T0314041

Mosquitoes of North America

Mosquitoes

OF NORTH AMERICA

(NORTH OF MEXICO)

STANLEY J. CARPENTER and **WALTER J. LaCASSE**

UNIVERSITY OF CALIFORNIA PRESS • Berkeley, Los Angeles, London

UNIVERSITY OF CALIFORNIA PRESS
BERKELEY AND LOS ANGELES
UNIVERSITY OF CALIFORNIA PRESS, LTD.
LONDON, ENGLAND
COPYRIGHT, 1955, BY
THE REGENTS OF THE UNIVERSITY OF CALIFORNIA
CALIFORNIA LIBRARY REPRINT SERIES EDITION, 1974
ISBN: 978-0-520-36661-9
LIBRARY OF CONGRESS CATALOG CARD NUMBER: 73-93048
DESIGNED BY JOHN B. GOETZ

Preface

Much progress has been made in our knowledge of the mosquitoes of North America since the comprehensive works of Howard, Dyar, and Knab (1912-1917), Dyar (1928), and Matheson (1944). Recent literature on the subject is vast and scattered. For many years there has been a need for a book that brings together much of this information and presents a concise and up-to-date account of the mosquitoes known to occur in this region. It is our sincere hope that this publication will fill this need.

This study contains pertinent information about the mosquitoes of America north of Mexico. Techniques used in collecting mosquitoes and preparing them for study; keys to the genera and species; descriptions of females, males, and larvae; accounts of distribution, bionomics and known medical importance; and numerous illustrations are included in an attempt to fill the needs of the systematic entomologist, the technician, and the field worker engaged in mosquito control.

Distribution records for each species have been compiled from published state and regional lists and are presented by states of the United States and provinces of the Dominion of Canada, listed in alphabetical order. As far as practical, the more complete and readily available references are cited. The numbers following each state and province refer to numbered references found in the bibliography.

The arrangement of genera and species follows rather closely that of Edwards (1932). The species are arranged alphabetically under subgenera or genera. The figures are distrib-uted throughout the text for the convenience of the reader. Plates of adult mosquitoes follow the text.

The illustrations of adult mosquitoes were drawn by members of the Taxonomic Entomology Section of the United States Army 406th Medical General Laboratory in Japan. The names of the artists and the plates for which each is responsible are listed as follows: Saburo Shibata—plates 2, 3, 5, 6, 8, 9, 10, 11, 12, 13, 14, 15, 16, 21, 22, 23, 24, 25, 26, 27, 30, 31, 32, 33, 34, 35, 36, 37, 38, 39, 40, 41, 42, 43, 45, 46, 47, 49, 50, 53, 55, 56, 58, 61, 63, 65, 66, 67, 69, 70, 72, 74, 80, 81, 85, 91, 92, 94, 95, 98, 99, 104, 106, 107, 109, 110, 111, 112, 113, 114, 115, 120, 122, 123, 125, 126, and 127; Kei Daishoji—plates 1, 4, 7, 18, 19, 20, 28, 29, 44, 48, 51, 57, 59, 60, 62, 64, 68, 71, 73, 75, 78, 79, 82, 84, 86, 87, 88, 90, 93, 97, 101, 102, 103, 105, 108, 116, 117, 118, 121, and 124; Kakuzo Yamazaki—plates 17, 52, 54, 76, 77, 89, 96, 100, and 119.

The generalized illustrations, 1 through 10, 12, *A*, 15 through 19, 21 and 22, and many of the illustrations of male terminalia and larvae were drawn by Elizabeth Kaston and originally published in "The Mosquitoes of The Southern United States East of Oklahoma and Texas" by Carpenter, Middlekauff, and Chamberlain, as Monograph No. 3 of the *American Midland Naturalist*. We are indeed grateful to Dr. John D. Mizelle, editor of the *American Midland Naturalist*, for granting us permission to copy these figures for this publication. We are also grateful to Dr. H. H. Roberts of the Philadelphia Academy of Sciences for the use of vi

Preface

figure 11 and to Dr. G. H. Penn of Tulane University for the use of figure 12, *B* and *C*.

The remainder of the illustrations of male terminalia and larvae were drawn by members of the Taxonomic Entomology Section of the 406th Medical General Laboratory; Eustorgio Mendez, graduate student in entomology at the University of California, Berkeley; and Corporal Reginald Jones of the Sixth Army Medical Laboratory.

Many acknowledgments are owing those who have assisted us during the preparation of this book. We are grateful to an Advisory and Editorial Committee, consisting of Dr. Alan Stone, chairman, and members Dr. Richard M. Bohart and Lt. Colonel Ralph W. Bunn, for their many helpful suggestions and assistance with technical problems, for helping us find specimen material, and for constructive criticism of the manuscript.

Special thanks are extended to the following members of the Army Medical Service for their coöperation and assistance: Colonel Robert L. Callison, M.C., Preventive Medicine Division, Office of the Surgeon General; Colonel Richard P. Mason, M.C., Commanding Officer, 406th Medical General Laboratory; Lt. Colonel Harold E. Shuey, M.C., Commanding Officer, Sixth

Army Medical Laboratory; Major Paul W. Oman, M.S.C., and Lt. John E. Scanlon, M.S.C., Entomology Section, 406th Medical General Laboratory; and Mr. W. C. Bentinck, Taxonomic Entomology Section, 406th Medical General Laboratory.

We also desire to thank the following entomologists who have contributed advice or specimens, or who have otherwise facilitated the preparation of this book: J. N. Belkin, R. E. Bellamy, G. H. Bradley, O. P. Breland, B. Brookman, R. B. Eads, F. F. Ferguson, R. F. Fritz, Pedro Galindo, C. M. Gjullin, R. A. Hedeen, K. L. Knight, A. W. Lindquist, E. C. Loomis, P. F. Mattingly, J. A. Mulrennan, W. B. Owen, H. D. Pratt, D. M. Rees, J. G. Rempel, E. S. Ross, H. H. Ross, J. A. Rowe, Harvey Scudder, G. E. Shewell, M. E. Smith, Ernestine Thurman, C. R. Twinn, Carl Venard, and J. R. Vockeroth.

Finally we wish to express our appreciation to Barbara Grier who meticulously typed the manuscript and assisted the senior author in checking the bibliography and proofreading the manuscript.

STANLEY J. CARPENTER
WALTER J. LaCASSE

February, 1954

Additional information on the taxonomy, biology, and geographical distribution of mosquitoes of North America (north of Mexico) since 1955 is summarized in the following articles:

Darsie, Richard F., Jr. 1973. A Record of Changes in Mosquito Taxonomy in the United States of America, 1955–1972. Mosquito Systematics, 5(2):187–193.

Carpenter, Stanley J. 1968. Review of Recent Literature on Mosquitoes of North America. California Vector Views, 15(8):71–98.

Carpenter, Stanley J. 1970. Review of Recent Literature on Mosquitoes of North America (Supplement I). California Vector Views, 17(6):39–65.

Contents

Life History

Mosquitoes are two-winged insects belonging to the order Diptera, family Culicidae. They are probably the best known group of insects because of their importance to man as pests and vectors of animal and human diseases and because they can be easily collected and studied in all their stages. They are widely distributed throughout the world, reaching their greatest variety in the tropical rain forests and probably their greatest abundance in the Arctic and Antarctic regions after the melting of snow in the spring and early summer.

A mosquito undergoes a complete metamorphosis, passing through four successive stages in its development, namely egg, larva, pupa and adult (imago). The principal characteristics of these four stages are described briefly.

Egg.—The female mosquito, at the time of oviposition, instinctively selects the habitat for the immature aquatic stages. Some species of *Anopheles* display a preference for large permanent bodies of water as their aquatic habitats; others are more general in their requirements and utilize a great variety of habitats such as small temporary pools or seepage areas.

The egg-laying habits of North American mosquitoes exhibit much diversity among the different genera. *Anopheles* deposit their eggs singly on the surface of still waters. The eggs float, and because of the surface tension of the water, usually arrange themselves in star-shaped patterns. The female of *Anopheles barberi* selects rot cavities in trees for ovipositing. *Psorophora* and most species of *Aedes* found

in this region deposit their eggs singly in moist depressions where they may remain dormant for months or even for several years. In northern latitudes eggs laid by *Aedes* in the summer usually do not hatch until flooded the following year. Most members of the subgenus *Finlaya* lay their eggs on the sides of tree cavities and depend on flooding by rainfall to submerge the eggs before hatching. Females of *Culex, Mansonia, Culiseta,* and *Uranotaenia* generally glue their eggs together in raftlike masses which are deposited upon the surface of the water.

Larva.—After the eggs of a mosquito have been in contact with water for a sufficient time, the larva cuts its way out of the egg by means of the egg breaker placed on the dorsal side of the head. During growth the larva sheds its skin four times; the stages between molts are called instars.

All mosquito larvae, except members of the genus *Mansonia*, must come to the surface at frequent intervals to obtain oxygen. *Mansonia* larvae and pupae attach themselves to the submerged roots and stems of plants and obtain oxygen from the plant tissue. The thorax and abdomen of the larva have two main tracheal trunks terminating in a pair of spiracles, situated dorsally on the eighth abdominal segment in the Anophelines and at the end of the dorsal siphon in the Culicines. When the larva is breathing, its spiracles must lie in the plane of the water surface. The Anopheline larva rests in a horizontal position, and the Culicine larva hangs head downward.

The food of mosquito larvae consists chiefly

1

of small plants and animals and particles of organic matter which are swept into the mouth by the mouth brushes or by actual nibbling. The *Anopheles* larva rotates its head 180°, and, with its mouth brushes upward, sweeps the surface film for food, whereas the Culicine larva obtains its food at various depths. The larvae of *Toxorhynchites* and the subgenus *Psorophora* are predacious and often feed on other species of mosquito larvae.

Many species of mosquitoes complete the larval stage in a week to 10 days, when conditions are favorable, but others require more time, even several months. Some of the *Psorophora* and even a few of the *Aedes* which utilize temporary pools as aquatic habitats may pass through the larval stage in 4 to 6 days.

Pupa.—The pupal stage appears with the fourth molt. Because it is lighter than water, the pupa rests at the surface until disturbed and then dives in a jerking, tumbling motion. Its buoyancy is owing to an air cavity between the wings of the future adult. A large pair of respiratory trumpets on the cephalothorax enables the pupa to break the surface film and obtain air. The pupal stage lasts but a few days (usually 3 or 4) for most species; however, it may last 2 weeks or more for some mosquitoes. At the end of this stage the pupa extends its abdomen nearly parallel to the water surface in preparation for the emergence of the adult.

Adult.—The adult swallows some of the air within the pupal skin and exerts internal pressure by muscular action to split the dorsum of the cephalothorax, thus enabling it to emerge. The adult slowly works its way out, using the cast skin as a float until its body can dry and harden.

Mosquitoes are best known because of their bloodsucking habits and because they often are vectors of important diseases of man and animals. However, not all mosquitoes suck human blood; some restrict their feeding to birds, amphibia, reptiles, and other nonmammalian hosts. Many species probably utilize juices of fruits and nectar of plants for food. The mouth parts of male mosquitoes are not developed for sucking blood.

Most of the *Anopheles* and many of the Culicines restrict their feeding to nighttime or the twilight hours of morning and evening, although some feed during the daytime in the shade and even in bright sunlight. *Culex* usually feed only at night or at dusk, and only a few species are avid feeders on man. *Culex erythrothorax* Dyar, found in the western United States, will attack man and warm-blooded animals indiscriminately during the daytime, even in bright sunlight, when their resting places are disturbed. Most of the *Aedes* and *Psorophora* readily attack man and other warm-blooded hosts and are noted for their bloodthirstiness.

The mating habits of many mosquitoes have been observed. They vary considerably among the different species. The males of many species form swarms late in the evening, often over some object such as a shrub, tree, or stone, and the females invade the swarm and emerge with males in the act of copulation. *Culiseta inornata* (Williston) has been observed to mate while resting on vegetation at the site of the larval habitat. Males and females of *Deinocerites cancer* Theobald have been seen in copula while resting in the upper part of crabholes in which the immature stages are passed. The males of the western treehole mosquito, *Aedes varipalpus* (Coquillett), often approach man or some warm-blooded animal and await the approach of the female mosquitoes for feeding, at which time mating takes place.

The females of most species of *Anopheles*, *Culex*, and *Culiseta* pass the winter in hibernation in protected places, whereas many others overwinter in the egg stage. This is especially true of *Aedes* and *Psorophora*. Some species are known to overwinter as larvae.

The flight habits of mosquitoes vary greatly with the different species. Several of the *Aedes*, particularly the salt-marsh species, are notorious wanderers, often migrating 40 miles or more from their aquatic habitats. *Anopheles* and *Culex* are usually weak fliers and do not move far from their aquatic habitats; however, many exceptions have been recorded by recapturing marked specimens.

Collecting

Adults.—Many species of *Anopheles* and *Culex*, and to a lesser degree members of other genera, rest during the daytime in relatively dark, humid shelters such as buildings, culverts, hollow trees, caves, and underneath rock ledges and overhanging banks along streams.

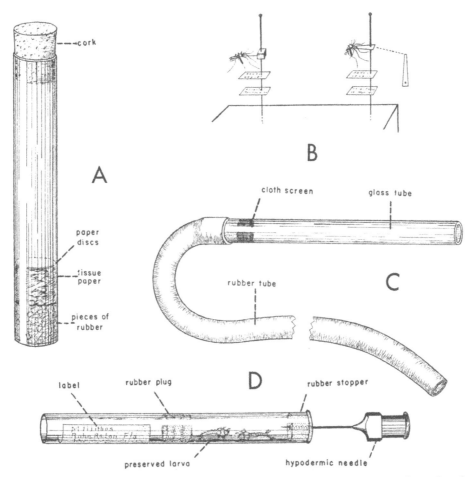

cork

A

paper
discs

tissue
paper

pieces of
rubber

B

cloth screen   glass tube

rubber tube

C

label   rubber plug   D   rubber stopper

preserved larva   hypodermic needle

Fig. 1. *A*, Chloroform tube for killing adult mosquitoes. *B*, Adult mosquitoes mounted on micropin and triangular paper point. *C*, Aspirator for collecting adult mosquitoes. *D*, Tube for preserving and storing mosquito larvae.

Specimens can be captured in these shelters with the aid of a chloroform or cyanide killing tube (fig. 1, *A*) or an aspirator (fig. 1, *C*). A flashlight is useful for finding specimens in dark shelters or in making nighttime collections.

Collecting mosquitoes at regular intervals from daytime shelters provides useful information on mosquito numbers and on control measures. Biting collections and landing-rate counts of mosquitoes attempting to feed on humans or animal attractants are often employed to obtain information on the species causing annoyance and their relative abundance. Artificial shelters in the form of nail kegs (661), small boxes (319), privy-type houses (126), and other similar structures have been installed and used for this purpose by workers in areas where suitable natural shelters are not available.

Several types of mosquito traps have been devised for mosquito surveys and for evaluating control measures. The better known traps now in use are the New Jersey light trap (519) shown in figure 2 and animal-baited traps. Care must be exercised in selecting sites for operating the New Jersey trap and in evaluating the results obtained, since many factors relating to light reactions of mosquitoes are unknown. The traps should not be placed in the vicinity of competing lights or in locations open to prevailing strong winds. They should be hung with the light about 5 to 6 feet from the ground. The traps can be operated one or more nights each week; they are usually equipped with an automatic time switch for starting and stopping. Seaman (653) describes a small light trap that can be operated from an automobile battery in the field. A rotary-type mechanical trap is recommended by Chamberlin and Lawson (149) and has been used successfully in many areas, particularly in the Arctic. The trap has rigid insect nets or cones equipped with collecting jars and attached to a horizontal rotating bar powered by a gasoline motor. A large trap, mounted on an automobile trailer, for collecting live mosquitoes for virus studies is described by Reeves and Hammon (590). A small portable trap made from a 50-pound lard can, equipped with an ingress funnel and baited with dry ice (carbon dioxide gas), is described

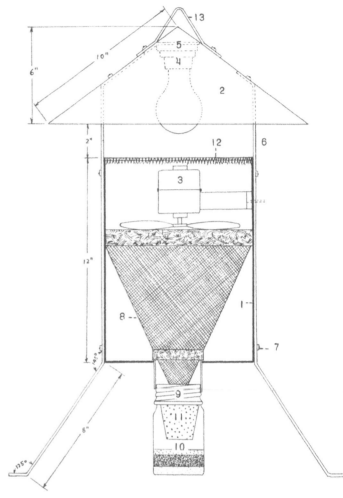

Fig. 2. Diagram of the New Jersey Mosquito Trap (adapted from Mulhern, 1942). *1*, Galvanized iron cylinder, 22 gauge, 9 inches inside diameter (cut away to show interior). *2*, Roof of galvanized iron. *3*, Motor and 8-inch fan. *4*, Porcelain receptacle for light bulb. *5*, Sponge rubber to absorb vibration. *6*, Three supporting ribs of band iron (⅛" by 1"). *7*, Screws and nuts for assembling trap. *8*, Screen funnel of 16 mesh bronze wire. *9*, Mason jar cap. *10*, Jar with cyanide for killing insects. *11*, Perforated paper cup. *12*, Galvanized screen ⅜" mesh. *13*, Loop for carrying or hanging trap.

by Bellamy and Reeves (54) and has been used in California for catching live mosquitoes.

Two common types of animal-baited traps, the Magoon trap (473) and the Egyptian trap (41), are in general use in the tropics for evaluating Anopheline mosquito densities. Each of these traps consists of a small portable stable for housing a small horse or other suitable animal bait and is equipped with an ingress baffle along the side walls.

Adult mosquitoes often remain among vegetation in moist shady places during the day, and specimens may be captured by sweeping with a net or by first disturbing them and then capturing the specimens in the net while they are in flight. It is frequently advisable to sweep through vegetation surrounding mosquito-breeding pools to obtain specimens of males and recently emerged females. A small midge net is particularly useful.

Larvae.—Mosquito larvae occur in various types of aquatic habitats, varying from lakes and marshes to small temporary pools and collections of water in treeholes, leaf axils of plants, and artificial containers. Every type of aquatic habitat should be examined when making a larval survey of mosquitoes in an area. It is important that adequate field notes be kept on all collections.

A useful tool for collecting mosquito larvae is a white-enameled dipper with a hollow handle into which a round stick or cane is inserted for greater length. For *Anopheles* the dips are made by skimming through the water surface in places where vegetation or floating debris offer protection for the larvae. The larvae and pupae of Culicine species are captured with an intercepting movement of the dipper since they generally sink to the bottom when disturbed. It is sometimes necessary to sit or stand quietly near the pool and capture the larvae and pupae as they come to the surface. A wide-mouth pipette is required for transferring the larvae and pupae from the dipper to the collection jars. A suction bulb attached to 2 or 3 feet of rubber hose is useful for collecting larvae from tree cavities, crabholes, and other inaccessible places.

Eggs.—Egg rafts of many species of Culicines are frequently encountered in the field and can be gathered and carried to the laboratory for hatching and rearing. Eggs of Anophelines may be collected by placing a white muslin bag over the hand and sweeping the hand through the water in spots where oviposition is likely to have taken place, or water can be taken with a dipper and poured through the muslin. Muslin fitted on embroidery hoops also may be used for this purpose. Anopheline eggs can often be detected on the water surface by examining carefully with a hand lens the water taken from likely breeding sites. Aitken (13) recommends egg sampling as a routine procedure in Anopheline surveys. Eggs of certain species of *Aedes* and *Psorophora* may be obtained in soil samples taken from possible breeding spots.

A satisfactory oviposition vial for obtaining eggs from gravid female mosquitoes can be prepared by pressing a layer of cotton, about ½ to 1 inch in depth, in the bottom of a 1-inch shell vial. Cover the cotton with a disk of filter paper and add sufficient water to moisten the the cotton and filter paper. A plug can be made by pushing a circular piece of wire screen into the top of the vial. This wire-screen plug will hold cotton soaked with glucose for feeding the enclosed mosquito.

Preparing
Specimens for Study

Adults. — Mosquitoes which have been reared should not be killed immediately but should be kept alive for 12 to 24 hours to permit the body to harden and thus avoid excessive shrinkage. When fresh specimens are available for pinning they should be mounted on micropins (*minuten nadeln*). The micropin is pushed into the thorax of the mosquito, preferably between the coxae (fig. 1, *B*). Dry specimens can be mounted on triangular paper points, using cellulose cement or shellac as an adhesive (fig. 1, *B*).

Reference collections of pinned adults should be stored in museum drawers or in Schmitt boxes or similar boxes of tight construction. Unmounted specimens can be packed for shipping or storage in pill boxes between layers of glazed cotton, cleansing tissue, or lens paper. Plain cotton is objectionable because of damage to specimens in removing them. Mosquitoes should always be handled with much care to avoid breakage and rubbing off of scales.

Male terminalia.—To prepare male terminalia for study, clip off the apical fourth of the abdomen with forceps or scissors and place it in a 10 to 20 per cent solution of potassium hydroxide in a small porcelain casserole. Heat the solution slowly almost to the boiling point and then transfer the specimen to water and rinse for several minutes. The specimen can be stored in 70 or 80 per cent alcohol or in glycerin in small vials.

A temporary mount of the male terminalia can be prepared by placing the specimen in a drop of glycerin or chloral gum on a slide. For a permanent mount the specimen should be dehydrated, cleared, and mounted in Canada balsam, clarite, or Euparal. We prefer balsam because of its permanency (see method for mounting mosquito larvae). Before mounting, the superfluous abdominal segments should be removed, and the specimen should be oriented dorsal side up with the dististyles extended.

It is often necessary to dissect the parts of the terminalia of *Anopheles* and *Culex* and mount them separately on the slide. This is also true of the claspette of *Aedes*. Some workers prefer to stain the terminalia before mounting; however, this is seldom necessary. Techniques for staining terminalia are described by Edwards (252) and Komp (441).

Pupae.—The pupal stage is best studied from the cast skin (exuviae). Unmounted specimens can be stored in 70 or 80 per cent alcohol. To mount the pupal skin, insert a dissecting needle between the junction of the metanotum and the cephalothorax proper and separate these two structures so as to leave the metanotum attached to the abdomen. The remainder of the cephalothorax is now open along the dorsal longitudinal mid-line and can be laid out flat and mounted with the outer side up. The specimen should be dehydrated, cleared, and then mounted, preferably in Canada balsam (see method for mounting mosquito larvae).

Larvae.—It is often desirable to isolate and rear single larvae to associate larval and pupal exuviae with the adult for taxonomic studies. Glass vials measuring about 1 inch by 3½ inches are satisfactory for this purpose. The exuviae can be removed with a pipette or a small spatula and preserved in 70 or 80 per cent alcohol in small vials (fig 1, *D*) or mounted if desired. The associated skins and emerged adult should always be given a corresponding identification number.

Full grown larvae may be killed and preserved in 70 or 80 per cent alcohol in small vials for storage or study. The live larvae may be placed directly into the preservative, but better specimens can usually be obtained by killing them in hot, but not boiling, water.

We have found it preferable to kill mosquito larvae in Peterson's KAAD solution and keep them in the solution overnight. The KAAD solution as described by Peterson (545) contains kerosene (1 part), isopropyl alcohol (7 to 9 parts), glacial acetic acid (1 part), and dioxan (1 part). The larvae are placed in a small dish and as much of the water as possible is withdrawn with a pipette; the KAAD solution is added and allowed to remain overnight. The KAAD solution is then withdrawn with a pipette, the larvae are rinsed in two or three changes of 70 or 80 per cent ethyl alcohol and stored in 70 or 80 per cent alcohol.

A convenient method of packing mosquito larvae for storage or shipping is to place them in dental procaine hydrochloride cartridges as shown in figure 1, *D*, and described by Carpenter (125) and Carpenter *et al.* (139).

It is often necessary to make permanent mounts of mosquito larvae for detailed studies. There are several known reliable mounting media for mosquito larvae but we prefer balsam which generally assures permanent mounts. Water-soluble chloral gum arabic media and polyvinyl alcohol have not proved satisfactory. To mount in Canada balsam, the larvae can be transferred from 70 or 80 per cent alcohol to cellosolve (ethylene glycol monoethyl ether) for 10 minutes before mounting. Small incisions should be made in the thorax and abdomen of the larvae with a sharp instrument before they are placed in cellosolve. The abdomen of a Culicine larva should be partly severed between the seventh and eighth segments with a dissecting needle or a small knife before mounting so that the siphon and anal segment will turn and lie flat. The balsam slides should be placed for a few days in an oven at about 150° F for hardening.

Eggs.—The study of mosquito eggs, in conjunction with other stages in the life cycle, is of particular importance when studying closely related species. Samples of mosquito eggs can be preserved by placing the filter paper disk, with the eggs from the oviposition vial, in a small vial containing a layer of cotton on the bottom, which has been saturated with 10 per cent formalin to provide formaldehyde fumes for preserving the eggs. A second layer of dry cotton is placed in the vial between the eggs and formalin-soaked cotton. The vial should be tightly corked and sealed with paraffin. It is seldom necessary to prepare permanent mounts of eggs for study; however, if it seems desirable, they can be mounted in balsam, clarite, or Euparal.

External Anatomy

ADULT CHARACTERS

The body of an adult mosquito comprises three distinct regions: the head, the thorax, and the abdomen. Each of these body regions has important characters which can be used in classification. Structures commonly used when identifying mosquitoes in their immature and mature stages are briefly described and illustrated in this study. The principal structures of an adult female mosquito are shown in figure 3.

HEAD. The large compound eyes occupy most of the lateral part of the head; ocelli are lacking. That part of the head posterior to the eyes is termed the occiput, and the part extending forward between the eyes, the vertex. The frons lies between the bases of the antennae and joins the anterior margin of the vertex. The occiput and vertex are clothed with erect and decumbent scales of various types and colors, often providing good characters for identification. Erect scales are narrow basally, and broadened and usually forked apically. Decumbent scales are either narrow and curved or broad and flat. A row of orbital bristles is present just behind the eyes. The frontal tuft is composed of setae arising from the anterior part of the vertex; it extends forward over the frons. It is well developed in *Anopheles*, and its color differences are often used as diagnostic characters for dark-winged species.

The clypeus is a short projection situated just anterior to the frons. It is longer than broad and has its distal margin rounded in the tribes Anophelini and Culicini, but is wider than long and has its distal margin trilobed in the tribe

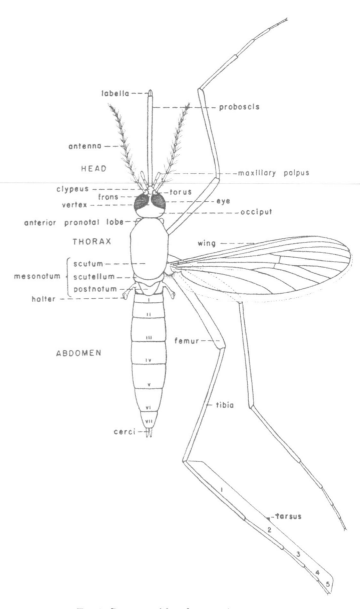

Fig. 3. Diagram of female mosquito.

8

Toxorhynchitini. It is usually bare but has scales in a few species.

The antennae are long, slender, 15-segmented structures arising on either side of the frons between the eyes. The first segment (scape) is small and is hidden beneath the large globular second segment (torus). The color of the integument and the scales of the torus often provide characters useful in specific determinations. The remaining thirteen segments are filamentlike and make up the flagellum. Each flagellar segment bears a whorl of hairs, short and sparse in the females and usually longer and more abundant (bushy) in the males. In this area the males of *Deinocerites* spp., *Uranotaenia lowii*, and *Wyeomyia* spp. have antennae similar to those of the females.

The palpi are 5-segmented, the first segment being very short. They exhibit sexual modifications and variations in some genera, subgenera, and species and are useful in classification. In Culicines the palpi of the female are usually smooth-scaled, more or less straight, and much shorter than the proboscis. The male Culicines usually have densely haired palpi, longer than the proboscis and with the distal segments angled upward and tapered to a point. The males are generally easy to recognize by their long bushy palpi and bushy antennae. The Anophelines of this region have the palpi about as long as the proboscis in the females, and as long or longer than the proboscis in the males. The male Anophelines have the two apical segments of the palpi flattened, angled upward, and rounded at the tip. Characteristics of the palpi of both Culicines and Anophelines are shown in figure 4.

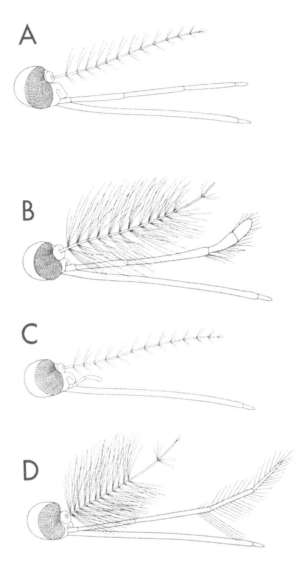

Fig. 4. Head and appendages of mosquitoes. *A*, Female *Anopheles. B*, Male *Anopheles. C*, Female Culicine. *D*, Male Culicine.

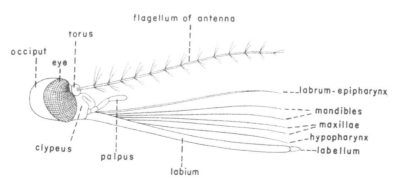

Fig. 5. Mouth parts of female mosquito.

The proboscis is composed of the greatly elongated lower lip known as the labium, with its enclosed piercing and sucking structures (fig. 5). The labium is sheathlike and terminates in a pair of small lobes (labellae). The labium never enters the wound during the feeding process. It serves as a protective sheath and guides and supports the piercing mouth parts, but is bent out of the way as the skin is pierced and penetrated.

The mouth parts are described as follows: the labrum-epipharynx, an elongated organ, inverted U-shape in cross section; the hypopharynx, lying directly beneath the labrum-epipharynx and forming a canal through which liquid food is drawn during feeding (the hypopharynx is traversed by a small salivary duct leading from the salivary glands); the paired mandibles, delicate in form and lying lateral to the labrum-epipharynx; and the paired maxillae, lying beneath and lateral to the hypopharynx and dentate apically. The mouth parts of the male are modified, the hypopharnx usually fused with the labium, the maxillae delicate and greatly reduced, and the mandibles, when present, poorly developed.

THORAX (fig. 6). The thorax provides many characters useful in classification. It is composed of three fused segments: the prothorax bearing the front pair of legs, the mesothorax bearing the wings and the middle pair of legs, and the metathorax bearing the halteres and the hind pair of legs.

Prothorax (fig. 6).—The prothorax is much

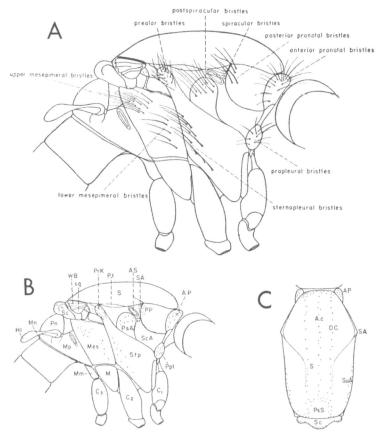

Fig. 6. Parts of thorax and pleural chaetotaxy of an adult *Psorophora. A,* Pleural chaetotaxy. *B,* Lateral view with sclerites and positions of bristles indicated. AP, anterior pronotum; AS, anterior spiracle; C_1, first coxa; C_2, second coxa; C_3, third coxa; Hl, halter; M, meron; Mes, mesepimeron; Mm, metameron; Mn, metanotum; Mp, metapleuron; Pn, Postnotum; PP, posterior pronotum; Ppl, propleuron; Prk, prealar knob; PsA, postspiracular area; Pt, paratergite; S, scutum; SA, spiracular area; Sc, scutellum; Sq, squama; SsA, subspiracular area; Stp, sternopleuron; WB, wing base. *C,* Dorsal view with positions of bristles indicated. Ac, acrostichal bristles; AP, anterior pronotal lobe; DC, dorsocentral bristles; PsS, prescutellar space; S, scutum; SA, scutal angle; Sc, scutellum; SuA, supra-alar bristles.

reduced and has on either side the anterior pronotum (AP), a lateral prominence back of the head; the posterior pronotum (PP), the area between the anterior pronotum and the spiracular area (the spiracular area lies just in front of the anterior spiracle and is set off from the posterior pronotum by a strong ridge); the propleuron (Ppl) just above the front coxa; and the prosternum, the region between the front coxae. The bristles and scales of these structures provide good taxonomic characters in some genera and species.

Mesothorax (fig. 6).—The mesothorax comprises the largest part of the thorax, as in all Diptera, and bears many excellent taxonomic structures. It is composed of two general areas, the dorsal mesonotum and the lateral pleura.

The mesonotum includes most of the dorsal surface of the thorax and consists of the following areas: the scutum (combined praescutum and scutum), the largest part of the mesonotum; the paratergite (Pt), a small area cut off from the lateral margin of the scutum; the scutellum (Sc), a trilobed or rounded structure behind the scutum; and the postnotum (Pn), a convex structure between the scutellum and the metanotum. Just in front of the middle and immediately above the anterior spiracle, the lateral margin of the scutum becomes somewhat prominent, forming the scutal angle (SA).

The type of scales on the scutum and their coloration provide characters often used in identification. In mosquitoes with unicolored scales on the scutum, the fineness or coarseness of the scales may be considered. The following setae are present on the scutum in rather definite areas or lines (fig. 6, *C*): the acrostichal bristles (Ac) in a median longitudinal row; the dorsocentral bristles (DC), a submedian row on either side of the acrostichals; and the supraalar group above and in front of the wing base. Conspicuous setae also may be present on the anterior margin of the scutum.

The paratergites bear scales in the *Aedes* but are usually bare in other genera. The scutellum is trilobed in most of the Culicines but is rounded posteriorly in *Toxorhynchites* and most Anophelines. Scales and marginal setae are present on the scutellum. The postnotum is similar in all mosquitoes but has a tuft of setae

in *Wyeomyia* that helps to distinguish the genus.

The mesopleural sclerites (fig. 6, *B*) occupy most of the side of the thorax and are described as follows: the postspiracular area (PsA), the region just behind the anterior spiracle; the subspiracular area (SsA), below the anterior spiracle and adjacent to the posterior pronotum and propleuron; the sternopleuron (Stp), a large ham-shaped sclerite below the postspiracular and subspiracular areas; the prealar area, the necklike upper part of the sternopleuron terminating in the prealar knob (PrK); the mesepimeron (Mes), a large subrectangular sclerite adjacent to the posterior margin of the sternopleuron; and the meron (M), a small triangular sclerite slightly above and immediately behind the second coxa, just below the mesepimeron.

The pleural sclerites are similar in shape throughout the subfamily and are seldom used in a systematic study of mosquitoes; however, their setae (bristles) and scales provide valuable diagnostic characters. The positions of the bristles are shown in figure 6, *A*. The presence or absence of postspiracular bristles in conjunction with the presence or absence of spiracular bristles is used as a generic character. The genus *Aedes*, for example, has postspiracular bristles but no spiracular bristles, whereas *Psorophora* has both. The number and position of the sternopleural bristles are used in some instances. They are sparse in *Wyeomyia* and *Uranotaenia* and abundant in *Aedes* and *Culex*.

The prealar bristles may be dense or sparse. The meron is always bare. The mesepimeron has two groups of bristles, the upper and lower mesepimerals. The upper mesepimerals in the upper corner of the mesepimeron are nearly always present. The lower mesepimerals are situated below the middle and near the anterior edge of the mesepimeron. *Orthopodomyia* may be separated from *Mansonia* on this basis, since lower mesepimerals are lacking on the former, but present on the latter. The number of lower mesepimeral bristles or their absence provides characters for differentiating certain *Aedes*.

The scales on the pleural sclerites are of more value for specific differentiation than the bristles. The scales are usually all broad and flat. In some genera and even in some species within a genus, the pleura are much more heavily scaled than in others. In some species of *Culex*, pleural scales are almost lacking. The presence or absence of scales on the postspiracular, subspiracular, and prealar areas and the extent of the patches of scales on the sternopleuron and the mesepimeron may provide means of distinguishing between closely related species. The absence or presence of the hypostigial spot (HS) and the number of scales it contains is used for differentiating certain *Aedes*.

Metathorax (fig. 6).—The metathorax, as in other Diptera, is greatly reduced. Its dorsal part (metanotum) is in the form of a narrow, usually indistinct, transverse band, connecting the postnotum with the first abdominal tergite. The metameron (Mm), a very small lateral sclerite, lies immediately above the hind coxa. The metapleuron (Mp) lies posterior to the mesepimeron and between the postnotum and the metameron. It is separated by a suture into the metepisternum, containing the posterior spiracle on its anterior margin, and the metepimeron, the narrow band bordering the first abdominal segment.

Legs.—Each segment of the thorax bears a pair of legs; the front legs arising from the prothorax, the middle legs from the mesothorax, and the hind legs from the metathorax. Each leg is composed of a coxa, trochanter, femur, tibia, and a 5-segmented tarsus (fig. 3). The fifth tarsal segment bears a pair of small tarsal claws or ungues. The claws may provide characters useful to the taxonomist. In most *Aedes* each claw of the front and middle legs

has a sharp tooth on the under side near the middle. The claws of the hind legs are similarly toothed in many species of *Aedes*. In the male the claws often have secondary sexual modifications. Between the bases of the claws is a small hairy empodium, apparently always present but inconspicuous. The pulvilli are a pair of small padlike structures that arise laterally near the base of the ungues or claws in *Culex*.

The legs are clothed with scales, hairs, and bristles. The scales are of the greatest diagnostic value and are usually rather broad, appressed, and imbricate but occasionally are long, slender and suberect as in the subgenus *Psorophora*. Ornamentation with contrasting dark and pale scales, usually in the form of bands on the tarsal segments, provides useful specific characters.

Mesothoracic wing.—The wing venation of Culicidae is rather uniform throughout the family, but does provide features for distinguishing some genera and species. The wings are elongate-oval in shape, each with two indentations near the base on the posterior margin. The small flaplike structure nearest the thorax is the squama. In most mosquitoes the squama is fringed, but this fringe is absent in *Toxorhynchites*, *Wyeomyia*, and *Uranotaenia* in this region. The squama is followed by the lobate alula.

The veins and cells of the wing are shown in figure 7. The veins are clothed with scales generally of two types. Those which lie close to the veins and are usually rather short and broad are termed squame scales; those which are sub-

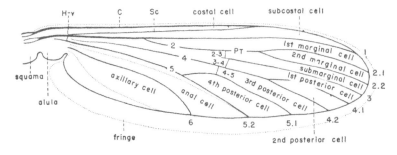

Fig. 7. Wing of mosquito, with venation shown. C, costa; Sc, subcosta; 1, first longitudinal vein; 2, second longitudinal vein (2.1, anterior branch; 2.2, posterior branch); 3, third longitudinal vein; 4, fourth longitudinal vein (4.1, anterior branch; 4.2, posterior branch); 5, fifth longitudinal vein (5.1, anterior branch; 5.2, posterior branch); 6, sixth longitudinal or anal vein; H-v, humeral cross vein; 2–3, 3–4 and 4–5, cross veins; Pt, petiole of vein 2.

erect and usually narrow are known as out-standing or plume scales. The entire posterior margin of the wing from the alula to the tip bears a close-set row of long slender fringe scales. The membranous areas of the wing bounded by the various veins and cross veins are known as cells (fig. 7) and are clothed with very fine hairs or microtrichia.

The length of the wing given in the descriptions represents the distance from the alula to the tip. The measurements are based on a few specimens of each species and are only an indication of the size of the wing.

Halter.—The metathoracic wings are represented by a pair of halteres which are not used in flight, except perhaps as balancing organs. The halter comprises three parts: the scabellum or base, the midhalter or stemlike part, and the capitellum or terminal knob. The color of the integument and scales of the halter, particularly of the capitellum, is occasionally used in separating species.

ABDOMEN. The abdomen is composed of ten segments, the first eight of which are distinct and unmodified. The ninth and tenth segments are greatly modified for sexual function in both males and females. The eighth, ninth, and tenth segments of the male are discussed separately under "Male terminalia." The characters of the female terminalia are rarely employed in taxonomic classifications of mosquitoes and are not included separately in this study.

Each of the distinct abdominal segments is made up of a large dorsal tergite and a smaller ventral sternite, connected laterally by intersegmental membranes. The tergites are collectively referred to as the dorsum and likewise the sternites as the venter. Both the dorsum and venter are clothed with scales in the Culicines, but are usually bare or have only a few scales in the Anophelines. Ornamentation of the dorsum, generally in the form of basal, apical, or lateral patches of scales may be present in some species, providing useful diagnostic characters. Scale patterns are generally not so evident on the venter. In *Aedes* and *Psorophora* the abdomen is tapered apically with the eighth segment withdrawn into the seventh. In other genera occurring in this region the abdomen does not taper appreciably, being bluntly rounded or broadly truncate at

the apex, although the eighth segment may be partly withdrawn in some cases.

MALE TERMINALIA. The terminal abdominal segments of the male mosquito are greatly modified for sexual purposes, thus exhibiting variations in structure of much taxonomic value. The term "male terminalia" is used here to include the anal and genital structures of the eighth, ninth, and tenth abdominal segments.

The male terminalia of the Culicinae undergo a rotation of 180° on the longitudinal axis shortly after the adult emerges so that the structures that were dorsal become ventral, and vice versa. References to structures of the male terminalia, however, are to the original positions before rotation, although they appear opposite on the mature specimen.

Important structures of the male terminalia include the eighth abdominal segment, ninth tergite, ninth sternite, proctiger, phallosome and supporting structures, basistyles, dististyles, and claspettes (figs. 8–11). These structures are described separately.

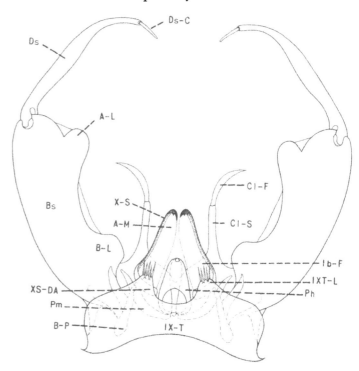

Fig. 8. Diagram of male terminalia of *Aedes*. A-L, Apical lobe. A-M, Anal membrane. B-L, basal lobe. B-P, Basal plate. Bs, Basistyle. Cl-F, Claspette filament. Cl-S, Claspette stem. Ds, Dististyle. Ds-C, Dististyle claw. Ib-F, Interbasal fold. IX-T, Ninth tergite. IXT-L, Lobe of ninth tergite. Ph, Phallosome. Pm, Paramere. X-S, Tenth sternite. XS-DA, Dorsal arm of tenth sternite.

Eighth abdominal segment.—This segment is usually unmodified and is relatively unimportant in the identification of *Aedes, Psorophora,* and most *Culex.* It bears diagnostic setae or spines dorsally in some genera, particularly *Mansonia* and *Wyeomyia.*

Ninth tergite (IX-T).—The ninth tergite and posterolateral lobes often provide good diagnostic characters. The extent of sclerotization and shape of the transverse band, and the shape, position, and armature of its lobes are significant.

Ninth sternite (IX-S).—The ninth sternite is usually unmodified in the mosquitoes of this region and relatively unimportant.

Proctiger.—The proctiger is made up of elements of the tenth abdominal or anal segment and varies considerably in the different genera. The tenth tergite is usually reduced. The tenth

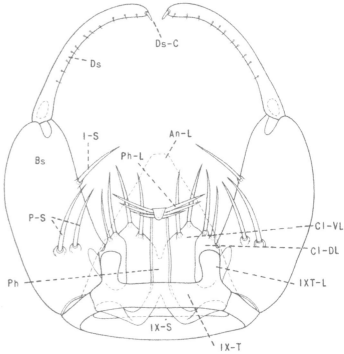

Fig. 10. Diagram of male terminalia of *Anopheles.* An-L, Anal lobe. Bs, Basistyle. Cl-DL, Dorsal lobe of claspette. Cl-VL, Ventral lobe of claspette. Ds, Dististyle. Ds-C, Dististyle claw. I-S, Internal spine. IX-S, Ninth sternite. IX-T, Ninth tergite. Ph, Phallosome. Ph-L, Leaflets of phallosome. P-S, Parabasal spines.

sternite (X-S) or paraproct is well developed in most genera and forms a pair of slender sclerotized supports for the anal membrane (A-M). It is vestigial or absent in *Anopheles* and *Uranotaenia,* which have the anal lobe (An-L) or membrane unsupported. The terminal armature of the tenth sternite provides diagnostic characters, particularly in the *Culex.*

Phallosome (Ph).—The phallosome (mesosome) is a chitinous, tubelike structure surrounding the penis. It lies just ventrad of the proctiger and is held in place by supporting structures, the basal plates (B-P) and parameres (Pm). These articulate with the basal processes of the tenth sternite, the phallosome, and with each other. Because of its range of variation in the different genera and subgenera it often provides reliable diagnostic characters.

Basistyle (Bs).—The basistyles are a pair of large, hollow processes arising from the ninth sternite. They may or may not possess apical,

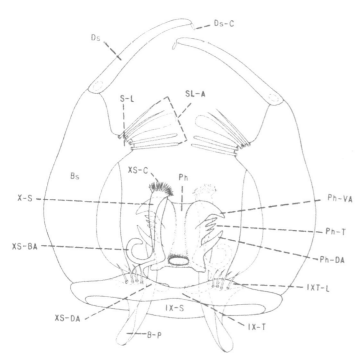

Fig. 9. Diagram of male terminalia of *Culex* (right plate of phallosome shown in bold lines for clearness, although it lies beneath the tenth sternite). B-P, Basal plate. Bs, Basistyle. Ds, Dististyle. Ds-C, Dististyle claw. IX-S, Ninth sternite. IX-T, Ninth tergite. IXT-L, Lobe of ninth tergite. Ph, Phallosome. Ph-DA, Dorsal arm of phallosome. Ph-T, Lateral teeth of phallosome. Ph-VA, Ventral arm of phallosome. S-L, Subapical lobe. SL-A, Appendages of subapical lobe. X-S, Tenth sternite. XS-BA, Basal arm of tenth sternite. XS-C, Crown of tenth sternite. XS-DA, Dorsal arm of tenth sternite.

subapical, or basal lobes on the inner surface. A subapical lobe (S-L) is present in *Culex* and bears important rods, spines, and leaflike appendages. An apical lobe (A-L) is often present in *Aedes*. A basal lobe (B-L) is present in several genera but is best developed in the *Aedes* where it may give rise to one or more large spines and characteristic smaller setae. The basal lobe is represented in *Anopheles* by the large parabasal spines (P-S); it is always absent in *Culex*, which bears instead the subapical lobe, probably homologous with the basal lobe of *Aedes*.

Dististyle (Ds).—The dististyle is an articulated appendage borne on the apical part of the basistyle. An articulated claw (Ds-C) is usually present at or near its apex. The shape of the distisyle, its point of origin on the basistyle, and the place of origin and shape of the claw provide useful diagnostic characters. The dististyle is greatly modified in form in *Wyeomyia* and some *Psorophora* species.

Claspette (Cl).—The connecting membranous projections between the bases of the basistyles, the interbasal folds (Ib-F), may bear a pair of structures ventrad of the phallosome, known as the claspettes. In *Anopheles* (fig. 10) the claspettes are represented by a pair of fleshy, spined, bilobed structures, each being incompletely divided into an outer or dorsal lobe (Cl-DL) and an inner or ventral lobe (Cl-VL). In *Aedes* (fig. 8) there is but one

lobe, which presumably corresponds to the ventral lobe of *Anopheles*, and it consists of a well-defined stem (Cl-S) and a filament (Cl-F).

PUPAL CHARACTERS

The pupae of Culicines and Anophelines resemble each other in many characteristics. The body comprises the enlarged anterior part or cephalothorax, consisting of the head and thorax, and the abdomen. The abdomen is slender, eight-segmented, and terminates in a pair of flattened paddles (fig. 12). Important diagnostic characters are found in the abdominal chaetotaxy, the paddles, and the respiratory trumpets.

CEPHALOTHORAX (fig 12, *B*). The designation of the setae of the cephalothorax follows that of Knight and Chamberlain (432) and Penn (539). The respiratory trumpet has a tubular part, the meatus, and an open part, the pinna. In the Anophelini the respiratory trumpets are short, truncate apically, and have a rather large oblique opening which terminates in a split. The respiratory trumpets in Culicines are variable, but are usually elongate or broadly conical and unsplit.

ABDOMEN (fig. 12, *C*). The chaetotaxy of mosquito pupae has been studied by many workers; one of the most recent comprehensive studies is that of Knight and Chamberlain (432) in which at least one species each of twenty-eight mosquito genera is illustrated. Descriptions of the pupae are not included in this study; however, the pupae of the Anopheline species in this region have been described and illustrated by Penn (540) and Darsie (165). Darsie (166) has recently described and illustrated pupae of the Culicine mosquitoes found in the northeastern United States.

The shape, position, or absence of hairs 7 and 8, and length and nature of the fringe of the paddles provide diagnostic characters (fig. 12, *C*). In the Anophelini there are two hairs near the posterior end of the paddle, the paddle hair 8 and the accessory paddle hair 9. In the genus *Culex* there is, in addition to hair 8, a median terminal accessory paddle hair 7. In the other genera of Culicini of this region, hair 8 is either single, branched distally, or absent. Hair 8 is absent in *Toxorhynchites*.

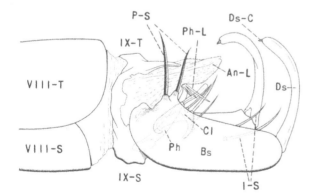

Fig. 11. Male Terminalia of *Anopheles* (Lateral aspect. Adapted from Ross and Roberts, 1943). An-L, Anal lobe. Bs, Basistyle. Cl, Claspette. Ds, Dististyle. Ds-C, Claw of dististyle. I-S, Internal spine. IX-S, Ninth sternite. IX-T, Ninth tergite. Ph, Phallosome. Ph-L, Leaflets of phallosome. P-S, Parabasal spines. VIII-S, Eighth sternite. VIII-T, Eighth tergite.

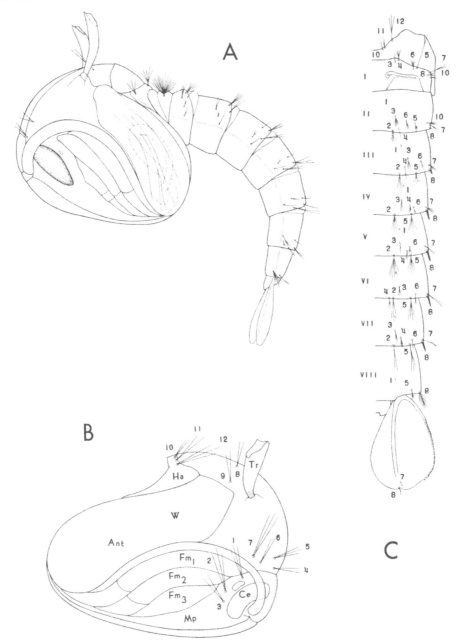

Fig. 12. *A*, Culicine pupa (lateral view). *B*, Generalized diagram of the cephalothorax of the mosquito pupa in lateral view (Redrawn from Penn, 1949). Ant, antenna. Ce, compound eye. Fm 1, 2, 3, femora of pro-, meso-, and metathorax. Ha, halter. Mp, mouth parts. Tr, respiratory trumpet. W, wing pad. *C*, Dorsal aspect of metanotum and abdomen of pupa of *Anopheles walkeri* Theobald (Redrawn from Penn, 1949).

Moorefield (518) and Carpenter (131) point out characters in the structure of the tenth abdominal segment (genital pouch) enabling us to determine sex in the pupal stage. In females, the genital pouch is generally short and broadly ovate in form (fig. 13, *B*), whereas in the male this pouch is usually much longer, more pointed, and bifurcate distally (fig. 13, *A*).

LARVAL CHARACTERS

The mosquito larva has three well-differentiated body regions, the head, the thorax, and the abdomen, each of which possesses variable

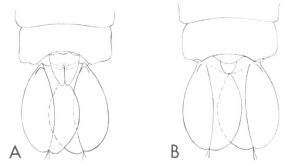

Fig. 13. Terminal segments of pupae of *Culiseta inornata* (Williston), showing sexual dimorphism. *A*, Male. *B*, Female. (Adapted from Moorefield, 1951.)

structures used by the taxonomist in classification. The principal morphological features are illustrated in figures 14–20. The terminology used here chiefly follows that employed by Belkin (48).

The head is flattened dorsoventrally and is formed of three large sclerites (figs. 15 and 16), a pair of lateroventral ocular sclerites (epicranial plates) and a single dorsal plate, the clypeus (frontoclypeus). A V-shaped epicranial suture is formed by the junction of these sclerites. The ocular sclerites bear the antennae and the imaginal and larval eyes. The clypeus has attached to its anterior border the preclypeus.

The mouth parts are ventral in position, but parts of the labrum extend anteriorly and are visible in a dorsal view. The labrum is composed of a median hairy palatum and lateral lobes bearing the mouth brushes. Other than

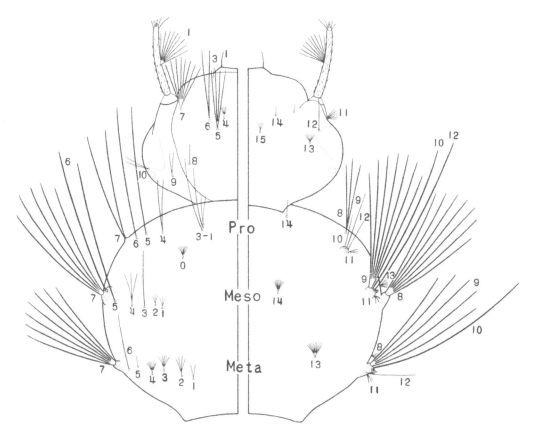

Fig. 14. Head and thorax of *Aedes* larva (dorsal and ventral aspects). *Antenna:* (1), antennal tuft. *Head:* (1), inner preclypeal spine; (3), outer clypeal hair; (4), postclypeal hair; (5), upper frontal hair; (6), lower frontal hair; (7), preantennal hair; (8), sutural hair; (9), transsutural hair; (10), supraorbital hair; (11), basal hair; (12), infraorbital hair; (13), subbasal hair; (14), postmaxillary hair; (15), submental hair. *Prothorax:* (0), accessory dorsal hair; (1–3), submedian or shoulder hairs; (4–7), dorsal hairs; (8), dorsolateral hair; (9–12), prothoracic pleural hairs; (14), median ventral hair. *Mesothorax:* (1–7), dorsal hairs; (8), dorsolateral hair; (9–12), mesothoracic pleural hairs; (13), ventrolateral hair; (14), median ventral hair. *Metathorax:* (1–7), dorsal hairs; (8), dorsolateral hair; (9–12), metathoracic pleural hairs; (13), median ventral hair.

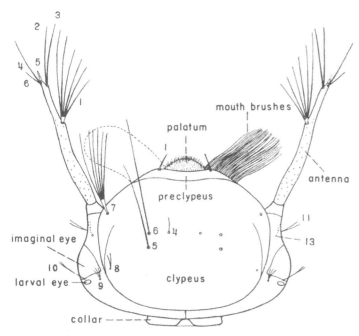

Fig. 15. Head of Culicine larva (dorsal view). *Antenna:*
(1), antennal tuft; (2), inner subapical hair; (3), outer
subapical hair; (4), terminal antennal hair; (5), papilla;
(6), fingerlike process. *Head:* (1), inner preclypeal spine;
(4), postclypeal hair; (5), upper frontal hair; (6), lower
frontal hair; (7), preantennal hair; (8), sutural hair; (9),
transsutural hair; (10), supraorbital hair; (11), basal hair;
(13), subbasal hair.

the labrum, the most prominent structures on
the ventral side of the head are the large
maxillae, which are occasionally armed with
characteristic spines, and the heavily sclerotized mandibles which bear teeth. The labium
forms the remainder of the floor of the mouth
and is composed of a proximal prementum, an
intermediate mentum, and a distal submentum.
The shape and size of the mentum and arrangement of its teeth provide diagnostic characters
in some species.

The thorax is made up of three fused segments, the pro-, meso-, and metathorax, distinguished only by hair groups, particularly
the pleurals which are present on each segment.

The abdomen is composed of nine segments,
the first seven of which are somewhat similar
and unmodified. The eighth segment bears the
breathing apparatus posterodorsally. The modifications of this respiratory organ provide excellent features for identification. The ninth or
anal segment bears several structures of taxo-

nomic importance, particularly in the Culicines.

HEAD (figs. 14 to 16). The principal taxonomic features of the antenna are its length,
shape, color, presence or absence of spicules
on the shaft, and the position and nature of the
antennal tuft or hair 1. The antenna has on its
distal end the inner subapical hair 2, the outer
subapical hair 3, the terminal antennal hair 4,
the papilla 5, and the fingerlike process 6 as
shown in figures 15 and 16, *B*. Hair 2 is known
as the dorsal sabre, and hair 3 the ventral sabre
in Anophelines.

The paired dorsal head hairs most frequently used in specific descriptions of larvae are as
follows: inner preclypeal hair or spine 1; inner
clypeal 2, well developed in Anophelines but
usually absent or minute in Culicines; outer
clypeal 3, well developed in Anophelines but
small in Culicines; postclypeal 4; inner or
upper frontal 5; mid or lower frontal 6; preantennal or outer frontal 7; sutural 8; transsutural 9; and supraorbital 10. The size, position, and number of branches of the head hairs
4–7, vary greatly in the genera and species of
Culicines and often provide excellent characters for classification.

THORAX (figures 14 and 17). The prothorax
has fifteen pairs of hairs, 0–14. The submedian
group 1–3 is frequently used in classifying
Anophelines and certain species of *Uranotaenia*. Prothoracic hairs 1–7 have been used
by some workers to separate certain species of
Aedes. The pleural groups 9–12 on all three
segments may be used for classifying Anophelines.

The mesothorax usually has fourteen pairs
of hairs, 1–14. Dorsal hairs 1 and 3 provide
means of distinguishing some of the species of
Culiseta found in this region. The metathorax
has thirteen pairs of hairs, 1–13. Dorsal hair 3
is palmate in some Anophelines. Most of the
larger hairs on the thorax of Anophelines are
pinnately branched. Anopheline larvae also
have situated anterodorsally on the thorax the
paired transparent retractile notched organs of
Nuttall and Shipley. When extended, these
organs make contact with the surface film,
probably helping support the larva and pre-
venting the thorax from rotating with the head
while feeding.

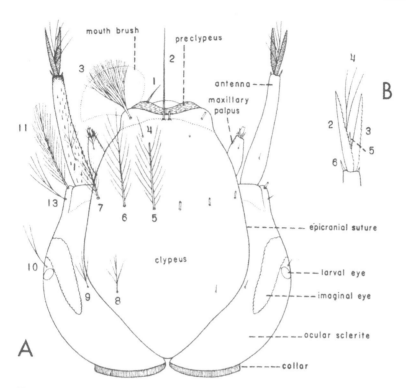

Fig. 16. Head of *Anopheles* larva. *A*, Dorsal view of head. (1), inner preclypeal spine; (2), inner clypeal hair; (3), outer clypeal hair; (4), postclypeal hair; (5), inner frontal hair; (6), middle frontal hair; (7), outer frontal or preantennal hair; (8), sutural hair; (9), transsutural hair; (10), supraorbital hair; (11), basal hair; (13), subbasal hair. *Antenna:* (1), antennal hair. *B*, Tip of antenna. (2), dorsal sabre; (3), ventral sabre; (4), terminal antennal hair; (5), papilla; (6), fingerlike process.

ABDOMEN. Important taxonomic structures found on abdominal segments I to VI of Anophelines are shown in figure 18. They include the accessory dorsal hair 0, palmate hair 1, antepalmate hair 2, dorsal hairs 3–5, upper and lower lateral hairs 6 and 7, anterior dorsolateral 8, the main tergal plate (MTP), and the accessory tergal plate (ATP). Lateral abdominal hairs 6 and 7 are used in the description of most of the Culicine larvae.

In Anopheline larvae, segment VIII bears the spiracular structures posterodorsally (siphon is absent), the pecten laterally, and hairs including pentad hairs 1–5 as illustrated (fig. 19). The bilateral pecten is a sessile sclerotized plate bearing both long and short teeth (fig. 19, *B*).

In Culicine larvae (fig. 20), segment VIII bears the siphon (air tube) posterodorsally, a bilateral comb composed of a row or patch of scales (absent in Toxorhynchitini), and the pentad hairs 1–5. A subventral longitudinal row of spines or teeth (pecten) extends bilater-

ally from the base of the siphon (absent in *Toxorhynchites, Orthopodomyia,* and *Wyeomyia*). The siphon also bears one or more pairs of siphonal tufts (sometimes obsolete in *Psorophora*). Lateral and subdorsal tufts are present in some genera and species. The dorsal preapical spine is situated dorsally near the distal end of the siphon or in a membrane beyond the tip of the siphon. The orifice of the siphon is surrounded by five valves: a single median or dorsal valve, a pair of small lateral valves, and a pair of large ventral or posterior valves, each of which usually bears small simple or branched hairs. A small sclerotized projection, the acus, may be present at the base of the siphon.

Features of the siphon of diagnostic value are its shape, relationship of its length to its basal diameter (siphonal index), nature of the pecten, and characteristics of the siphonal tuft or tufts. The siphonal index is obtained by comparing the length of the siphon, excluding the acus and the valves, with its diameter at the

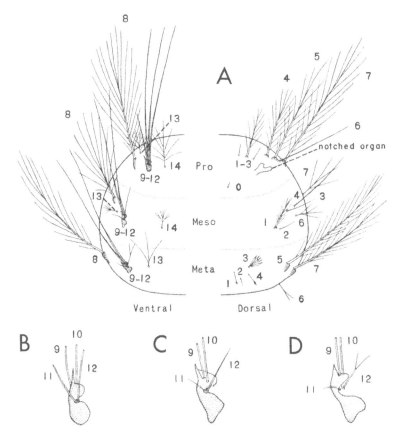

Fig. 17. Thorax of *Anopheles* larva. *A*, Ventral and dorsal views of prothorax, mesothorax and metathorax. *Prothorax:* 1–3, submedian or shoulder hairs. 4–7, dorsal hairs. 8, dorsolateral hair. 9–12, prothoracic pleural hairs. 13, ventrolateral hair. 14, median ventral hair. *Mesothorax:* 1–7, dorsal hairs. 8, dorsolateral hair. 9–12, mesothoracic pleural hairs. 13, ventrolateral hair. 14, median ventral hair. *Metathorax:* 1–2, dorsal hairs. 3, metathoracic palmate hair. 4–7, dorsal hairs. 8, dorsolateral hair. 9–12, metathoracic pleural hairs. 13, median ventral hair. *B, C,* and *D,* Bases of pleural hairs.

base. A siphon is referred to as "inflated" when it is considerably wider near the middle than at the base.

The ninth or anal segment, in both the Culicines and Anophelines, bears the following prominent structures: a sclerotized dorsal saddle which may or may not completely encircle the anal segment; a lateral hair or tuft arising on either side near the posterior margin of the saddle; a dorsal brush composed of upper (inner) and lower (outer) caudal hairs arising from the dorsoapical angle on either side of the anal segment; the ventroapical ventral brush composed of a staggered row of hair tufts, the bases of which may be sclerotized to form the barred area or grid; and two or four papilliform posterior anal gills. Tufts of the ventral brush arising from the barred area are gener-

ally referred to as the cratal tufts and those arising before the grid as precratal tufts.

EGG CHARACTERS

The shell of the mosquito egg comprises three layers: the innermost thin vitelline membrane surrounding the yolk; the intermediate endochorion, the more or less sclerotized opaque outer shell; and the exochorion, a thin transparent layer covering the endochorion and marked with small protuberances and reticulations. The endo- and exochorion together make up the chorion. In the new-laid egg the endochorion is also transparent but soon becomes opaque. Mosquito eggs are thus white when laid but gradually become dark brown or black.

The anterior pole of the egg bears the micropylar apparatus surrounding a minute opening, the micropyle. The micropyle permits entrance of the sperm cells from the spermathecae of the female during oviposition.

The eggs of Culicine mosquitoes are usually elongate-oval in shape (fig. 21). The larger end, containing the head of the developing em-

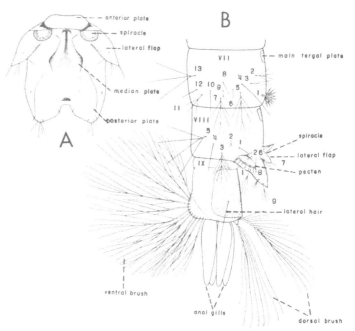

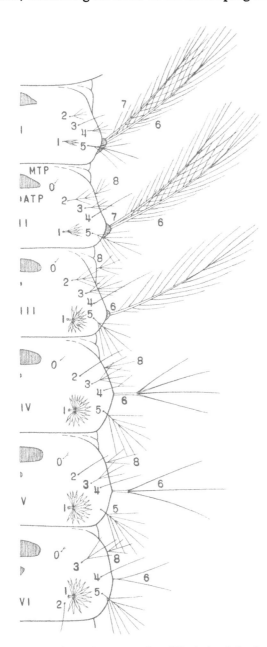

Fig. 19. Terminal segments of abdomen of *Anopheles* larva. *A*, Spiracular apparatus (dorsal view). *B*, Terminal segments VII to IX (lateral view). *Segment VII:* (1–5), dorsal hairs; (6–7), upper and lower lateral hairs; (8), anterior dorsolateral hair; (9), posterior ventrolateral hair; (10–13), ventral hairs. *Segment VIII:* (1), first pentad; (2), second pentad; (3), third pentad; (4), fourth pentad; (5), fifth pentad. *Spiracular Lobe:* (1), postspiracular hair; (2), pecten hair; (6), proximal dorsal valve hair; (7), distal dorsal valve hair; (8), proximal ventral valve hair; (9), distal ventral valve hair.

bryo, is usually rounded, though the posterior end is bluntly pointed. The eggs are laid singly by some mosquitoes, whereas certain others lay them in raftlike masses (fig. 21, *A*). The shape of the individual eggs is rather characteristic for various genera. The nature of the markings of the exochorion and the manner in which the eggs are laid are also useful in classification.

The eggs of Anopheline mosquitoes are boat-shaped, flattened, or slightly concave dorsally and convex ventrally (fig. 21). They are deposited separately on the water. The exochorion is modified to form a projecting frill, which partly or entirely surrounds the upper part, and a pair of air-filled lateral or dorsolateral floats. These structures, together with the bosses and reticulations of the exochorion, provide characters for separating allied species and races of *Anopheles* in some parts of the world.

Fig. 18. Abdominal segments I to VI of *Anopheles* larva (dorsal view). MTP, main tergal plate. ATP, accessory tergal plate. 0, accessory dorsal hair. 1, palmate hair. 2, antepalmate hair. 3–5, dorsal hairs. 6 and 7, upper and lower lateral hairs. 8, anterior dorsolateral hair.

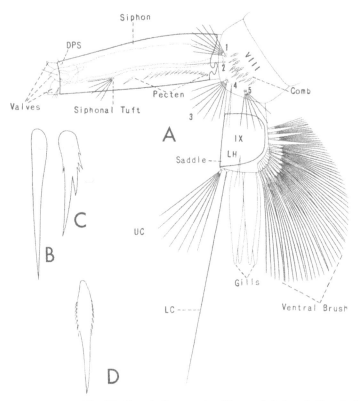

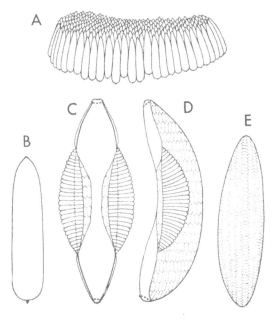

Fig. 21. Eggs of mosquitoes. *A*, Egg raft of *Culex*. *B*, Single egg of *Culex*. *C*, Egg of *Anopheles* (dorsal view). *D*, Egg of *Anopheles* (lateral view). *E*, Egg of *Aedes aegypti*.

Fig. 20. Terminal segments of larva of *Aedes*. *A*, Terminal segments. *Segment VIII:* (1), first pentad; (2), second pentad; (3) third pentad; (4), fourth pentad; (5), fifth pentad. *Segment IX (Anal Segment):* LH, lateral hair; LC, lower caudal hair of dorsal brush; UC, upper caudal tuft of dorsal brush. *Siphon:* DPS, dorsal preapical spine. *B* and *C*, pecten teeth. *D*, Comb scale.

Internal Anatomy
of the Female Mosquito

Since this present study deals primarily with systematics and biology of mosquitoes, detailed descriptions of internal anatomy will not be given. However, brief discussions of certain internal organs including the salivary glands, the principal divisions of the alimentary canal, the reproductive organs, and the fat body are included. These organs are shown in figure 22.

hypopharynx and forces the salivary fluid into the wound.

Alimentary canal.—The alimentary canal is composed of three principal divisions, the fore-gut, the mid-gut, and the hind-gut. In the insect embryo the fore-gut and hind-gut arise as invaginations of the body wall and are therefore lined with cuticle. The mid-gut is the main

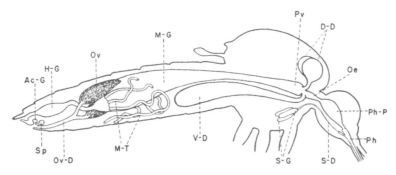

Fig. 22. Internal anatomy of female mosquito (adapted from Marshall, 1938). Ac-G, accessory gland; D-D, dorsal diverticula; H-G, hind-gut; M-G, mid-gut; M-T, malphigian tubes; Oe, oesophagus; Ov, ovary; Ov-D, oviduct; Ph, pharynx; Ph-P, pharyngeal pump; Pv, proventriculus; S-D, salivary duct; S-G, salivary glands; Sp, spermatheca; V-D, ventral diverticulum.

Salivary gland.—The paired salivary glands (S-G) are situated within the prothorax above the forelegs on either side of the oesophagus. Each gland is composed of three tubular lobes (acini). A salivary duct (S-D) passes from each gland forward through the neck of the mosquito, merging in the head to form the common salivary duct. The salivary duct traverses the entire length of the hypopharynx. A salivary pump is situated near the base of the

digestive part of the canal and has an epithelial lining which carries on secretion and absorption.

The fore-gut consists of the pharynx (Ph) and the oesophagus (Oe). The pharynx has a conspicuous pharyngeal pump (Ph-P) equipped with muscles which enable the chamber to be alternately expanded and contracted rapidly, thus pumping food up through the food channel and forcing it into the oesophagus. The

23

oesophagus is a simple tube leading from the pharynx to the mid-gut. Three slender tubes lead from the hind part of the oesophagus to the oesophageal diverticula, which consist of two dorsal diverticula (D-D) and a much larger ventral diverticulum (D-V). The proventriculus (Pv) is a two-way sphincter valve. By selective action of this valve it is said that blood sucked up by the female mosquito is admitted directly to the mid-gut whereas other foods, such as fruit juices, are temporarily shunted to the diverticula.

The mid-gut (M-G) is a straight tube narrowed anteriorly and expanded posteriorly. The narrow anterior part is situated within the thorax whereas the expanded posterior part, generally referred to as the stomach, is situated mostly within the first five abdominal segments. The malphigian tubes (M-T), five in number, arise from the mid-gut near its junction with the hind-gut. These tubes absorb waste matter from the blood and empty it into the gut for evacuation through the hind-gut.

The hind-gut varies in size and shape as it passes posteriorly to the anal opening. Its parts are referred to as the ileum-colon and rectum. Within the rectum are six rectal papillae which are said to absorb water from the excrement and retain it for further use.

Reproductive organs. — The reproductive system has two ovaries (Ov) situated in the abdomen dorsal to the hind part of the mid-gut. An ovarian duct leads from each ovary to the common oviduct (Ov-D), which is expanded distally to form the atrium. The sperm ducts leading from the spermathecae (Sp) and the duct from the accessory glands (Ac-G) also open into the atrium. In impregnated females the spermathecae are filled with spermatozoa which pass through the sperm ducts and fertilize the eggs as they pass through the atrium during oviposition. Culicine mosquitoes have three spermathecae, and Anophelines have one. In mosquitoes which lay their eggs in rafts, the accessory gland provides the secretion which cements the eggs together.

Fat body.—The fat body of insects is made up of masses of cells in which fat is stored in the form of globules. This stored nutriment may be used by the female mosquito during certain periods in her life such as hibernation or during the reproductive period. The fat body becomes well developed in female mosquitoes which go into complete hibernation.

Family Culicidae

The principal diagnostic features of the family Culicidae are found in the wing venation, which shows only slight modfications in the subfamilies and genera. Other family characters include the small rudimentary first segment of the antenna, the more or less enlarged second antennal segment, the completely divided pronotum, the combined praescutum and scutum, and the long legs usually lacking spurs on the tibiae. The larvae possess a complete head capsule and have only one pair of functional abdominal spiracles (absent in *Chaoborus*). The larvae are without exception essentially aquatic.

The Culicidae family consists of three subfamilies, Dixinae, Chaoborinae, and Culicinae. A rather full account of the genera and species of the subfamily Culicinae occurring in North America, north of Mexico, is given in the following pages. The main distinguishing characters of the larvae, pupae, and adults of the three subfamilies are presented in the following keys.

KEYS TO THE SUBFAMILIES
Adults

1. Antenna with a 14-segmented flagellum; wings without scales, vein Sc ending before or near the base of vein 2; mouth parts short, not fitted for piercing ....*Dixinae*
   Antenna with a 13-segmented flagellum; wings with scales at least on fringe, vein Sc extending beyond base of vein 2................................2
2. Mouth parts short; wings with veins clothed for the most part with long hairs, scales almost confined to the fringe...................*Chaoborinae*
   Mouth parts prolonged into the form of a proboscis; wings with scales on veins......*Culicinae*

Pupae

1. Paddles fused and incapable of movement
   Dixinae and *Chaoborinae* (in part)
   Paddles free and movable.....................................2
2. Respiratory trumpet on the cephalothorax either almost closed, or if open, with the spiracular opening near its middle..................*Chaoborinae*
   Respiratory trumpet on the cephalothorax open at the tip and with the spiracular opening at its base..*Culicinae*

Larvae

1. Thorax narrow and with distinct segmentation
   Dixinae
   Thorax distinctly broader than abdomen and without distinct segmentation 2
2. Antennae prehensile, with long and strong apical spines ....*Chaoborinae*
   Antennae not prehensile.........................*Culicinae*

Subfamily Culicinae

ADULT. Mouth parts in the form of a proboscis. Antenna with the first segment small, cylindrical and hidden beneath the enlarged globular second segment; flagellum with thirteen filamentous segments, usually plumose in males (sometimes alike or nearly so in both sexes), basal hair whorls on all segments except the first. Compound eyes uniform; ocelli absent. Thorax clothed dorsally with scales of various forms and setae. Pleura usually with patches of flat appressed scales. Abdomen clothed with scales (except in *Anophelini*). Legs clothed with scales which are usually flat and closely appressed. Wings with long slender fringe scales on the posterior margin. Veins clothed with scales of various sizes, shapes and colors. Vein Sc extending beyond the base of vein 2. *Male terminalia*. Always undergoes a rotation of 180° on its longitudinal axis shortly after emergence from the pupa and has variations in structure which are of much taxonomic value.

PUPA. Respiratory trumpets of various shapes and lengths, open at tip, spiracle located at or near base, surface never with hexagonal reticulations. First abdominal segment usually with a pair of large dendritic tufts. Eighth segment with a pair of free and movable paddles, each provided with a strong midrib and thickened on the outer margin but not on the inner margin. Pupa active and always rests with the cephalothorax horizontal and the abdomen somewhat curved underneath.

LARVA. Body with three distinct regions: head, thorax, and abdomen—each of which possesses variable structures. Posterior spiracles always functional. Mouth brushes well developed. Thoracic segments all more or less completely fused, always wider than the abdomen. Lacks ventral pseudopods, and without eversible groups of hooks on the ninth abdominal segment. Anal gills usually four in number and never retractile.

CLASSIFICATION. Following the classification given by Edwards (251), the subfamily Culicinae is divided into three tribes: Anophelini, Toxorhynchitini, and Culicini. Important characters for separating the three tribes in the larval, pupal, and adult stages are presented in the keys. Lane and Cerqueira (451) raised the Sabethines to tribal rank, and several other workers have followed this procedure.

KEYS TO TRIBES

ADULTS

1. Abdomen without scales, or at least the sternites largely bare; palpi of female nearly as long as the proboscis (except in *Bironella*)
   Anophelini, p. 28

   Abdomen densely scaled both dorsally and ventrally; palpi of female short (in some *Toxorhynchitini* they may be nearly two-thirds as long as the proboscis) 2

2. Proboscis rigid, the basal half thickened and the apical half strongly curved downward and backward; scutellum rounded on posterior margin ....*Toxorhynchitini*, p. 58

   Proboscis not rigid, of nearly uniform thickness, not curved downward and backward on apical half; scutellum trilobed on posterior margin
   Culicini. p. 62

Pupae

1. Lateral seta 8 spinelike on abdominal segments
   III to VI and placed almost exactly at the
   apical corners of the segments

   Anophelini, p. 28

   Lateral seta 8 on abdominal segments III to VI
   each consisting of a single hair or a branched
   hair and placed well before the apical corners
   of the segments..2
2. Outer part of the swimming paddles elongated
   lobelike and extending beyond tip of the midrib and inner part of paddle

   Toxorhynchitini, p. 58

   Outer part of the swimming paddles not lobelike
   and not conspicuously longer than the midrib
   and inner part of paddle...............*Culicini*, p. 62

Larvae

1. Eighth abdominal segment without an elongate
   dorsal siphon............................*Anophelini*, p. 28

   Eighth abdominal segment with an elongate dorsal siphon which is as long or longer than
   broad 2
2. Mouth brushes prehensile, each composed of
   about ten stout rods........*Toxorhynchitini*, p. 58

   Mouth brushes composed of 30 or more hairs
   (prehensile only in the subgenus *Psorophora*)

   Culicini, p. 62

Tribe Anophelini

ADULT. Clypeus longer than broad, rounded anteriorly. Palpi long in both sexes except in the genus *Bironella*. Thorax usually only slightly arched. Scutellum evenly rounded posteriorly and with marginal setae arranged in an unbroken row, except in *Chagasia* which has the scutellum trilobed. Abdomen either without scales or with loosely applied scales; the sternites at least are largely bare in scaly species. Legs long and slender, uniformly covered with closely applied scales, without distinct tibial bristles, without pulvilli. Wings usually with distinct markings. Squama with fringe. While the adult is in the resting position, the proboscis, thorax, and abdomen are held nearly in a straight line. *Male terminalia*. Phallosome tubular, with or without one or more pairs of leaflets distally. Basistyle short, usually without distinct basal lobe. Dististyle long, slender, and bearing a short terminal spine.

PUPA. Respiratory trumpets short and with wide opening. Abdominal segments III–VII with seta 8 spinelike and placed at or near the posterolateral corners of each segment; segment VIII with seta 8 stout and plumose. Paddles with hair 8 at tip of midrib, and a smaller hair 7 on midrib anterior to tip. The pupae of the Nearctic Anopheline mosquitoes north of Mexico have been described and illustrated by Penn (540).

LARVA. Head usually broader than long. While feeding, the head can be rotated through an arc of 180° so that the ventral side is uppermost. Most of the larger body hairs are pinnately branched. Mouth brushes are composed of a large number of simple hairs. The thorax has a pair of notched organs of Nuttall and Shipley. Abdomen with a series of palmate float hairs occurring in pairs on some or all of the first seven abdominal segments. Eighth abdominal segment without a dorsal respiratory siphon. A lateral chitinous plate with a posterior row of strong teeth (pecten) is present on the posterodorsal surface of the eighth abdominal segment. Dorsal brush of anal segment bilaterally consists of two irregularly branched hairs. The ventral brush is composed of a fan-shaped group of irregularly branched hairs.

EGGS. The eggs of Anophelini are usually boat-shaped and are equipped with dorsolateral or lateral floats. Both ends of the egg are pointed, the end corresponding to the head of the larva being somewhat broader and blunter than the other.

CLASSIFICATION. Edwards (251) recognized three genera: *Chagasia*, found in tropical America; *Bironella*, found in New Guinea and adjacent islands; and *Anopheles*, practically world-wide in distribution.

Genus ANOPHELES Meigen[1]

Anopheles Meigen, 1818, Syst. Beschr. Zweifl. Ins., 1:10.

Edwards (251) divides the genus *Anopheles* into four subgenera, based largely on characters of the male terminalia: *Anopheles*, found throughout the Old and New Worlds; *Nyssorhynchus*, found principally in Central and South America; *Stethomyia*, known from Cen-

[1]Consult Dyar (1928) or Edwards (1932) for synonyms.

28

tral and South America; and *Myzomyia*, found throughout the tropical and subtropical regions of the Old World. Two subgenera, *Anopheles* and *Nyssorhynchus*, are represented in this area.

The adult and larval characters given for the tribe Anophelini are for the most part applicable to the genus *Anopheles*. Salient characters of the male terminalia are as follows: phallosome tubular, tip with one or more pairs of leaflets or bare; claspette usually consisting of a dorsal and a ventral lobe; basistyle without basal and apical lobes. The basistyle may bear parabasal, accessory, and internal spines. Subgenus *Anopheles* has two parabasal spines, no accessory spines, and one internal spine. Subgenus *Nyssorhynchus* has one parabasal spine, two accessory spines, and one internal spine. The dististyle is long, curved, and not swollen medially.

We have followed Vargas and Matheson (729) in recognizing *Anopheles earlei* Vargas, *A. freeborni* Aitken, and *A. occidentalis* Dyar and Knab as distinct species and not involving them in the *maculipennis* group of European authors. *Anopheles bradleyi* King and *A. georgianus* are here recognized as distinct species.

KEYS TO THE SPECIES

ADULT FEMALES

1. Wings with areas of pale scales....................2
    Wings entirely dark-scaled (except for a silver- or bronze-colored apical fringe spot in *A. earlei* and *A. occidentalis*).............6

2 (1). Hind tarsus with a broad white band; wing costa with four or more spots of pale scales.....................*albimanus* Wied., p. 55
    Hind tarsus entirely dark; wing costa with no more than two spots of pale scales........3

3 (2). Wing costa with a pale spot at outer third opposite tip of subcosta; vein 6 with one or two areas of dark scales.........................4
    Wing costa dark except for pale spot at extreme tip; vein 6 with three areas of pale scales (2 in male) *crucians* Wied., p. 35; *bradleyi* King, p. 34; *georgianus* King, p. 42

4 (3). Wing veins 3 and 5 entirely dark-scaled; vein 6 with apical half and basal fourth dark-scaled; palpi unbanded
    punctipennis (Say), p. 48
    Wing veins 3 and 5 with long areas of pale scales; vein 6 with apical half dark, basal half white; palpi banded..........................5

5 (4). Wing vein 4 predominantly pale-scaled; terminal segment of palpus white-scaled
    p. pseudopunctipennis Theob., p. 46
    Wing vein 4 with stem predominantly dark-scaled; apical part of terminal segment of palpus dark-scaled
    p. franciscanus McCracken, p. 45

6 (1). Wings without spots of dark scales; setae on scutum at least half as long as width of scutum; scutum shiny on rubbed specimens; small species......*barberi* Coq., p. 32
    Wings with spots of dark scales, more or less distinct; setae on scutum rarely half as long as width of scutum; scutum dull on rubbed specimens; medium-sized species 7

7 (6). Tip of wing with a silver- or copper-colored fringe spot 8
    Tip of wing without pale fringe spot...........9

8 (7). Scales on stem of wing vein 2, between fork and dark spot, raised..*earlei*[2] Vargas, p. 38
    Scales on stem of wing vein 2, between fork and dark spot, rather closely appressed
    occidentalis D. and K., p. 43

9 (7). Frontal tuft with some pale setae; wing with four distinct dark spots; palpi entirely dark 10
    Frontal tuft entirely dark; wing usually with indistinct dark spots; segments of palpi with or without distinct white apical rings 11

10 (9). Wing scales on basal part of vein 5, before the fork, truncate at apices with serrate edges; generally occurs west of the Rocky Mountains..............*freeborni* Aitken, p. 39
    Wing scales on basal part of vein 5, before the fork, predominantly broadly rounded at apices without serrate edges; occurs east of the Rockies..*quadrimaculatus* Say, p. 50

11 (9). Knob of halter usually pale-scaled; palpi with narrow white apical segmental rings; knee spots present......*walkeri* Theob., p. 53
    Knob of halter entirely dark-scaled; palpi with faint apical segmental white rings or entirely dark; knee spots reduced or absent......................*atropos* D. and K., p. 31

MALE TERMINALIA

1. Basistyle with one internal spine, two accessory spines, and one parabasal spine (subgenus *Nyssorhynchus*)
    albimanus Wied., p. 55
    Basistyle with one internal spine, no accessory spines, and two parabasal spines (subgenus *Anopheles*)...............................2

2 (1). Leaflets of phallosome extremely delicate and hard to discern, or absent....................3
    Leaflets of phallosome distinct......................4

[2]*Anopheles earlei* generally occurs in the northern United States and southern Canada east of the Rockies, whereas *Anopheles occidentalis* is found in a narrow strip along the Pacific coast.

Tribe Anophelini

3 (2). Phallosome without leaflets; small species
    barberi Coq., p. 33
  Phallosome with extremely delicate leaflets appearing as fine lines (hard to discern); medium-sized species
    p. franciscanus McCracken, p. 45

4 (2). Leaflets of phallosome serrate nearly to tips; dorsal lobe of claspette bearing three overlapping bladelike apical filaments
    p. pseudopunctipennis Theob., p. 47
  Leaflets of phallosome smooth, although a few coarse basal teeth may be present; dorsal lobe of claspette bearing spines or short rods, not bladelike............................5

5 (4). Claspette with dorsal and ventral lobes fused forming a single fleshy conical lobe on each side........*crucians* Wied., p. 36; *bradleyi* King, p. 34; *georgianus* King, p. 42
  Claspette with dorsal and ventral lobes distinct, at least apically................................6

6 (5). External spine of dorsal lobe of claspette capitate or bluntly rounded at apex..........7
  External spine of dorsal lobe of claspette acuminate at apex......................................9

7 (6). One or more leaflets of phallosome with a few coarse basal teeth; lobes of ninth tergite usually somewhat constricted medially, widened and rounded apically
    quadrimaculatus Say, p. 51
  Leaflets of phallosome without basal teeth; lobes of ninth tergite narrow and pointed or rounded apically, not widened............8

8 (7). Anteapical pair of phallosome leaflets not more than one-half the length of the apical pair......................*atropos* D. and K., p. 31
  Anteapical pair of phallosome leaflets more than one-half as long as the apical pair
    walkeri Theob., p. 53

9 (6). Lobes of ninth tergite wide and short, with the apex somewhat expanded
    earlei Vargas, p. 38
  Lobes of ninth tergite long and narrow
    freeborni[3] Aitken, p. 40; *occidentalis* D. and K., p. 43; *punctipennis* (Say), p. 49

Larvae (Fourth Instar)

1. Frontal head hairs 5–7 large, plumose; lateral hair 6 of abdominal segments 1–III plumose 2
  Frontal head hairs 5–7 small, simple; lateral hair 6 of abdominal segments I–VI plumose..........................*barberi* Coq., p. 33

2 (1). Palmate hair 1 well developed on abdominal segments I–VII, the individual leaflets with smooth margins....*albimanus* Wied., p. 56
  Palmate hair 1 obsolete or greatly reduced on abdominal segments I and II..................3

_____
[3]Lobes of ninth tergite are usually a little narrower in *A. freeborni* than in *A. occidentalis* and *A. punctipennis.*

3 (2). Outer clypeal hair 3 simple; hair 9 of meso- and metathoracic pleural groups short and stout, no more than half as long as hair 10 4
  Outer clypeal hair 3 feathered or branched; hairs 9 and 10 of meso- and metathoracic pleural groups nearly equal in length........5

4 (3). Spiracular apparatus with the tip of the posterior plate prolonged into two long black upturned tails; antepalmate hair 2 on abdominal segment IV single
    p. pseudopunctipennis Theob., p. 47
  Spiracular apparatus with tip of posterior plate rounded, not prolonged; antepalmate hair 2 on abdominal segment IV 1- to 3-branched..*p. franciscanus* McCracken, p. 45

5 (3). Outer clypleal hair 3 sparsely branched (five to ten branches)....*atropos* D. and K., p. 31
  Outer clypeal hair 3 strongly dichotomously branched (more than twenty-five branches) 6

6 (5). Inner clypeal hair 2 with sparse minute feathering toward tip; prothoracic hair 1 with three to five strong branches near base............................*walkeri* Theob., p. 53
  Inner clypeal hair 2 simple (sometimes forked beyond middle); prothoracic hair 1 short, single or branched beyond middle 7

7 (6). Accessory hair 0 and hair 2 on abdominal segments IV and V both well developed, each with four to nine branches
    crucians Wied., p. 36
  Accessory hair 0 absent or rudimentary......8

8 (7). Palmate hair 1 on abdominal segments III–VII about equally developed (palmate on segment VII may be somewhat smaller); individual leaflets of palmates III and VII serrated on margins 9
  Palmate hair 1 on abdominal segments IV–VI about equally developed, those on segments III and VII being rudimentary or only about one-half to two-thirds as large; individual leaflets of palmates III and VII usually with smooth margins...................10

9 (8). Inner clypeal hairs 2 generally forked beyond middle....................*earlei* Vargas, p. 38
  Inner clypeal hairs 2 simple........................11

10 (8). Palmate hair 1 functional on abdominal segments IV–VI only, those on segments III and VII being rudimentary; antepalmate hair 2 with three to six branches on segments IV and V (rarely with two branches)..............*georgianus* King, p. 42
  Palmate hair 1 functional on abdominal segments III–VII, although those on segments III and VII are reduced to about one-half to two-thirds the size of those on segments IV–VI; antepalmate hair 2 single or double on segments IV and V (sometimes triple)..........................*bradleyi* King, p. 34

11 (9). Basal tubercles of inner clypeal hairs 2 separated by at least the diameter of one tubercle............*quadrimaculatus* Say, p. 51
Basal tubercles of inner clypeal hairs 2 usually separated by less than the diameter of one tubercle....................................12

12(11). Antepalmate hair 2 on abdominal segments IV and V usually single
occidentalis D. and K., p. 43
Antepalmate hair 2 on abdominal segments IV and V usually double or triple
freeborni[4] Aitken, p. 40;
punctipennis (Say), p. 49

Anopheles (Anopheles) atropos
Dyar and Knab

Anopheles atropos Dyar and Knab, 1906, Proc. Biol. Soc. Wash., *19*:160.

ADULT FEMALE (pl. 1). Medium-sized species. *Head:* Proboscis long, dark brown; palpi about as long as the proboscis, entirely dark or with very narrow pale rings at apices

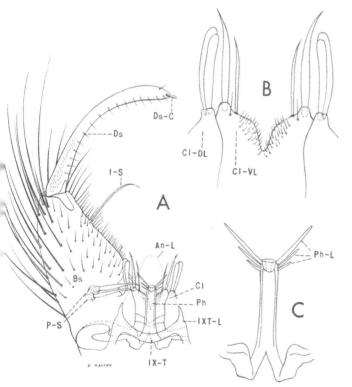

Fig. 23. *Anopheles atropos* Dyar and Knab. *A*, Male terminalia. *B*, Claspettes. *C*, Phallosome.

[4]No reliable characters are known for separating the fourth instar larvae of these two species. In *A. freeborni* the frontal hairs of the head are usually associated with three or five spots, whereas *A. punctipennis* from the western United States usually has these hairs inserted in a transverse dark band.

of distal segments, with raised scales on basal part. Occiput densely clothed with dark erect forked scales on median and posterior regions; frontal tuft dark. *Thorax:* Integument of scutum dark brown to dark reddish brown; scutum clothed with short golden-brown hairs and larger black setae, the black setae are more numerous laterally. Scutellum crescent-shaped, clothed with dark brown setae. *Abdomen:* Integument dark brown to black, densely clothed with dark hairs. *Legs:* Legs dark; knee spots reduced or absent. *Wing:* Length about 3.5 to 4.0 mm. Scales narrow, entirely dark, some of the scales often arranged in four darker spots, usually indistinct. *Halter:* Knob dark-scaled.

ADULT MALE. Coloration similar to that of the female. *Terminalia* (fig. 23). Ninth tergite (IX-T) with band narrow, moderately sclerotized; lobes (IXT-L) slender, tapered, pointed, or rounded at tip. Tenth sternite absent. Anal lobe (An-L) large, triangular, spiculate. Phallosome (Ph) cylindrical, broad and furcate at base; apex with three or four pairs of ligulate subequal leaflets (Ph-L) without coarse basal teeth, the apical pair about twice as long as the anteapical pair (fig. 23, *C*). Claspette (fig. 23, *B*) broad, consisting of a dorsal lobe (Cl-DL) bearing one to several (usually two or three) large rounded or slightly capitate spines and a ventral lobe (Cl-VL) with a large broad pointed apical spine and one or two smaller subapical spines. Basistyle (Bs) conical, less than twice as long as mid-width, clothed with long and short setae and few or no scales; two parabasal spines (P-S) and one internal spine (I-S) present. Dististyle (Ds) slightly longer than basistyle, curved, with a row of small papillated setae along dorsal surface; numerous minute nonpapillated hairs on basal third or half; claw (Ds-C) short, blunt.

LARVA (fig. 24). *Head:* Antenna sparsely spiculate; antennal tuft 1 small, branched distally, inserted on basal third of shaft; terminal hair 4 with several branches. Inner clypeal hairs 2 usually sparsely feathered at tip (occasionally simple), basal tubercles may or may not be separated by more than the diameter of a single tubercle; outer clypeal 3 with five to ten short branches on distal half; postclypeal 4 small, simple or branched beyond base; frontal hairs 5–7 large, plumose; sutural 8 and trans-

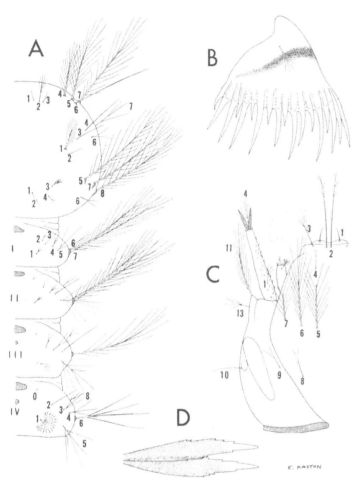

Fig. 24. Larva of *Anopheles atropos* Dyar and Knab. *A*, Thorax and abdomen. *B*, Pecten. *C*, Head. *D*. Leaflets of palmate hair.

sutural 9 simple or 2- or 3-branched beyond base. *Thorax:* Prothoracic hairs 1 and 3 short, simple; hair 2 larger, variable (with two to several branches). Prothoracic pleural group 9–12 with hairs 9, 10, and 12 long, simple; hair 11 about one-fourth as long as 9, 10, or 12, simple or bifid. Mesothoracic pleural group 9–12 with hairs 9 and 10 long, simple, subequal; hair 11 minute, simple; hair 12 simple, short, but much longer than hair 11. Metathoracic pleural group 9–12 with hairs 9 and 10 long, simple, subequal; hair 11 minute, simple; hair 12 is 2- or 3-branched, short, but much longer than hair 11. Metathoracic palmate hair 3 small, with transparent leaflets. *Abdomen:* Accessory dorsal hair 0 obsolete; palmate hair 1 rudimentary on segments I and

II, partly developed on segment III and well developed on segments IV–VII; leaflets with serrations on apical half. Antepalmate hair 2 long, simple (rarely bifid) on segments IV and V. Lateral hair 6 long, plumose on segments I–III. Pecten as illustrated (fig. 24, *B*).

DISTRIBUTION. The Atlantic and Gulf coastal regions of the United States from New Jersey to Texas. It has been found also in Cuba (143) and Jamaica (361). *United States:* Alabama (322); Florida (135, 418); Georgia (135, 514); Louisiana (418); Maryland (64, 161); Mississippi (135, 418); New Jersey (674); North Carolina (420, 646); South Carolina (135, 742); Texas (418); Virginia (418).

BIONOMICS. The larvae develop in permanent salt-water pools or in marshes where the salt content varies from about 1 to 12 per cent. The larvae are often found in very shallow pools. The adults may occur in rather large numbers in salt marshes near the larval habitat. Griffitts (322) observed that the females attack man during the day, even in bright sunlight, and also at night. He noticed that the adults entered human habitations in considerable numbers in one area in Louisiana; however, they rarely enter buildings in many other areas. Komp (438) describes the females as vigorous biters along the Mississippi coast where they attack any exposed part of the human body, even in broad daylight. He also observed adults flying some distance from the shore to attack people in boats. Komp observed males in a swarm hovering in the protection of low bushes near the beach on Ship Island, Mississippi. The adults are readily attracted to artifical light, more so than are most other Anophelines.

MEDICAL IMPORTANCE. Mayne and Griffitts (503) were able to infect females of *A. atropos* which had fed on carriers of *Plasmodium vivax*; however, the species is believed to be unimportant as a natural vector of malaria.

Anopheles (Anopheles) barberi
Coquillett

Anopheles barberi Coquillett, 1903, Can Ent., 35:310.

ADULT FEMALE (pl. 2). Small species. *Head:* Proboscis long, dark; palpi about as long as the proboscis, dark. Occiput clothed with numerous erect forked scales, the ones on the central part and vertex pale; frontal tuft dark. *Thorax:*

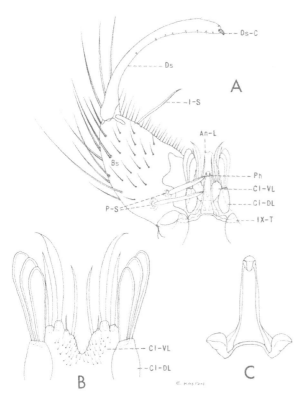

Fig. 25. *Anopheles barberi* Coquillett. *A*, Male terminalia. *B*, Claspettes. *C*, Phallosome.

one of which is rather broad and the other is slender. Basistyle (Bs) a little longer than broad, clothed with long and short setae, without scales; two strong parabasal spines (P-S) and one internal spine (I-S) present. Dististyle (Ds) much longer than basistyle, curved, with a few small papillated setae on the distal half, nonpapillated hairs absent; claw (Ds-C) short, blunt.

LARVA (fig. 26). *Head:* Antenna short, dark and glabrous; antennal hair 1 small, simple, inserted near middle of shaft; terminal hair 4 rather long, simple. Inner clypeal hairs 2 simple, widely spaced; outer clypeal 3 shorter than inner clypeal 2, usually simple, occasionally bifid; postclypeal 4 simple or bifid, longer than outer clypeal; frontal hairs 5–7 small, simple; sutural 8 and transsutural 9 small, simple. *Thorax:* Integument sparsely spiculate. Prothoracic hair 1 long, usually sparsely feathered (feathering varies from weak to strong and may be along most of the shaft or restricted to apex); hair 2 usually a little

Integument of scutum brown, shiny; scutum clothed with long dark setae which are at least half as long as the width of the scutum. Scutellum crescent-shaped, clothed with dark setae. *Abdomen:* Integument brown, rather densely clothed with dark-brown hairs. *Legs:* Legs entirely dark. *Wing:* Length about 3.0 mm. Scales slightly broadened, uniformly dark. *Halter:* Knob dark-scaled.

ADULT MALE. Coloration similar to that of the female. *Terminalia* (fig. 25). Ninth tergite (IX-T) membranous dorsally, sclerotized laterally; lobes represented by the rounded posterior part of each lateral sclerotized area. Tenth sternite absent; anal lobe (An-L) large, triangular, finely spiculate. Phallosome (Ph) cylindrical, broad at base, tapering to a blunt, rounded point at apex, without leaflets (fig. 25, *C*). Claspette (fig. 25, *B*) broad, consisting of a dorsal lobe (Cl-DL) bearing three overlapping spatulate filaments and a ventral lobe (Cl-VL) bearing one large, broad, pointed, apical spine and two smaller subapical spines,

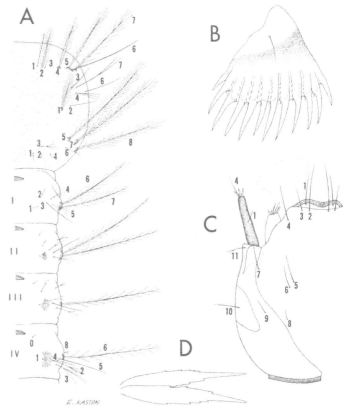

Fig. 26. Larva of *Anopheles barberi* Coquillett. *A*, Thorax and abdomen. *B*, Pecten. *C*, Head. *D*, Leaflets of palmate hair.

34

Tribe Anophelini

longer than hair 1, with many lateral branches; hair 3 short, simple. Prothoracic pleural group 9–12 with hair 9 long, barbed; hairs 10 and 12 long, simple; hair 11 minute. Mesothoracic pleural group 9–12 with hair 9 long, barbed; hair 10 long, simple; hair 11 obsolete; hair 12 short, simple, or branched apically. Metathoracic pleural group 9–12 with hair 9 long, barbed; hair 10 long, simple; hair 11 obsolete; hair 12 short, simple or branched apically. Metathoracic palmate hair 3 obsolete. *Abdomen:* Accessory dorsal hair 0 obsolete. Palmate hair 1 rudimentary on segment I, well developed on segments II–VII, leaflets with serrations on apical half. Antepalmate hair 2 long, branched apically, and inserted lateral to palmate 1 on segments IV and V. Lateral hair 6 long, plumose on segments I–VI (lateral branches rather short). Pecten as illustrated (fig. 26, *B*).

DISTRIBUTION. Eastern United States, north to New York and west to Nebraska and Texas. It has been found also in northern Mexico (730). *United States:* Alabama (135, 418); Arkansas (88, 121); Delaware (167); District of Columbia (318); Florida (135, 418); Georgia (135, 419); Illinois (615); Indiana (133, 152); Iowa (400); Kansas (14); Kentucky (418, 570); Louisiana (418, 759); Maryland (161, 213); Mississippi (135, 546); Missouri (169, 324); Nebraska (14); New Jersey (418); New York (3); North Carolina (135, 646); Ohio (400); Oklahoma (628); Pennsylvania (756); South Carolina (135, 742); Tennessee (135, 658); Texas (463, 634); Virginia (176, 178).

BIONOMICS. The larvae are found in rot cavities in trees of many kinds, in stumpholes, and occasionally in artificial containers such as wooden tubs and tin cans in or near wooded areas. When found in artificial containers, the larvae are generally associated with leaves and other plant debris. The larvae usually dive when disturbed but soon rise to the surface where they are easily seen because of their shiny black appearance. According to Matheson (492) the winter is passed in the north in hibernation as early larval instars, sometimes frozen in solid ice in the treeholes. The larval development is quite rapid following the onset of warm weather in the spring in the north, and

the first adults usually emerge in June. Larvae may be found any time during the year in the south when the habitats are flooded.

The adults are occasionally found resting during the daytime underneath bridges, in culverts, and in buildings in or near wooded areas. Both sexes are attracted to artificial light, but the males are more often found in light-trap collections than the females. Thibault (683) states that *A. barberi* enters houses readily and is a persistent biter, although the feeding female is nervous and seldom finishes a blood meal at one sitting.

MEDICAL IMPORTANCE. Stratman-Thomas and Baker (678) infected *A. barberi* with *Plasmodium vivax* and demonstrated transmission of the malaria to another patient. No data are avalaible as to its infection in nature; however, it is rarely abundant and is probably unimportant as a natural vector of malaria.

Anopheles (Anopheles) bradleyi King

Anopheles crucians var. *bradleyi* King, 1939, Amer. Jour. Trop. Med., *19*:468.

ADULT FEMALE. Very similar to *A. crucians* (pl. 3) and *A. georgianus*, but frequently has stem of vein 5 entirely white-scaled.

ADULT MALE. Coloration similar to that of the female, but with only two dark spots present on vein 6. *Terminalia.* Similar to that of *A. crucians* (fig. 28) and *A. georgianus*.

LARVA (fig. 27). *Head:* Antenna brown, spiculate (spicules coarse and stout, thickened and darkly pigmented at base); antennal tuft 1 with several branches, inserted at basal third of shaft; terminal hair 4 with several branches. Inner clypeal hairs 2 simple, the distance between the basal tubercles is variable but is usually less than the diameter of one of the tubercles; outer clypeal 3 densely dichotomously branched; postclypeal 4 small, single, or rarely branched on apical half; frontal hairs 5–7 large, plumose; sutural 8 and transsutural 9 usually 3- to 5-branched (hair 9 longer than hair 8 and reaching beyond base of hair 7). *Thorax:* Prothoracic hair 1 simple or branched distally; hair 2 stout and with many branches; hair 3 short, simple. Prothoracic pleural group 9–12 with hairs 9, 10, and 12 long, simple, subequal; and hair 11 short, simple. Meso-

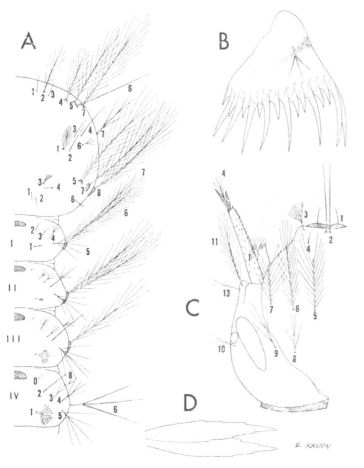

Fig. 27. Larva of *Anopheles bradleyi* King. *A*, Thorax and abdomen. *B*, Pecten. *C*, Head. *D*, Leaflets of palmate hair.

thoracic pleural group 9–12 with hairs 9 and 10 long, subequal; hair 11 minute; and hair 12 about one-third as long as hair 9 or 10. Metathoracic pleural group 9–12 with hairs 9 and 10 long, simple; hair 11 minute, simple; and hair 12 short, double or triple. Metathoracic palmate hair 3 small, with transparent leaflets. *Abdomen:* Accessory dorsal hair 0 obsolete on segments IV and V; palmate hair 1 well developed and of equal size on segments IV–VI, leaflets serrated beyond middle; palmates usually about one-half to two-thirds as large on segments III and VII, leaflets usually smooth; palmates weakly developed on I and II. Antepalmate hair 2 usually simple on segments IV and V, sometimes double or rarely triple. Hair 5 on segment I is 4- or 5-branched (rarely more) and about twice the size of hair 4. Lateral hair 6 long, plumose on segments I–III. Pecten as illustrated (fig. 27, *B*).

DISTRIBUTION. The Atlantic and Gulf coastal regions of the United States from New York to Texas. It has also been found along the Gulf coast of Mexico from Tamaulipas to Quintana Roo (730). *United States:* Alabama (135, 418); Delaware (167); Florida (135, 418); Georgia (135, 516); Louisiana (418); Maryland (418, 735); Mississippi (135, 546); New Jersey (165, 269); New York (31); North Carolina (135, 646); South Carolina (135, 742); Texas (532); Virginia (176, 178).

BIONOMICS. The larvae are found in pools of brackish water near the coast, preferably in pools containing *Chara* and other aquatic grasses. King *et al.* (419) state that the larvae of *A. bradleyi* have been taken together with *A. atropos* in water of rather high salt concentration and with the larvae of *A. crucians* in nearly fresh water. Vogt (735) observed that the larvae of *A. bradleyi* were distributed in a study area in Maryland with little regard to salinity, but were generally found in situations where emergent vegetation was present, except during the spring and fall when larvae could also be found in exposed water surfaces. Very little is known of the habits of the adults, since they cannot be distinguished with certainty at this time from *A. crucians* which also occurs in the range of *A. bradleyi*.

MEDICAL IMPORTANCE. Boyd *et al.* (80) were able to infect *A. bradleyi* with *Plasmodium falciparum*, but nothing is known of its ability to transmit malaria in nature.

Anopheles (Anopheles) crucians
Wiedemann

Anopheles crucians Wiedemann, 1828, Ausser. Zweifl., Ins., *1*:12.
Anopheles crucians var. *crucians*, King, 1939, Amer. Jour. Trop. Med., *19*:470.

ADULT FEMALE (pl. 3). Indistinguishable from *A. bradleyi* and *A. georgianus*. Medium-sized species. *Head:* Proboscis long, black; palpi a little shorter than the proboscis, black, with raised scales on basal part, segment 3 with a few white scales basally, segment 4 white-ringed basally and apically, segment 5 entirely white. Occiput clothed with numerous erect forked scales, those of central part white, others dark; scales of vertex narrow, white;

frontal tuft white. *Thorax:* Integument of scutum mottled gray, brown, and black, a pair of dark gray submedian longitudinal stripes present; scutum clothed with numerous short yellowish hairs, a few narrow whitish scales on anterior promontory and longer black setae on lateral fossae. Scutellum crescent-shaped, clothed with yellowish hairs and long brown setae. *Abdomen:* Integument of abdomen dark brown to black, clothed with numerous yellow to dark-brown hairs. *Legs:* Legs dark, femora and tibiae tipped with white. *Wing:* Length about 4.0 mm. Wing with white to yellowish-white scales, arranged on the veins in contrasting lines and spots (costa dark except for pale spot at extreme tip; vein 6 with three spots of dark scales; stem of vein 5 partly or entirely dark-scaled). *Halter:* Knob black-scaled.

ADULT MALE. Coloration similar to that of the female, but with only two dark spots on wing vein 6. *Terminalia* (fig. 28). Similar to that of *A. bradleyi* and *A. georgianus*. Ninth tergite (IX-T) with band narrow, sclerotized; lobes (IXT-L) very long and slender, rounded or bluntly pointed apically. Tenth sternite absent; anal lobe (An-L) large, triangular, spiculate. Phallosome (Ph) cylindrical, broad and furcate at base; apex with three or four pairs of ligulate leaflets (Ph-L), the apical pair longer, one or more pairs of leaflets with coarse basal teeth. Claspette (Cl) with the dorsal and ventral lobes fused to form a single conical

lobe on each side; spines variable in number, usually consisting of one or two large pointed apical spines, one long slender spine arising from inner surface near middle, and one to three large pointed spines arising from outer surface (fig. 28, *B*). Basistyle (Bs) nearly twice as long as mid-width, clothed with long and short setae and many scales; two parabasal spines (P-S) and one internal spine (I-S) present. Dististyle (Ds) longer than basistyle, with a row of small papillated setae along dorsal surface and lacking minute nonpapillated hairs; claw (Ds-C) short, blunt.

LARVA (fig. 29). *Head:* Antenna spiculate; antennal tuft 1 inserted at basal third of shaft, with several branches; terminal hair 4 with several branches. Inner clypeal hairs 2 simple, with basal tubercles separated by less than the diameter of one of the tubercles; outer clypeal 3 densely dichotomously branched; postclypeal 4 small, single or branched distally; frontal hairs 5–7 large, plumose; sutural 8 and transsutural 9 usually 4- or 5-branched. *Thorax:* Prothoracic hair 1 short, simple or sparsely branched toward apex; hair 2 long, stout, with many branches; hair 3 short, simple. Prothoracic pleural group 9–12 with hairs 9, 10, and 12 long, simple; and hair 11 short, usually simple. Mesothoracic pleural group 9–12 with hairs 9 and 10 long, simple; hair 12 about one-third as long as hair 9 or 10, simple; and hair 11 minute, simple. Metathoracic pleural group 9–12 with hairs 9 and 10 long, simple; hair 12 short, branched beyond base; and hair 11 minute, simple. Metathoracic palmate hair 3 small, with transparent leaflets. *Abdomen:* Accessory dorsal hair 0 well developed (4- to 9-branched) on segments IV and V; palmate hair 1 rudimentary on segments I and II, well developed and nearly equal in size on III–VII (occasionally smaller on III and VII); leaflets with serrations on apical third. Antepalmate hair 2, 4- to 9-branched on segments IV and V, about equal in size to hair 0. Lateral hair 6 long, plumose on segments I–III. Pecten as illustrated (fig. 29, *B*).

DISTRIBUTION. Southeastern United States, north to Massachusetts and west to Kansas and New Mexico. Its range extends south to Mexico (730), Guatemala (102), British Honduras (446), Honduras (156), Nicaragua (444),

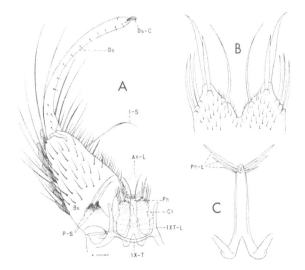

Fig. 28. *Anopheles crucians* Wiedemann. *A*, Male terminalia. *B*, Claspettes. *C*, Phallosome.

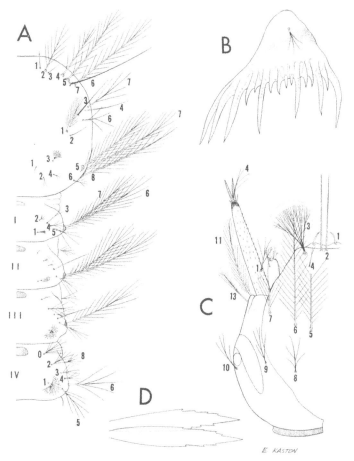

Fig. 29. Larva of *Anopheles crucians* Wiedemann. *A*, Thorax and abdomen. *B*, Pecten. *C*, Head. *D*, Leaflets of palmate hair.

BIONOMICS. The larvae are found in ponds, lakes, swamps, and semipermanent and permanent pools. They are generally associated with aquatic vegetation, either surface or emergent types, usually under partly shaded conditions. This mosquito seems to prefer acid water in its larval habitat and reaches its greatest abundance in the acid waters of the cypress swamps in Georgia and Florida. Larvae may be found throughout most of the winter in the southern states. Frohne and Hart (295) observed that *A. crucians* survives the winter wholly in the aquatic stages at Manning, South Carolina, undergoing retarded larval development.

The adults occasionally enter houses but are principally outdoor biters, attacking mostly at night, but sometimes they do attack in the daytime, especially on cloudy days or in shade. The adults rest during the day underneath houses or bridges, in hollow trees, culverts, and other similar shelters. When in the resting position the body forms an angle with the resting surface of nearly 90°. The adults are readily attracted to light traps.

MEDICAL IMPORTANCE. Although the species has been found infected with malaria in nature, it is not usually considered to be an important vector. However, Sabrosky *et al.* (641) dissected about 700 specimens of *A. crucians* from Santee Swamps, South Carolina, in 1945, and found a gland infection rate of 3.28 per cent, which was much higher than that found in *A. quadrimaculatus* taken in this area. We wish to point out that this finding, together with the fact that the species is active earlier and later in the season than *A. quadrimaculatus*, may indicate the need for further evaluation of its importance as a vector of malaria.

Anopheles (Anopheles) earlei Vargas

Anopheles earlei Vargas, 1943, Bol. Of. Sanit. Pan-Amer., 22:8.

Anopheles earlei, Vargas, 1943, Rev. Inst. Salub. Enf. Trop., 5:215.

Anopheles earlei, Vargas and Matheson, 1948, Rev. Inst. Salub. Enf. Trop., 9:27.

The larvae, pupae, and male terminalia provide satisfactory characters for distinguishing *Anopheles earlei* from the closely related *A. occidentalis* of the Pacific coast region. There

Cuba (443), Jamaica (361), and Puerto Rico (700). Pritchard and Pratt (561) did not find *A. crucians* in Puerto Rico and expressed the opinion that the single larva reported by Tulloch was actually *Anopheles grabhamii* Theobald. *United States:* Alabama (135); Arkansas (88,121); Connecticut (494); Delaware (120, 167); District of Columbia (213, 318); Florida (135, 419); Georgia (135, 419); Illinois (615); Indiana (152); Iowa (14); Kansas (14); Kentucky (134, 570); Louisiana (419, 759); Maryland (64, 161); Massachusetts (270, 701); Mississippi (135, 546); Missouri (169, 324); New Jersey (643); New Mexico (20); New York (31, 418); North Carolina (135, 419); Ohio (731a); Oklahoma (368, 628); Pennsylvania (298, 756); Rhode Island (35, 270); South Carolina (135, 742); Tennessee (135, 658); Texas (418, 634); Virginia (176, 178).

are striking differences in the pupae of the two species; *A. occidentalis* has a very long slender spine in the posterolateral corner of segment VII, whereas this spine is short and blunt in *A. earlei*.

ADULT FEMALE (pl. 4). Similar to *A. occidentalis*. Medium-sized species. *Head:* Proboscis long, dark brown; palpi about as long as the proboscis, dark brown. Occiput with many brown erect forked scales, those on the vertex yellowish white; frontal tuft with some silvery setae. *Thorax:* Integument of scutum with a broad median frosted stripe, brown laterally; the frosted stripe clothed with short pale-yellow hairs, the lateral areas with longer and dark setae. Scutellum crescent-shaped, clothed with yellow hairs and larger brown setae. *Abdomen:* Integument dark brown, with frosted areas, clothed with pale-yellow to brown hairs. *Legs:* Legs dark brown, tinted with blue; femora and tibiae narrowly tipped with yellowish white. *Wing:* Length about 5.0 mm. Scales narrow, dark brown; some of the scales arranged in distinct darker spots; scales on stem of vein 2, between fork and dark spot, raised. Fringe at apex of wing with a silver- or bronze-colored spot. *Halter:* Knob dark-scaled.

ADULT MALE. Coloration usually paler than that of the female. *Terminalia* (fig. 30). Ninth tergite (IX-T) with band narrow, sclerotized; lobes (IXT-L) short and broad, with apex somewhat expanded. Tenth sternite absent; anal lobe (An-L) large, subtriangular, spicu-

late. Phallosome (Ph) broad and furcate at base, apex with three or four pairs of smooth, ligulate leaflets (Ph-L), the distal ones longer. Claspette (fig. 30, *B*) broad, consisting of a dorsal lobe (Cl-DL) bearing two or three acuminate spines and a large, pilose ventral lobe (Cl-VL) bearing a long pointed apical spine and a short slender and a long slender spine subapically. The lobes of the claspette are more or less distinct apically. Basistyle (Bs) a little less than twice as long as broad, clothed with long and short setae; two stout parabasal spines (P-S) and one internal spine (I-S) present. Dististyle (Ds) a little longer than the basistyle, narrower medially; claw (Ds-C) short, blunt.

LARVA (fig. 31). *Head:* Antenna spiculate, darker at apex; antennal tuft 1 inserted at basal fourth or third of shaft, with several branches; terminal hair 4 with several branches. Inner clypeal hairs 2 with two to five branches distally, rarely simple; usually separated at their bases by the width of one tubercle; outer clypeal 3 densely dichotomously branched; postclypeal 4 short, rarely single, 2- to 5-branched, rarely more; frontal hairs 5–7 large, plumose; sutural 8 about 7- to 9-branched; transsutural 9 usually 5- or 6-branched. *Thorax:* Prothoracic hair 1 short, usually simple; hair 2 long, with about ten to twelve branches; hair 3 short, simple. Prothoracic pleural group 9–12 with hairs 9, 10, and 12 long, simple; hair 11 short, simple. Mesothoracic pleural group 9–12 with hairs 9 and 10 long, simple; hair 11 minute, simple; hair 12 short, simple. Metathoracic pleural group 9–12 with hairs 9 and 10 long, simple; hair 11 simple; hair 12 short, simple or 2-branched beyond middle. Metathoracic palmate hair 3 small, with transparent leaflets. *Abdomen:* Accessory dorsal hair 0 obsolete or minute, simple on segments IV and V. Palmate hair 1 rudimentary on segments I and II, well developed and of about equal size on III–VII; leaflets with serrations beyond middle. Antepalmate hair 2 usually 2- to 6-branched on segments IV and V. Lateral hair 6 long, plumose on segments I–III. Pecten as illustrated (fig. 31, *B*).

DISTRIBUTION. This species, described by Vargas (718) but only recently recognized as valid, occurs across the northern United States

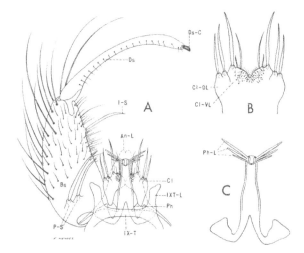

Fig. 30. *Anopheles earlei* Vargas. *A*, Male terminalia. *B*, Claspettes. *C*, Phallosome.

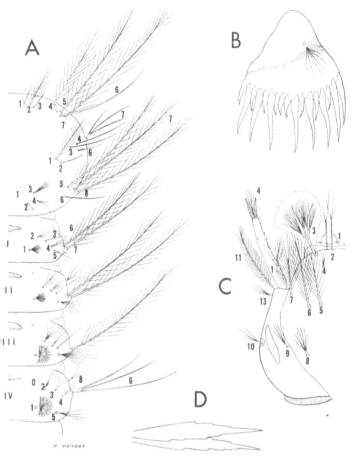

Fig. 31. Larva of *Anopheles earlei* Vargas. *A*, Thorax and abdomen. *B*, Pecten. *C*, Head. *D*, Leaflets of palmate hair.

and Canada to Alaska, whereas *A. occidentalis* appears to be restricted to a rather narrow strip along the Pacific coast, from California north to Canada and possibly the lower panhandle of Alaska. The following records, many of which were earlier reported as *A. occidentalis*, are probably *A. earlei*. *United States:* Colorado (343); Connecticut (494); Idaho (12, 344); Iowa (12, 625); Maine (270, 557); Massachusetts (12, 270); Michigan (12, 557); Minnesota (533, 557); Montana (557, 631); Nebraska (12); New Hampshire (66, 270); New York (31, 557); North Dakota (12, 553); South Dakota (14); Vermont (29, 270); Wisconsin (12, 718); Wyoming (557). *Canada:* Alberta (12, 708); British Columbia (162a); Labrador (348); Manitoba (12, 708); New Brunswick, Nova Scotia, Ontario and Quebec (536, 708); Saskatchewan (12, 708). *Alaska* (Stone, personal communication[5]).

BIONOMICS. The larvae seem to prefer cold clear water in the shallow margins of semi-permanent and permanent ponds overgrown with emergent and floating vegetation. They are occasionally found in woodland pools, marshes, open bogs, and along the margins of sluggish streams. The females have been observed to attack man mostly at dusk or shortly after dark. The females overwinter in hibernation, and Matheson (492) observed that the overwintering females will attack man, even in bright sunshine, after coming out of hibernation. Owen (533) observed the females in caves in Minnesota during the winter months. Owen also observed that the females readily entered cabins in Minnesota, and in one instance noticed that they gained entrance by coming down the chimney. Pratt (557) claims that the females invade houses and feed on man at all hours during the night, rivaling *Mansonia perturbans* (Walker) in bloodthirstiness.

Rozeboom (631) observed, while rearing the species in the laboratory, that the males became active and began to swarm when his arm was thrust into the cage for feeding the females. The males were especially active at dusk, and pairing seemed always to take place during flight. After union the female would come to rest on the side of the cage with the male suspended head down. Contact was maintained for 3 to 6 seconds.

MEDICAL IMPORTANCE. The species occurs outside the endemic malaria areas in North America and is not known to be a vector of the disease in nature.

Anopheles (Anopheles) freeborni Aitken

Anopheles maculipennis freeborni Aitken, 1939, Pan-Pac. Ent., *15*:192.
Anopheles maculipennis Meigen (of authors, in part).
Anopheles maculipennis freeborni, Aitken, 1945, Univ. Calif. Publ. Entom., 7:296.

ADULT FEMALE (pl. 5). Similar to *A. quadrimaculatus*. Medium-sized species. *Head:* Proboscis long, dark; palpi about as long as the proboscis, dark, with scales slightly raised on basal part. Occiput with dark erect forked scales, those on median area yellowish white; vertex with long narrow pale scales; frontal

[5]Dr. Alan Stone says "Dr. Robert Matheson has examined *Anopheles* from Alaska and found them to be entirely *earlei*."

tuft with some pale setae. *Thorax:* Integument of scutum brown to black, median area with more or less distinct grayish-brown median and submedian stripes; prescutellar space grayish brown; scutum clothed with pale-yellow to golden-brown hairs, more numerous medially. Scutellum crescent-shaped, clothed with golden-brown hairs and long dark setae. *Abdomen:* Integument of abdomen brown to black, clothed with yellowish-brown hairs. *Legs:* Legs dark-scaled, femora and tibiae tipped with pale scales. *Wing:* Length about 4.5 mm. Scales narrow, dark, some of the scales forming four darker spots. Scales on basal part of vein 5, before the fork, truncate at apices, with serrate edges. Scales on distal third of wing less dense than in *A. quadrimaculatus;* wing with conspicuous bare spaces between the veins on apical third. *Halter:* Knob dark-scaled.

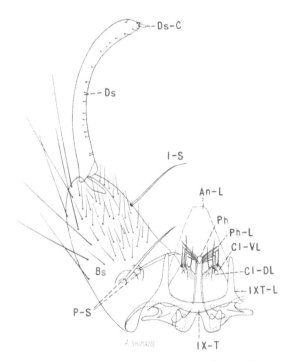

Fig. 32. Male terminalia of *Anopheles freeborni* Aitken.

ADULT MALE. Coloration similar to that of the female, wing spots less distinct. *Terminalia* (fig. 32). Similar to *A. occidentalis* and *A. punctipennis* but usually has lobes of ninth tergite a little narrower. Ninth tergite (IX-T) with band narrow, sclerotized; lobes (IXT-L) long and narrow. Tenth sternite absent; anal

lobe (An-L) large, subtriangular, densely spiculate. Phallosome (Ph) cylindrical, broad and furcate at base; apex with three or four pairs of ligulate leaflets (Ph-L), the distal pairs longer. Claspette broad, consisting of a dorsal lobe (Cl-DL) bearing two or three acuminate apical spines and a pilose ventral lobe (Cl-VL) bearing one or two long pointed apical spines, and a short slender and a long slender spine subapically. The lobes are more or less distinct apically. Basistyle (Bs) conical, a little less than twice as long as mid-width, clothed with long and short setae; two stout parabasal spines (P-S) and one internal spine (I-S) present. Dististyle (Ds) a little longer than the basistyle, with a row of small papillated setae on inner dorsal surface and a small patch of minute nonpapillated hairs near base; claw (Ds-C) short, blunt.

LARVA (fig. 33). Appears to be indistinguishable from *A. punctipennis* by any constant characters. *Head:* Antenna darker at apex, spiculate; antennal tuft 1, 4- to 8-branched, usually inserted a little beyond basal third of shaft; terminal hair 4 with several branches. Inner clypeal hairs 2 simple, with basal tubercles separated by less than the diameter of one tubercle; outer clypeal 3 densely dichotomously branched; postclypeal 4 small, usually 2-branched near base or middle; frontal hairs 5–7 large, plumose (usually associated with three dark spots just posterior to points of insertion of hairs 5–7); sutural 8 and trans-sutural 9 about 6- to 9-branched. *Thorax:* Prothoracic hair 1 short, with several weak branches beyond base; hair 2 long, stout, with many branches; hair 3 short, simple. Prothoracic pleural group 9–12 with hairs 9, 10, and 12 long, simple; and hair 11 short, simple. Mesothoracic pleural group 9–12 with hairs 9 and 10 long, simple; hair 11 minute, simple; and hair 12 simple, about one-third as long as hair 9 or 10. Metathoracic pleural group 9–12 with hairs 9 and 10 long, simple; hair 11 minute, simple; and hair 12 short, double or triple. Metathoracic palmate hair 3 small. *Abdomen:* Accessory dorsal hair 0 obsolete on segments IV and V; palmate hair 1 rudimentary on segments I and II, well developed and of about equal size on III–VII; leaflets with serrations on apical half. Antepalmate hair 2

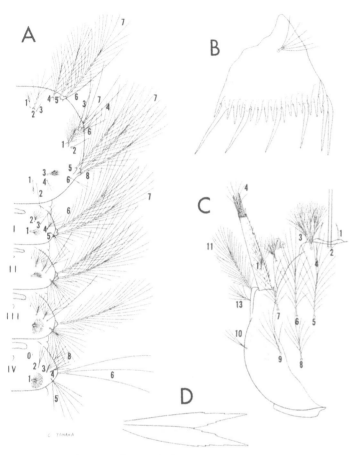

Fig. 33. Larva of *Anopheles freeborni* Aitken. *A*, Thorax and abdomen. *B*, Pecten. *C*, Head. *D*, Leaflets of palmate hair.

532); Utah (575); Washington (67, 672); Wyoming (578). *Canada:* British Columbia (12, 708).

BIONOMICS. The larvae are found in clear seepage water, in roadside pools, in rice fields, and in other similar habitats. Pools that are sunlit, at least part of the day, seem to be preferred, although larvae are occasionally found in rather densely shaded pools. Heavy production of larvae often occurs in matted algal growths in water along the margins of rice fields.

The adults rest during the daytime in the dark, cool corners of sheds, houses, and basements as well as underneath bridges and in culverts. The females enter houses readily and feed on man. They are more active at dusk and during the night but occasionally do attack man during the daylight hours in dense shade or on cloudy days. Hibernation of the females generally takes place in outbuildings, cellars, homes, and other similar shelters where they are semiactive, often moving from one shelter to another during the winter. Ryckman and Arakawa (637) found adults hibernating in wood rats' nests in California. Freeborn (285) observed that there is a dispersal flight of overwintering females in the Sacramento Valley in February for oviposition.

A prehibernation flight takes place in the fall, and at this time the females may invade areas even 10 or 12 miles from the aquatic habitats of their immature stages. Rosenstiel (611) did not observe a spring dispersal of this species in the Sacramento Valley but claims that blood feeding by *A. freeborni* was observed at that time. He describes fall population movements of *A. freeborni* as consisting of a congregative phase during the last 3 weeks of August, followed in September and October by a dispersion phase during which the mosquitoes moved in large numbers away from their breeding grounds. When the females emerge from their winter shelters on warm sunshiny days during February, they bite viciously even in bright sunlight. The species reaches its greatest abundance in the hot interior valleys of California. It occurs from below sea level in the edges of Death Valley to approximately 7,000 feet elevation in Utah.

MEDICAL IMPORTANCE. It is the principal

usually double on segments IV and V (sometimes single or triple, rarely 4-branched on segment IV). Lateral hair 6 long, plumose on segments I–III. Pecten as illustrated (fig. 33, *B*).

DISTRIBUTION. Southwestern Canada and western United States, including the Great Basin and the central valleys of California, Oregon, and Washington. It is generally found west of the Continental Divide, but also occurs east of the divide in southern Colorado, in New Mexico, and in the Rio Grande Valley as far south as El Paso, Texas. It has been reported also from Mexico (106, 730). The species probably reaches its greatest abundance in the Sacramento and San Joaquin valleys in California. *United States:* Arizona (12); California (12, 287); Colorado (12, 343); Idaho (12, 672); Montana (12, 550); Nevada (12); New Mexico (12, 273); Oregon (12, 672); Texas (12,

vector of malaria in the arid and semiarid western United States. Burgess and Young (117) demonstrated that *A. freeborni* showed greater susceptibility than *A. quadrimaculatus* to infection by a domestic strain (St. Elizabeth) of *Plasmodium vivax*. Western equine encephalitis virus has been isolated from wild-caught *A. freeborni* in the Yakima Valley, Washington (332).

Anopheles (Anopheles) georgianus King

Anopheles crucians var. *georgianus* King, 1939, Amer. Jour. Trop. Med., *19*:462.

ADULT FEMALE. Indistinguishable from *A. crucians* (pl. 3) and *A. bradleyi*. Bellamy and Repass (55) found that the eggs of *A. georgianus* are usually distinct from those of *A. crucians*.

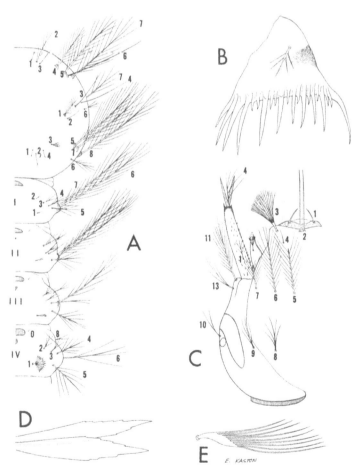

Fig. 34. Larva of *Anopheles georgianus* King. *A*, Thorax and abdomen. *B*, Pecten. *C*, Head. *D*, Leaflets of palmate hair. *E*, Palmate hair III.

ADULT MALE. Coloration similar to that of the female. *Terminalia:* Indistinguishable from *A. crucians* (fig. 28) and *A. bradleyi*.

LARVA (fig. 34). *Head:* Antenna spiculate (spicules slender, not pigmented, and with little thickening at base); antennal tuft 1 inserted on basal third of shaft, with several branches; terminal hair 4 with several branches. Inner clypeal hairs 2 simple, with basal tubercles rarely separated by more than the diameter of one of the tubercles; outer clypeal 3 densely dichotomously branched; postclypeal 4 small, single or branched apically; frontal hairs 5–7 large, plumose; sutural 8, 3- to 5-branched; transsutural 9 usually 3- or 4-branched, rarely reaching base of hair 7. *Thorax:* Prothoracic hair 1 short, about 3- to 6-branched apically; hair 2 large, stout, with many branches; hair 3 short, simple. Prothoracic pleural group 9–12 with hairs 9, 10, and 12 long, simple, subequal; and hair 11 short, simple or bifurcate. Mesothoracic pleural group 9–12 with hairs 9 and 10 long, simple, subequal; hair 11 simple, minute; and hair 12 about one-third as long as 9 or 10, 2- or 3-branched beyond middle. Metathoracic pleural group 9–12 with hairs 9 and 10 long, simple; hair 11 simple, minute; and hair 12 short, with several branches beyond base. Metathoracic palmate hair 3 small, with transparent leaflets. *Abdomen:* Accessory dorsal hair 0 obsolete on segments IV and V. Palmate hair 1 well developed and of equal size on segments IV–VI, with leaflets serrated on apical half; palmates very small and rudimentary on segments I–III and VII, with leaflets smooth. Antepalmate 2, 3- to 6-branched (rarely double) on segments IV and V. Hair 5 on segment I, 5- or 6-branched (rarely more), nearly twice as large as hair 4. Lateral hair 6 long, plumose on segments I–III. Pecten as illustrated (fig. 34, *B*).

DISTRIBUTION. This species is apparently restricted to the southeastern United States. *United States:* Alabama (135, 516); Florida and Georgia (135, 516); Louisiana (420, 759); Mississippi (135, 516); North Carolina (135, 646); South Carolina (516, 742).

BIONOMICS. The larvae are found in hoofprints in seepage areas in pastures, in potholes, and in sluggish swamp streams. Bellamy (51) first reported this mosquito from fifteen collections taken in four counties in Georgia. An analysis of these collections made by Bellamy points to pastureland seepage areas with acid waters as the typical habitat. Freeborn (286) stated that the larvae of *A. georgianus* are frequently associated with small accumulations of water favoring the growth of the small pitcher plant, *Sarracenia purpurea*. The adults of *A. georgianus* cannot at this time be distinguished from those of *A. crucians*; therefore nothing is known of their habits. Bellamy and Repass (55) collected adults of *A. georgianus* in the field in Georgia and were able to obtain eggs from which larvae were hatched.

MEDICAL IMPORTANCE. Nothing is known of the importance of this species in relation to the transmission of malaria, but it is not believed to be a natural vector of the disease.

Anopheles (Anopheles) occidentalis
Dyar and Knab

Anopheles occidentalis Dyar and Knab, 1906, Proc. Biol. Soc. Wash., *19*:159.

Anopheles maculipennis occidentalis, Aitken (in part), 1945, Univ. Calif. Publ. Entom., 7:282.

ADULT FEMALE (pl. 6). The female is similar to that of *A. earlei*. Medium-sized species. *Head:* Proboscis long, dark brown; palpi about as long as the proboscis, dark brown. Occiput clothed with dark erect forked scales, those on vertex white; frontal tuft with some silvery setae. *Thorax:* Integument of scutum with a broad median frosted stripe, dark brown laterally; the frosted area clothed with short, pale-yellow hairs, the darker lateral areas with larger dark setae. Scutellum crescent-shaped, clothed with yellow hairs and longer brown setae. *Abdomen:* Integument dark brown and frosted, densely clothed with pale-yellow to brown hairs. *Legs:* Legs dark-scaled; femora and tibiae narrowly tipped with pale scales. *Wing:* Length about 4.0 mm. Scales narrow, dark; some of the scales arranged in distinct darker spots; scales on stem of vein 2, between fork and dark spot, rather closely appressed (not raised as in *A. earlei*). Fringe at apex of wing with a silver- or bronze-colored spot. *Halter:* Knob dark-scaled.

ADULT MALE. Coloration similar to that of the female but is usually paler. *Terminalia*

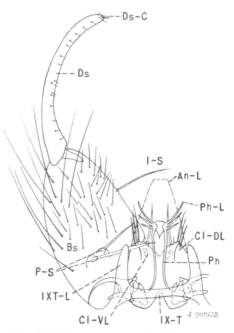

Fig. 35. Male terminalia of *Anopheles occidentalis* Dyar and Knab.

(fig. 35). Similar to *A. punctipennis* and *A. freeborni*. Ninth tergite (IX-T) with band narrow, sclerotized; lobes (IXT-L) long and narrow. Tenth sternite absent; anal lobe (An-L) large, subtriangular, densely spiculate. Phallosome (Ph) cylindrical, broad and furcate at base; apex with three or four pairs of ligulate leaflets (Ph-L), the more distal pairs longer. Claspette broad, consisting of a dorsal lobe (Cl-DL) with two (rarely one or three) acuminate spines and a pilose ventral lobe (Cl-VL) with a long pointed apical spine and a short slender and a long slender spine subapically. The lobes are more or less distinct apically. Basistyle (Bs) conical, slightly less than twice as long as mid-width, clothed with long and short setae; two (rarely three) stout parabasal spines (P-S) and one internal spine (I-S) present. Dististyle (Ds) slightly longer than basistyle, with a rather dense patch of minute nonpapillated hairs near base; claw (Ds-C) short, blunt.

LARVA (fig. 36). *Head:* Antenna spiculate, darker on distal third; antennal tuft 1 inserted at basal fourth or third of shaft, with several branches; terminal hair 4 with several branches. Inner clypeal hairs 2 simple, the dis-

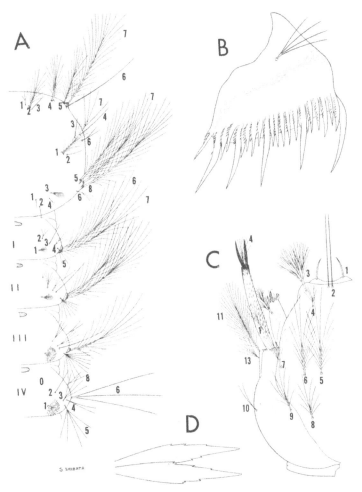

Fig. 36. Larva of *Anopheles occidentalis* Dyar and Knab.
A, Thorax and abdomen. *B*, Pecten. *C*, Head. *D*, Leaflets of
palmate hair.

hairs 9 and 10 long, simple; hair 11 minute, simple; hair 12 short, simple or 2-branched beyond middle. Metathoracic palmate hair 3 small, with transparent leaflets. *Abdomen:* Accessory dorsal hair 0 obsolete or minute, simple on segments IV and V. Palmate hair 1 rudimentary on segments I and II, well developed and of about equal size on III–VII; leaflets with serrations beyond middle. Antepalmate hair 2 usually simple on segments IV and V. Lateral hair 6 long, plumose on segments I–III. Pecten as illustrated (fig. 36, *B*).

DISTRIBUTION. *Anopheles occidentalis* occurs in a narrow strip along the Pacific coast from Baja California, Mexico (106), northward in the coastal areas of California, Oregon, Washington, Canada, and possibly the panhandle of Alaska. The exact range of this species is not known at this time, since it has so long been confused with an eastern form of the "*Maculipennis* complex" now known as *A. earlei. United States:* California (287); Oregon (672); Washington (557). *Canada:* British Columbia and Yukon (557). *Alaska:* (557[6]).

BIONOMICS. The larvae occur in a variety of aquatic habitats, including permanent pools, lagoons, ponded creeks, hillside seepages, watering troughs, and quiet protected streams. Impounded water habitats seem to be preferred. Larvae are often found in the shallow margins of ponds where they live in the dense shade of cattails, duckweeds, watercress, grass, or algae. The larvae are frequently found in heavy concentrations in shallow hillside seepage areas along the cool fog-bound Pacific coast, where they may occur in water of less than one inch in depth. The species tolerates more shade in its larval habitats than does *A. freeborni.* Aitken (12) believes it is quite probable that larval development may continue at a slow rate during the winter months along the California coast. The adults are seldom found far from their larval habitats, and they rarely feed on humans.

MEDICAL IMPORTANCE. Moore *et al.* (517) were able to infect this species with *Plasmo-*

tance between the basal tubercles less than the diameter of one of the tubercles; outer clypeal 3 densely dichotomously branched; postclypeal 4 short, 4- to 10-branched near base; frontal hairs 5–7 large, plumose; sutural 8 and transsutural 9 about 5- to 9-branched. *Thorax:* Prothoracic hair 1 rather short, often split beyond middle into two to five branches; hair 2 long, stout, with about eight to ten branches; hair 3 short, simple. Prothoracic pleural group 9–12 with hairs 9, 10, and 12 long, simple; hair 11 short, simple. Mesothoracic pleural group 9–12 with hairs 9 and 10 long, simple; hair 11 minute, simple; hair 12 short, simple. Metathoracic pleural group 9–12 with

---

[6]Although Pratt (1952) shows the range of *A. occidentalis* extending well along the coastal areas of Alaska, Dr. Alan Stone says he believes the *Anopheles* of Alaska is *earlei* with the possible exception of the lower panhandle.

dium vivax, and transmission of the parasites to a neurosyphilitic patient was accomplished. It is not known to transmit malaria in nature.

Anopheles (Anopheles) pseudopunctipennis franciscanus McCracken

Anopheles franciscanus McCracken, 1904, Ent. News, *15*:12.

Anopheles boydi Vargas, 1939, "Medicina," Rev. Mexico, *19*:356.

Anopheles pseudopunctipennis var. *willardi* Vargas, 1941, Bull. Brooklyn Ent. Soc., *36*:74.

Anopheles pseudopunctipennis franciscanus, Aitken, 1945, Univ. Calif. Publ. Entom., *7*:327.

ADULT FEMALE. Medium-sized species. Similar to *A. pseudopunctipennis pseudopunctipennis* (pl. 7) but differs as follows: Terminal segment of the palpi has the apical part dark-scaled; wing vein 2 is entirely dark-scaled except for a subapical white patch on fork 2.1 and a small patch on stem near the cross vein 2–3; vein 4 has stem predominantly dark-scaled. *Head:* Proboscis long, dark; palpi about as long as the proboscis, black, with apices of segments 2, 3, and 4 narrowly ringed with white, apical part of terminal segment dark. Occiput clothed with erect forked scales, those on median area white, the others dark brown; scales of vertex white; frontal tuft white. *Thorax:* Integument of scutum with a broad median longitudinal frosted stripe extending the full length of the scutum, dark brown laterally; the frosted stripe clothed with narrow yellowish-white scales and pale-yellow hairs, the darker lateral areas with larger dark-brown setae. Scutellum crescent-shaped, clothed with yellow hairs and long brown setae. *Abdomen:* Integument brown to black, clothed with golden-brown hairs. *Legs:* Legs dark, tips of femora and tibiae pale. *Wing:* Length about 4.0 mm. Scales black and pale yellow, arranged on the veins in contrasting lines and spots. *Halter:* Knob dark-scaled.

ADULT MALE. Coloration similar to that of the female. *Terminalia.* Closely resembles *A. pseudopunctipennis pseudopunctipennis* (fig. 37), but has the leaflets of the phallosome, when observed, appearing as delicate lines; if serrations are present on the leaflets, they are difficult to discern. Ninth tergite (IX-T) with band narrow; lobes (IXT-L) stout, about as long as broad, blunt at tips. Tenth sternite absent; anal lobe (An-L) large, subtriangular, spiculate. Phallosome (Ph) cylindrical, broad and furcate at base; leaflets (Ph-L) when present, extremely delicate and hard to discern. Claspette broad, consisting of a dorsal lobe (Cl-DL) bearing three very broad, flat, leaflike spines with rounded, bent tips; and a ventral lobe (Cl-VL) usually bearing two long slender spines and a much shorter weak spine. Basistyle (Bs) about twice as long as mid-width, clothed with long and short setae; two parabasal spines (P-S) and one internal spine (I-S) present. Dististyle (Ds) longer than the basistyle, with a row of small papillated setae on inner dorsal surface; minute nonpapillated setae often absent on basal part; claw (Ds-C) short, blunt.

LARVA. Similar to *A. pseudopunctipennis pseudopunctipennis* (fig. 38) but differs in having the antennal hair 1 single or branched; antepalmate hair 2, 1- to 3-branched on abdominal segment IV; and the posterior lobes of the spiracular plates rounded, without long slender, sclerotized projections. *Head:* Antenna smooth on outer surface, spiculate on inner surface; antennal hair 1 single or branched, inserted on basal third of shaft; terminal hair 4 with several branches. Inner clypeal hairs 2 long, simple, with basal tubercles separated by more than the width of one tubercle; outer clypeal 3 long, simple; post-clypeal 4 long, simple; frontal hairs 5–7 large, plumose; sutural 8 and transsutural 9 with several branches arising from a rather stout shaft. *Thorax:* Prothoracic hair 1 short, 2- to 4-branched beyond base; hair 2 long, stout, with many lateral branches; hair 3 simple, about twice as long as hair 1. Prothoracic pleural group 9–12 with hairs 9, 10, and 12 long, simple; and hair 11 much shorter, simple. Mesothoracic pleural group 9–12 with hair 9 short, stout; hair 10 stout, about twice as long as hair 9; hair 11 minute; and hair 12 weak, usually double or triple, about two-thirds as long as hair 9. Metathoracic pleural group 9–12 with hair 9 short, stout; hair 10 stout, a little more than twice as long as hair 9; hair 11 obsolete; and hair 12 weak, usually branched, about half as long as hair 9. Metathoracic palmate hair 3 absent. *Abdomen:* Accessory dorsal hair 0 absent on segments I–VII. Palmate hair 1 rudimentary on segments I and II, well developed on III–VII; leaflets long, slender, with serrations beyond middle. Antepalmate hair 2 single to triple on segment IV, usually single on V. Lateral hair 6 long, plumose on segments I to III. Posterior spiracular plates without a long slender sclerotized projection or "tail" arising from the inner caudal margin.

DISTRIBUTION. The distribution of *A. pseudopunctipennis franciscanus* is not well known at this time, since it has been confused with other members of the *"pseudopunctipennis* complex." At present it is known to occur in the southwestern United States, as far north as Oregon and east to western Texas and Oklahoma. It also occurs in northern Mexico (106). The *"pseudopunctipennis* complex" needs much study, but for the present, two forms are recognized in this area. *United States:* Arizona (12, 286); California (12, 287); Colorado (343); Kansas (286); Nevada and New Mexico (12, 286); Oklahoma (321); Oregon (672); Texas and Utah (12, 286); Wyoming (578).

BIONOMICS. The immature stages of this mosquito are found in shallow pools along the margins of receding streams during the dry season. The larvae are almost invariably associated with abundant growth of green algae in pools open to sunlight. The larvae have been found also on occasions in brackish water and in artificial containers. According to Freeborn (286) the larvae have been found in salt water equivalent to 34 and 60 per cent sea water near San Diego, California.

Belkin *et al.* (50) made a rather intensive study of the behavior of the adults of this mosquito. They found that the males, and the females to a lesser extent, utilized various types of diurnal shelters near their larval habitats. Small numbers of blooded females were found resting in farm buildings within a half-mile radius of known larval habitats. The adults were entirely crepuscular and were attracted in moderate numbers to artificial light. The females displayed a preference for feeding on larger mammals, including horses, cows, and sheep, and a lesser preference for feeding on smaller mammals and fowl. It is generally known as a field mosquito in California; it rarely comes indoors and seldom attacks man. Belkin *et al.* (50) describe swarms of approximately 5,000 males of *franciscanus* over a porch in California, and mating of males and females was observed. The paired mosquitoes fell to the roof of the porch and separated after a few seconds.

MEDICAL IMPORTANCE. Barber *et al.* (26) were able to infect this species with *Plasmodium vivax* in New Mexico; however, because of its habits it is not regarded as an important vector in nature.

Anopheles (Anopheles) pseudopunctipennis pseudopunctipennis Theobald

Anopheles pseudopunctipennis Theobald, 1901, Mon. Culic., 2:305.

Anopheles peruvianus Tamayo and Garcia, 1907, Mem. Munic. Lima, p. 35.

Anopheles (Proterorhynchus) argentinus Brethes, 1912, Bol. Ins. Ent. y Pat. Veg., 1:15.

Anopheles tucumanus Lahille, 1912, Ann. Mus. Nac. B. A., 23:253.

Anopheles pseudopunctipennis pseudopunctipennis, Aitken, 1945, Univ. Calif. Pub. Entom., 7:327.

ADULT FEMALE (pl. 7). Medium-sized species. *Head:* Proboscis long, dark; palpi about as long as the proboscis, black, with apices of segments 2, 3, and 4 narrowly ringed with white, base of segment 4 more broadly ringed with white, terminal segment yellowish white. Occiput clothed with erect forked scales, those on central part white, others dark brown; scales of vertex white; frontal tuft long, white. *Thorax:* Integument of scutum with a median broad frosted stripe, dark brown laterally; the frosted area clothed with narrow yellowish-white scales and pale-yellow hairs, the darker lateral areas with longer dark setae. Scutellum crescent-shaped, clothed with yellow hairs and long brown setae. *Abdomen:* Integument dark brown to black, clothed with golden-brown hairs. *Legs:* Legs dark, tips of femora and tibiae pale. *Wing:* Length about 4.0 mm. Scales black and pale yellow, arranged on the veins in contrasting lines and spots as follows: costa all dark to junction of subcosta, with pale spots at outer third at junction of subcosta and near apex; vein 1 with six alternating white and dark areas, beginning with white base; vein 2 with base of stem white, followed by alternating dark and white areas; fork 2.1 with basal half dark, apical half white, tip dark; fork 2.2 dark, tip white; vein 3 white at base, followed by a broad dark area, a white area and a dark tip; vein 4 white to fork, except for two central small dark patches; forks 4.1 and 4.2 dark, followed by an equal area of white, tips white; vein 5 white at base, followed by a dark area, then white to fork; fork 5.1 with basal fifth dark, apical two-thirds dark, tip white; fork 5.2 with basal two-thirds white, apical third dark, tip white; vein 6 with basal half white, apical half dark, tip white. *Halter:* Knob dark-scaled.

ADULT MALE. Coloration similar to that of the female. *Terminalia* (fig. 37). Ninth tergite (IX-T) with band narrow, sclerotized; lobes (IXT-L) stout, about as long as broad, blunt or obtuse-angulate at tips. Tenth sternite absent; anal lobe (An-L) large, subtriangular, spiculate. Phallosome (Ph) cylindrical, broad and furcate at base; apex with two or three pairs of leaflets (Ph-L), slender, curved, and serrated (fig. 37, *C*). Claspette (fig. 37, *B*) broad, consisting of a dorsal lobe (Cl-DL) bearing three overlapping bladelike apical filaments; and a ventral lobe (Cl-VL) bearing two long slender spines and a short weak spine. Basistyle (Bs)

conical, nearly twice as long as mid-width, clothed with long and short setae; two parabasal spines (P-S) and one internal spine (I-S) present. Dististyle (Ds) longer than basistyle, with a row of small papillated setae on inner dorsal surface; usually without a patch of minute nonpapillated hairs on basal part; claw (Ds-C) short, blunt.

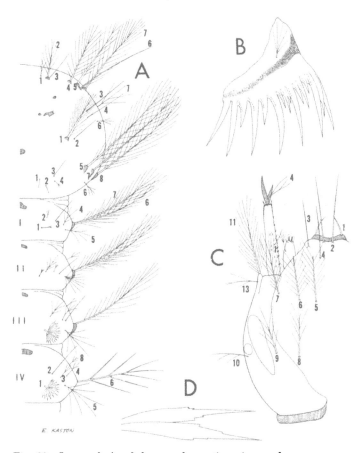

Fig. 38. Larva of *Anopheles pseudopunctipennis pseudopunctipennis* Theobald. *A*, Thorax and abdomen. *B*, Pecten. *C*, Head. *D*, Leaflets of palmate hair.

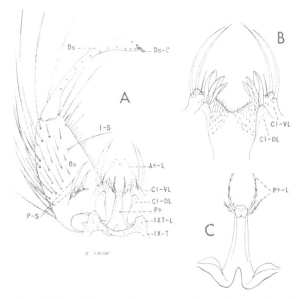

Fig. 37. *Anopheles pseudopunctipennis pseudopunctipennis* Theobald. *A*, Male terminalia. *B*, Claspettes. *C*, Phallosome.

LARVA (fig. 38). *Head:* Antenna smooth on outer surface, spiculate on inner surface; antennal hair 1 usually single, inserted before middle of shaft; terminal hair 4, 2- to 4-branched. Inner clypeal hairs 2 long, simple, with basal tubercles separated by more than the diameter of one of the tubercles; outer clypeal 3 simple, nearly as long as the inner clypeals 2; postclypeal 4 long, simple; frontal hairs 5–7 large, plumose; sutural 8 and transsutural 9 with several branches arising from a stout shaft, reaching bases of frontal hairs. *Thorax:* Prothoracic hair 1 short, usually 2- to 4-branched beyond base; hair 2 long, stout, with many lateral branches; hair 3 simple, about twice as long as hair 1. Prothoracic pleural group 9–12 with hairs 9, 10, and 12 long, simple; and hair 11 much shorter, simple. Mesothoracic pleural group 9–12 with hair 9 short, stout; hair 10 stout, about twice as long as hair 9; hair 11 minute; and hair 12 weak, usually 2- or 3-branched toward apex, about half to two-thirds as long as hair 9. Metathoracic pleural group 9–12 with hair 9 short, stout; hair 10 stout, about twice as long as hair 9; hair 11 obsolete; and hair 12 weak, usually 3- or 4-branched, about half as long as hair 9. Metathoracic palmate hair 3 absent. *Abdomen:* Accessory dorsal hair 0 absent on segments I–VII. Palmate hair 1 rudimentary on segments I and II, well developed on III–VII; leaflets long, slender, with serrations beyond middle. Antepalmate hair 2 usually single on segments IV and V. Lateral hair 6 long, plumose on segments I–III. Posterior spiracular plates each with a long slender sclerotized projection or "tail" arising from the inner caudal margin. Pecten as illustrated (fig. 38, *B*).

DISTRIBUTION. This is probably the most widespread of the New World Anophelines, ranging from the southern United States southward to Argentina. Its distribution largely coincides with the dry western mountains and plateaus, and it is not generally common in the lowlands. Its distribution is not well known in the United States, since many of the earlier records have been confused with *A. pseudopunctipennis franciscanus* in the western states. *United States:* Arkansas (88, 122); Colorado (14); Kansas (14, 360); Louisiana (286, 759); Mississippi (418); Missouri (531); New Mexico (273); Oklahoma (321, 628); Tennessee (418); Texas (249, 634).

BIONOMICS. The larvae apparently require a great deal of sunlight for their development, and the species thus displays a preference for pools and eddies in drying streams, particularly those in full sunlight containing mats of *Spirogyra*. The larvae can also be found along the margins of slow moving streams during the dry season, where they generally rest against leaves, sticks, or other bits of floating debris. The larvae are frequently abundant in shallow, drying streams in mountainous areas during the dry season when the streams are not subject to flushing by heavy rainfall.

The habits of the adults appear to differ markedly in the different regions. In some areas, from Mexico south, the species is exceedingly domestic, enters houses readily, and attacks man. Bordas and Downs (74), while studying the behavior pattern of this species in Mexico, observed that the adults enter houses, apparently for shelter, in the early morning shortly after daybreak. The mosquitoes leave the houses at dusk. Most feeding takes place outside the houses, although some females will bite indoors. These workers observed that the females feed freely outdoors on cattle, horses, burros, and human beings.

MEDICAL IMPORTANCE. This species is regarded as an important vector of malaria in some areas and is relatively unimportant in others. It is not known to be a major vector in the United States.

Anopheles (Anopheles) punctipennis (Say)

Culex punctipennis Say, 1823, Jour. Acad. Nat. Sci. Phila., 3:9.

Anopheles perplexens Ludlow, 1907, Can. Ent., 39:267.

Anopheles punctipennis stonei Vargas, 1941, Rev. Soc. Mexico Hist. Nat., 2:175.

ADULT FEMALE (pl. 8). Medium-sized species. *Head:* Proboscis long, black; palpi about as long as the proboscis, dark-scaled, with raised scales on basal part. Occiput clothed with erect forked scales, those on central part white, others dark; scales of vertex narrow, white; frontal tuft white. *Thorax:* Integument of scutum with a broad median frosted stripe, dark brown laterally; the frosted area clothed with short pale-yellow hairs, the darker lateral areas with larger dark setae. Scutellum crescent-shaped, clothed with yellow hairs and long brown setae. *Abdomen:* Integument dark brown to black, clothed with pale and dark hairs. *Legs:* Legs dark-scaled, femora and tibiae tipped with pale scales. *Wing:* Length about 4.0 mm. Scales black and pale yellow, in contrasting lines and spots (costa with a pale spot at outer third opposite tip of subcosta; vein 6 with basal fourth and apical half dark-scaled; veins 3 and 5 entirely dark-scaled). *Halter:* Knob dark-scaled.

ADULT MALE. Coloration similar to that of the female. *Terminalia* (fig. 39). Similar to *A. freeborni* and *A. occidentalis*. Ninth tergite (IX-T) with band narrow, sclerotized; lobes (IXT-L) about three times as long as broad, often slightly constricted medially, rounded apically. Tenth sternite absent; anal lobe (An-L) large, triangular, spiculate. Phallosome (Ph) cylindrical, broad and furcate at base; apex with three or four pairs of ligulate leaflets (Ph-L), the more distal pairs longer; one or more pairs of leaflets with a few coarse basal teeth. Claspette (fig. 39, *C*) broad, consisting of a dorsal lobe (Cl-DL) bearing one or two long stout acuminate spines; and a ventral lobe (Cl-VL) bearing a long stout acuminate apical spine (spine broader than the apical spine of the dorsal lobe), and a slender short and slender long spine subapically. The lobes are fused basally but are more or less distinct apically. Basistyle (Bs) conical, nearly twice as long as mid-width, clothed with long and short setae and a few scales; two parabasal spines (P-S) and one internal spine (I-S) present. Dististyle (Ds) a little longer than the basistyle, with a row of small papillated setae on inner dorsal surface, usually without a patch of minute non-papillated hairs on basal part; claw (Ds-C) short, blunt.

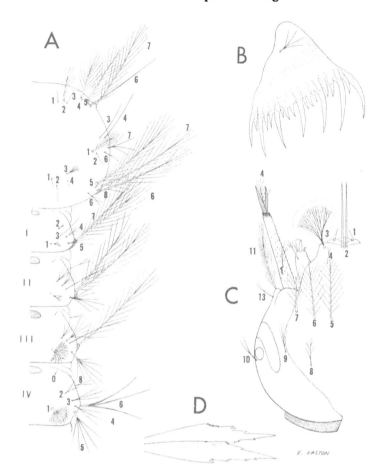

Fig. 40. Larva of *Anopheles punctipennis* (Say). *A*, Thorax and abdomen. *B*, Pecten. *C*, Head. *D*, Leaflets of palmate hair.

LARVA (fig. 40). Appears to be indistinguishable from *A. freeborni* by any constant characters. *Head:* Antenna pale, darker at apex, spiculate (spicules usually clear and without pigment); antennal tuft 1 inserted near basal third of shaft, with several branches; terminal hair 4 with several branches. Inner clypeal hairs 2 simple, with basal tubercles separated by less than the diameter of one tubercle; outer clypeal 3 densely dichotomously branched; postclypeal 4 small, branched; frontal hairs 5–7 large, plumose (usually inserted in a transverse dark band in California specimens); sutural 8 and transsutural 9 usually 5- to 7-branched (transsutural rarely reaching beyond base of hair 7). *Thorax:* Prothoracic hair 1 short, with several weak branches beyond base; hair 2 long, stout, with many branches; hair 3 short, simple. Prothoracic pleural group 9–12 with hairs 9, 10, and 12 long, simple; and hair

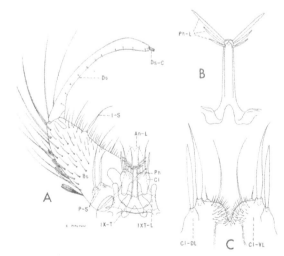

Fig. 39. *Anopheles punctipennis* (Say). *A*, Male terminalia. *B*, Phallosome. *C*, Claspettes.

11 short, simple. Mesothoracic pleural group 9–12 with hairs 9 and 10 long, simple; hair 11 minute, simple; and hair 12 simple, about one-third as long as hair 9 or 10. Metathoracic pleural group 9–12 with hairs 9 and 10 long, simple; hair 11 minute, simple; and hair 12 short, 2- or 3-branched. Metathoracic palmate hair 3 small or obsolete. *Abdomen:* Accessory dorsal hair 0 obsolete on segments IV and V; palmate hair 1 rudimentary on segments I and II, well developed and of about equal size on III–VII (may be somewhat smaller on VII); leaflets with serrations on apical half. Antepalmate hair 2 usually double (occasionally single or triple) on segments IV and V. Hair 5 on segment I is 3-branched (rarely 4) and only slightly larger than hair 4. Lateral hair 6 long, plumose on segments I–III. Pecten as illustrated (fig. 40, *B*).

DISTRIBUTION. This species is found throughout most of the United States and southern Canada (rare throughout much of the Rocky Mountain region). Its range extends south to the tablelands of Mexico (481, 730). *United States:* Alabama (135); Arkansas (88, 121); California (287); Colorado (343); Connecticut (270, 494); Delaware (120, 167); District of Columbia (318); Florida (135, 419); Georgia (135, 608); Idaho (344, 672); Illinois (615); Indiana (152, 345); Iowa (625); Kansas (360, 418); Kentucky (134, 570); Louisiana (419, 759); Maine (45, 270); Maryland (64, 161); Massachusetts (270); Michigan (391, 639); Minnesota (533); Mississippi (135, 546); Missouri (6, 530); Montana (475, 550); Nebraska (14, 680); New Hampshire (66, 270); New Jersey (213, 350); New Mexico (418); New York (31); North Carolina (135); North Dakota (553); Ohio (418, 482); Oklahoma (628); Oregon (672); Pennsylvania (298, 756); Rhode Island (35, 270); South Carolina (135, 742); South Dakota (14); Tennessee (135, 658); Texas (249, 418); Vermont (270); Virginia (176, 178); Washington (67, 672); West Virginia (213); Wisconsin (163, 171); Wyoming (578). *Canada:* British Columbia (352); Manitoba (536, 708); New Brunswick (708); Nova Scotia (536, 708); Ontario (212, 708); Quebec (536, 708).

BIONOMICS. The larvae of *A. punctipennis*

are found in a large variety of aquatic habitats, including ponds, temporary pools, springs, pools in intermittent streams, borrow pits, roadside puddles, wheel ruts in muddy roads, hog wallows, eddies along the margins of flowing streams, and in rain-water barrels and other artificial containers. The species seems to prefer cool, clear water, particularly in hill streams.

The females feed mostly after dusk but will attack man during the daytime in dense woodlands or in their daylight resting places. This mosquito is generally regarded as an outdoor species and seldom enters dwellings in large numbers to feed. The adults rest during the daytime in hollow trees, culverts, underneath overhanging banks of streams, rock ledges, and in other similar, dark moist shelters. The species reaches its greatest abundance in the southern United States in the early spring and late fall, being somewhat less abundant in midsummer. A more even seasonal population is generally found in the northern part of its range. The females overwinter in hibernation in buildings, cellars, hollow trees, and other well-protected shelters. During the winter months the females occasionally enter living and sleeping quarters and bite man.

MEDICAL IMPORTANCE. Although this mosquito is readily infected with malaria parasites in the laboratory, it is not regarded as an important natural vector of the disease.

Anopheles (Anopheles) quadrimaculatus Say

Anopheles quadrimaculatus Say, 1824, Keating's Narr. Exp. St. Peter's River, 2:356.

Anopheles guttulatus Harris, 1835, Hitchcock's Rept. Geol. Min. Bot. Zool. Mass., p. 595.

Anopheles annulimanus Van der Wulp, 1867, Tijds. voor. Ent., 10:129.

ADULT FEMALE (pl. 9). Medium-sized species. *Head:* Proboscis long, dark; palpi about as long as the proboscis, dark, with raised scales on basal part. Occiput clothed with numerous dark erect forked scales, those on the median area and vertex pale; frontal tuft with some pale setae. *Thorax:* Integument of scutum brown to black; scutum clothed with numerous pale-yellow to golden-brown hairs, more numerous medially, longer laterally.

Scutellum crescent-shaped, clothed with yellow hairs and long dark setae. *Abdomen:* Integument of abdomen dark brown to black, densely clothed with yellowish-brown hairs. *Legs:* Legs dark-scaled; femora and tibiae tipped with pale scales. *Wing:* Length about 4.5 mm. Scales narrow, entirely dark, some of the scales forming four rather distinct darker spots. Scales on basal part of vein 5, before the fork, predominantly broadly rounded at apices, without serrate edges. Scales on distal third of wing dense and laid down in a more regular manner than in *A. freeborni;* wing membrane between the veins almost covered with scales. *Halter:* Knob, dark-scaled.

ADULT MALE. Coloration similar to that of the female. *Terminalia* (fig. 41). Ninth tergite (IX-T) with band narrow, sclerotized; lobes (IXT-L) about two to three times as long as broad, usually somewhat constricted medially, rounded at apex. Tenth sternite absent; anal lobe (An-L) large, triangular, spiculate. Phallosome (Ph) cylindrical, broad and furcate at base; apex with three or four pairs of ligulate subequal leaflets (Ph-L), the apical pair longer; one or more pairs of the leaflets with a few coarse basal teeth (fig. 41, *C*). Claspette (fig. 41, *B*) broad, consisting of a dorsal lobe (Cl-DL) bearing one to five stout, bluntly rounded or capitate spines (often fused to one another); and a ventral lobe (Cl-VL) usually bearing two (occasionally one or three) large acuminate apical spines of unequal length; the largest of these spines is rarely as broad as the blunt spine or fused spines of the dorsal lobe. The lobes are fused basally but are more or less distinct apically. Basistyle (Bs) conical, less than twice as long as mid-width, clothed with long and short setae and few or no scales; two parabasal spines (P-S) and one internal spine (I-S) present. Dististyle (Ds) as long or longer than the basistyle, with a row of small papillated setae on dorsal surface, usually without nonpapillated setae on basal part; claw (Ds-C) short, blunt.

LARVA (fig. 42). *Head:* Antenna spiculate; antennal tuft 1 inserted near middle of shaft, with several branches; terminal hair 4 with several branches. Inner clypeal hairs 2 simple, with basal tubercles separated by at least the diameter of one of the tubercles; outer

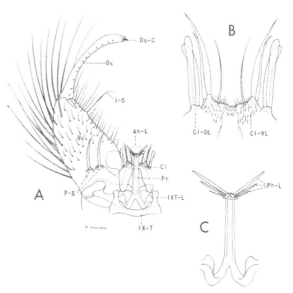

Fig. 41. *Anopheles quadrimaculatus* Say. *A*, Male terminalia. *B*, Claspettes. *C*, Phallosome.

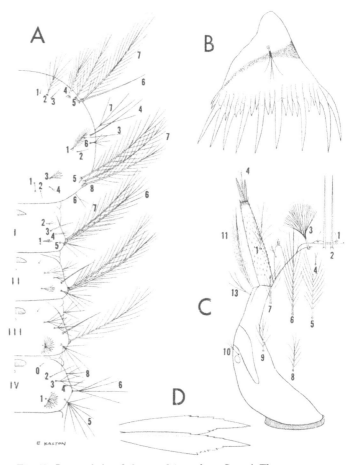

Fig. 42. Larva of *Anopheles quadrimaculatus* Say. *A*, Thorax and abdomen. *B*, Pecten. *C*, Head. *D*, Leaflets of palmate hair.

clypeal 3 densely branched dichotomously; postclypeal 4 small, usually branched distally; frontal hairs 5–7 large, plumose; sutural 8 usually 8- to 10-branched; transsutural 9 usually 6- to 8-branched. *Thorax:* Prothoracic hair 1 short, simple or weakly branched at tip; hair 2 long, stout, with many branches; hair 3 short, simple. Prothoracic pleural group 9–12 with hair 9 long, simple or 2-branched before middle; hair 10 and 12 long, simple; and hair 11 short, simple. Mesothoracic pleural group 9–12 with hairs 9 and 10 long, simple; hair 11 minute, simple; and hair 12 simple, about one-third as long as hair 9 or 10. Metathoracic pleural group 9–12 with hairs 9 and 10 long, simple; hair 11 minute, simple; and hair 12 short, 2- to 4-branched. Metathoracic palmate hair 3 small, with transparent leaflets. *Abdomen:* Accessory dorsal hair 0 obsolete on segments IV and V; palmate hair 1 well developed on abdominal segments III–VII, rudimentary on segments I and II; leaflets with serrations on apical half. Antepalmate hair 2 single, occasionally double on segments IV and V. Lateral hair 6 long, plumose on segments I–III. Pecten as illustrated (fig. 42, *B*).

DISTRIBUTION. Eastern and central United States, north to southern Canada. It also occurs in Mexico (730). This species reaches its greatest abundance in southeastern United States. *United States:* Alabama (135); Arkansas (88, 121); Connecticut (270, 494); Delaware (120, 167); District of Columbia (318); Florida (135, 419); Georgia (135, 608); Illinois (615); Indiana (152, 345); Iowa (625); Kansas (14, 530); Kentucky (134, 570); Louisiana (419, 759); Maine (270); Maryland (64); Massachusetts (270); Michigan (391, 639); Minnesota (163, 533); Mississippi (135, 546); Missouri (6, 14); Nebraska (530, 680); New Hampshire (66, 270); New Jersey (350); New York (31); North Carolina (135, 419); North Dakota (553); Ohio (418, 482); Oklahoma (368, 628); Pennsylvania (298, 756); Rhode Island (35, 270); South Carolina (135, 742); South Dakota (14, 531); Tennessee (135, 658); Texas (249, 418); Vermont (29, 270); Virginia (176, 178); Wisconsin (163, 171). *Canada:* Ontario (367, 409); Quebec (708).

BIONOMICS. The larvae are found in permanent fresh water in sluggish streams, canals, ponds, and lakes containing surface-growing or emergent vegetation or floating debris, and only occasionally in pools of a temporary nature. The eggs are deposited by the female singly upon the surface of the water, but tend to arrange themselves in rather definite patterns. The larvae feed almost entirely upon small organisms at the water surface; they do not appear to make any selection of their food, but ingest any material small enough to be swept into the mouth by the mouth brushes. Bradley (82), while working at Mound, Louisiana, concluded that flagellates, diatoms, and green algae, which make up a large proportion of the surface plankton in that area, constituted the principal food of the larvae. During the summer season the larval period is rather short, about 12 to 20 days depending on temperature and food supply. The pupal period varies from 2 to 6 days. The number of generations per season varies in different regions. Boyd (76) believed that there were seven to eight generations in North Carolina, whereas Hurlbut (390) records nine to ten generations in the Tennessee Valley.

The females of *A. quadrimaculatus* are active feeders on man and on wild and domesticated animals. The adults are active principally at night and rest in dark corners in buildings, underneath houses, in stables, in hollow trees, and other shelters during daylight hours. They are very active for a short period after dusk, but during the remainder of the night their activities are probably limited mostly to flights in search of a blood meal except for another active period at dawn when they shift to daytime resting places. The overwintering of inseminated females in hibernation in the Tennessee Valley is described by Hinman and Hurlbut (366) and by Hess and Crowell (359). The females normally do not feed during the daytime, but they will seek blood meals in daylight on warm days in the early spring, after coming out of hibernation, and occasionally in dark buildings and on cloudy days.

Keener (411) studied the habits of *A. quadrimaculatus* in a laboratory colony and found that if blood meals were available to the females, oviposition took place about 3 days after emergence. Each female laid from nine to twelve batches of eggs during her lifetime, and each batch varied from 194 to 263 eggs. The effective flight range of this species often varies, probably depending to some extent on the proximity to suitable hosts upon which the females can feed and on the number of adults of the species produced in an area. It is usually regarded as 1 mile or less, under average conditions; however, much longer flights have been recorded by the recapture of marked specimens. Hybridization between *A. quadrimaculatus* and *A. freeborni*, under laboratory conditions, has been reported by Burgess (114). Keener (411) describes the mating of males and females in small rearing cages. They mate in flight, usually fall to the floor of the cage, and separate after about 10 to 15 seconds.

MEDICAL IMPORTANCE. The species occurs in large numbers throughout the southeastern United States, except in southern Florida, and is regarded as the most important vector of malaria in this region. The susceptibility of this mosquito to experimental infection with *Plasmodium vivax*, *P. falciparum*, and *P. malariae* has been well established, and gland infections in wild-caught specimens have been reported by numerous workers.

Anopheles (Anopheles) walkeri Theobald

Anopheles walkeri Theobald, 1901, Mon. Culic.,
1:299.

ADULT FEMALE (pl. 10). Medium-sized species. *Head:* Proboscis long, dark; palpi about as long as the proboscis, dark-scaled except for a narrow white ring at apex of each segment, basal part with raised scales. Occiput clothed with dark-brown erect forked scales; the vertex with a few pale recumbent scales; frontal tuft dark brown. *Thorax:* Integument of scutum dark brown to black; scutum clothed with short golden-brown hairs medially, longer dark setae laterally. Scutellum crescent-shaped, clothed with golden-brown hairs and long dark setae. *Abdomen:* Integument dark brown to black, densely clothed with yellow to brown hairs. *Legs:* Legs dark-scaled, femora and tibiae tipped with pale scales. *Wing:* Length about 4.0 to 4.5 mm. Scales narrow, entirely dark; some of the scales forming four darker spots, more or less distinct. *Halter:* Knob usually pale-

scaled. Stone (676) points out that specimens from Canada and the states bordering Canada usually have the knobs brownish and the scales distinctly darkened.

ADULT MALE. Coloration similar to that of the female. *Terminalia* (fig. 43). Ninth tergite (IX-T) with band narrow, moderately sclerotized; lobes (IXT-L) slender, tapered, pointed, or rounded at tip. Tenth sternite absent; anal lobe (An-L) large, triangular, spiculate. Phallosome (Ph) cylindrical, broad and furcate at base; apex with three to five pairs of ligulate leaflets (Ph-L), the anteapical pair more than half as long as the apical pair (usually about three-fourths as long); none of the leaflets with coarse basal teeth. Claspette (fig. 43, C) broad, consisting of a dorsal lobe (Cl-DL) bearing one or two large, blunt, or slightly capitate apical spines; and a ventral lobe (Cl-VL) bearing a large acuminate apical spine and a long slender subapical spine. Basistyle (Bs) conical, less than twice as long as mid-width, clothed with long and short setae and few or no scales; two parabasal spines (P-S) and one internal spine (I-S) present. Dististyle (Ds) a little longer than the basistyle, with a row of small papillated setae along dorsal surface and usually with a patch of minute nonpapillated hairs on basal sixth of dististyle; claw (Ds-C) short, blunt.

LARVA (fig. 44). *Head:* Antenna spiculate antennal tuft 1 inserted near basal third of

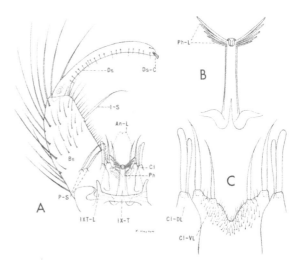

Fig. 43. *Anopheles walkeri* Theobald. *A*, Male terminalia. *B*, Phallosome. *C*, Claspettes.

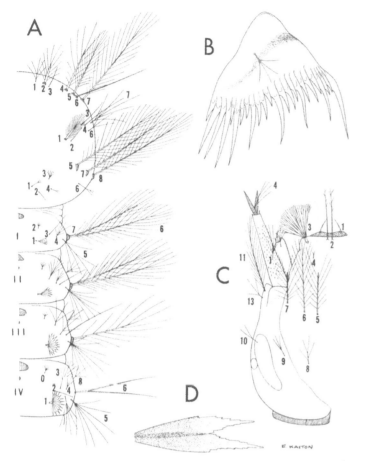

Fig. 44. Larva of *Anopheles walkeri* Theobald. *A*, Thorax and abdomen. *B*, Pecten. *C*, Head. *D*, Leaflets of palmate hair.

shaft, several branched; terminal hair 4 with several branches. Inner clypeal hairs 2 with sparse, minute feathering toward tips, basal tubercles rarely separated by more than the diameter of one tubercle; outer clypeal 3 densely branched dichotomously; postclypeal 4 small, branched distally; frontal hairs 5–7 large plumose; sutural 8 and transsutural 9 with several branches. *Thorax:* Prothoracic hair 1, 3- to 5-branched; hair 2 longer and with many branches; hair 3 short, simple. Prothoracic pleural group 9–12 with hairs 9, 10, and 12 long, simple; and hair 11 short, simple. Mesothoracic pleural group 9–12 with hairs 9 and 10 long, simple; hair 11 minute, simple; and hair 12 simple, about one-third as long as hair 9 or 10. Metathoracic pleural group 9–12 with hairs 9 and 10 long, simple; hair 11 minute, simple; and hair 12 short, usually simple. Metathoracic palmate hair 3 small, with

transparent leaflets. *Abdomen:* Accessory dorsal hair 0 comparatively well developed on segments IV and V (much smaller than antepalmate hair 2), with three to seven branches. Palmate hair 1 well developed, nearly equal in size on segments III–VII; leaflets serrated on apical half; palmate hairs rudimentary on segments I and II. Antepalmate hair 2 usually single on segments IV and V, sometimes double or triple. Lateral hair 6 long, plumose on segments I–III. Pecten as illustrated (fig. 44, *B*).

DISTRIBUTION. Eastern United States from the Gulf of Mexico north to southern Canada and west to North Dakota, Nebraska, and Texas. It has been taken also in Mexico (730). *United States:* Alabama (136); Arkansas (683); Connecticut (494); Delaware (120, 167); District of Columbia (318); Florida (135, 419); Georgia (419, 516); Illinois (615); Indiana (152, 418); Iowa (530, 625); Kansas (14, 286); Kentucky (570); Louisiana (418, 759); Maine (45, 270); Maryland (64, 161); Massachusetts (28, 270); Michigan (391, 639); Minnesota (163, 533); Mississippi (420, 516); Missouri (14); Nebraska (14, 680); New Hampshire (66, 270); New Jersey (350, 418); New York (31); North Carolina (135, 514); North Dakota (553); Ohio (418); Pennsylvania (298, 756); Rhode Island (35, 436); South Carolina (84, 742); South Dakota (14); Tennessee (18, 658); Texas (532); Vermont (29, 270); Virginia (176, 418); Wisconsin (163, 171). *Canada:* British Columbia (536, 708); Manitoba and New Brunswick (708); Nova Scotia, Ontario and Quebec (536, 708).

BIONOMICS. Hurlbut (387) found that *A. walkeri* produces winter eggs, and he believes that this mosquito hibernates in the egg stage in the northern part of its range. The larvae of *A. walkeri* occur in fresh-water marshes containing emergent or floating vegetation or debris. Bang *et al.* (18) observed that the larvae occur in the Reelfoot Lake area in Tennessee in almost any body of water in which there is thick emergent vegetation. These workers concluded that cut-grass (*Z. zaniopsis miliacea*), shaded by willows or button bushes, offered the most ideal larval habitat for the species.

The adults are known to enter dwellings at night to feed on humans and then retire to secretive daytime hiding places. The senior author has collected the females in Georgia in a swamp where they were attacking and biting at midday in bright sunlight. Bang *et al.* (18) found the adults resting in dark, extremely moist situations, particularly around the shaded bases of cut-grass and other shore-line shrubbery. The adults are readily attracted to light traps. These workers captured specimens of *A. walkeri* in a light trap 1½ to 2 miles from the nearest known aquatic habitat available for the immature stages.

MEDICAL IMPORTANCE. This mosquito appears to be a potential vector of malaria on the basis of successful experimental transmission of the disease and the finding of a single wild-caught specimen by Bang *et al.* (17) containing oöcysts and a gland infection.

Anopheles (Nyssorhynchus) albimanus
Wiedemann

Anopheles albimanus Wiedemann, 1821, Dipt. Exot., p. 10.
Anopheles (Nyssorhynchus) albimanus, Komp, 1942, Nat. Inst. Hlth. Bull., *179:*67, 115, 154.

ADULT FEMALE (pl. 11). Medium-sized species. *Head:* Proboscis long, black. Palpi about as long as the proboscis, dark, with the apices of segments 2 and 3 narrowly marked with white, all of segment 5 white, basal part with raised scales. Occiput with white erect forked scales dorsally, dark-scaled laterally; vertex white; frontal tuft white. *Thorax:* Integument of scutum dark gray to dark brown, with a pair of small submedian black spots near middle and a posteromedian black spot on prescutellar space; scutum clothed with gray scales and pale setae, the setae more numerous laterally. Scutellum crescent-shaped, clothed with gray scales and yellowish-brown setae. *Abdomen:* Integument dark brown to black, clothed dorsally with long setae and pale scales, the scales more numerous on middle of the segments; ovate dark scales present laterally on segments III–VII. *Legs:* Femora dark, sparsely speckled with pale scales, tipped with white; femora II and III each with a pale-scaled subapical spot on one side. Tibiae dark, sparsely speckled with white, tipped with white. Tarsi marked as follows: front tarsus with segment 1 narrowly white ringed at apex, 2 and 3

white apically, 4 and 5 entirely dark; middle tarsus with narrow white rings at apices of segments 1 and 2, 3–5 entirely dark; hind tarsus with segment 1 entirely dark except for a few white scales at apex, 2 usually with basal half dark and apical half white, 3 and 4 entirely white, 5 dark basally and white apically. *Wing:* Length 3.5 to 4.0 mm. Scales black and yellowish white, arranged on the veins in contrasting lines and spots (pl. 11). *Halter:* Knob dark-scaled.

ADULT MALE. Coloration similar to that of the female. *Terminalia* (fig. 45). Ninth tergite (IX-T) membranous dorsally, sclerotized laterally; lobes absent. Tenth sternite absent; anal lobe (An-L) subtriangular, not spiculate, lateral margins pigmented. Phallosome (Ph) cylindrical, broad and furcate at base, rounded apically and without leaflets (fig. 45, *B*). Ventral lobes of claspette (Cl-VL) fleshy, fused on the midline, each rounded at apex and with a large, ovoid lobe on ventral side. Dorsal lobes of claspette (Cl-DL) widely separated, each with three or four sickle-shaped apical filaments set close together (fig. 45, *C*). Basistyle (Bs) conical, about two to two and a half times as long as mid-width, clothed with scales and long and short setae; one stout recurved parabasal spine (P-S), two stout recurved accessory spines (Ac-S) and one slender straight internal spine (I-S) present. Dististyle (Ds) a little shorter than the basistyle, with a row of small papillated setae on inner dorsal surface, without minute nonpapillated setae; claw (Ds-C) bluntly pointed.

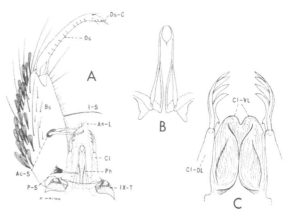

Fig. 45. *Anopheles albimanus* Wiedemann. *A,* Male terminalia. *B,* Phallosome. *C,* Claspettes.

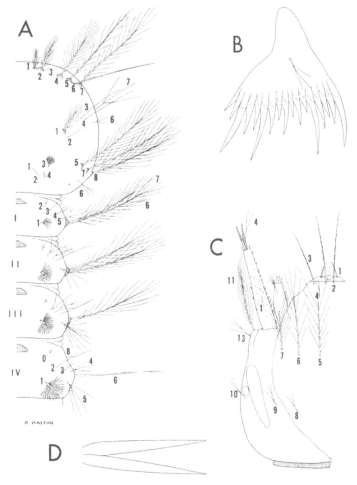

Fig. 46. Larva of *Anopheles albimanus* Wiedemann. *A*, Thorax and abdomen. *B*, Pecten. *C*, Head. *D*, Leaflets of palmate hair.

LARVA (fig. 46). *Head:* antenna spiculate (spicules on inner margin at middle of shaft much coarser); antennal tuft 1 short, 2- to 4-branched, inserted on basal third of shaft; terminal hair 4, 2- to 4-branched. Inner clypeal hairs 2 and outer clypeal 3 finely barbed, basal tubercles nearly equidistant; postclypeal 4 single or branched apically; frontal hairs 5–7 large, plumose; sutural 8 usually 2- to 4-branched; transsutural 9 usually 2- to 5-branched. *Thorax:* Prothoracic hairs 1–3 arising from a common sclerotized base; hair 1 stout, about half as long as hair 2, shaft thickened with many lateral branches; hair 2 long, strong, shaft thickened and with many lateral branches; hair 3 small, simple. Prothoracic pleural group 9–12 with hairs 10 and 12 long,

simple; hair 9 long, with several lateral branches; and hair 11 about one-half as long as hair 10 or 12, bifid or trifid. Mesothoracic pleural group 9–12 with hairs 9 and 10 long, simple; hair 11 minute; and hair 12 short (about one-fourth as long as 9 or 10), simple. Metathoracic pleural group 9–12 with hair 9 long, with several weak lateral branches; hair 10 long, simple; hair 11 minute; and hair 12 short, usually bifid. Metathoracic palmate hair 3 with numerous slender transparent leaflets. *Abdomen:* Accessory dorsal hair 0 small, branched, present on segments II–VII. Palmate hair 1 well developed on segments I–VII (smaller on segments I and VII); leaflets slender, with smooth margins. Antepalmate hair 2 long, simple on segments IV and V. Lateral hair 6 long, plumose on segments I–III. Pecten as illustrated (fig. 46, *B*).

DISTRIBUTION. Along the Rio Grande River in Texas (463, 634), Mexico, the West Indies, Central America, and South America, south to Ecuador, Colombia, and Venezuela. A collection, consisting of 131 adult specimens, was taken at Key West, Florida, by Dr. G. N. Mac-Donnell, during 1904 and reported by King (414). One fourth instar larva was collected by Ernest Erb at Boca Raton, Florida, during May, 1944, and reported by Carpenter (129) and Carpenter *et al.* (136). Pritchard *et al.* (563) report the finding of a single fourth instar larva of this species on the northwestern side of Big Pine Key on January 13, 1946. An Anopheline survey of the Florida Keys was then made, but no other specimens of *A. albimanus* were found. However, Pritchard *et al.* (563) state that the species may possibly be endemic to the Florida Keys where conditions are such that it maintains itself only in erratic and extremely limited numbers. Burgess (115) reports the successful establishment of a laboratory colony of this species, developed from larvae collected in the Florida Keys. Eads (247) reports the capture of several female specimens in light traps operated near Corpus Christi, Texas, on the Gulf Coast approximately 150 miles from Brownsville, where the species is known to be well established.

BIONOMICS. The larvae of *A. albimanus* are found in a wide variety of water collections, either fresh or brackish. The species seems to require a great deal of sunlight in its larval habitats and is rarely found in densely shaded pools. During the rainy season it is abundant in rain-filled pools throughout most of its range. It also utilizes seepage pools, irrigation ditches, and the quiet shallow margins of ponds, streams, and lakes for its immature stages. Water-filled hoof prints in cow pastures are favorite habitats during the rainy season. Larval densities are frequently very high during the dry season in shallow areas of lakes where surface growths of vegetation are sufficient to protect the larvae and pupae from predatory fish.

The adults are strong fliers and feed readily on man and domestic animals, particularly horses and cattle. Horse-baited traps are favorite devices for measuring adult populations in the American tropics. Pritchard and Pratt (561) found the New Jersey light trap to be superior to an animal-baited trap for sampling adult populations of *A. albimanus* in Puerto Rico. Carpenter and Peyton (140) and others have observed that the adults are not readily attracted to artificial light in Panama and that the light trap is generally inferior to a horse-baited trap as a sampling device in that area. The females are domestic and are commonly found in houses at night, but return to the jungle early in the morning. They do not seem to congregate in diurnal shelters as most Anophelines do, but probably remain dispersed in the jungle during the daytime.

MEDICAL IMPORTANCE. The species is generally regarded as the principal lowland vector of malaria throughout much of Central America, northern South America, and the West Indies.

Tribe Toxorhynchitini

The tribe Toxorhynchitini consists of large usually brilliantly colored diurnal, nonblood-sucking mosquitoes, occurring mostly in tropical regions. The tribe includes the single genus *Toxorhynchites*.

Genus TOXORHYNCHITES Theobald[7]

Megarhinus Robineau-Desvoidy, 1827, Mem. Soc. d'Hist. Nat., *3*:403.

Toxorhynchites Theobald, 1901, Mon. Culic., *1*:244.

Ankylorhynchus Lutz, 1904, in Bourroul, Mosq. do Brasil, p. 53.

Lynchiella Lahille, 1904, Act. y Trab. 2 Con. Med. Lat.-Amer., *2*:13.

Worcesteria Banks, 1906, Phila. Jour. Sci., *1*:779.

Teromyia Leicester, 1908, Culic. Malaya, p. 49.

Toxorhynchites, Stone, 1948, Proc. Ent. Soc. Wash., *50*:161.

ADULT. Clypeus broader than long, anterior margin slightly trilobed. Proboscis rigid on basal half, flexible, and strongly curved downward on distal half. Palpi of females vary in length in different species from about one-sixth to about two-thirds as long as the proboscis. Antennae of male plumose. Scutellum evenly rounded. All species large, usually with flat scales of metallic luster on head, thorax, abdomen, and legs. Pleura heavily scaled. Wing scales sparse, short, and broad. Second marginal cell of wing much shorter than its stem; a slight emargination present on hind margin of wing opposite vein 5.2, and a submarginal V-shaped thickening of the membrane between veins 5.1 and 5.2. Squama without fringe. *Male terminalia*. Tenth sternite with not more than two or three terminal teeth. Phallosome composed of two elongated plates, heavily sclerotized. Basistyle conical; basal lobe bearing spines; apical lobe absent. Claspette absent. Dististyle long, slender, not swollen medially.

PUPA. Respiratory trumpets short in species occurring in this region. Dendritic tufts large on abdominal segment I. Outer part of paddle extends beyond tip of midrib.

LARVA. Very large and predacious. Mouth brushes consist of about ten strong, curved, flattened setae, not serrated but with hooked tips. Head nearly square; frontoclypeus divided by a transverse suture into an anterior part and a much larger posterior part comprising most of the dorsal head surface. Mentum much broader than long. Antenna smooth, with a small or minute tuft inserted beyond middle. Hairs of thorax and abdomen arising from sclerotized plates. Comb and pecten absent.

EGGS. Ovate, almost round, surface minutely spinose. Eggs are laid singly, usually on the surface of the water.

BIONOMICS. The larvae are found in small collections of water in the leaf bases of *Bromeliaceae*, in treeholes, bamboo sections, and occasionally in artificial water containers. The larvae are predacious on other mosquito larvae with which they are associated in their aquatic habitats. Both the males and females fly by day and feed on plant juices but do not suck blood.

CLASSIFICATION. Edwards (251) recognized three groups, A, B, and C, distinguished by the

---

[7]Stone (1948) has shown that the name *Megarhinus* Robineau-Desvoidy is preoccupied, and *Toxorhynchites* Theobald thus becomes the available name for the genus.

form of the palpi of the females. Groups A and B occur in the Americas, but group C is confined to the Old World. This tropical genus is represented in the United States by two closely related geographical subspecies, *T. rutilus rutilus* (Coquillett) and *T. rutilus septentrionalis* (Dyar and Knab), the only known character separating them being a color difference of the males.

DISTRIBUTION. The genus is largely confined to the tropical regions of both the Old and New Worlds, with a few species occurring in the warmer parts of the north temperate zone.

KEY TO THE SPECIES

ADULT MALES

Front tarsus with segment 2 and the basal part of segment 3 with white scales..........*rutilus rutilus* (Coq.), p. 59

Front tarsus entirely dark-scaled........*rutilus septentrionalis* (D. and K.), p. 60

Toxorhynchites rutilus rutilus (Coquillett)

Megarhinus rutila Coquillett, 1896, Can. Ent., *28*:44.
Toxorhynchites rutilus rutilus, Jenkins, 1949, Proc. Ent. Soc. Wash., *51*:225.

ADULT FEMALE. A very large brilliantly ornamented mosquito indistinguishable from *T. rutilus septentrionalis* (pl. 12). *Head:* Proboscis long, black, the apical half strongly curved downward, a few metallic-blue scales intermixed. Palpi about two-thirds as long as the proboscis, metallic dark blue to violet, with golden-yellow scales laterally on second and third segments. Occiput with appressed dark and metallic-blue and coppery scales dorsally; the eye margins and lateral parts of the occiput with flat, silvery to golden-yellow scales. Tori black, with dense gray pubescence on inner and dorsal surfaces. *Thorax:* Anterior pronotal lobe with broad flat scales, those on the dorsal part brilliant metallic blue. Integument of scutum dark brown; a median stripe on anterior two-thirds of scutum and the sides primarily gold-scaled with blue reflection; remainder of scutum clothed with fine dark purplish-brown scales. Pleura shingled with flat golden-yellow scales. Scutellum with closely appressed purplish-brown scales, those on posterior and lateral margins gold, with bluish

reflection. *Abdomen:* Abdomen clothed dorsally with brown to metallic blue-green scales, with golden-yellow scales laterally. *Legs:* Femora dark brown, with purple reflection, golden yellow basally and on posterior surface; white knee spots present. Tibiae primarily clothed with dark brown and purple scales. Front and middle tarsi with apex of segment 1, all of 2 and 3, and all but apex of 4 white; hind tarsus with apex of segment 3, all of 4, and all but tip of 5 white. *Wing:* Length about 6.0 mm. Scales rather broad and sparse, dark purple.

ADULT MALE. Coloration similar to that of the female but with segment 2 and the basal part of segment 3 of the front and middle tarsi and segment 5 of the hind tarsus silvery white in typical specimens. These segments are all dark purple in typical specimens of *T. rutilus septentrionalis*. However, all types of intergrades occur in the zone of overlap of the range of the two subspecies. *Terminalia.* Indistinguishable from *T. rutilus septentrionalis* (fig. 47). Ninth tergite (IX-T) with band well developed; lobes (IXT-L) slightly elevated and widely separated, each bearing many long setae. Tenth sternite (X-S) heavily sclerotized and toothed apically. Phallosome (Ph) bottle-shaped, composed of two separate lateral plates connected on the dorsum near base by a narrow bridge; apical part slender, heavily sclerotized, pointed, each plate with a row of short teeth directed ventrally. Claspette absent. Basistyle (Bs) about twice as long as basal width, clothed with long and short setae and numerous scales; basal lobe (B-L) rounded, bearing three or four large strong setae and several smaller ones; apical lobe absent. Dististyle (Ds) nearly as long as basistyle, slender, with many small setae on distal half; claw (Ds-C) slender, inserted a little before apex.

LARVA. Indistinguishable from *T. rutilus septentrionalis* (fig. 48). Large larva. Head quadrate, about as long as broad. Antenna smooth, about one-third as long as the head; antennal tuft short, multiple, inserted on outer fourth or fifth of shaft, preceded by a pair of single hairs at outer third or fourth. Mouth brushes composed of about ten stout, prehensile, smooth, curved rods. Head hairs: Postclypeal 4 short, multiple; upper frontal 5, lower frontal 6, and preantennal 7 single and about equal in length. Thorax and abdomen with many of the hairs and spines arising from sclerotized bases. Eighth abdominal segment without a comb but with a large sclerotized lateral plate bearing on its posterior margin two long barbed spines and two short branched hairs. Siphonal index 1.5 to 2.0; pecten absent; siphonal tuft multiple, barbed, inserted near base of siphon. Anal segment ringed by the saddle; posterolateral margin of saddle fringed with a row of intermixed short and long spines; lateral hair represented by a stout, barbed spine, inserted on distal margin of the saddle, a little shorter than the saddle; dorsal brush bilaterally consisting of a lower caudal tuft of about four or five subequal branches and an upper caudal tuft of about eight to ten subequal branches; ventral brush with individual hairs feathered, confined to the barred area; gills bulbous, much shorter than the saddle.

DISTRIBUTION. Known only from the extreme southeastern United States. *United States:* Florida (38, 397, 652); Georgia (397, 400); South Carolina (137, 397). Basham *et al.* (38) list collections of this species from four counties in Florida.

BIONOMICS. The larvae have been found in rot cavities in trees in South Carolina and Florida. Basham *et al.* (38) found the larvae in holes in water oak, scrub oak, live oak, orange, pecan, and pine trees, and in various types of artificial containers in Florida. Seabrook and Duffey (652) were unable to find the larvae of this mosquito in rot cavities in trees in Palm Beach County, Florida, but found them in one of the arboreal bromeliads (*Tillandsia utriculata* Linn.) where they were associated with the larvae of *Wyeomyia mitchellii* and *W. vanduzeei.* The habits and time required for development of the immature stages to adults are apparently similar to those of *T. rutilus septentrionalis.*

Toxorhynchites rutilus septentrionalis
(Dyar and Knab)

Megarhinus septentrionalis Dyar and Knab, 1907, Jour. N.Y. Ent. Soc., *15*:12.

Megarhinus herrickii Theobald, 1907, Mon. Culic., 4:131.

Toxorhynchites rutilus septentrionalis, Jenkins, 1949, Proc. Ent. Soc. Wash., *51*:225.

ADULT FEMALE (pl. 12). A very large, brilliantly ornamented mosquito indistinguishable from *T. rutilus rutilus.*

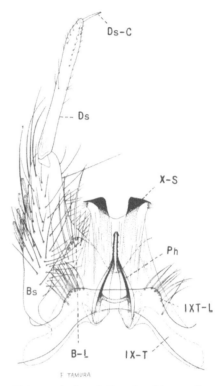

Fig. 47. Male terminalia of *Toxorhynchites rutilus septentrionalis* (Dyar and Knab).

ADULT MALE. Coloration similar to that of the female, but in typical specimens the front and middle tarsi have all segments dark purple, and the hind tarsus has segment 4 white. However, all types of intergrades occur in the zone of overlap of the two subspecies. *Terminalia* (fig. 47). Indistinguishable from *T. rutilus rutilus.*

LARVA (fig. 48). Indistinguishable from *T. rutilus rutilus.*

DISTRIBUTION. Eastern United States, north to New Jersey and Pennsylvania and west to the great plains of Kansas, Oklahoma, and Texas. *United States:* Alabama (397, 400); Arkansas (121, 397); Delaware (167); District of Columbia (213, 318); Florida (38, 397); Georgia (135, 397); Illinois (397, 615); Kansas (360, 397); Kentucky (397, 570); Louisiana (397, 759); Maryland (213, 397); Mississippi (135, 397); Missouri (6,

397); New Jersey (397); North Carolina
(135, 397); Ohio (397, 482); Oklahoma
(397, 628); Pennsylvania (90, 397); South
Carolina (135, 742); Tennessee (397, 658);
Texas (96, 397); Virginia (176, 397); West
Virginia (213, 397).

BIONOMICS. The adults are among the largest
and most colorful mosquitoes found in this
region. The larvae are found principally in rot
cavities of trees of many kinds and occasionally
in water in artificial containers such as rain
barrels, wooden tubs, tin buckets and cans,
stone and glass jars, and in rockholes. The larvae are predacious, feeding on mosquito larvae
and other small aquatic animals with which
they are associated in their habitats. Breland
(96) made a rather comprehensive study of
the biology and immature stages of this mosquito in Texas. He found the pure white oval
eggs floating on the water surface in a tree
cavity near Austin, Texas, hatched them, and
studied all the larval instars.

While collecting the eggs, Breland observed
a female dipping up and down above the water
surface in an aerial dance, presumably in the
act of depositing eggs. Michener (510) reports
finding 18 eggs of this mosquito floating on the
water surface in a small treehole at Camp
Shelby, Mississippi. The eggs were brilliant
white, but when placed in alcohol the white of
the egg immediately disappeared, leaving the
chorion transparent and slightly brownish.
Some workers believe that the white color of
the eggs may be produced by fine air bubbles.
Michener also observed that when hatching, the
egg is split lengthwise into two halves which
frequently remain attached to one another by
a narrow ribbon of shell. Breland (96) also
observed that *Toxorhynchites* larvae did not
attack anything that was not moving at the time.
He noticed that the mandibles alone, not the
modified mouth brushes, were used to seize and
hold prey.

The larvae are frequently associated with
the immature stages of *Anopheles barberi* Coq.,
Aedes triseriatus (Say), and *Orthopodomyia*
spp. in their aquatic habitats. The larval period
is quite long, usually requiring several weeks,
but has been observed to last as long as 6
months. The pupal stage is usually 4 to 7 days,

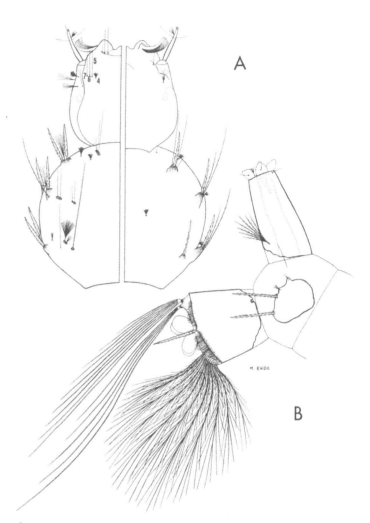

Fig. 48. Larva of *Toxorhynchites rutilus septentrionalis*
(Dyar and Knab). *A*, Head and thorax (dorsal and ventral).
B, Terminal segments.

but has been known to last 23 days (400). It
is said that this mosquito passes the winter in
the larval stage in some parts of its range, but
Breland believes that it passes the winter as an
adult in some areas, particularly around
Austin, Texas. Michener (510) is of the opinion that it passes the winter as mature larvae in
southern Mississippi, and he noticed that all
larvae collected during the winter in southern
Mississippi were full grown. The males and
females fly by day and feed on the nectar of
flowers and other plant juices. The adults have
been observed resting on vines and tree trunks,
often in the sunlight, near the habitats of their
immature stages.

Tribe Culicini

ADULT. Clypeus longer than broad, rounded anteriorly. Proboscis slender, flexible, not hooked. Thorax rather strongly arched. Scutellum trilobed, each lobe bearing a distinct group of setae. Abdomen clothed with broad flat scales. Wing with cross vein 3–4 not bent at right angles, no spur extending basally from angle of vein 3. Wing without a V-shaped thickening of the membrane between veins 5.1 and 5.2. Second marginal cell about as long as its petiole (much shorter in *Uranotaenia*). *Male terminalia*. The structures are of various forms but always with well-developed tenth sternites or paraprocts; phallosome never with leaflets at its tip.

PUPA. Abdominal segments II to VI with lateral hairs inserted some distance from the posterolateral corners of each segment. Paddles usually with a hair tuft 8 at tip of midrib; when a second hair 7 is present, as in *Culex*, it is placed near hair 8, not anterior to it. Outer part of paddle never produced distinctly beyond end of midrib.

LARVA. Head rarely rotatable. Mouth brushes composed of a large number of hairs, often reduced in number in the predacious species, but never less than about thirty. Hairs of body never pinnately branched as in Anophelini. Thorax broad and without eversible organs. Abdomen without float hairs. Eighth abdominal segment with a siphon. Lateral comb always present on the eighth abdominal segment.

EGGS. Variously shaped in the different genera and species but never stoutly oval or rounded as in Toxorhynchitini or equipped with lateral floats as in Anophelini.

CLASSIFICATION. The tribe Culicini is represented in this region by nine genera included in the following keys:

KEYS TO THE GENERA

ADULTS

1. Second marginal cell of wing short, less than half as long as its petiole......*Uranotaenia* L. A., p. 73
   Second marginal cell of wing as long or longer than its petiole 2

2. Postnotum with a tuft of setae; wing squama without fringe of hairs..........*Wyeomyia* Theob., p. 63
   Postnotum bare; wing squama with a fringe of hairs 3

3. Spiracular bristles present 4
   Spiracular bristles absent 5

4. Postspiracular bristles present; tip of abdomen pointed.......................*Psorophora* R. D., p. 112
   Postspiracular bristles absent; tip of abdomen blunt....................................*Culiseta* Felt, p. 82

5. Postspiracular bristles present...........................6
   Postspiracular bristles absent..............................7

6. Wing scales very broad; tip of abdomen blunt
   (in part) *Mansonia* Blanch., p. 104
   Wing scales narrow (rarely moderately broad); tip of abdomen pointed......*Aedes* Meigen, p. 137

7. Antennae much longer than the proboscis; first flagellar segment longer than the next three segments combined......*Deinocerites* Theob., p. 325
   Antennae not longer than the proboscis or only slightly so, first flagellar segment about as long as each succeeding segment...................................8

8. Scutum bicolorous, with narrow longitudinal lines of white scales; penultimate segment of front tarsi very short, only about as long as wide
   Orthopodomyia Theob., p. 98
   Scutum unicolorous, lacking white longitudinal

62

lines; penultimate segment of front tarsi much longer than wide................................9

9. Wing scales very broad, brown and white mixed (in part) *Mansonia* Blanch., p. 104
Wing scales narrow, uniformly dark
Culex Linn., p. 269

LARVAE (FOURTH INSTAR)

1. Distal half of siphon strongly attenuated, adapted for piercing roots of aquatic plants
Mansonia Blanch., p. 000
Siphon cylindrical or fusiform, not adapted for piercing roots of aquatic plants.....................2

2. Siphon with a pecten................................3
Siphon without a pecten................................8

3. Head longer than wide; eighth abdominal segment with a prominent sclerotized plate bearing the comb on its posterior border
Uranotaenia L. A., p. 73
Head at least as wide as long; eighth abdominal segment without a prominent sclerotized plate (small plate present in some *Psorophora* species) 4

4. Head with a prominent triangular pouch on each side; anal segment with divided dorsal and ventral sclerotic plates, membranous laterally
Deinocerites Theob., p. 325
Head lacking a prominent triangular pouch on each side; anal segment completely ringed by the dorsal saddle or membranous only ventrally 5

5. Siphon with a pair of large basal siphonal tufts, or, if tufts are small, the comb scales are arranged in a single row and barlike
Culiseta Felt, p. 82
Siphon without a pair of basal siphonal tufts........6

6. Siphon with several pairs of siphonal tufts or single hairs....................*Culex* Linn., p. 269
Siphon with one pair of median or subapical siphonal tufts (sometimes vestigial) or one pair of single hairs................................7

7. Anal segment completely ringed by the saddle and pierced on the midventral line by tufts of the ventral brush................*Psorophora* R. D., p. 112
Anal segment not completely ringed by the saddle, or if ringed, not pierced on the midventral line by tufts of the ventral brush
Aedes Meigen, p. 137

8. Anal segment with a prominent median ventral brush consisting of a close-set row of tufts; comb with a row of short pointed scales and a second row of much longer pointed scales
Orthopodomyia Theob., p. 98
Anal segment without a median ventral brush, but with a pair of ventrolateral tufts; comb scales arranged in a single row
Wyeomyia Theob., p. 64

Genus WYEOMYIA Theobald

Wyeomyia Theobald, 1901, Mon. Culic., 2:267.
Dendromyia Theobald, 1903, Mon. Culic., 3:313.
Menolepis Lutz, 1905, Imp. Med., 13:269.
Prosopolepis Lutz, 1905, Imp. Med., p. 312.
Miamyia Dyar, 1919, Ins. Ins. Mens., 7:116.
Dinomyia Dyar, 1919, Ins. Ins. Mens., 7:117.
Triamyia Dyar, 1919, Ins. Ins. Mens., 7:120.
Pentemyia Dyar, 1919, Ins. Ins. Mens., 7:122.
Heliconiamyia Dyar, 1919, Ins. Ins. Mens., 7:123.
Diphalangarpe Dyar, 1919, Ins. Ins. Mens., 7:126.
Cleobonnea Dyar, 1919, Ins. Ins. Mens., 7:134.
Decamyia Dyar, 1919, Ins. Ins. Mens., 7:135.
Calladimyia Dyar, 1919, Ins. Ins. Mens., 7:137.
Dodecamyia Dyar, 1919, Ins. Ins. Mens., 7:138.
Hystatomyia Dyar, 1919, Ins. Ins. Mens., 7:140.
Shropshirea Dyar, 1922, Ins. Ins. Mens., 10:97.
Eunicemyia Dyar and Shannon, 1924, Jour. Wash. Acad. Sci., 14:482.
Janicemyia Dyar and Shannon, 1924, Jour. Wash. Acad. Sci., 14:482.
Phyllozomyia Dyar, 1924, Ins. Ins. Mens., 12:112.
Technicomyia Dyar, 1925, Ins. Ins. Mens., 13:20.
Nunezia Dyar, 1928, Mosq. Amer., p. 50.
Cruzmyia Lane and Cerqueira, 1942, Arq. Zool. Estad., 7:579.
Davismyia Lane and Cerqueira, 1942, Arq. Zool., Estad., 7:582.
Antunesmyia Lane and Cerqueira, 1942, Arq. Zool., Estad., 7:587.

ADULT. Proboscis variable in length, usually somewhat swollen at tip. Palpi short. Palpi and antennae of males and females similar; however, the plumes are sometimes better developed in males than in females. Scales of head broad and flat. Thorax dull-colored, never conspicuously ornamented. Spiracular bristles always present. Pronotal lobes large. Pleura with broad appressed scales. Upper mesepimeral bristles present; lower mesepimerals absent. Postnotum with a tuft of setae. Abdomen blunt, almost without hairs except on segment I and toward tip. Tarsal claws simple in both sexes. Wing without fringe on squama, one to three small hairs rarely present. Wing membrane with distinct microtrichia. Fork of vein 2 usually much longer than its petiole. *Male terminalia.* Tenth sternite with 2 or 3 terminal teeth. Phallosome simple. Basistyle and dististyle variable; dististyle usually with swollen head.

Tribe Culicini

PUPA. Respiratory trumpets variable, usually short. Dendritic tufts on abdominal segment I well developed. Lateral tufts on segments VII and VIII large and multiple. Paddles short, narrowed distally, without fringe and apical hair.

LARVA. Antenna usually short, smooth; antennal tuft consisting of a single, double, or triple hair inserted beyond the middle of the shaft. A small dorsal saddle on the anal segment (never completely ringed). Comb scales of eighth segment usually in a single regular row. Pecten absent or represented by a few teeth. Hairs of siphon single or branched, arising from all aspects, in no definite pattern. Ventral brush of anal segment represented by bilateral ventrolateral tufts.

EGGS. Deposited singly, not modified in structure.

BIONOMICS. The immature stages of this genus are found in small collections of water in living or dead plants such as rot cavities in trees, bamboo stems, leaves of pitcher plants, and flower bracts of *Heliconia*. The adults are mostly diurnal and are rarely encountered except in the forest near the aquatic habitats of their immature stages.

CLASSIFICATION. The pleural chaetotaxy, length of the proboscis, character of the wing scales, the presence or absence of scales on the postnotum, and ornamentation of the scutum are the chief means usually employed for defining subgenera. Four species belonging to a single subgenus, *Wyeomyia*, are known to occur in this region.

DISTRIBUTION. The members of the genus *Wyeomyia* are largely confined to the tropical and subtropical regions of the New World. Two species, *W. haynei* and *W. smithii*, range far into the north temperate zone in North America.

KEYS TO THE SPECIES

ADULT FEMALES

1. Anterior pronotal lobe entirely silver-white scaled; hind tarsal segments streaked with white basally on under side (immature stages found in bromeliads)..............*vanduzeei* D. and K., p. 70
   Anterior pronotal lobe almost entirely covered with dark scales, with purplish reflection; hind tarsal segments either streaked with white basally on under side or entirely dark........................2

2. Middle tarsal segments 3 and 4 (occasionally 5) and apex of 2 and hind tarsal segments (at least 1 to 3) each streaked with white on one side (immature stages found in bromeliads)
   mitchellii (Theob.), p. 66
   Middle and hind tarsal segments entirely dark (immature stages found in pitcher plants).............3

3. Scutellum with a patch of silvery scales
   haynei Dodge, p. 64
   Scutellum with all scales dark
   smithii (Coq.), p. 68

MALE TERMINALIA

1. Basal part of stem of dististyle slender, longer than the apical capitate part
   mitchellii (Theob.), p. 67
   Basal part of stem of dististyle stout, shorter than the apical capitate part.......................2

2. The branches of the capitate head about equal in length...........................*vanduzeei* D. and K., p. 70
   The branches of the capitate head unequal in length...............................*haynei* Dodge, p. 65;
   smithii (Coq.), p. 68

LARVAE (FOURTH INSTAR)

1. Upper frontal head hair 5 multiple, lower frontal 6 double; ventrolateral tufts of anal segment short, fan-shaped, each with about twelve subequal branches............*mitchellii* (Theob.), p. 67
   Upper frontal head hair 5 and lower frontal 6 single; ventrolateral tufts of anal segment each with about three to six branches........................2

2. Siphonal index about 6.0; several of the siphonal hairs double or triple; ventrolateral tufts of anal segment each with one or two long and three or four shorter branches
   vanduzeei D. and K., p. 71
   Siphonal index about 4.0 to 5.0; siphon with long single hairs on all aspects; ventrolateral tufts of anal segment each with about two or three branches 3

3. Dorsal pair of anal gills much reduced, about one-third the size of the ventral pair; lateral hair of anal segment usually double
   haynei Dodge, p. 65
   Dorsal pair of anal gills absent or represented by a pair of small swellings; lateral hair of anal segment usually triple........*smithii* (Coq.), p. 69

Wyeomyia (Wyeomyia) haynei Dodge

Wyeomyia haynei Dodge, 1947, Proc. Ent. Soc. Wash., *49*:117.

ADULT FEMALE (pl. 13). Similar to *W. smithii* but has a small patch of silvery scales on the scutellum. Very small species. *Head:* Proboscis long, black, slightly swollen apically; palpi very short, dark. Occiput with dark broad flat scales with metallic blue-green reflection dorsally; the vertex with broad silver-white scales. Scales of gena broad, appressed, silver white. Tori dark brown to black, naked. *Thorax:* Anterior pronotal lobe covered with broad appressed violaceous scales; scales of posterior pronotum, pleuron and coxa broad, appressed, silver white, occasionally faintly tinted with blue. Integument of scutum dark brown; scutum densely matted with large elliptical dark gray-brown scales, somewhat blue-green in certain lights. Scutellum with dark scales and brown setae on the lobes; a patch of silvery scales on the middle lobe. *Abdomen:* Abdomen usually laterally compressed. Tergites clothed with dark-brown scales with coppery to blue-green reflection. Venter yellowish white to silver white. The dark of the dorsum and the white of the venter meet laterally in a straight longitudinal line. *Legs:* Femora and tibiae dark scaled on anterior surface, with metallic blue-green reflection, posterior surface largely pale. Knee spots absent. Hind and front tarsi entirely dark; middle tarsus with segments 3 and 4 and apex of 2 white on one side. *Wing:* Length about 2.5 mm. Scales narrow, dark.

ADULT MALE. Coloration, palpi, and antennae similar to those of the female. *Terminalia* (fig. 49). Lobes of ninth tergite (IXT-L) slightly elevated, separated by about twice the width of one lobe, connected by a narrow band; each lobe bearing about three or four long stout spines. Tenth sternite (X-S) sclerotized apically, with apical teeth and two or three small setae near apex. Phallosome (Ph) longer than broad, narrower at base, broadly rounded toward apex, open ventrally, closed dorsally. Claspette absent. Basistyle (Bs) about three times as long as mid-width, narrowed beyond middle, with scales and short setae on outer aspect, three long strong dorsal setae arising near base. Basal lobe (B-L) triangular in outline, bearing several small setae (this structure may not be a true basal lobe but a modification of the interbasal fold); apical lobe absent. Dististyle (Ds) nearly as long as the basistyle and consisting of a short stout stem and a greatly enlarged apical head, head somewhat longer than stem. For structural details see figure 49.

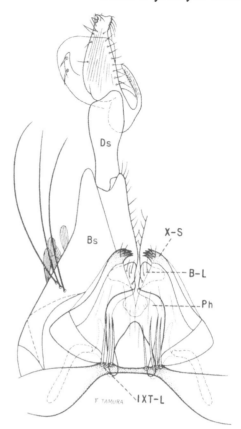

Fig. 49. Male terminalia of *Wyeomyia haynei* Dodge.

LARVA (fig. 50). Head nearly as long as broad. Antenna small, about one-third as long as the head, smooth; antennal hair single, inserted on outer third of shaft, reaching beyond tip. Head hairs: Postclypeal 4, upper frontal 5, lower frontal 6, and preantennal 7 single, nearly as long as the antenna; hair 5 shorter than hair 6. Lateral abdominal hair 6 long, two or more branched on segments I–VI. Comb of eighth segment of about 11 to 28 scales in a single row, the scales becoming progressively smaller ventrally; individual scale elongate and evenly fringed with small delicate spinules, those on apex somewhat stronger. Siphonal index about 4.0; siphon with long single sparsely barbed hairs on all aspects; pecten absent; dorsal preapical spine stout, about as long as apical width of the siphon. Anal segment with the saddle reaching about halfway down the sides; lateral hair very long, barbed, usually 2-branched; dorsal brush bilaterally consisting of two long barbed upper and two long barbed lower caudal hairs; ventrolateral tuft usually

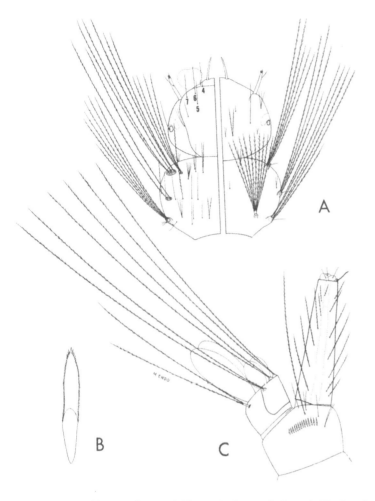

Fig. 50. Larva of *Wyeomyia haynei* Dodge. *A*, Head and thorax (dorsal and ventral). *B*, Comb scale. *C*, Terminal segments.

2-branched on either side; four gills, bulbous, ventral pair longer than the saddle, dorsal pair much reduced, usually about one-third the size of the ventral pair.

DISTRIBUTION. The exact range of this species cannot be determined until additional collections are made in areas where the host plant occurs; however, it is probably limited to the southern United States. *United States:* Alabama and North Carolina (175); South Carolina (175, 743). Collections of *Wyeomyia* reported as *smithii* from the District of Columbia and Maryland (213) may also be this species.

BIONOMICS. The larvae occur in the leaves of the pitcher plant, *Sarracenia purpurea*. According to Dodge (175), the larvae of *W. haynei*

are found in the southern subspecies of this plant, *Sarracenia purpurea venosa* Raf. The habits of both the larvae and adults appear to be similar to those of *W. smithii*.

Wyeomyia (Wyeomyia) mitchellii
(Theobald)

Dendromyia mitchellii Theobald, 1905, Mosq. or Cul. of Jamaica, p. 37.
Wyeomyia violescens Dyar and Knab, 1906, Proc. Biol. Soc. Wash., *19*:138.
Wyeomyia guatemala Dyar and Knab, 1906, Proc. Biol. Soc. Wash., *19*:139.
Wyeomyia adelpha Dyar and Knab, 1906, Proc. Biol. Soc. Wash., *19*:140.
Wyeomyia ochrura Dyar and Knab, 1906, Jour N.Y. Ent. Soc., *14*:227.
Wyeomyia glaucocephala Dyar and Knab, 1906, Proc. Biol. Soc. Wash., *19*:140.
Wyeomyia homothe Dyar and Knab, 1907, Jour. N.Y. Ent. Soc., *15*:211.
Wyeomyia ablabes Dyar and Knab, 1908, Proc. U.S. Nat. Mus., *35*:66.
Wyeomyia ablechra Dyar and Knab, 1908, Proc. U.S. Nat. Mus., *35*:66.
Wyeomyia antoinetta Dyar and Knab, 1909, Smithson. Misc. Coll., *52*:263.
Wyeomyia rolonca Dyar and Knab, 1909, Proc. Ent. Soc. Wash., *11*:173.
Wyeomyia (Wyeomyia) mitchellii, Lane and Cerqueira, 1942, Arq. Zool. Estad., *3*:556.

ADULT FEMALE (pl. 14). Very small species. *Head:* Proboscis long, black, slightly swollen apically; palpi very short, dark. Occiput with dark broad flat scales with metallic blue-green reflection dorsally; the vertex with broad silver-white scales. Scales of gena broad, appressed, silver white, a narrow line of white scales extended along the margin of the eye. Tori brown, naked. *Thorax:* Anterior pronotal lobe covered with broad appressed dark violaceous scales; scales of posterior pronotum, pleuron and coxa broad, appressed, silver white, faintly tinted with blue. Scutum and scutellum densely matted with large elliptical dark scales, appearing brown to metallic blue-green in different lights. A small patch of pale scales present on anterior margin of the scutum. Scutellum with dark-brown setae on the lobes. *Abdomen:* Abdomen usually somewhat laterally compressed. Tergites clothed with dark-brown scales with coppery to blue-green reflection. Venter yellowish white to silver white. The dark of the dorsum and the white of the venter meet laterally in a straight longitudinal line. *Legs:* Femora and tibiae dark-scaled, with metallic blue-green reflection on anterior surface, the posterior surface largely pale. Knee spots absent. Front tarsus entirely dark; middle tarsus with segments 3 and 4 (occasionally 5) and apex of 2 white on one side; each segment of hind tarsus basally streaked with white on posterior surface, obsolete or absent on the more distal segments. *Wing:* Length about 2.5 mm. Scales narrow, dark.

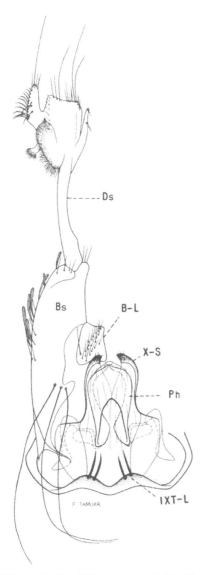

Fig. 51. Male terminalia of *Wyeomyia mitchellii* (Theobald).

ADULT MALE. Coloration, palpi, and antennae similar to those of the female. *Terminalia* (fig. 51). Lobes of ninth tergite (IXT-L) small, slightly elevated, widely separated, connected by a narrow band; each lobe bearing two short stout spines. Tenth sternite (X-S) with apical teeth and two or three small setae near apex. Phallosome (Ph) longer than broad, narrower at base, broadened and rounded toward apex, open ventrally, closed dorsally. Claspette absent. Basistyle (Bs) nearly three times as long as mid-width, a little narrower beyond middle, with scales and short setae on outer aspect, three very long strong dorsal setae arising near base. Basal lobe (B-L) situated near middle of basistyle, bearing a spine and numerous small setae (this structure may not be a true lobe but a modification of the interbasal fold); apical lobe absent. Dististyle (Ds) about as long as basistyle, consisting of a long slender stem and a greatly enlarged apical head; the head about two-thirds as long as the stem. For structural details see figure 51.

LARVA (fig. 52). Head nearly as long as broad. Antenna small, about one-third as long as the head, smooth; antennal tuft small, triple, inserted on outer third of shaft, reaching slightly beyond tip. Head hairs: Postclypeal 4 single, about as long as hair 5; upper frontal 5 usually 3- or 4-branched, shorter than hair 6; lower frontal 6 long, double; preantennal 7 long, usually 3- to 5-branched. Lateral abdominal hair 6 long, usually triple on segment I, double on II–VI. Comb of eighth segment of numerous scales, variable in number, arranged in a single row, the scales becoming progressively somewhat smaller ventrally; individual scale elongate, with numerous fine lateral spinules extending nearly to apex. Siphonal index 4.0 to 5.0; pecten absent; siphon with long, usually single barbed hairs on all aspects and a few small 2-branched tufts subapically; dorsal preapical spine stout, about as long as apical width of the siphon. Anal segment with the saddle reaching about three-fourths down the sides; lateral hair very long, single, barbed; dorsal brush bilaterally consisting of a long barbed upper and a long barbed lower caudal hair; ventrolateral tuft of about twelve short, barbed, subequal branches on either side; four gills,

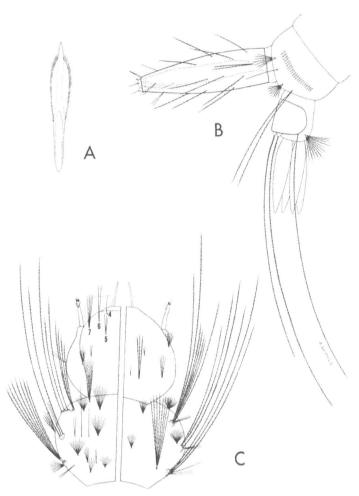

Fig. 52. Larva of *Wyeomyia mitchellii* (Theobald). *A*, Comb scale. *B*, Terminal segments. *C*, Head and thorax (dorsal and ventral).

varying from about one to two times as long as the saddle, blunt.

DISTRIBUTION. Southern Florida (135, 213, 562); Central America and the West Indies.

BIONOMICS. The larvae of this mosquito are found in water collections at the bases of the leaves of epiphytic *Bromeliaceae*. Larvae may be found any time during the year when the host plants contain water. In addition to the normal larval habitats, Galindo *et al.* (302) found the larvae of this species once in a tree-hole and once in a bamboo section in Panama. Hill and Hill (361) found the larvae in bamboo stumps, wild cocoes, and both terrestial and arboreal bromeliads in Jamaica.

The females are often encountered in the jungle and will occasionally feed on man. Galindo *et al.* (302) were able to rear typical males and females of this species from eggs laid by females which had been captured and fed on humans. Seabrook and Duffey (652) observed that the females bite readily in Florida when their haunts are invaded. Hill and Hill (361) state that the adults occasionally enter houses in Jamaica.

Wyeomyia (Wyeomyia) smithii
(Coquillett)

Aedes smithii Coquillett, 1901, Can. Ent., *33*:260.

ADULT FEMALE (pl. 15). Very similar to *W. haynei*. The only apparent difference in the female of the two species is in the scales of the scutellum. *Wyeomyia haynei* has a patch of silvery scales on the scutellum, but the corresponding scales are all dark in *W. smithii*. A very small species. *Head:* Proboscis long, black, slightly expanded apically; palpi very short, dark. Occiput with dark, broad, flat scales with metallic blue-green reflection dorsally; the vertex with broad silver-white scales. Scales of gena broad, appressed silver white. Tori with integument dark, naked. *Thorax:* Anterior pronotal lobe with broad appressed violaceous scales; posterior pronotum, pleuron, and coxa with broad appressed silver-white scales, often faintly tinted with blue. Integument of scutum dark brown; scutum and scutellum densely matted with large elliptical dark grayish-brown scales, somewhat metallic in some lights. Integument of scutellum light brown, the lobes bearing brown setae. *Abdomen:* Abdomen usually laterally compressed. Tergites clothed with dark-brown scales with coppery to blue-green reflection. Venter yellowish white to silver white. The dark of the dorsum and the white of the venter meet laterally in a straight longitudinal line. *Legs:* Femora and tibiae dark-scaled with metallic blue-green reflection on anterior surface, posterior surface pale. Hind and front tarsi entirely dark; middle tarsus with segments 3 and 4 and apex of 2 white on one side. *Wing:* Length about 2.5 mm. Scales narrow, dark.

ADULT MALE. Coloration similar to that of the female. *Terminalia* (fig. 53). The male terminalia of this species appears to be indistinguishable from that of *W. haynei*. Speci-

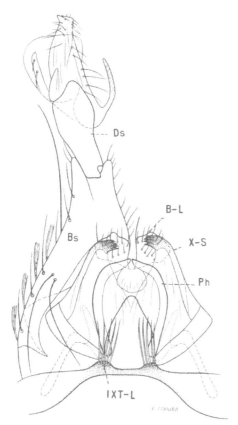

Fig. 53. Male terminalia of *Wyeomyia smithii* (Coquillett).

broad. Antenna small, less than half as long as the head, smooth; antennal hair single, inserted on outer fourth of shaft, reaching beyond tip. Head hairs: Postclypeal 4, upper frontal 5, lower frontal 6, and preantennal 7 long, single; hair 5 shorter than hair 6. Lateral abdominal hair 6 long, two or more branched on segments I–VI. Comb of eighth segment consists of a single row of 4 to 16 (usually 7 to 12) scales; individual scale elongate and evenly fringed with small delicate spinules. Siphonal index about 4.0 to 5.0; siphon with long single hairs on all aspects; pecten absent; dorsal preapical spine as long as apical diameter of the siphon. Anal segment with the saddle extending about two-fifths to three-fifths down the sides; lateral hair very long, usually 3-branched (occasionally 2- or 4-branched, often varying in number on the two sides); dorsal brush bilaterally consisting of long 2-branched (rarely 3-branched) upper and lower caudal tufts; ventrolateral tuft usually 3-branched on either side; two gills,

mens of *W. smithii* from Forked River, New Jersey, have two long strong dorsal setae arising from near base of basistyle. Lobes of ninth tergite (IXT-L) slightly elevated, separated by nearly twice the width of one lobe, each bearing about four to six long stout spines. Tenth sternite (X-S) sclerotized and bearing teeth apically, with two or three small subapical setae. Phallosome (Ph) longer than broad, narrowed basally, open ventrally, closed dorsally, broadly rounded at apex. Claspette absent. Basistyle (Bs) nearly three times as long as mid-width, with scales and short setae on outer aspect, with two or three long strong dorsal setae arising near base. Basal lobe (B-L) triangular-shaped, bearing several small setae (this structure may not be a true basal lobe); apical lobe absent. Dististyle (Ds) nearly as long as the basistyle, with a short stout stem and a greatly enlarged apical head, head longer than the stem. For structural details see figure 53.

LARVA (fig. 54). Head about as long as

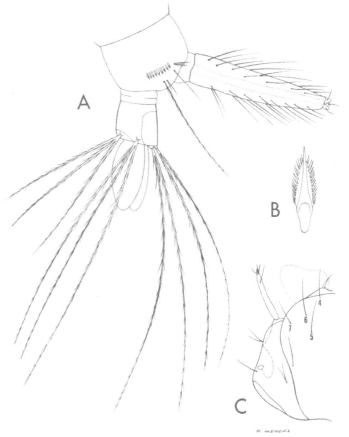

Fig. 54. Larva of *Wyeomyia smithii* (Coquillett). *A*, Terminal segments. *B*, Comb scale. *C*, Head.

large, bulbous, about two to two and a half times as long as the saddle.

DISTRIBUTION. Southeastern Canada and the northeastern United States, west to Minnesota. *United States:* Connecticut (494); Delaware (167); Illinois (213, 615); Maine (45, 564); Massachusetts (213, 701); Michigan (391); Minnesota (533); New Hampshire (66, 213); New Jersey (175, 213); New York (31, 175); Ohio (731a); Rhode Island (436); Wisconsin (171, 175). This species is also reported from the District of Columbia and Maryland (213) but these specimens may be *W. haynei*, the recently described southern form. *Canada:* Labrador (289, 348); Manitoba (598, 708); Nova Scotia (289, 708); Ontario (175, 289).

BIONOMICS. The larvae occur in the leaves of the pitcher plant, *Sarracenia purpurea.* Larvae may be found during any season of the year, even in winter, and are apparently able to withstand freezing and thawing when the water in the leaves of the plant freezes and thaws. Owen (533) found that, although the larvae can withstand freezing, they are destroyed by temperatures near −14°C. Haufe (348) observed the larvae in the leaves of *Sarracenia purpurea* L. at Goose Bay, Labrador, throughout the summer. The larvae were active in the larger leaves exposed to the sun during the day, even when temperatures were near freezing and ground pools were still covered with snow and ice.

The eggs are apparently deposited by the female in the younger leaves before water collects in them; however, they may be deposited on the sides of older leaves above the water level. The female of this species is not known to bite humans. Smith (662) gives an excellent account of this species and claims that he has never observed the females attempting to bite. The female rests with the head pointed downward at an angle between the front legs and with the hind legs curled high above the abdomen.

Wyeomyia (Wyeomyia) vanduzeei
Dyar and Knab

Wyeomyia vanduzeei Dyar and Knab, 1906, Proc. Biol. Soc. Wash., *19*:138.
Wyeomyia fratercula Dyar and Knab, 1906, Proc. Biol. Soc. Wash., *19*:139.
Wyeomyia sororcula Dyar and Knab, 1906, Proc. Biol. Soc. Wash., *19*:139.
Wyeomyia argyrura Dyar and Knab, 1908, Proc. U.S. Nat. Mus., *35*:70.
Wyeomyia conchita Dyar and Knab, 1909, Smithson. Misc. Coll., *52*:264.
Wyeomyia (Wyeomyia) vanduzeei, Lane and Cerqueira, 1942, Arq. Zool. Estad., *3*:550.

ADULT FEMALE (pl. 16). Very small species. *Head:* Proboscis long, black, slightly swollen apically; palpi very short, dark. Occiput with broad, dark, appressed scales with metallic blue-green reflection dorsally; the vertex with broad silver-white scales. Scales of gena broad, appressed, silver white, a line of white scales extending along the margin of the eye. Tori brown, naked. *Thorax:* Anterior pronotal lobe, posterior pronotum, pleuron, and coxa covered with broad appressed silver-white scales. Scutum and scutellum densely matted with large elliptical scales appearing coppery brown to metallic blue-green in different lights. A small patch of pale scales on anterior margin of the scutum. Scutellum with dark-brown setae on the lobes. *Abdomen:* Abdomen usually laterally compressed. Tergites clothed with dark-brown scales with coppery to metallic blue-green reflection. Venter yellowish white to silver white. The dark of the dorsum and the white of the venter meet laterally in a straight longitudinal line. *Legs:* Femora and tibiae dark-scaled with metallic blue-green reflection on anterior surface, the posterior surface largely pale. Knee spots absent. Front tarsus entirely dark; middle tarsus with segments 3 and 4 (occasionally 5) and apex of 2 white on one side; each segment of hind tarsus basally streaked with white on posterior surface. *Wing:* length about 2.5 mm. Scales narrow, dark.

ADULT MALE. Coloration, palpi, and antennae similar to those of the female. *Terminalia* (fig. 55). Lobes of ninth tergite (IXT-L) small, slightly elevated, separated by about the width of one lobe; each lobe bearing about three stout spines. Tenth sternite (X-S) moderately sclerotized, with apical teeth and a few short setae near apex. Phallosome (Ph) open ventrally, closed dorsally, longer than broad, narrower at base, somewhat broadened and rounded apically; each plate of phallosome with three small teeth at apex and bearing several small subapical setae. Claspette absent. Basistyle (Bs)

frontal 5, lower frontal 6, and preantennal 7 single; hair 6 a little longer than hair 5. Lateral abdominal hair 6 long, multiple on segments I and II, single on III–VI. Comb of eighth segment of numerous scales in a single row, the scales becoming progressively smaller ventrally; individual scale elongate, fringed laterally and apically with minute spinules. Siphonal index about 6.0; pecten absent; siphon bearing a row of six small 1- or 2-branched tufts dorsally, a larger 2- or 3-branched tuft at basal fourth and two or three small 1- or 2-branched subventral tufts at apical third or fourth of siphon. Dorsal preapical spine stout, nearly as long as apical width of the siphon. Anal segment with the saddle extending nearly three-fourths down the sides; lateral hair long, double, sparsely barbed; dorsal brush bilaterally consisting of a long 2-branched, sparsely barbed, upper and lower

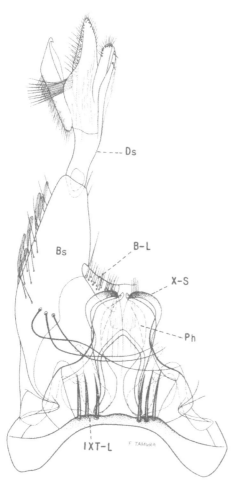

Fig. 55. Male terminalia of *Wyeomyia vanduzeei* Dyar and Knab.

about three times as long as mid-width, with scales and short setae on outer aspect, three very long strong dorsal setae arising near base. Basal lobe (B-L) triangular, bearing a spine and numerous small setae (this structure may not be a true basal lobe but a modification of the interbasal fold); apical lobe absent. Dististyle (Ds) nearly as long as the basistyle and consists of a short stout stem and a greatly enlarged apical head; the head about one and a half times as long as the stem. For structural details see figure 55.

LARVA (fig. 56). Head nearly as long as broad. Antenna about one-third as long as the head, smooth; antennal hair small, single, inserted on outer third of shaft, extending slightly beyond tip. Head hairs: Postclypeal 4, upper

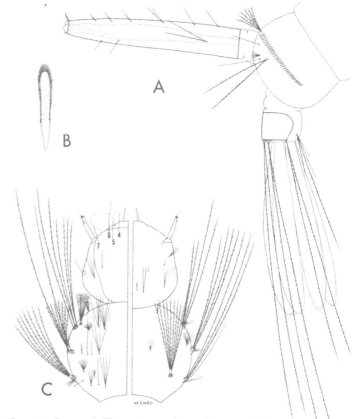

Fig. 56. Larva of *Wyeomyia vanduzeei* Dyar and Knab. *A*, Terminal segments. *B*, Comb scale. *C*, Head and thorax (dorsal and ventral).

caudal tuft; a ventrolateral tuft of one or two long and three or four shorter hairs on either side; four gills, swollen, four times as long as saddle or longer, bluntly rounded.

DISTRIBUTION. Southern Florida (135, 213, 419); Central America and the West Indies.

BIONOMICS. The habits of *W. vanduzeei* are apparently very similar to those of *W. mitchellii* with which it is commonly associated in its aquatic stages in the leaves of epiphytic *Bromeliaceae*. Seabrook and Duffey (652) observed that the females readily bite man in Florida.

Genus URANOTAENIA Lynch Arribálzaga

Uranotaenia Lynch Arribálzaga, 1891, Rev. Mus. de La Plata, *1*:375.

Anisocheleomyia Theobald, 1905, Entomologist, *38*:52.

Pseudouranotaenia Theobald, 1905, Jour. Econ. Biol., *1*:33.

Pseudoficalbia Theobald, 1912, Trans. Linn. Soc. Zool., *15*:89.

ADULT. Proboscis variable in length, usually swollen at tip. Palpi very short in both sexes. Antennae of male more or less plumose. Occiput clothed with broad appressed scales. Pronotal lobes well separated. Posterior pronotal, spiracular, prealar, lower mesepimeral and upper mesepimeral bristles reduced in number; postspiracular bristles absent. Pleuron with scales usually limited to one or two patches or stripes. Abdomen of female blunt. Pulvilli absent. Squama without fringe. Fork of vein 2 much shorter than its stem. Wing membrane with minute microtrichia, not visible under low magnification. *Male terminalia.* Tenth sternites vestigial, not supporting the anal membrane. Phallosome divided into a pair of heavily sclerotized strongly toothed plates. Claspettes absent. Basistyle short, stout.

PUPA. Respiratory trumpets variable. Dendritic tufts large on abdominal segment I. Paddles ovate, serrate.

LARVA. Head usually longer than broad, heavily sclerotized. Antenna short, smooth, or nearly so. Upper and lower frontal head hairs 5 and 6 stout and spinelike in many species. Comb of eighth segment consisting of a single row of scales on the posterior margin of a sclerotized plate. Siphon short or of medium length, with a pecten and a single pair of siphonal tufts inserted beyond the pecten near the middle of siphon. Anal segment completely ringed by the dorsal saddle.

EGGS. Eggs elongate, deposited in boat-shaped masses on the surface of the water.

BIONOMICS. The larvae are found mostly in ground pools, swamps, and grassy margins of lakes; some occur in rot cavities in trees, potholes, and other small containers. The larvae are frequently associated with *Anopheles* in the field, with which they may be confused by the collector since they rest nearly horizontal with the surface of the water. The members of this genus are of little economic importance and rarely appear to feed on man.

CLASSIFICATION. No important modifications of adult structures, including the male terminalia, are known in the genus. However, ornamentation is varied and usually provides means for identifying the species.

DISTRIBUTION. Occurs throughout the tropics. A few species are found in temperate North America.

KEYS TO THE SPECIES

ADULT FEMALES

1. Hind tarsi with apex of third and entire fourth and fifth segments white-scaled....*lowii* Theob., p. 76
   Tarsi entirely dark-scaled......................................2
2. Scutum with a narrow median longitudinal stripe of iridescent blue scales
   sapphirina (O. S.), p. 77
   Scutum brown, without a narrow median longitudinal stripe of iridescent blue scales...............3
3. Scutum with a distinct lateral line of iridescent blue scales running anteriorly from base of wing, narrowly broken about midway, the anterior half cutting a dark-brown patch
   syntheta Dyar and Shannon, p. 79
   Scutum with a faint or obsolete lateral line of iridescent blue scales extending anteriorly from base of wing, widely separated about midway
   anhydor Dyar, p. 74

MALE TERMINALIA

1. Dististyle stout, only about twice as long as broad, bearing about fifteen to twenty stout curved spines...................*lowii* Theob., p. 76
   Dististyle slender, about four or five times as long as wide, bearing a single short subapical spinelike claw 2
2. Dististyle distinctly swollen at apical third; each plate of phallosome bearing three or four subapical teeth......................*anhydor* Dyar, p. 74;
   syntheta D. and S., p. 80
   Dististyle not swollen at apical third; each plate of phallosome bearing a single large subapical thornlike projection....*sapphirina* (O. S.), p. 78

LARVAE (FOURTH INSTAR)

1. Head hairs upper frontal 5 and lower frontal 6 stout and spinose...2
   Head hairs upper frontal 5, 2- or 3-branched, lower frontal 6 single (hairs 5 and 6 coarse but not stout) ....*anhydor* Dyar, p. 75;
   syntheta D. and S., p. 80

2. Upper lateral abdominal hair 6 double on segments
   I and II, the lower 7 single; prothoracic hair 3,
   4- to 8-branched, barbed, more than half as long
   as hairs 1 and 2......................*lowii* Theob., p. 76

   Upper lateral abdominal hair 6 triple on segments
   I and II, the lower 7 single; prothoracic hair 3
   8- to 10-branched, smooth, much less than half
   as long as hairs 1 and 2

   sapphirina (O. S.), p. 78

Uranotaenia anhydor Dyar

Uranotaenia anhydor Dyar, 1907, Proc. U.S. Nat.
Mus., *32*:128.

Uranotaenia anhydor, Brookman and Reeves, 1953,
Ann. Ent. Soc. Amer., *46*:225 (male illustrated).

The larvae, male terminalia, and adult females of *U. anhydor* Dyar and *U. syntheta*
Dyar and Shannon appear to be quite similar,
and the study of additional material, when
available, may reveal that we are dealing with
two subspecies or possibly with a single species.

ADULT FEMALE (pl. 17). Similar to *U. syntheta* Dyar and Shannon. Very small species.
Head: Proboscis long, dark-scaled, swollen
apically; palpi very short, dark. Broad dorsal
region of occiput clothed with bronze-brown
appressed scales and brown erect forked scales;
lateral region of occiput with broad appressed
bluish-white scales, these scales are continued
dorsally along the eye margin forming a small
patch on the vertex. Tori with integument tan
or light brown, a few black scales on inner surface. *Thorax:* Integument of scutum light
brown; scutum clothed with narrow bronze-
brown scales and three longitudinal rows of
dark bristles. A short oblique row of iridescent
bluish scales along the lateral margin on either
side of anterior half of scutum and a short
longitudinal row of similar iridescent bluish
scales above the paratergite on either side
(these lines are often faint or obsolete). Scutellum with brown setae on the lobes, middle
lobe with narrow bronze-brown scales. Pleura
light brown, with some iridescent bluish scales
on the anterior pronotal lobes and the sternopleura. A single spiracular bristle present;
postspiracular bristles absent. *Abdomen:* Tergites clothed with dark scales with bronzy or
dark-blue reflection and small basolateral segmental white patches. Venter with yellow or

light-brown scales. Eighth segment blunt,
largely retracted within the seventh. *Legs:* Femora with dark bronze-brown scales, posterior
surface pale; knee spots pale. Front and middle
tibiae dark, with pale scales on posterior surface; hind tibiae mostly dark. Tarsi dark.
Wing: Length about 2.5 mm. Scales rather
broad, brown, with a line of pale scales at base
of vein 1; second marginal cell much shorter
than its petiole.

ADULT MALE. Coloration and palpi similar
to the female. *Terminalia* (fig. 57). Similar to
U. syntheta. Ninth tergite (IX-T) broad, widely
and deeply emarginate anteriorly, convex;
lobes long, fingerlike, blunt at apex, separated
by at least twice the width of one lobe (these
structures do not seem to be true lobes of the
ninth tergite). Tenth sternite vestigial, not supporting the anal membrane (A-M). Phallosome
(Ph) consisting of two sclerotized plates narrowly open dorsally but with two angular
projections touching at middle, broadly open
ventrally but joined by an arched bridge near
middle; each plate with a large median blunt
dorsal projection, with 3 or 4 pointed teeth
directed ventrolaterally. Claspette absent.

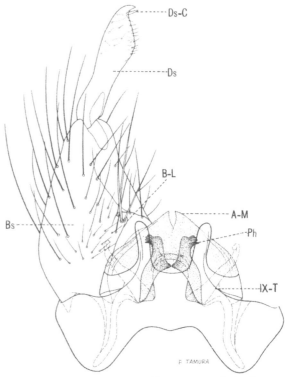

Fig. 57. Male terminalia of *Uranotaenia anhydor* Dyar.

Basistyle (Bs) a little longer than wide, clothed with long setae; basal lobe (B-L) represented by a rounded area on inner surface of basistyle, bearing two or three long rather strong setae and a few smaller ones; apical lobe absent. Disistyle (Ds) a little shorter than the basistyle, expanded at apical third, tapered and strongly curved at apex and bearing small papillated setae; claw (Ds-C) short, slender, inserted a little before the apex.

LARVA (fig. 58). Very similar to *U. syntheta* Dyar and Shannon. Head about as broad as long. Antenna short, dark, with a few coarse spicules on basal half; antennal hair small, single, inserted on basal third of shaft. Head hairs: Postclypeal 4 large, multiple; upper frontal 5 double or triple, sparsely barbed, shorter than hair 6; lower frontal 6 long, single, coarse, sparsely barbed; preantennal 7 large, multiple, sparsely barbed. Prothoracic hairs: 1–2 long, single, barbed; 3 medium, 3- to 7-branched, barbed; 4 long, 3- to 5-branched; 5–6 long, single; 7 long, 3- to 5-branched. Lateral abdominal hair 6 long and coarse on segments I and II, short, fine, multiple on III–VI. Comb of eighth segment consisting of a row of about 8 to 11 scales on the distal margin of a large transverse sclerotized plate; individual scale broadly thorn-shaped, fringed with minute lateral spinules to near distal third. Siphonal index about 4.5; siphon slightly upcurved; pecten of about 14 to 20 teeth reaching near middle of siphon; siphonal tuft large, multiple, sparsely barbed, inserted near end of pecten, longer than basal diameter of siphon. Anal segment longer than wide, ringed by the saddle, a row of spines on apical margin of the saddle; lateral hair represented by a large 4- or 5-branched tuft; dorsal brush bilaterally consisting of two long lower caudal hairs and a shorter multiple upper caudal tuft; ventral brush long, sparse, confined to the barred area; gills shorter than the saddle, pointed.

DISTRIBUTION. Known from San Diego County, California (183, 284, 382, 654), and Baja California, Mexico (106). Brookman and Reeves (106) state that Dr. C. B. Philip has recently taken a collection of females in western Nevada, of what appears to be *U. anhydor*.

BIONOMICS. Dyar (183) described a single

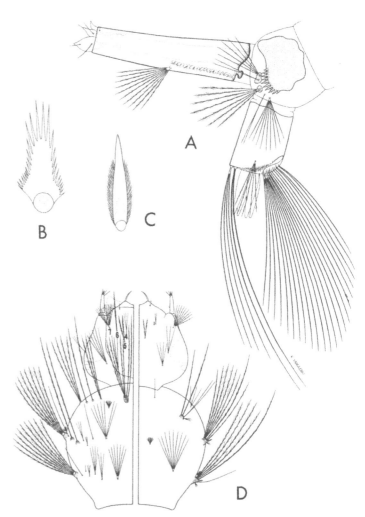

Fig. 58. Larva of *Uranotaenia anhydor* Dyar. *A*, Terminal segments. *B*, Pecten tooth. *C*, Comb scale. *D*, Head and thorax (dorsal and ventral).

larva collected in a swamp at Sweetwater Junction, San Diego County, California, in June, 1906. Larvae were again collected by Dyar in a pond at Old Town, San Diego County, in May, 1916, and adults were reared. Freeborn (284) states that a female specimen of this species was taken in a large packing box placed near a fresh-water creek at Camp Kearney near San Diego in May, 1918. Seaman (654) collected a larva in a hoof print in a grassy area in the river bed of San Luis Rey River 4 miles upstream from Bonsall, San Diego County, California. Two pupae were taken by Brookman and Reeves (106) in Baja California, Mexico, and one male and one female were reared.

Uranotaenia lowii Theobald

Uranotaenia lowii Theobald, 1901, Mon. Culic., 2:339.

Uranotaenia continentalis Dyar and Knab, 1906, Jour. N.Y. Ent. Soc., *14*:187.

Uranotaenia minuta Theobald, 1907, Mon. Culic., 4:559.

ADULT FEMALE (pl. 18). Very small species. *Head:* Proboscis long, dark-scaled, swollen apically; palpi very short, dark-scaled. Occiput clothed with dark bronzy broad appressed scales, margined anteriorly and laterally with broad iridescent pale-blue scales, a patch of pale-blue scales on vertex. Tori light brown, somewhat darker on inner surface. *Thorax:* Integument of scutum light brown except for a dark broad median longitudinal stripe, and a darkly pigmented spot just anterior to the wing base; scutum clothed with narrow dark-brown scales on dorsal surface; a line of iridescent bluish scales on the lateral dark spot. Patches of pale-blue iridescent scales on the anterior pronotal lobes, on the bases of the coxae, and on the sternopleura. Scutellum with dark-brown scales and dark setae. A single spiracular bristle present; postspiracular bristles absent. Sternopleuron with a large darkly pigmented spot on upper half. *Abdomen:* Tergites dark-brown-scaled with a metallic luster; the third, fifth, and sixth segments each with a rather large apicolateral patch of pale-blue iridescent scales, the second segment often similarly marked. Sternites with pale-yellow scales. Eighth segment blunt, largely retracted within the seventh. *Legs:* Front and middle legs dark-scaled; posterior surface of femora pale; knee spots pale; a small white patch on outer surface of each tibia at apex. Hind leg similarly marked but with tarsal segments 4 and 5, and apex of segment 3 white-scaled. *Wing:* Length nearly 2.5 mm. Scales rather broad, dark brown, with a small patch of iridescent pale-blue scales on base of veins 1 and 5; second marginal cell much shorter than its stem.

ADULT MALE. Coloration, palpi, and antennae similar to those of the female. *Terminalia* (fig. 59). Ninth tergite (IX-T) broad, convex, anterior margin deeply emarginate; the lobes pointed, longer than their basal width (these do not seem to be true lobes of the ninth ter-

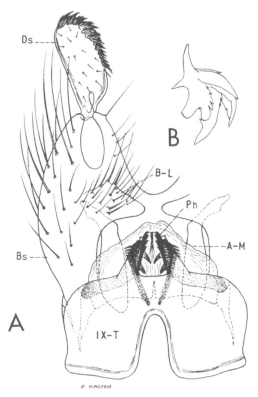

Fig. 59. *Uranotaenia lowii* Theobald. *A*, Male terminalia. *B*, Phallosome (lateral view).

gite). Tenth sternite vestigial, not supporting the anal membrane (A-M). Phallosome (Ph) composed of two sclerotized plates, open ventrally, closed dorsally; each plate with three large thornlike dorsal projections and three very large sickle-shaped serrated ventrolateral processes. Claspette absent. Basistyle (Bs) a little longer than broad, clothed with long setae; basal lobe (B-L) elevated and rounded, crowned with numerous prominent setae; apical lobe absent. Dististyle (Ds) greatly swollen beyond base, about twice as long as broad, about half as long as basistyle; apex bluntly rounded, bearing about fifteen to twenty short broad heavily sclerotized spines.

LARVA (fig. 60). Head longer than broad. Antenna short, about one-fourth as long as the head, sparsely spiculate; antennal hair small, single, inserted near basal third of shaft. Head hairs: Postclypeal 4 multiple, shorter than upper frontal 5 and lower frontal 6; upper frontal 5 and lower frontal 6 stout, spinelike, dark, as long as the antenna; preantennal 7 multiple, sparsely barbed. Prothoracic hairs:

1–2 long, single, barbed; 3 more than half as long as 1 and 2, about 4- to 8-branched, barbed; 4 long, single to triple; 5 long, double or triple; 6 long, single; 7 long, 3- or 4-branched. Lateral abdominal hair 6 double on segments I and II; hair 7 single. Comb of eighth segment with about 6 to 9 scales on the distal margin of a large sclerotized plate; individual scale thorn-shaped, with minute lateral spinules on basal half to three-fourths. Siphonal index about 4.0; siphon slightly upcurved; pecten of about 10 to 17 evenly spaced teeth not reaching middle of siphon; individual tooth broad, fringed laterally and apically with fine spines; siphonal tuft multiple, inserted near end of pecten. Anal segment completely ringed by the saddle; a row of spines on apical margin of the saddle;

lateral hair represented by a multiple tuft, shorter than the saddle; dorsal brush bilaterally consisting of two long lower caudal hairs and a long 2- or 3-branched upper caudal tuft; ventral brush sparse, confined to the barred area, the two proximal tufts very short; gills shorter than the saddle, bluntly pointed.

DISTRIBUTION. Southern United States, Mexico, the West Indies, Central and South America. *United States:* Alabama (135, 420); Arkansas (84, 88); Florida (135, 419); Georgia (135, 420); Louisiana (420, 759); Mississippi (135, 420); North Carolina (135, 514); South Carolina (135, 742); Texas (463, 634).

BIONOMICS. The larvae of *U. lowii* are found in the grassy shallow margins of ponds and lakes, usually exposed to sunlight. Larvae may be found throughout the year in Florida, but farther north they are limited to the summer and early fall months. The larval habits are quite similar to those of *U. sapphirina*. The female is not known to bite man. Remington (597) conducted a rather intensive study of the feeding habits of this species in Louisiana and observed that the females did not feed on man and reptiles but did feed on several kinds of amphibians used in the study. The adults are readily attracted to light traps and have been occasionally captured in the daytime by sweeping through shaded vegetation. Michener (510) states that the scarcity of larvae in comparison with adult catches in light traps at Camp Shelby, Mississippi, suggests that the adults may fly considerable distances.

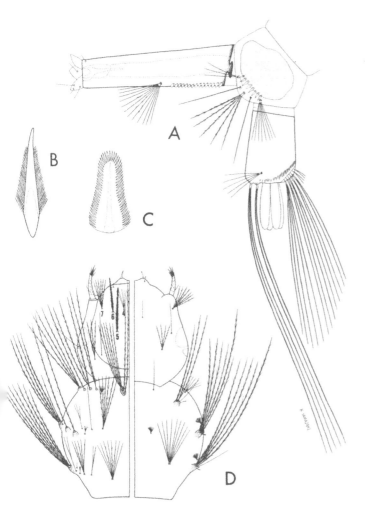

Fig. 60. Larva of *Uranotaenia lowii* Theobald. *A,* Terminal segments. *B,* Comb scale. *C,* Pecten tooth. *D,* Head and thorax (dorsal and ventral).

Uranotaenia sapphirina (Osten Sacken)

Aedes sapphirinus Osten Sacken, 1868, Trans. Amer. Ent. Soc., 2:47.

Uranotaenia coquilletti Dyar and Knab, 1906, Jour. N.Y. Ent. Soc., *14*:186.

ADULT FEMALE (pl. 19). Very small species. *Head:* Proboscis long, dark-brown scaled, often with a few iridescent blue scales sprinkled on basal half, swollen apically; palpi very short, dark-scaled. Occiput clothed with dark bronze, broad appressed scales, margined anteriorly and laterally with broad iridescent bluish scales, vertex with a patch of blue scales. Tori light brown, naked. *Thorax:* Integument of scutum light brown; scutum clothed with narrow golden to dark-brown scales; a narrow median line of broad iridescent bluish scales extending nearly the entire length of the scutum and covering most of middle lobe of the scutellum; a similar line of iridescent bluish scales on lateral margin of scutum reaching from the base of the wing to the scutal angle; patches of iridescent scales also present on anterior pronotal lobe and on the mid-part of the sternopleuron. A single spiracular bristle present; postspiracular bristles absent. *Abdomen:* Tergites clothed with brown scales with metallic luster; apices of third, fifth, and often the sixth segment with a rounded median patch of white scales. Sternites with dingy pale-brown scales. Eighth segment blunt, largely retracted within the seventh. *Legs:* Legs dark-brown-scaled except for a small patch of bluish-white scales at apices of femora and tibiae and yellowish scales on the posterior surface of the femora. *Wing:* Length about 2.5 to 2.7 mm. Scales rather broad, brown, with a row of iridescent bluish scales on the stem of vein 5 and a much shorter row near the base of vein 1; second marginal cell much shorter than its petiole.

ADULT MALE. Coloration and palpi similar to that of the female. *Terminalia* (fig. 61).

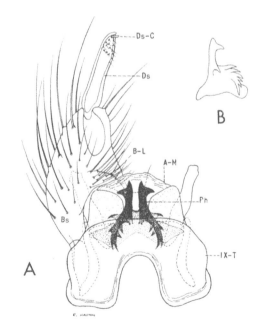

Fig. 61. *Uranotaenia sapphirina* (Osten Sacken). *A,* Male terminalia. *B,* Phallosome (lateral view).

Ninth tergite (IX-T) broad, convex, deeply emarginate on anterior margin; the lobes are large, quadrate, and separated by less than the width of one lobe, although these structures do not seem to be true lobes of the ninth tergite. Tenth sternite vestigial, not supporting the anal membrane (A-M). Phallosome (Ph) composed of two heavily sclerotized plates, open ventrally, closed dorsally at base; each plate slender, bluntly pointed apically, bearing a large subapical thornlike projection and a row of about six or seven stout, curved processes attached near base. Claspette absent. Basistyle (Bs) about four-fifths as broad as long, clothed with numerous long setae; basal lobe (B-L) conical, appressed to inner surface of the basistyle, bearing two or three long, rather strong setae and several smaller ones; apical lobe absent. Dististyle (Ds) nearly three-fourths as long as the basistyle, four or five times as long as broad, bearing several very short papillated setae on distal third; claw (Ds-C) short, slender, inserted before apex.

LARVA (fig. 62). Head longer than broad. Antenna short, about one-fourth as long as the head, dark, sparsely spiculate; antennal hair small, single, inserted at basal third of shaft. Head hairs: Postclypeal 4 double, about as long as upper and lower frontals 5 and 6; upper frontal 5 and lower frontal 6 stout, spinelike, dark, as long as the antenna; preantennal 7 multiple, sparsely barbed. Prothoracic hairs: 1–2 long, single, barbed; 3 multiple, 7- to 10-branched, weakly barbed, less than half as long as hairs 1 and 2; 4–5 long, double or triple; 6 long, single; 7 long, usually triple. Lateral abdominal hair 6 triple on segments I and II; hair 7 single. Comb of eighth segment with 7 to 10 scales on the distal margin of a large sclerotized plate; individual scale thorn-shaped, with small lateral spinules on basal half. Siphonal index 3.5 to 4.5; siphon slightly upturned; pecten of about 12 to 15 evenly spaced teeth not reaching middle of siphon; individual tooth broad, fringed on both sides with spinules; siphonal tuft multiple, inserted near end of pecten or slightly beyond. Anal segment completely ringed by the saddle; a row of spines on apical margin of the saddle; lateral hair represented by a multiple tuft, shorter than the saddle; dorsal brush bilater-

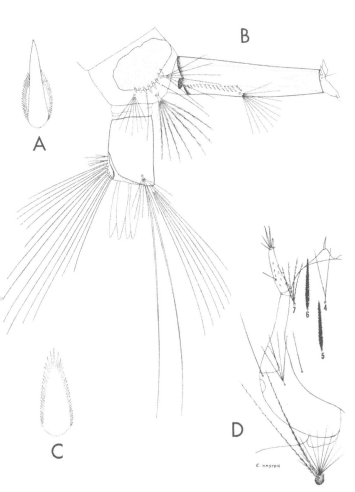

Fig. 62. Larva of *Uranotaenia sapphirina* (Osten Sacken). *A*, Comb scale. *B*, Terminal segments. *C*, Pecten tooth. *D*, Head.

Michigan (391, 529); Minnesota (533); Mississippi (135, 546); Missouri (6, 324); Nebraska (680); New Hampshire (66, 270); New Jersey (350); New Mexico (273); New York (31); North Carolina (135, 646); North Dakota (553); Ohio (213, 482); Oklahoma (628); Pennsylvania (298, 756); Rhode Island (35, 270); South Carolina (135, 742); South Dakota (14); Tennessee (135, 658); Texas (463, 634); Vermont (270); Virginia (176, 178); Wisconsin (171). *Canada:* Ontario (409, 708); Quebec (708).

BIONOMICS. The larvae occur commonly in permanent pools and ponds and lakes that contain emergent or floating vegetation exposed to sunlight. This mosquito is commonly associated with *Anopheles quadrimaculatus* Say in its larval habitats in the southern United States. Larvae may be found throughout most of the year in the southern part of its range, but farther north they are present mostly during the summer and early fall months.

Very little information is available on the feeding habits of the adults of the *Uranotaenia.* It is the opinion of most workers that the females of *U. sapphirina* rarely, if ever, bite humans. The adults are seldom seen during the daytime but have been observed resting in damp situations in culverts, hollow trees, and among vegetation near the larval habitats. This species overwinters as adult females in shelters, particularly hollow trees. The adults are readily attracted to artificial light.

Uranotaenia syntheta Dyar and Shannon

Uranotaenia syntheta Dyar and Shannon, 1924, Ins. Ins. Mens., *12*:189.

Uranotaenia syntheta, Dampf, 1943, Rev. Soc. Mexico Hist. Nat., *4*:147.

Uranotaenia syntheta, Porter, 1946, Amer. Mid. Nat., *35*:535.

ADULT FEMALE (pl. 20). Similar to *U. anhydor* Dyar, but appears to have a more distinct lateral line of iridescent blue scales extending anteriorly from the base of the wing, narrowly separated midway. In *U. anhydor* this lateral line is faint or obsolete and is widely separated midway. Very small species. *Head:* Proboscis long, dark-scaled, swollen apically; palpi very short, dark scaled. Occiput clothed dorsally

ally consisting of two long lower caudal hairs and a shorter multiple upper caudal tuft; ventral brush sparse, confined to the barred area, the two proximal tufts small; gills usually a little shorter than the saddle, pointed.

DISTRIBUTION. Southeastern Canada and the eastern United States, west to North Dakota and New México. It also occurs in Mexico and the West Indies. *United States:* Alabama (135, 419); Arkansas (88, 121); Connecticut (494); Delaware (120, 167); District of Columbia (318); Florida (135, 419); Georgia (135, 608); Illinois (615); Indiana (152, 345); Iowa (530, 625); Kansas (14, 530); Kentucky (134, 570); Louisiana (419, 759); Maryland (161, 213); Massachusetts (270);

with bronzy dark-brown appressed scales and dark erect forked scales; lateral region of occiput with broad appressed bluish-white scales, these scales continuing dorsally along the eye margin and forming a small patch on the vertex. Tori tan or light brown, usually a little darker on inner surface. *Thorax:* Integument of scutum light brown; scutum clothed with small narrow reddish-brown and dark-brown scales. A narrow lateral line of iridescent bluish scales extending forward from the wing base to near the anterior margin of the scutum; this line is broken into two segments at the scutal angle, the anterior segment bisecting a dark patch. Patches of iridescent bluish scales also present on anterior pronotal lobes and on the sternopleura. Scutellum with narrow brown scales and dark setae on the lobes. A single spiracular bristle present; postspiracular bristles absent. *Abdomen:* Tergites brown-scaled with metallic luster. Venter clothed with dingy pale-brown scales. *Legs:* Femora brown, posterior surface pale; knee spots whitish. Tibiae dark, streaked with pale scales on posterior surface, tipped with whitish scales. Tarsi dark. *Wing:* Length about 2.5 mm. Scales rather broad, brown, a line of pale scales on base of vein 1; second marginal cell much shorter than its petiole.

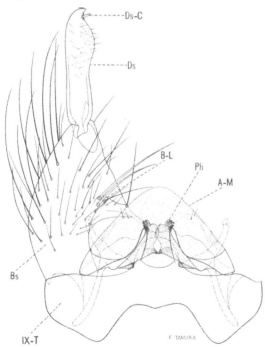

Fig. 63. Male terminalia of *Uranotaenia syntheta* Dyar and Shannon.

ADULT MALE. Coloration and palpi similar to the female. *Terminalia* (fig. 63). Similar to *U. anhydor* Dyar. Ninth tergite (IX-T) broad, deeply and widely emarginate anteriorly, convex; lobes bluntly pointed, longer than their basal width, although these do not seem to be true lobes of the ninth tergite. Tenth sternite vestigial, not supporting the anal membrane (A-M). Phallosome (Ph) composed of two sclerotized plates narrowly open dorsally but with two angular projections touching at middle, broadly open ventrally but joined by an arched bridge near middle; each plate bluntly pointed apically and with about 3 coarse blunt teeth directed ventrolaterally. Claspette absent. Basistyle (Bs) a little longer than broad, clothed with long setae; basal lobe (B-L) represented by a rounded area on inner surface of basistyle bearing two or three long rather strong setae and several smaller ones; apical lobe absent. Dististyle (Ds) about two-thirds to three-fourths as long as the basistyle, about four or five times as long as broad, roundly expanded on inner margin at apical third, bearing short setae on distal half, tapered and strongly curved at apex; claw (Ds-C) short, slender, inserted a little before apex.

LARVA (fig. 64). Similar to *U. anhydor* Dyar. Head slightly longer than broad. Antenna short, about one-fourth as long as the head, dark, spiculate; antennal hair small, single, inserted at basal fourth of shaft. Head hairs: Postclypeal 4 large, 4- or 5-branched; upper frontal 5 usually double, coarse, sparsely barbed, about two-thirds as long as lower hair 6; lower frontal 6 long, single, coarse, sparsely barbed; preantennal 7 large, multiple, barbed. Prothoracic hairs: 1–2 long, single, barbed; 3 nearly two-thirds as long as 1 and 2, about 4- to 6-branched, barbed; 4 long, double or triple; 5–6 long, single; 7 long, usually 3- or 4-branched. Lateral abdominal hair 6 long and coarse on segments I and II, short, fine and multiple on III–VI. Comb of eighth segment consists of a row of 7 to 10 scales on the distal margin of a large transverse sclerotized plate; individual scale thorn-shaped, with minute lateral spinules on basal half to two-thirds. Siphonal index 4.5 to 5.0; pecten of about 16 to 20 teeth reaching near middle of siphon; siphonal tuft about 7- to 10-

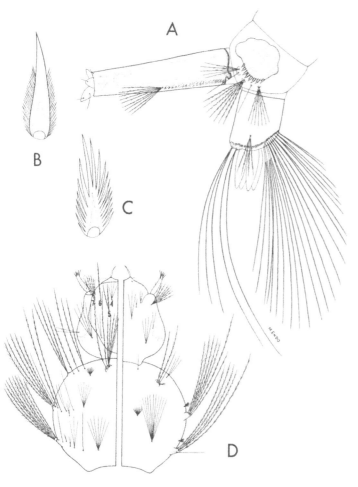

Fig. 64. Larva of *Uranotaenia syntheta* Dyar and Shannon. *A*, Terminal segments. *B*, Comb scale. *C*, Pecten tooth. *D*, Head and thorax (dorsal and ventral).

branched, barbed, inserted near middle of siphon at end of pecten, longer than basal

diameter of siphon. Anal segment ringed by saddle; a row of spines on apical margin of the saddle; lateral hair represented by a 2- to 5-branched barbed tuft, a little shorter than the saddle; dorsal brush bilaterally consisting of two long lower caudal hairs and an upper caudal tuft usually 3-branched; ventral brush sparse, confined to the barred area; gills a little shorter to a little longer than the saddle, pointed.

DISTRIBUTION. Mexico (164, 481) and the southwestern United States. The species apparently reaches its greatest abundance in the Rio Grande Valley and west Texas, but its range also extends to western and southern Oklahoma. *United States:* New Mexico (20, 273); Oklahoma (321); Texas (93, 249).

BIONOMICS. Porter (551) described the larva from a single specimen collected in still water along the margin of a shallow grassy ditch at Fort Worth, Texas. Breland (93) made several collections of larvae of this species in Palmetto State Park, near Lulling, Texas, where the larvae occurred in small depressions along a stream containing masses of water hyacinths. Dampf (164) described the male and states that the species was rather abundant in some areas of Mexico. He found it to be primarily a winter species on the Mexican tableland, most abundant in the month of November, and with scanty spring and summer generations. The adults have been frequently captured in light traps in several areas in Texas. Eads *et al.* (249) reported light-trap collections of this species from eighteen counties in west Texas.

Genus CULISETA Felt

Theobaldia Neveu-Lemaire (not *Theobaldius*Nevill),
  1902, C. R. Soc. Biol., *54*:1331.
Culicella Felt, 1904, N.Y. State Mus. Bull., *79*:391c.
Culiseta Felt, 1904, N.Y. State Mus. Bull., *79*:391c.
Theobaldinella Blanchard, 1904, Les Moust., p. 390.
Pseudotheobaldia Theobald, 1907, Mon. Culic.,
  4:150.
Climacura Howard, Dyar, and Knab, 1915, Carnegie
  Inst. Wash. Pub., *159*, *3*:452.
Allotheobaldia Brolemann, 1919, Ann. Ent. Soc.
  France, *88*:90.
Theomyia Edwards, 1930, Bull. Ent. Res., *21*:303.

ADULT. Proboscis moderately long. Palpi short in female, variable in length in male but usually as long as the proboscis. Antennae of male always plumose and shorter than the proboscis. Spiracular bristles present; postspiraculars absent. Tip of abdomen of female bluntly rounded, the eighth segment not retractile. Pulvilli absent. Squama fringed. A tuft of setae present on under side of wing at base of subcosta. *Male terminalia.* Tenth sternites with a few terminal teeth. Phallosome usualy divided. Basistyle rather long; basal lobe present; apical lobe small in some species, absent in others. Dististyle simple and with terminal claw. Claspettes absent.

PUPA. Respiratory trumpets short. Dendritic tufts on abdominal segment I large; lateral tufts of segments VII and VIII inconspicuous. Paddles finely serrate and with one hair at outside tip of midrib.

LARVA. Head wider than long. Comb of eighth segment present. A pair of siphonal tufts inserted near base of siphon; pecten present.

EGGS. In most species the eggs are rounded at one end and somewhat pointed at the opposite end. They are usually laid on the surface of the water in boat-shaped masses, but according to Edwards (251), there are exceptions. The eggs in subgenera *Climacura* and *Culicella* have been found to be laid singly in captivity.

BIONOMICS. The larvae are found for the most part in ground pools, rarely in other situations. Some of the species, especially those of the subgenus *Culiseta*, are severe biters.

DISTRIBUTION. Almost world-wide, but largely confined to the temperate regions of America, Europe, and Asia.

CLASSIFICATION. Three subgenera, *Culiseta*, *Culicella*, and *Climacura* are represented in this region. These are very distinct in the larval form, but less readily defined on adult characters.

KEYS TO THE SPECIES

ADULT FEMALES

1. Hind tarsi with pale rings on some segments
   (very narrow in some species)..................2
   Hind tarsi entirely dark 5

2 (1). Hind tarsal rings broad, that of segment 2
   covering one-fourth to one-third of the segment; cross veins with scales......................3
   Hind tarsal rings narrow, that of segment 2
   covering about one-tenth of the segment;
   cross veins without scales..........................4

3 (2). Femora each with a narrow subapical white-scaled ring....*maccrackenae* D. and K., p. 91
   Femora without subapical white-scaled rings
     alaskaensis (Ludlow), p. 83

4 (2). Wing with dense patches of dark scales
     incidens (Thomson), p. 87
   Wing without dense patches of dark scales,
   uniformly scaled
     morsitans (Theob.), p. 93

5 (1). Wing costa with mixed white and dark scales
     inornata (Will.), p. 89
   Wing costa entirely dark-scaled.....................6

6 (5). Points of origin of cross veins 3–4 and 4–5
   separated by more than the length of either
   cross vein; spiracular bristles dark
     melanura (Coq.), p. 95
   Points of origin of cross veins 3–4 and 4–5
   separated by less than the length of either
   cross vein; spiracular bristles yellow
     impatiens (Walker), p. 85

MALE TERMINALIA

1. Lobes of ninth tergite bluntly rounded, very
   dark and bearing many short stout spines
     inornata (Will.), p. 90
   Lobes of ninth tergite slightly elevated, lightly
   sclerotized and bearing long slender
   setae 2

2 (1). Basistyle with a small apical lobe.....................3
   Basistyle without an apical lobe.....................6

3 (2). Eighth tergite with a long even row of about
   thirty to forty short stout spines on median
   apical margin....*impatiens* (Walker), p. 85
   Eighth tergite with one stout median apical
   spine or a median apical clump of several
   spines 4

4 (3). Eighth tergite with a single stout spine on median apical margin
                        maccrackenae D. and K., p. 92
Eighth tergite with a median row or clump
of several short stout spines on apical
margin 5

5 (4). Apical lobe of basistyle broadly rounded and
bearing numerous setae of rather uniform
size.................*alaskaensis* (Ludlow), p. 83
Apical lobe of basistyle small, bearing several
setae, one of which is much longer and
stronger............*incidens* (Thomson), p. 88

6 (2). Eighth tergite with a broadly rounded median
apical lobe bearing a group of about eight
to ten long slender setae
                        moristans (Theob.), p. 94
Eighth tergite without a broadly rounded median apical lobe and lacking a group of
setae........................*melanura* (Coq.), p. 95

LARVAE (FOURTH INSTAR)

1. Antennae long, with tuft inserted at distal
third or fourth of shaft; pecten not followed
by an even row of single hairs (followed by
a row of small tufts in *C. melanura*)..........2
Antennae short, with tuft inserted near middle
of shaft; pecten followed by an even row
of long hairs..3

2 (1). Comb of eighth segment consisting of a single
row of about twenty-five long barlike
scales; pecten followed by a row of about
twelve subequal tufts
                        melanura (Coq.), p. 96
Comb of eighth segment consisting of many
scales in a patch; siphon without a row of
small tufts beyond pecten
                        morsitans (Theob.), p. 94

3 (1). Saddle of anal segment with a conspicuous
group of coarse short spicules at apex;
mesothoracic hair 1 large, 2- to 4-branched
                        maccrackenae D. and K., p. 92
Saddle of anal segment smooth, without
coarse spicules at apex, mesothoracic hair
1 small, single or multiple...........................4

4 (3). Upper frontal and lower frontal head hairs 5
and 6 similar in size and number of
branches............*impatiens* (Walker), p. 86
Lower frontal head hair 6 with fewer branches
than hair 5 and usually somewhat longer..5

5 (4). Lateral hair of anal segment rather strong, as
long or longer than the saddle
                        inornata (Will.), p. 90
Lateral hair of anal segment fine, shorter than
the saddle..6

6 (5). Mesothoracic hair 1 small, usually single
                        incidens (Thomson), p. 88
Mesothoracic hair 1 small, usually multiple
                        alaskaensis (Ludlow), p. 84

Culiseta (Culiseta) alaskaensis (Ludlow)

Theobaldia alaskaensis Ludlow, 1906, Can. Ent.,
        38:326.
Culiseta siberiensis Ludlow, 1919, Ins. Ins. Mens.,
        7:151.
Theobaldia arctica Edwards, 1920, Bull. Ent. Res.,
        10:136.

ADULT FEMALE (pl. 21). Large species.
Head: Proboscis long, dark-scaled, lightly
sprinkled with pale scales on basal half; palpi
short, dark, speckled with pale scales. Occiput
with narrow curved whitish scales and dark
erect forked scales dorsally, with broad appressed white scales laterally. Tori brown,
darker on inner surface, with yellowish-white
scales on inner surface. *Thorax:* Integument of
scutum black; scales on dorsal surface narrow,
dark brown, with many whitish scales intermixed and tending to form paired spots; anterior and lateral margins and prescutellar
space with white scales predominating. Scutellum with narrow whitish scales and dark
setae on the lobes. Posterior pronotum with
curved yellowish scales dorsally becoming
white ventrally. Pleura with sparse patches of
rather narrow curved grayish-white scales.
Spiracular bristles yellow; postspiracular
bristles absent. *Abdomen:* First tergite with a
median patch of white scales; remaining tergites dark, with rather broad basal white bands,
the second tergite with a pale median stripe.
Venter white-scaled, a few dark scales intermixed. *Legs:* Femora dark brown, with pale
scales intermixed, posterior surface pale; knee
spots white. Tibiae dark brown, with scattered
pale scales, tipped with white at apices. Tarsi
dark, first segment sprinkled with white; with
basal white rings on segments 1 to 4, broader
on hind tarsi, often poorly defined on front and
middle legs. *Wing:* Length about 6.0 to 7.0 mm.
Scales narrow, dark, some pale scales intermixed on costa, subcosta and vein 1; scales
dense and dark, producing spots at origin of
vein 2, at cross veins, at forks of veins 2 and 4,
and at middle of vein 6; cross veins with scales;
points of origin of cross veins 3–4 and 4–5 on
vein 4 separated by less than the length of either
cross vein. A dense tuft of yellowish setae present at base of subcosta on under side of wing.

ADULT MALE. Coloration similar to that of
the female. Palpi with light markings at the

joints of the segments and at the middle of the long segment. *Terminalia* (fig. 65). Eighth tergite (VIII-T) with about two to ten short stout spines on median apical margin. Lobes of ninth tergite (IXT-L) slightly raised, separated by less than the width of one lobe, each bearing an irregular row of long slender setae. Tenth sternite (X-S) heavily sclerotized and with several dark teeth at apex. Phallosome (Ph) long, cylindrical, narrowed and hooked at apex, consisting of two sclerotized plates open dorsally and ventrally. Basistyle (Bs) about two and a half times as long as mid-width, clothed with many long setae on outer aspect and with short setae on inner aspect; basal lobe (B-L) small, conical, sclerotized, bearing two (sometimes three) stout spines apically and many shorter setae; apical lobe (A-L) small, broadly rounded, sclerotized, bearing many setae. Disistyle (Ds) about two-thirds as long as the basistyle, curved and tapered apically, bearing small papillated setae; claw (Ds-C) short, stout, bifurcate.

LARVA (fig. 66). Antenna less than half as long as the head, spiculate; antennal tuft multiple, inserted near middle of shaft, reaching near tip. Head hairs: Postclypeal 4 small, multiple; upper frontal 5, 5- to 7-branched, barbed, shorter than hair 6; lower frontal 6, 3- or 4-branched, barbed, the middle branch usually longer and stronger than the others; preantennal 7 multiple, barbed, reaching slightly beyond insertion of antennal tuft; sutural 8 and transsutural 9 small, double or triple; supraorbital 10 small, single, or branched distally. Prothoracic hairs: 1–2 long, single; 3 medium, 2- to 4-branched; 4 medium, 3- to 5-branched; 5–6 long, single; 7 long, 3- to 5-branched. Mesothoracic hair 1 small, multiple. Lateral abdominal hair 6 multiple on segments I and II, usually double on III–VI. Comb of eighth segment of many scales in a patch; individual scale rounded apically and fringed with subequal spinules. Siphonal index 2.5 to 3.5; pecten of numerous teeth on basal fifth of siphon, closely followed by an even row of hairs extending to near apical fourth of siphon; siphonal tuft large, multiple, barbed, inserted within the pecten near base of siphon. Anal segment completely ringed by the saddle; lateral hair 2- to 5-branched, shorter than the saddle; dorsal brush bilaterally consisting of a lower caudal tuft of three or four unequal branches and a shorter multiple upper caudal tuft; ventral brush well developed, with two to five precratal tufts; gills as long or a little longer than the saddle, pointed.

DISTRIBUTION. Western North America from Colorado north to Alaska, northern Europe, and Siberia. It is usually restricted to the higher mountains in the southern part of its range. *United States:* Colorado (226); Montana (475); Wyoming (578). *Canada:* Alberta (212, 708); British Columbia (289, 352); Labrador (348); Manitoba (289, 708); Northwest Territories (289); Quebec (289, 402); Yukon (212, 289). *Alaska:* (396, 670).

BIONOMICS. The species apparently overwinters as adult females. The eggs are deposited in raftlike masses on the surface of the water. The larvae have been found in late spring and summer in open pools, in river beds and adjacent pools, and in pools in sluggish streams with a heavy growth of vegetation. Mail (475) found larvae in a pool formed by melting snow

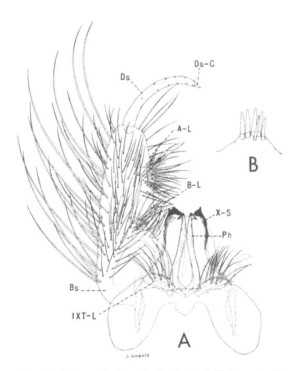

Fig. 65. *Culiseta alaskaensis* (Ludlow). *A*, Male terminalia.
B, Apical margin of eighth tergite.

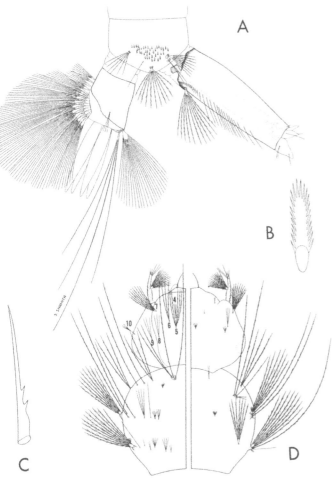

Fig. 66. Larva of *Culiseta alaskaensis* (Ludlow). *A,* Terminal segments. *B,* Comb scale. *C,* Pecten tooth. *D,* Head and thorax (dorsal and ventral).

in a woodland in Montana. Stackelberg (668) reports finding the larvae in small open pools in Russia where they were associated with the larvae of *Aedes excrucians* and *A. flavescens.* Tulloch (699) found the larvae in permanent bodies of water near Fairbanks, Alaska.

Information on the bloodsucking habits of this species is scanty; however, Natvig (523) describes observations recorded by Martini where the females are said to have been very annoying in bright sunshine at 2:00 P.M. near Hamburg, Germany. Mail (475) collected the adults in the vicinity of grassy marshes in river valleys in Montana but observed that they were usually found near timber. Frohne (290) noted that the females attack man in shaded protected situations in Alaska.

Culiseta (Culiseta) impatiens (Walker)

Culex impatiens Walker, 1848, List Dipt. Brit. Mus., 1:5.

Culex pinguis Walker, 1866, Lord's Nat. in Vanc. and B.C., 2:337.

Culex absobrinus Felt, 1904, N.Y. State Mus. Bull., 79:318.

ADULT FEMALE (pl. 22). Large species. *Head:* Proboscis long, dark-scaled, sparsely speckled with pale scales; palpi short, dark, lightly speckled with pale scales. Occiput with narrow curved yellowish-white scales and dark erect forked scales dorsally, with broad yellowish appressed scales laterally. Tori brown, darker on inner surface; a few pale scales sometimes present on inner surface. *Thorax:* Integument of scutum dark brown; scutum clothed with narrow reddish-brown and yellowish scales in a variable pattern; two fine yellowish lines extending posteriorly from median patches of yellowish scales; prescutellar space margined with yellowish scales. Scutellum with narrow yellowish scales and dark setae on the lobes. Pleura with patches of dingy-white scales. Numerous yellow spiracular bristles present; postspiracular bristles absent. *Abdomen:* First tergite with a small median patch of pale scales; remaining tergites bronze-brown, with basal white bands. Venter yellowish-scaled, with dark scales intermixed. *Legs:* Femora dark brown, yellowish on posterior surface; knee spots white. Tibiae dark, streaked posteriorly with pale scales. Tarsi dark. *Wing:* Length about 5.0 to 6.0 mm. Scales narrow, brown; scales denser at base of vein 3 and fork of vein 4; cross veins without scales; points of origin of cross veins 3–4 and 4–5 on vein 4 separated by less than the length of either cross vein. A dense tuft of yellowish setae present at base of costa on under side of wing.

ADULT MALE. Coloration similar to that of the female. *Terminalia* (fig. 67). Eighth tergite (VIII-T) with an even row of twenty to fifty short stout spines on median apical margin. Lobes of ninth tergite (IX-T-L) slightly elevated, separated by less than the width of one lobe, each bearing several long slender setae. Tenth sternite (X-S) heavily sclerotized apically, with several strong dark teeth at apex.

Phallosome (Ph) cylindrical, gradually tapered to apex, formed of two sclerotized plates, open dorsally and ventrally. Basistyle (Bs) about twice as long as mid-width, bearing many long setae on outer aspect and short setae on inner aspect; basal lobe (B-L) small, conical, bearing one large spinelike seta and several smaller ones; apical lobe (A-L) represented by a small prominence, bearing a few moderately long setae, one or more of which are longer and stronger. Disistyle (Ds) a little more than half as long as the basistyle, tapered and curved, bearing small papillated setae; claw (Ds-C) short, stout, bifurcate.

LARVA (fig. 68). Antenna about half as long as the head, spiculate; antennal tuft multiple, inserted near middle of the shaft, reaching near tip. Head hairs: Postclypeal 4 small, usually double; upper frontal 5 and lower frontal 6 multiple, barbed, similar in size and number of branches; preantennal 7 multiple, barbed, reaching near insertion of antennal tuft; sutural 8 and transsutural 9 small, single

or double; supraorbital 10 small, single or branched distally. Prothoracic hairs: 1–2 long, single; 3 medium, 3- or 4-branched; 4 medium, 3- to 6-branched; 5 long, single; 6 long, single or double; 7 long, 3- to 7-branched. Mesothoracic hair 1 small, single or double. Lateral abdominal hair 6 usually multiple on segments I and II, double on III–VI. Comb of eighth segment with many scales in a patch; individual scale rounded apically and fringed with subequal spinules. Siphonal index 2.5 to 2.8; pecten of numerous teeth on basal fifth or fourth of siphon, followed by an even row of hairs extended to near apical fourth of siphon; siphonal tuft large, multiple, barbed, inserted within the pecten near base of siphon. Anal segment completely ringed by the saddle; lateral hair 2- to 4-branched, shorter than the saddle; dorsal brush bilaterally consisting of a lower caudal tuft of three or four unequal branches and a shorter multiple upper caudal tuft; ventral brush well developed, with two to five precratal tufts; gills longer than the saddle, pointed.

DISTRIBUTION. Northern United States, Canada, and Alaska. *United States:* California (287); Colorado (226, 343); Idaho (344, 672); Iowa (625); Maine (45, 213); Massachusetts (270); Michigan (391); Missouri (14); Montana (213, 475); Nebraska (14, 680); New Hampshire (213, 458); New York (31); Oregon (672); Utah (575); Vermont (213); Washington (67, 672); Wisconsin (171); Wyoming (578). *Canada:* Alberta (213, 598); British Columbia (352); Labrador (289, 348); Manitoba (466); New Brunswick (289, 708); Northwest Territories (289, 708); Nova Scotia (289); Ontario (212, 289); Quebec (289, 402); Yukon (212, 289). *Alaska:* (213, 396, 673).

BIONOMICS. The females overwinter and are among the first mosquitoes on wing in the spring. Frohne (292) collected hibernators of this species near Juneau, Alaska, late in March, and he believes that they may become active during warm spells, even during January, February, or March. The eggs are laid in rafts on the water usually in deep well-shaded pools. Frohne (290) found the larvae of this species to be extremely abundant in a grossly polluted

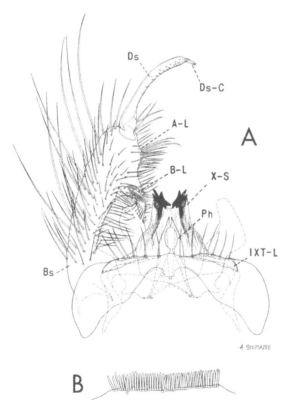

Fig. 67. *Culiseta impatiens* (Walker). *A*, Male terminalia. *B*, Apical margin of eighth tergite.

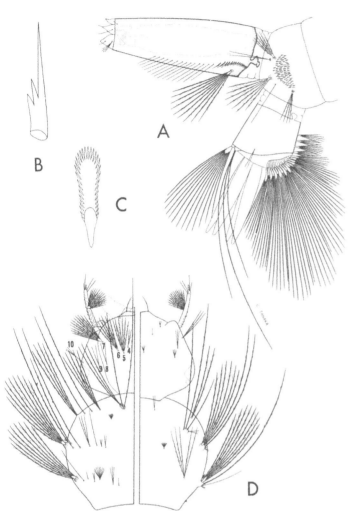

Fig. 68. Larva of *Culiseta impatiens* (Walker). *A*, Terminal segments. *B*, Pecten tooth. *C*, Comb scale. *D*, Head and thorax (dorsal and ventral).

impoundage in Anchorage, Alaska. The larvae are often found in pools filled by melting snow. Hearle (351) found that there is more than one generation a season in British Columbia.

Rees (575) observed that the females feed only about dusk in Utah, and at times are vicious biters but are generally timid and easily frightened away. The senior author collected several females of *C. impatiens* which attacked him in an evergreen forest in Sierra County, California, near midday in June, 1953. Other workers have observed that the females readily attack man in sheltered situations in many other areas where the adults are numerous.

Culiseta (Culiseta) incidens (Thomson)

Culex incidens Thomson, 1868, Kongl. Sven. Freg. Eug. Resa, 6, Dipt., p. 443.

Culex particeps Adams, 1903, Kans. Univ. Sci. Bull., 2:26.

ADULT FEMALE (pl. 23). Rather large species. *Head:* Proboscis long, dark-scaled, with a few pale scales intermixed, the pale scales more numerous on the basal half and underneath; palpi short, dark, speckled with pale scales. Occiput with yellowish-white lanceolate scales and numerous dark erect forked scales on dorsal region; eyes margined with white scales; lateral region of occiput with broad white scales. Tori brown or black, with a few white scales on inner surface. *Thorax:* Integument of scutum dark brown; scutum clothed dorsally with narrow dark-brown and golden scales, arranged in poorly defined lines and spots; anterior margin, lateral margins, and prescutellar space with broader yellowish or yellowish-white scales. Scutellum with lobes densely clothed with yellowish-white scales and dark setae. Posterior pronotum with a mixture of narrow curved white and dark-brown scales dorsally, with broad white scales ventrally. Pleura with medium-sized patches of grayish-white to creamy-white scales. Spiracular bristles yellow; postspiraculars absent. *Abdomen:* First tergite with a small median patch of yellowish-white scales; remaining tergites bronze-brown- or black-scaled, with basal segmental bands of yellowish-white scales. Venter yellowish-white-scaled, with black scales intermixed. *Legs:* Femora dark brown, posterior surface yellow to near apex; yellowish-white knee spots present. Tibiae dark, streaked with white scales on posterior surface, tipped with pale scales apically. Tarsi dark, with narrow basal pale bands, more distinct on hind tarsi, often indistinct or absent on distal segments of front and middle tarsi. *Wing:* Length about 5.5 to 6.0 mm. Points of origin of cross veins 3–4 and 4–5 on vein 4 separated by less than the length of either cross vein; cross veins without scales; wing scales narrow, dark; veins 2, 4, 5.1, and 6 with dense patches of dark scales. A tuft of yellowish setae present at base of subcosta on under side of wing; a small area of white scales usually present on basal part of costa.

ADULT MALE. Coloration similar to that of the female. Palpi with whitish rings at base of joints and at middle of long joint. *Terminalia* (fig. 69). Eighth tergite (VIII-T) bearing about four to ten short stout spines on median apical margin. Lobes of ninth tergite (IXT-L) slightly raised, separated by less than the width of one lobe, each lobe bearing a row of rather long setae. Tenth sternite (X-S) heavily sclerotized, with about 4 strong dark teeth on apex. Phallosome (Ph) cylindrical, abruptly narrowed subapically, pointed at apex, consisting of two sclerotized plates narrowly open ventrally, broadly open dorsally except near tip. Claspette absent. Basistyle (Bs) stout, about two and a half times as long as mid-width, bearing long setae on outer aspect and shorter setae on inner aspect; basal lobe (B-L) small, conical, bearing two strong spines apically and numerous setae subapically; apical lobe (A-L) represented by a small prominence bearing several setae, one of which is distinctly stronger and longer. Dististyle (Ds) about two-thirds as long as the basistyle, bearing small papillated setae; claw (Ds-C) short, stout, bifurcate.

LARVA (fig. 70). Antenna about half as long as the head, sparsely spiculate; antennal tuft multiple, barbed, inserted near middle of shaft, reaching near tip. Head hairs: Postclypeal 4 double or triple, much shorter than hair 5; upper frontal 5 multiple, barbed, with more branches than hair 6; lower frontal 6 multiple, barbed, reaching slightly beyond anterior margin of head; preantennal 7 multiple, barbed, reaching slightly beyond insertion of antennal tuft; sutural 8 double, nearly equal in size to hair 4; transsutural 9 usually 3- or 4-branched; supraorbital 10 single or double. Prothoracic hairs: 1 long, 2- to 5-branched; 2 long, single; 3 medium, 2- to 4-branched; 4 medium, 4- to 7-branched; 5–6 long, single; 7 long, 2- to 6-branched. Mesothoracic hair 1 small, usually single. Lateral abdominal hair 6 multiple on segments I and II, double on III–VI. Comb of eighth segment with many scales in a patch; individual scale rounded apically and fringed with subequal spinules. Siphonal index about 3.0; pecten of about 11 to 15 teeth on basal fifth of siphon, followed by an even row of hairs extending to near apical fourth of siphon; siphonal tuft large, multiple, barbed, inserted within pecten near base of siphon. Anal segment completely ringed by the saddle; lateral hair 1- to 5-branched, usually triple, shorter than the saddle; dorsal brush bilaterally consisting of a lower caudal tuft of three or four unequal branches and a shorter multiple upper caudal tuft; ventral brush well developed, with two or three precratal tufts; gills usually a little longer than the saddle, pointed.

DISTRIBUTION. Western North America from southern California to Alaska. *United States:* Arizona (213); California (287); Colorado (226, 343); Idaho (344, 672); Michigan (391); Montana (213, 475); Nebraska (680); Nevada (213); New Mexico (20, 213); North Dakota (14); Oklahoma (321); Oregon (213, 672); Texas (249, 532); Utah (575); Washington (67, 672); Wyoming (578). *Canada:* Alberta (212, 708); British Columbia (352); Northwest Territories and Nova Scotia (289); Yukon (212, 289). *Alaska:* (213, 291).

BIONOMICS. The larvae of *C. incidens* develop in a wide variety of aquatic habitats, including ground pools, springs, pools formed by melting snow in the mountains, rain barrels, reservoirs, hoof prints, brackish water along the coast, and often in grossly polluted pools.

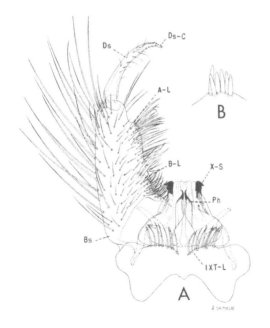

Fig. 69. *Culiseta incidens* (Thomson). *A*, Male terminalia. *B*, Apical margin of eighth tergite.

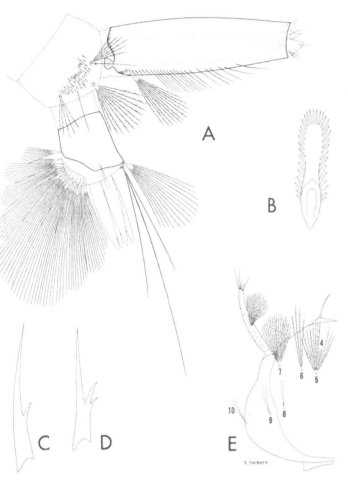

Fig. 70. Larva of *Culiseta incidens* (Thomson). *A*, Terminal segments. *B*, Comb scale. *C* and *D*, Pecten teeth. *E*, Head.

They are often encountered in stagnant garden pools, residual pools in stream beds, and in animal-drinking troughs; however, the preferred habitat seems to be clear or semiclear water with at least partial shade. The females hibernate overwinter in the colder regions; however, the species remains active throughout the winter months along the California coast.

The females are reported to be extremely annoying in some areas, whereas in other communities they are timid biters, rarely attacking man. Hubert (384) describes the successful colonization of this species in a room-sized enclosure, and observed the mating flights. Mating flights were induced by turning on the light in the room. They took place just above the level of the floor. As females flew into the swarm they were immediately seized in copu-lation, the females pulling the males in flight to the wall where they came to rest momentarily.

MEDICAL IMPORTANCE. Successful experimental transmission of St. Louis encephalitis virus (330), western equine virus (331), and Japanese B encephalitis virus (592) indicates that *C. incidens* may have potential medical importance.

Culiseta (Culiseta) inornata (Williston)

Culex inornatus Williston, 1893, U.S. Dept. Agr. Div. Ornith. and Mam., No. Amer. Fauna, 7:253.

Culex magnipennis Felt, 1904, N.Y. State Mus. Bull., 79:278.

ADULT FEMALE (pl. 24). Large species. *Head:* Proboscis dark-scaled, speckled with pale scales; palpi short, dark, speckled with white scales. Occiput with narrow pale yellow or yellowish-white scales and dark erect forked scales dorsally, with broad whitish scales laterally. Tori dark brown, a few white scales on inner surface. *Thorax:* Integument of scutum brown; scutum clothed with narrow, golden-brown and pale-yellow intermixed scales on dorsal surface; anterior and lateral margins and prescutellar space pale-yellow-scaled. Scutellum with narrow pale yellow scales and dark setae on the lobes. Posterior pronotum sparsely clothed with broad brown and white scales. Pleura with large, rather sparse patches of cream-white scales. Yellow spiracular bristles present; postspiracular bristles absent. *Abdomen:* First tergite with a median patch of yellowish-white scales; tergites II to VII dark brown, each with a yellowish-white basal band which widens laterally to cover the full length of each tergite; eighth tergite entirely pale-scaled. Venter clothed with pale-yellow scales. *Legs:* Legs dark brown. Femora, tibiae, and often the first and second tarsal segments speckled with pale scales, pale on posterior surface; pale knee spots present. *Wing:* Length about 5.0 to 6.0 mm. Points of origin of cross veins 3–4 and 4–5 on vein 4 separated by less than the length of either cross vein; scales narrow, dark, and rather sparse on all veins but costa, subcosta, and vein 1, where they are broader, more numerous, and intermixed with white scales. A dense tuft of yellowish setae present at base of subcosta on under side of wing.

ADULT MALE. Coloration similar to that of the female. The palpi differ from those of most male Culicines in lacking long hairs and having the terminal segment broad and flattened. *Terminalia* (fig. 71). Lobes of ninth tergite (IXT-L) sclerotized, bluntly rounded, bearing numerous short stout spines. Tenth sternite (X-S) heavily sclerotized, with a broad sclerotized striated structure near base and toothed at apex. Phallosome (Ph) broadest near base, tapering to apex, open dorsally and ventrally, formed of two plates which are heavily sclerotized except for an apical hyaline part. Claspette absent. Basistyle (Bs) about twice as long as mid-width, bearing many long setae on outer aspect and shorter setae on inner aspect; basal lobe (B-L) conical, bearing several spinelike setae apically and numerous smaller setae subapically; apical lobe absent. Dististyle (Ds) about half as long as the basistyle, stout, bearing many small papillated setae; claw (Ds-C) short, strong, bifurcate.

LARVA (fig. 72). Antenna nearly half as long as the head, sparsely spiculate. Antennal tuft multiple, inserted near middle of shaft, reaching near tip. Head hairs: Postclypeal 4 small, multiple; upper frontal 5 multiple, barbed, a little shorter and with more branches than hair 6; lower frontal 6 multiple, barbed; preantennal 7 long, multiple, barbed; sutural 8 usually double, smaller than hair 4; transsutural 9 usually 3- or 4-branched; supraorbital 10 single to triple. Prothoracic hairs: 1 very long, usually single; 2 long, single; 3 long, double, or triple; 4 medium, 3- to 6-branched; 5–6 long, single; 7 long, usually 3- or 4-branched. Mesothoracic hair 1 small, single. Lateral abdominal hair 6 multiple on segments I and II, double on III–VI. Comb of eighth segment of many scales in a patch; individual scale rounded apically and fringed with subequal spinules. Siphonal index about 3.5; pecten of about 12 to 20 teeth on basal fourth or fifth of siphon, closely followed by an even row of long hairs extending to near apical fourth of siphon; siphonal tuft large, multiple, barbed, inserted within the pecten near base of siphon. Anal segment completely ringed by the saddle; lateral hair 1- to 5-branched (usually double), as long or longer than the saddle; dorsal brush bilaterally consisting of a lower caudal tuft of three or four unequal branches and a shorter multiple upper caudal tuft; ventral brush large, with one or two precratal tufts; gills as long or longer than the saddle, bluntly pointed.

DISTRIBUTION. United States, southern Canada, and northern Mexico. *United States:* Alabama (135, 213); Arizona (213); Arkansas (88, 121); California (287); Colorado (343); Delaware (120, 167); District of Columbia (213, 318); Florida (135, 213); Georgia (135, 420); Idaho (344, 672); Illinois (615); Indiana (152, 345); Iowa (625); Kansas (213, 530); Kentucky (134, 570); Louisiana (213, 759); Maryland (64, 161); Massachusetts (213, 701); Michigan (391); Minnesota (533); Mississippi (135, 546); Missouri (213, 324); Montana (213, 475); Nebraska (680); Nevada (213); New Hampshire (66); New Jersey (643); New Mexico (20, 213); New York (31); North Carolina (135, 646); North Dakota (553); Ohio (731a); Oklahoma (628); Oregon (213, 672); Pennsylvania (298, 756); South Carolina (135, 742); South Dakota (14, 213); Tennessee (135, 658);

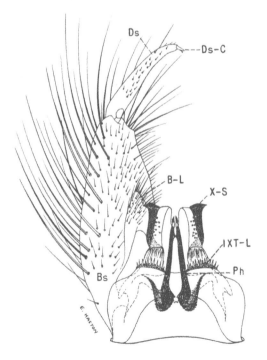

Fig. 71. Male terminalia of *Culiseta inornata* (Williston).

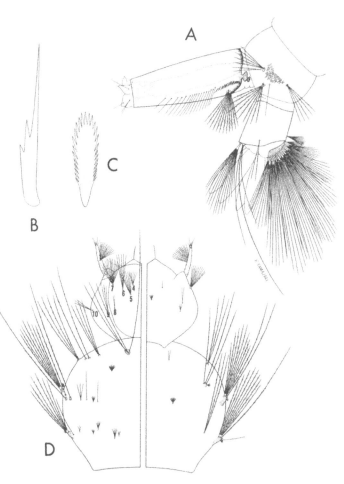

Fig. 72. Larva of *Culiseta inornata* (Williston). *A*, Terminal segments. *B*, Pecten tooth. *C*, Comb scale. *D*, Head and thorax (dorsal and ventral).

and spring months, whereas in the north the females pass the winter in hibernation, and the immature stages are found only during the spring, summer, and fall months. The senior author has, on several occasions during the winter months in Arkansas, collected the larvae from pools covered with ice.

The females are seldom persistent biters but do occasionally attack man. Mail (475) states that the females seem to prefer feeding on larger mammals and are at times very troublesome to livestock in timbered areas in the early spring in Montana. Precipitin tests made by Reeves and Hammon (591) in the Yakima Valley, Washington, seem to support this conclusion, indicating that the females feed principally on larger mammals such as horses and cows and rarely on man. The adults are readily attracted to artificial light and are often captured in light traps during cold weather when few other night-flying insects are active. The senior author observed males and females in copulation at the edge of a ground pool in Arkansas during 1941. Approximately twenty-five pairs were united by the tips of their abdomens and were resting on the branch of a fallen tree. Similar mating habits are described by Howard, Dyar, and Knab (382). The continuous rearing of this species in the laboratory is described by Owen (534) and McLintock (468).

MEDICAL IMPORTANCE. Western equine encephalitis virus has been isolated from wild-caught *C. inornata* in the Yakima Valley, Washington (332). Reeves and Hammon (592) demonstrated successful experimental transmission of Japanese B encephalitis virus with this mosquito.

Texas (213, 463); Utah (575); Virginia (176, 178); Washington (67, 672); Wisconsin (171); Wyoming (578). *Canada:* Alberta (598); British Columbia (352); Manitoba (466, 708); Northwest Territories (289); Ontario (708); Saskatchewan (212, 464); Yukon (289).

BIONOMICS. The larvae of *C. inornata* are found in ground pools, ditches, and occasionally in artificial water containers, often grossly polluted. They occur in brackish water in the coastal marshes in California where they are sometimes associated with the larvae of *Aedes squamiger* (Coq.). The larvae are somewhat less partial to shade and cool water than *C. incidens.* In the southern United States, breeding is generally restricted to the fall, winter,

Culiseta (Culiseta) maccrackenae
Dyar and Knab

Culiseta maccrackenae Dyar and Knab, 1906, Proc. Biol. Soc. Wash., *19*:133.

Culiseta dugesi Dyar and Knab, 1906, Proc. Biol. Soc. Wash., *19*:134.

ADULT FEMALE (pl. 25). Rather large species. *Head:* Proboscis long, dark-scaled, with pale scales intermixed; palpi short, dark, with a few pale scales intermixed, tipped with white, segments banded with white scales. Occiput with narrow pale yellow or yellowish-white scales and dark erect forked scales on dorsal region, with broad grayish-white scales laterally. Tori brown, with a few grayish-white scales on inner surface. *Thorax:* Integument of scutum dark brown; scutum mottled with patches of golden and dark bronze-brown scales, with some white scales intermixed; the white scales are more numerous on the anterior and lateral margins and around the prescutellar space. White scales sometimes arranged more or less in the form of narrow lines. Scutellum densely clothed with rather broad silver-white or cream-white scales, the lobes bearing brown setae. Posterior pronotum sparsely clothed with broad brown and white scales. Pleura with medium-sized patches of grayish-white scales. Spiracular bristles yellow; post-spiraculars absent. *Abdomen:* First tergite with a small median patch of pale scales; remaining tergites dark-scaled, each with a white basal band slightly extended medially; eighth tergite almost entirely pale-scaled. Venter largely pale-scaled. *Legs:* Femora dark brown, speckled with pale scales, pale scales more numerous on posterior surface; each with a preapical ring of pale scales; pale knee spots present. Tibiae dark, speckled with white, streaked with white on posterior surface. Tarsi dark, each segment with a broad white basal ring, narrower on more distal segments, often indistinct or absent on segment 5 of front and middle tarsi. *Wing:* Length about 5.0 to 5.5 mm. Points of origin of cross veins 3–4 and 4–5 on vein 4 separated by less than the length of one cross vein; cross veins with scales; wing scales brown, with a few white scales intermixed on costa, subcosta, and vein 1; scales dense and dark so as to produce spots at base of vein 2, fork of veins 2.1 and 2.2, on cross veins, base of vein 5.1 and on apical part of vein 6. A tuft of yellowish setae present at base of subcosta on under side of wing.

ADULT MALE. Coloration similar to that of the female. Palpi with white bands at bases of segments and at middle of long segment. *Terminalia* (fig. 73). Eighth tergite (VIII-T) bearing a single short stout spine on median apical margin. Lobes of ninth tergite (IXT-L) slightly elevated, separated by less than the width of one lobe, each bearing numerous long setae. Tenth sternite (X-S) heavily sclerotized apically, with about 3 or 4 dark teeth at apex. Phallosome (Ph) cylindrical, tapered to apex, consisting of two elongated, sclerotized plates, touching on dorsal side. Claspette absent. Basistyle (Bs) about two and a half times as long as mid-width, bearing long setae on outer aspect and shorter setae on inner aspect; basal lobe (B-L) conical, bearing three stout pointed apical spines and numerous fine setae on sides; apical lobe (A-L) represented by a slight prominence bearing several setae. Dististyle (Ds) a little more than half as long as the basistyle, tapered and slightly curved beyond middle, bearing many small papillated setae; claw (Ds-C) short, stout, bifurcate.

LARVA (fig. 74). Antenna nearly half as long as the head, sparsely spiculate; antennal tuft multiple, barbed, inserted near middle of shaft, reaching near tip. Head hairs: Postclypeal 4 3- to 5-branched, weakly barbed, nearly as long as hair 5; upper frontal 5 multiple, barbed, with more branches than hair 6; lower frontal 6 multiple, barbed; preantennal 7 multiple, barbed, extending a little beyond insertion of the antennal tuft; sutural 8 single or double; transsutural 9 multiple, barbed, longer than hair 6; supraorbital 10 single or double. Prothoracic hairs: 1 long, 3- to 5-branched; 2 long, single; 3 long, 3- to 6-branched; 4 long, 3- to 7-branched; 5–6 long, single; 7 long, usually 4- to 6-branched. Mesothoracic hair 1 large, conspicuous, 2- to 4-branched. Lateral abdominal hair 6 multiple on segments I and II, double on III–VI. Comb of eighth segment of numerous scales in a patch; individual scale rounded apically and fringed with subequal spinules. Siphonal index about 3.0; pecten of about 12 to 18 teeth on basal fourth of siphon, followed by an even row of hairs extending to near apical fourth of siphon; siphonal tuft large, multiple, barbed, inserted in a sclerotized area near base of siphon. Anal segment completely ringed by the saddle; dorsoapical part of the saddle with prominent heavy short dark spicules; lateral hair of anal segment 1- to 5-branched, shorter than the saddle; dorsal brush bilaterally consisting of a lower caudal tuft of three or four unequal branches and a shorter multiple upper caudal tuft; ventral brush large, with three or four precratal tufts;

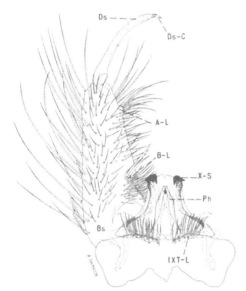

Fig. 73. Male terminalia of *Culiseta maccrackenae* Dyar and Knab.

gills usually a little longer than the saddle, pointed.

DISTRIBUTION. Mexico (106, 481), Costa Rica (237), and along the coast of California and southern Oregon. *United States:* California (284, 287); Oregon (672).

BIONOMICS. The larvae develop in cool, rather clear, shaded pools. The pools in which the immature stages occur often contain prolific growths of filamentous algae and bottom deposits of dead leaves and other plant debris. The larvae spend much of their time at the bottom, and when disturbed will hide under leaves and other bottom debris. Successive broods occur throughout the entire year, even during the winter months, in the California coastal region. The females will feed on man in the shade any time during the day but feed more actively after sundown. The species is seldom found in sufficient numbers to be of much importance. The adults are readily attracted to light traps.

Culiseta (Culicella) morsitans (Theobald)

Culex morsitans Theobald, 1901, Mon. Culic., 2:8.
Culex dyari Coquillett, 1902, Jour. N.Y. Ent. Soc., 10:192.
Culex brittoni Felt, 1905, Ent. News, 16:79.

ADULT FEMALE (pl. 26). Medium to rather

large species. *Head:* Proboscis long, dark-scaled; palpi short, dark. Occiput with narrow yellowish-white scales and dark erect forked scales dorsally, with broad yellowish-white scales laterally. Tori brown, inner surface darker and bearing a few brown setae. *Thorax:* Integument of scutum brown, with two rather broad submedian reddish-brown bare stripes, two shorter bare stripes one on either side of the prescutellar space; scutum clothed with narrow fine golden-brown scales except for coarser narrow white scales medially, along anterior and lateral margins, on either side of posterior bare stripes, around the prescutellar space, and in front of wing bases. Scutellum with narrow white lanceolate scales and dark setae on the lobes. Posterior pronotum sparsely clothed with narrow curved dark-brown scales. Pleura with small patches of grayish-white scales. Numerous yellow spiracular bristles present; postspiraculars absent. *Abdomen:* First tergite dark; remaining tergites dark,

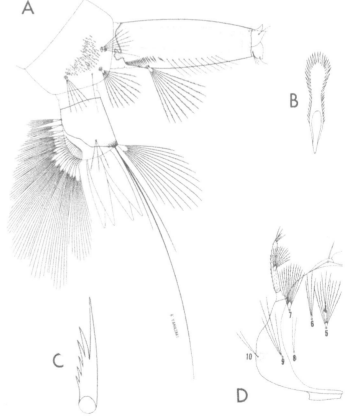

Fig. 74. Larva of *Culiseta maccrackenae* Dyar and Knab. *A*, Terminal segments. *B*, Comb scale. *C*, Pecten tooth. *D*, Head.

with broad appressed bronze-brown scales and rather narrow basal bands of yellowish-white scales. Venter largely pale-scaled, and with many pale setae. *Legs:* Femora dark brown, posterior surface pale; knee spots pale. Tibiae bluish-brown-scaled, each with a streak of pale scales on posterior surface, tipped with white apically. Tarsi dark, except for rather faint pale rings involving both ends of joints, more pronounced on basal segments, indistinct or absent on distal segments. *Wing:* Length about 5.0 to 5.5 mm. Points of origin of cross veins 3–4 and 4–5 on vein 4 separated by more than the length of either cross vein; scales narrow, dark brown; cross veins without scales. A dense tuft of dark setae present at base of subcosta on under side of wing.

ADULT MALE. Coloration similar to that of the female. Palpi long and with ventral brushes at apex of second and on all of third and fourth segments. *Terminalia* (fig. 75). Eighth tergite (VIII-T) with a group of about eight to ten long slender spines on a broadly rounded lobe on median apical margin. Lobes of ninth tergite (IXT-L) only slightly elevated, each bearing many long weak setae. Tenth sternite (X-S) heavily sclerotized apically, with 2 (rarely 3) teeth at apex. Phallosome (Ph) short, bulbous, tapered from outer third and pointed at apex, narrowly open ventrally, broadly open at base

and closed for a short distance at apex on dorsal side. Claspette absent. Basistyle (Bs) about two and a half times as long as mid-width, clothed with long setae on outer aspect, with short setae on inner aspect; basal lobe (B-L) prominent, conical, with about three to five strong apical spines and several weaker subapical setae; apical lobe absent. Dististyle (Ds) about two-thirds as long as the basistyle, glabrous, a few minute papillated setae present; claw (Ds-C) short, stout, bifurcate.

LARVA (fig. 76). Antenna as long as the head, curved, spiculate; antennal tuft large, multiple, barbed, inserted at outer third or fourth of shaft, reaching well beyond tip. Head hairs: Postclypeal 4 small, single; upper frontal 5 usually 4- to 6-branched, shorter than hair 6; lower frontal 6 very long, double, sparsely barbed; preantennal 7 long, multiple, barbed; sutural 8 and transsutural 9 small, usually double or triple; supraorbital 10 usually single or branched. Prothoracic hairs: 1–2 very long, usually single; 3 long, usually double; 4 long, double; 5–6 very long, single; 7 long, usually 3- or 4-branched. Lateral abdominal hair 6 usually multiple on segments I and II, single on III–VI. Comb of eighth segment of numerous scales in a patch; individual scale somewhat expanded outwardly and fringed laterally and apically with subequal spinules. Siphonal index about 6.0 to 7.0; pecten of a few teeth on basal fifth or fourth of siphon, distal teeth detached; siphonal tuft large, usually 4- or 5-branched, inserted within the pecten near base of siphon. Anal segment much longer than wide, completely ringed by the saddle; lateral hair single, a little shorter than the saddle; dorsal brush bilaterally consisting of a lower caudal tuft of three unequal branches and a short multiple upper caudal tuft; ventral brush well developed, with about six or seven precratal tufts; gills variable in length, usually as long or longer than the saddle, slender, and pointed.

DISTRIBUTION. Northern United States, Alaska, Canada, and Europe. *United States:* Colorado (343); Connecticut (270, 494); Delaware (167); Idaho (344); Illinois (615); Iowa (625); Kentucky (65); Maine (45); Massachusetts (213, 270); Michigan (391. 529); Minnesota (533); New Hampshire (66.

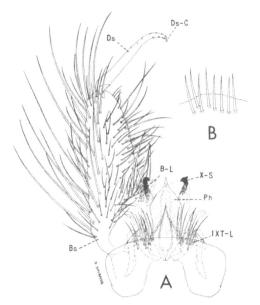

Fig. 75. *Culiseta morsitans* (Theobald). *A*, Male terminalia. *B*, Apical margin of eighth tergite.

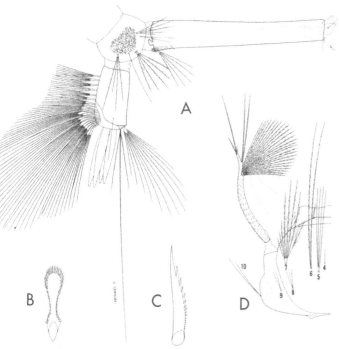

Fig. 76. Larva of *Culiseta morsitans* (Theobald). *A*, Terminal segments. *B*, Comb scale. *C*, Pecten tooth. *D*, Head.

213); New Jersey (213, 350); New York (31); North Dakota (553); Ohio (482); Oregon (672); Pennsylvania (298, 756); Rhode Island (35, 270); South Dakota (237); Washington (67, 672); Wisconsin (171). *Canada:* Alberta (212, 708); British Columbia (212, 708); Labrador (348); Manitoba (708); New Brunswick (536, 708); Northwest Territories (289); Nova Scotia (536, 708); Ontario (212, 289); Prince Edward Island (710); Quebec (536,708); Saskatchewan (598,708); Yukon (212, 708). *Alaska:* (396,670).

BIONOMICS. The larvae occur in rather large temporary cold rain-water pools in both shaded and open marshes. Jenkins (396) found the larvae of *C. morsitans* in central Alaska in temporary *Carex* marshes, along lake margins, in coastal marshes and in bogs. Marshall (480) observed that in England the females lay their eggs during the late summer in dried-up hollows or above the water level of partly filled pools and, after hatching, the larvae often reach the fourth instar by November and overwinter in that stage. In Denmark, Wesenberg-Lund (749) made similar observations on the bionomics of this species. Macan (460) tells of

finding the larvae of *C. morsitans* in a treehole in England. The females rarely, if ever, feed on man. Wesenberg-Lund (749) in Denmark and Marshall (480) in England report that they never have been able to induce laboratory-reared females to feed on man. Natvig (523) cites a reference where Eckstein claims to have induced a female of *C. morsitans* to feed on the blood of a greenfinch, indicating that *C. morsitans* probably feeds on birds.

Culiseta (Climacura) melanura
(Coquillett)

Culex melanurus Coquillett, 1902, Jour. N.Y. Ent. Soc., *10*:193.

ADULT FEMALE (pl. 27). Medium-sized species. *Head:* Proboscis long, slender, dark-scaled; palpi short, dark. Occiput with narrow yellowish scales and dark erect forked scales dorsally, with a patch of broad whitish scales laterally. Tori brown, darker on inner surface. *Thorax:* Integument of scutum dark brown to reddish-brown; scutum clothed with fine dark bronze-brown scales, paler on prescutellar space. Scutellum with narrow golden-brown scales and dark setae on the lobes. Pleura with patches of dingy-white scales. A few dark spiracular bristles present; postspiracular bristles absent. *Abdomen:* Tergites dark brown to black-scaled, with bronze to purplish reflection, and with small yellowish-white basal patches laterally (not visible from above); faint narrow yellowish-white bands sometimes present on some segments. Venter with dingy-white or yellowish scales, usually blended with a few dark scales. *Legs:* Legs entirely dark-scaled except for pale posterior surface of femora. *Wing:* Length 3.5 to 4.0 mm. Points of origin of cross veins 3–4 and 4–5 on vein 4 separated by more than the length of either cross vein; cross veins without scales; longitudinal veins densely clothed with dark, slightly broadened, ligulate scales. A tuft of dark setae arising from base of subcosta on under side of wing.

ADULT MALE. Coloration similar to that of the female. Palpi long, the terminal segments slender, bearing numerous long hairs. *Terminalia* (fig. 77). Lobes of ninth tergite (IXT-L) slightly elevated, each bearing several long weak setae. Tenth sternite (X-S) heavily sclerotized apically, with 2 or 3 strong recurved teeth at apex. Phallosome (Ph) short, bulbous, open ventrally, closed dorsally beyond middle, rounded at apex. Claspette absent. Basistyle (Bs) about two and a half times as long as mid-width, bearing numerous long setae on outer aspect and with short setae on inner aspect; basal lobe (B-L) conical, bearing four or five stout spines apically and numerous smaller setae subapically; apical lobe absent. Dististyle (Ds) two-thirds as long as basistyle, glabrous, a few minute papillated setae sometimes present; claw (Ds-C) short, stout, bifurcate.

LARVA (fig. 78). Antenna nearly as long as the head, spiculate; antennal tuft large, multiple, barbed, inserted in a notch at outer fourth of shaft and extending much beyond tip. Head hairs: Postclypeal 4 short, single; upper frontal 5 multiple, nearly half as long as hair 6; lower frontal 6 long, single, sparsely barbed; preantennal 7 multiple, barbed; sutural 8 and transsutural 9 short, double, or triple; supraorbital 10 single. Prothoracic hairs: 1–2 very long, single; 3 long, double or triple; 4 short, usually double or triple; 5 very long, single;

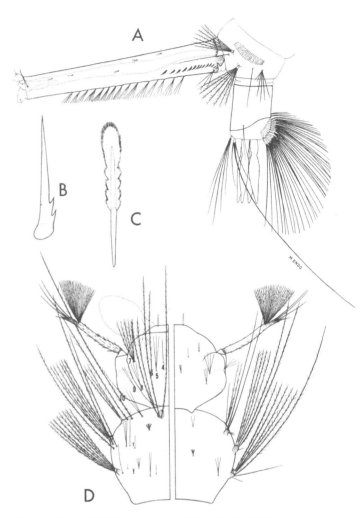

Fig. 78. Larva of *Culiseta melanura* (Coquillett). *A*, Terminal segments. *B*, Pecten tooth. *C*, Comb scale. *D*, Head and thorax (dorsal and ventral).

6 long, single; 7 medium, usually double or triple. Lateral abdominal hair 6 multiple on segments I and II, double on III to V, usually single on VI. Comb of eighth segment of about 25 barlike scales in a single row; individual scale long, slender, pointed basally, rounded, and fringed apically with spinules. Siphonal index about 6.0 to 7.0; pecten of numerous teeth on basal third of siphon; a small 2- or 3-branched siphonal tuft at base of siphon; a median row of about ten to sixteen short multiple ventral tufts beginning within the pecten and extending to outer fourth of siphon; a row of five or six minute 2-branched (rarely 3-branched) dorsolateral tufts present. Anal segment completely ringed by the saddle;

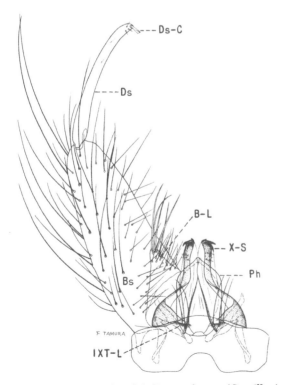

Fig. 77. Male terminalia of *Culiseta melanura* (Coquillett).

lateral hair single or double, usually a little shorter than the saddle; dorsal brush bilaterally consisting of a lower caudal tuft of 3 hairs (1 long and 2 short) and a shorter multiple upper caudal tuft of about 5 to 8 subequal hairs; ventral brush well developed, with one or two precratal tufts; gills about as long or longer than the saddle, pointed.

DISTRIBUTION. Eastern and central United States, west to Colorado and Texas. *United States:* Alabama (135, 419); Arkansas (213, 375); Colorado (343); Delaware (120, 167); District of Columbia (318); Florida (135, 419); Georgia (135, 213); Iowa (14, 531); Kentucky (570); Louisiana (419, 759); Maine (45, 270); Maryland (64, 161); Massachusetts (270); Michigan (391); Minnesota (14); Mississippi (135, 546); Missouri (14, 122); Nebraska (531); New Hampshire (66, 213); New Jersey (213, 350); New York (31); North Carolina (135, 646); Ohio (731a); Oklahoma (628); Pennsylvania (298, 756); Rhode Island (35, 270); South Carolina (135, 742); Tennessee (135, 658); Texas (532); Virginia (176, 213); Wisconsin (171).

BIONOMICS. The larvae are found more frequently in small permanent bodies of water, particularly in swamps. Matheson (492) states that the species overwinters as larvae under the ice. Very little is known of the adult habits of this species except that the adults are frequently captured in light traps in the southern United States and to a lesser extent in diurnal resting stations.

MEDICAL IMPORTANCE. Chamberlain *et al.* (147) recovered the virus of eastern equine encephalitis from wild-caught female specimens of *C. melanura* which were captured in a light trap in a swamp near Ponchatoula, Louisiana.

Tribe Culicini

Genus ORTHOPODOMYIA Theobald

Orthopodomyia Theobald, 1904, Entomologist, *37*: 236.

Bancroftia Lutz, 1904, *in* Bourroul, Mosq. do Brasil, p. 59.

Pneumaculex Dyar, 1905, Proc. Ent. Soc. Wash., 7:45.

Thomasina Newstead and Carter, 1911, Ann. Trop. Med., 4:553.

ADULT. Medium-size mosquito, usually with conspicuous ornamentation. Proboscis long. palpi of male about as long as the proboscis, about one-third as long as the proboscis in female. Antennae rather long in both sexes, plumose in male. Posterior pronotal bristles present; spiracular and postspiracular bristles absent. Abdomen of female blunt-tipped, eighth segment not retractile. Squama fringed. Fork of wing vein 2 much longer than its petiole. *Male terminalia.* Apical margin of eighth tergite with a rounded median part. Tenth sternite with several teeth at tip. Claspette absent. Basistyle with small basal lobe, armed with spines and setae; apical lobe absent. Dististyle long, simple, with terminal spine.

PUPA. Respiratory trumpets short, widening gradually from the base, opening large. Dendritic tufts on abdominal segment I large; lateral tufts on segments VII and VIII rather large and multiple. Paddles with smooth margin, and a small hair at tip of midrib.

LARVA. The body is pink, reddish, or often greenish, depending upon the food. Antenna usually tapered from near base, with multiple tuft inserted before middle. Siphon without a pecten; a single pair of siphonal tufts present, inserted a little before or near middle of siphon. Comb of eighth segment usually consisting of two rows of thornlike scales, those of the posterior row much longer. Anal segment completely ringed by the dorsal saddle in fourth instar larvae.

EGGS. The eggs of several species are brown or black and have prominent longitudinal flanges and a coarse pattern of irregular polygons. According to Howard, Dyar, and Knab (382), the eggs of the best-known species are laid singly on the sides of the treehole at the edge of the water line. Each egg is covered with a gelatinous brown wrinkled membrane re-sembling a veil. They hatch in two or three days and the little larvae descend into the water.

BIONOMICS. The larvae are found mostly in treeholes, cut or broken bamboo sections, leaf bases of *Bromeliaceae*, and occasionally in artificial containers. The species are sylvan and are not known to bite man.

CLASSIFICATION. The genus is represented in this region by three species: *O. alba, O. californica,* and *O. signifera.* Only minor differences have been observed in the adult males and females of these three species, but they can be readily distinguished in the larval stage. Edwards (251) divides the genus into two groups, based on ornamentation of the adults: Group A (*Orthopodomyia*) and Group B (*Bancroftia*). The three species occurring in this region belong to the latter group.

DISTRIBUTION. The genus is represented by a few species occurring in North and South America, Europe, and the Orient; absent from Africa and Australia.

KEY TO THE SPECIES

LARVAE (FOURTH INSTAR)

1. Siphonal tuft 2- to 4-branched and less than three-fourths as long as that part of the siphon beyond point of insertion of the tuft; abdominal segment VIII without a sclerotized plate; pentad hair 3 of eighth abdominal segment small, less than half as long as the saddle of the anal segment .... *alba* Baker, p. 99

   Siphonal tuft with more than four branches and more than three-fourths as long as that part of the siphon beyond point of insertion of the tuft; abdominal segment VIII with a large dorsal sclerotized plate; pentad hair 3 of eighth abdominal segment large, usually about as long as the saddle of the anal segment 2

2. Siphonal tuft about as long as that part of the siphon beyond point of insertion of the tuft .... *signifera* (Coquillett), p. 102

   Siphonal tuft near twice the length of that part of the siphon beyond the point of insertion of the tuft .... *californica* Bohart, p. 100

Orthopodomyia alba Baker

Orthopodomyia alba Baker, 1936, Proc. Ent. Soc. Wash., *38*:1.

ADULT FEMALE. The adult female appears to be indistinguishable from that of *O. signifera* (pl. 28) and is very similar to that of *O. cali-*

fornica. The bands on hind tarsal segments 1–2 and 2–3 are about as broad as in *O. signifera*, but they are more evenly placed on the joints.

ADULT MALE. Coloration similar to that of the female. *Terminalia.* The male terminalia of this species appears to be indistinguishable from that of *O. signifera* (fig. 81) and *O. californica.*

LARVA (fig. 79). Antenna less than half as long as the head, swollen basally, and tapered apically, smooth; antennal tuft multiple, short, barbed, inserted near basal fourth of shaft. Head hairs: Postclypeal 4 rather large, multiple, barbed; upper frontal 5 and lower fron-

tal 6 large, multiple, barbed; preantennal 7 rather large, multiple, barbed; sutural 8 long, single; transsutural 9 multiple, barbed, about as strong as preantennal 7; supraorbital 10 single or double, lightly barbed. Lateral abdominal hair 6 long, usually double or triple (occasionally 4- or 5-branched) on segments I and II. Abdominal segment VI–VIII without sclerotized dorsal plates. Comb of eighth segment of two rows of scales, the anterior row with about 12 to 17 short scales and the posterior row with about 8 to 12 long scales; individual scale long, pointed, and fringed on basal half with small spines, small scales on ventral part of anterior row often fringed apically with subequal spines. Pentad hair 3 of eighth abdominal segment small, barbed, 2- to 5-branched. Siphonal index about 2.0 to 2.5; pecten absent; siphonal tuft 2- to 4-branched, barbed, inserted near basal third of siphon, usually about half as long as that part of siphon beyond insertion of tuft. Anal segment with the saddle variable (in some specimens it fades ventrally, not meeting on the midventral line, whereas in others it appears to ring the segment); a small transverse linear sclerotized area near base of the segment; lateral hair 2- to 4-branched, shorter than the saddle; dorsal brush bilaterally consisting of a long lower caudal hair and a shorter multiple upper caudal tuft; ventral brush rather large, confined to the barred area; gills shorter than the saddle, the dorsal pair longer than the ventral pair, blunt.

DISTRIBUTION. Eastern and central United States. The records given here are based on larval collections. *United States:* Alabama (400, 657); Illinois (615); Kentucky (400, 423); Louisiana (341); Mississippi (510, 514); Missouri (324); New Jersey (448a); New York (16, 31); North Carolina (135, 646); Texas (92, 755); Virginia (178, 400).

BIONOMICS. The larvae are found in water-filled rot cavities in trees and occasionally in artificial containers where they are often associated with the larvae of *O. signifera*. They have been reported as occurring in rot cavities in elm, silver maple, pecan, sweet gum, and Texas oak. Wilkins and Breland (755) say that there seems to be a definite tendency for the mosquito to deposit its eggs in rot cavities

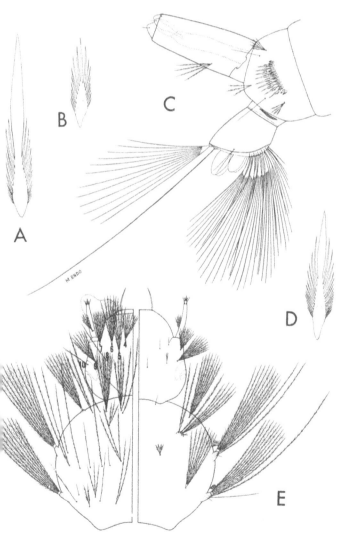

Fig. 79. Larva of *Orthopodomyia alba* Baker. *A, B,* and *D,* Comb scales. *C,* Terminal segments. *E,* Head and thorax (dorsal and ventral).

that have relatively small external openings, which they believe may limit the occurrence of the species in a given area.

Very little information is available on the feeding habits of the adults in nature. Wilkins and Breland (755) observed that reared caged adults sought out dark corners of the cage, avoiding the light. According to Baker (16) and Matheson (492), the species overwinters in the larval stage in the north, but Breland is of the opinion that it overwinters in the adult stage in Texas. Baker (16) found that second and third instar larvae of *O. alba* survived after having been frozen in solid ice for a week. Wilkins and Breland (755) also found the larvae to be quite resistant to low temperatures in Texas as long as the insects were not completely frozen.

Orthopodomyia californica Bohart

Orthopodomyia californica Bohart, 1950, Ann. Ent. Soc. Amer., *43*:399.

ADULT FEMALE. The female of this species is very similar to that of *O. signifera* (pl. 28), except for the following: Specimens of *O. signifera* generally have a larger white ring at the joints between hind tarsal segments 1–2 and 2–3 with the ring covering a greater part of 1 than of 2, and 2 than of 3 in both sexes. In *O. californica* the hind tarsal ring usually covers an equal part of segments 1 and 2, and 2 and 3 in the female. The male does not appear to be so definite in this respect.

ADULT MALE. Coloration similar to that of the female. *Terminalia.* The male terminalia of this species appears to be indistinguishable from that of *O. signifera* (fig. 81).

LARVA (fig. 80). Antenna about half as long as the head, swollen basally and tapered apically, smooth; antennal tuft multiple, barbed, inserted near basal third of shaft, not reaching tip. Head hairs: Postclypeal 4 short, multiple; upper frontal 5 and lower frontal 6 large, multiple, barbed; preantennal 7 medium-sized, multiple, barbed; sutural 8 long, single; transsutural 9 and supraorbital 10 single or branched distally. Lateral abdominal hair 6 short and multiple on segments I and II. A large sclerotized dorsal plate present on seg-

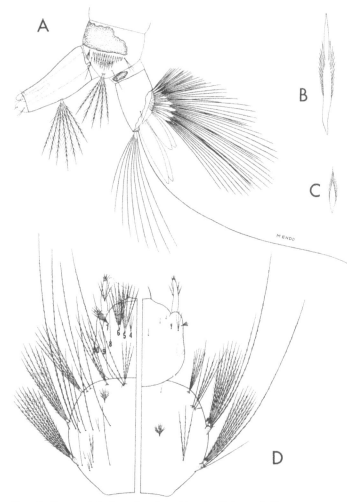

Fig. 80. Larva of *Orthopodomyia californica* Bohart. *A,* Terminal segments. *B,* and *C,* Comb scales. *D,* Head and thorax (dorsal and ventral).

ment VIII, often present on VI and VII. Comb of eighth segment consisting of two rows of scales, the anterior row with about 15 to 21 short scales and the posterior row with about 7 to 9 long scales; individual scale long, pointed and fringed on basal half with small spines, small scales on ventral part of anterior row are fringed apically with subequal spines; pentad hair 3 of eighth abdominal segment large, multiple, barbed. Siphonal index about 2.5 to 3.0; pecten absent; siphonal tuft 5- to 9-branched, barbed, inserted near middle of siphon, individual hairs varying from about 1.7 to 2.3 times as long as that part of the siphon beyond point of insertion of the tuft. Anal segment completely ringed by the saddle; a small linear sclerotic plate at base of segment; lateral hair single or double, shorter than the saddle; dorsal brush bilaterally consisting of a long lower caudal hair and a shorter multiple upper caudal tuft; ventral brush rather large, confined to the barred area; gills usually as long or longer than the saddle, the dorsal pair longer than the ventral pair, blunt.

DISTRIBUTION. Known only from California. This species was earlier reported from Solano County as "*Culex* (?) *signifer*" by Ludlow (459) and from Riverside and San Bernardino counties as *Orthopodomyia signifera* by Reeves (584). Freeborn and Bohart (287) list known records of *O. californica* from six counties in California.

BIONOMICS. Reeves (584) found the larvae during February at Riverside in rot cavities in cottonwood trees where they were associated with the larvae of *Aedes varipalpus*. Collections reported by Reeves (584), Bohart (72), and Freeborn and Bohart (287) indicate that breeding may be continuous during the warmer months. Bohart (72) expresses the opinion that the species overwinters in both the egg stage and as fourth instar larvae. The larvae develop slowly. Very little is known about the habits of the adults.

Orthopodomyia signifera (Coquillett)

Culex signifer Coquillett, 1896, Can. Ent., *28*:43.

Mansonia waverleyi Grabham, 1907, Can. Ent., *39*:25.

ADULT FEMALE (pl. 28). Medium-sized species. *Head:* Proboscis long, dark, with numerous white scales forming narrow longitudinal lines dorsally. Palpi about one-third as long as the proboscis, dark-scaled, with a white-scaled line dorsally. Occiput with rather narrow white scales and long dark erect forked scales dorsally; eye margins and lateral parts of occiput white-scaled. Tori dark; torus and first four or five flagellar segments of antenna white-scaled on inner surface. *Thorax:* Integument of scutum dark brown to black; scutum clothed with numerous long dark setae and small reddish-brown scales, except for three paired narrow longitudinal lines of silver-white scales. Scutellum with long narrow silver-

white scales and dark setae. Pleura with patches and lines of silver-white scales. *Abdomen:* First tergite with a median patch of broad pale scales; second tergite with the pale basal band wide and often projecting medially nearly to the apical margin of the segment; remaining tergites dark bronze-brown-scaled dorsally, with more or less distinct narrow basal bands of pale scales. Sternites with mixed dark and pale scales, the dark ones predominant toward apex. *Legs:* Femora with posterior surface pale-scaled; anterior surface brown, with white scales intermixed. Pale knee spots present. Tibiae dark, speckled with white, tipped with white apically. Front tarsi with segments 1 and 2 faintly marked with pale scales at joints or all segments dark brown. Middle tarsi with base and apex of segment 1 and base of segment 2 narrowly ringed with white, remaining segments usually entirely dark brown. Hind tarsi with all segments broadly ringed with white apically and basally, rings broadest on first and second segments, narrower on third and fourth, segment 5 entirely white on one side (rings on segments 1–2 and 2–3 covering a greater part of segment 1 than of 2, and 2 than of 3). *Wing:* Length about 4.0 mm. Scales broad, ovate, intermixed dark brown and white; white scales concentrated on base of vein 1, on basal half of vein 6, and at point of cross veins.

ADULT MALE. Coloration similar to that of the female. *Terminalia* (fig. 81). Eighth tergite (VIII-T) with a median posterior tonguelike projection overlapping the emarginate margin of the ninth tergite (IX-T); lobes of ninth tergite represented by a few setae. Tenth sternite (X-S) heavily sclerotized and toothed apically. Phallosome (Ph) open ventrally, closed dorsally toward apex by a membranous connection; each plate broad on basal half, narrowed distally, and with several short heavily sclerotized subapical teeth. Claspette absent. Basistyle (Bs) about three times as long as mid-width, clothed with scales and long and short setae, one or two large curved spines arising from inner face of basistyle near middle; basal lobe (B-L) conical, appressed to the basistyle, bearing a few large apical spines and numerous smaller subapical setae; apical lobe absent. Dististyle (Ds) about three-fifths as long as

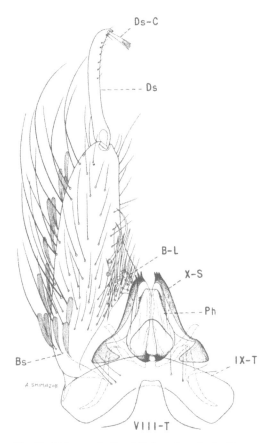

Fig. 81. Male terminalia of *Orthopodomyia signifera*
(Coquillett).

basistyle, bluntly pointed at apex, with several small papillated setae on distal half; claw (Ds-C) short, split distally into numerous comblike teeth.

LARVA (fig. 82). Antenna about half as long as the head, swollen basally and tapered apically, smooth; antennal tuft multiple, barbed, inserted near basal third of shaft, not reaching tip. Head hairs: Postclypeal 4 short, multiple, barbed; upper frontal 5 and lower frontal 6 large, multiple, barbed; preantennal 7 medium-sized, multiple, barbed; sutural 8 long, single; transsutural 9 and supraorbital 10 single or branched distally. Lateral abdominal hair 6 short, multiple on segments I and II. A large sclerotized dorsal plate present on segment VIII, often present on VI and VII. Comb of eighth segment of two rows of scales, the anterior row with about 17 to 23 short scales, and the posterior row with about 6 to 10 long scales; individual scale long, pointed, and

fringed on basal half with small spines, small scales on ventral part of anterior row are fringed apically with subequal spines; pentad hair 3 of eighth abdominal segment large, 5- to 12-branched, barbed. Siphonal index about 3.0 to 3.5; pecten absent; siphonal tuft 5- to 12-branched, barbed, inserted a little beyond basal third of siphon, individual hairs varying from about three-fourths as long to a little longer than that part of the siphon beyond point of insertion of the tuft. Anal segment completely ringed by the saddle; a small linear sclerotic plate at base of segment; lateral hair single, often feathered at tip, nearly as long as the saddle; dorsal brush bilaterally consisting of a long lower caudal hair and a shorter multiple upper caudal tuft; ventral brush rather large, confined to the barred area; gills usually a little shorter than the saddle, the dorsal pair longer than the ventral pair, blunt.

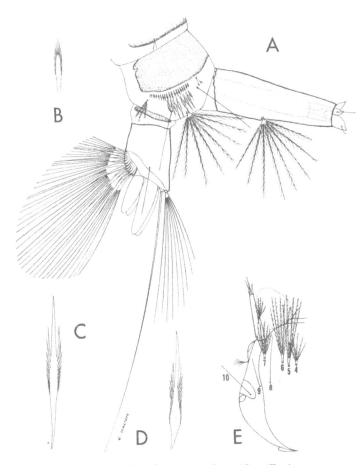

Fig. 82. Larva of *Orthopodomyia signifera* (Coquillett). *A*, Terminal segments. *B*, *C*, and *D*, Comb scales. *E*, Head.

DISTRIBUTION. Southern United States, north to Massachusetts and west to North Dakota and New Mexico. It has been found also in Jamaica (361) and Puerto Rico (561). *United States:* Alabama (135, 419); Arkansas (88, 121); Connecticut (494); Delaware (120, 167); District of Columbia (213, 318); Florida (135, 419); Georgia (135, 213); Illinois (615); Indiana (152, 345); Iowa (626); Kansas (360, 531); Kentucky (570); Louisiana (419, 759); Maryland (64, 161); Massachusetts (701); Mississippi (135, 546); Missouri (213, 324); Nebraska (680); New Jersey (213, 350); New Mexico (273); New York (31); North Carolina (135, 646); North Dakota (14); Ohio (485); Oklahoma (213, 628); Pennsylvania (298, 756); Rhode Island (436); South Carolina (135, 742); Tennessee (135, 658); Texas (213, 463); Virginia (176, 178).

BIONOMICS. The larvae occur in water-filled rot cavities in trees of many types and occasionally in artificial, particularly wooden, containers. Howard, Dyar, and Knab (382) claim that the eggs are laid singly or in twos or threes close to the water line on the side of the water container, and that after hatching, the small larvae descend into the water. However, Horsfall (375) states that since oviposition takes place on the water, the rise and fall of water in the treeholes does not affect the abundance of the larvae. The winter is passed in the larval stage, at least in the southern United States. Baker (16) reports that the larvae of this species are killed with complete freezing of the water in their habitats in New York, though the larvae of *Orthopodomyia alba* survive freezing. The larvae are slow in developing, and two or more broods are frequently found together in a single habitat.

The adults are often found resting on tree trunks near the larval habitat or inside the treehole in which the immature stages develop. Most workers state that this mosquito has never been known to bite man, but it is known to take avian blood.

Tribe Culicini

Genus MANSONIA Blanchard

Taeniorhynchus Lynch Arribálzaga, 1891, Rev. Mus. de La Plata, *1*:374.

Panoplites Theobald (not Gould), 1901, Mon. Culic., 2:173.

Mansonia Blanchard, 1901, C.R. Soc. Biol., *53*:1045.

Coquillettidia Dyar, 1905, Proc. Ent. Soc. Wash., 7:45.

Mansonioides Theobald, 1907, Mon. Culic., 4:498.

Rhynchotaenia Brethes, 1911, Ann. Mus. Nac. Buenos Aires, *13*:470.

Pseudotaeniorhynchus Theobald, 1911, Novae Culicidae, p. 19.

ADULT. Proboscis moderate, not swollen at tip. Palpi of male usually as long or longer than the proboscis, about one-fourth to a little more than one-third as long as the proboscis in the female. Antenna of male plumose. Posterior pronotal bristles present; spiraculars absent; postspiraculars present or absent. Postnotum bare. Pleura usually with small patches of scales. Pulvilli absent. Wings with many or all the scales broad; squama fringed. *Male terminalia.* Variable in structure. Basal lobe either appressed to basistyle or long and slender, usually bearing blunt spines or modified filaments. Apical lobe absent. Dististyle variable in shape.

PUPA. Respiratory trumpets long, the tip of each in the form of a chitinized spine for piercing the roots of plants. Abdominal segment I without dendritic tufts; remaining segments either with some stout single setae or all setae minute. Paddles long, narrow, tip emarginate, without a terminal seta, margins with small serrations.

LARVA. Antenna long, with a large multiple tuft inserted before middle and two long filaments inserted before tip. Anal segment long, completely ringed by the dorsal saddle. Comb of eighth segment of a few scales in a single row. Siphon short, with a pair of siphonal tufts inserted near middle; pecten absent. Siphon attenuated and modified to form a sawlike apparatus for piercing underwater stems and roots of aquatic plants.

EGGS. Each elongate, usually with a long process. The eggs of the subgenera *Coquillettidia* and *Rhynchotaenia* are laid in masses on the surface of the water; those of the subgenera *Mansonia* and *Mansonioides* are laid in small groups on the undersurface of leaves of floating water plants.

BIONOMICS. Shortly after hatching, the larvae attach themselves by means of the modified siphon to roots or stems of aquatic plants from which they obtain oxygen. The pupae also carry on respiration in a similar manner, piercing the plants with their modified trumpets, rising to the surface when ready for emergence. According to Edwards (251), holarctic species hibernate as larvae in the mud. The females are troublesome biters.

CLASSIFICATION. Satisfactory characters for separating the species are found in the adult female, male terminalia, and larva. Edwards (251) recognized four subgenera: *Mansonia, Rhynchotaenia, Coquillettidia,* and *Mansonioides.* Representatives of two subgenera, *Mansonia* and *Coquillettidia,* are found in this region.

DISTRIBUTION. Practically world-wide, except for the Arctic and Antarctic regions.

KEYS TO THE SPECIES

ADULT FEMALES

1. Postspiracular bristles absent; segment 1 of hind tarsus with a median pale ring
   perturbans (Walker), p. 109

   Postspiracular bristles present; segment 1 of hind tarsus without a median pale ring.....................2

2. Palpus a little more than one-third as long as proboscis, with fourth segment about twice as long as third; spines of eighth tergite clumped posteriorly (visible only when dissected)
   titillans (Walker), p. 107

   Palpus less than one-third as long as proboscis, with fourth segment about one and a half times as long as third; spines of eighth tergite more or less uniformly spaced
   indubitans Dyar and Shannon, p. 105

MALE TERMINALIA

1. Dististyle without a toothlike branch on inner margin.........................*perturbans* (Walker), p. 109

   Dististyle with a toothlike branch on inner margin 2

2. Dististyle not convoluted..*titillans* (Walker), p. 107
   Dististyle somewhat convoluted
   indubitans Dyar and Shannon, p. 105

1. Lateral spine of maxilla smooth; anal segment either without precratal tufts or with one or two small tufts piercing the saddle
                perturbans (Walker), p. 000
   Lateral spine of maxilla strongly serrate on one side; anal segment with about four precratal tufts piercing the saddle..................................2
2. Comb scale long, slender, thornlike, with a few small spinules on basal part
                titillans (Walker), p. 000
   Comb scale rather broad and fringed apically with several stout subequal spines
        indubitans Dyar and Shannon, p. 000

Mansonia (Mansonia) indubitans
Dyar and Shannon

Mansonia indubitans Dyar and Shannon, 1925, Jour. Wash. Acad. Sci., *15*:41.

ADULT FEMALE (pl. 29). Medium-sized species. *Head:* Proboscis dark, speckled with pale scales and with a narrow white ring near apical third; palpi a little less than one-third as long as the proboscis (segment 4 about one and a half times as long as segment 3), dark-scaled, speckled with white, with small terminal segment white. Occiput with pale lanceolate scales and numerous dark-brown or black erect forked scales. Tori brown on outer surface, darker on inner surface; a few white scales on dorsal and inner surfaces. *Thorax:* Integument of scutum mottled chocolate brown and dark brown; scutum clothed with dark-brown lanceolate scales intermixed with golden lanceolate scales; the prescutellar space surrounded by golden scales. Scutellum with golden scales and brown setae on the lobes. Postspiracular bristles present. *Abdomen:* First tergite with median area pale-scaled; remaining tergites predominantly dark-scaled, a few scattered whitish scales laterally and with white scales apically. Sternites with intermixed white and brown scales, the white scales more numerous apically. Eighth segment blunt, largely retracted within the seventh; eighth tergite (fig. 83, *B*) with numerous short stout spines more or less regularly spaced and arranged in a curved posterior row and a somewhat irregular anterior row. Apical margin of seventh tergite without a close-set row of small spines. *Legs:* Femora and tibiae dark brown-scaled, speckled with pale scales; posterior surface of middle and

hind femora predominantly pale-scaled. Tarsal segments 1 to 4 of front and middle legs each with a narrow basal white band; all segments of hind tarsus with white basal bands. *Wing:* Length about 3.5 mm. Scales mixed brown and white, very broad.

ADULT MALE. Coloration similar to that of female. *Terminalia* (fig. 83). Ninth tergite (IX-T) with lobes indistinct. Tenth sternite (X-S) heavily sclerotized apically, dorsal arm long, curved dorsally. Phallosome (Ph) open ventrally, closed dorsally near apex, broadly rounded at apex. Claspette absent. Basistyle (Bs) about three times as long as mid-width, clothed with scales and long and short setae, a group of strong setae at apex and several very long strong setae near base; basal lobe (B-L) long and slender, extending a little beyond

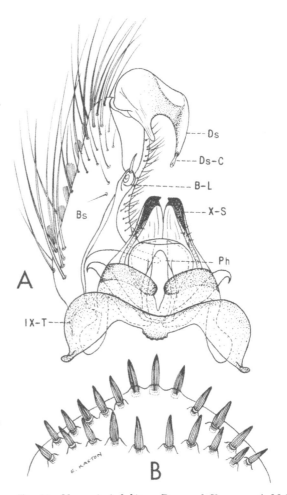

Fig. 83. *Mansonia indubitans* Dyar and Shannon. *A,* Male terminalia. *B,* Eighth abdominal tergite of female.

apical third of basistyle, basal two-thirds of lobe narrow and stemlike, fused to the basistyle, apical third swollen and bearing a short stout bluntly pointed terminal spine; apical lobe absent. Dististyle (Ds) about two-fifths as long as basistyle, basal third broad and bearing a slender inner branch, remainder of dististyle abruptly narrowed beyond origin of inner branch, curved, slightly convoluted, tapered distally; claw (Ds-C) short, stout.

LARVA (fig. 84). Head much broader than long. Antenna nearly twice as long as the head, whiplike, sparsely spiculate on basal half; antennal tuft large, multiple, barbed, arising from a notch on basal third; a pair of long smooth setae inserted a little beyond middle of shaft and extending a little beyond tip; apex of antenna with a short spine, a seta, and a membranous papilla. Head hairs: Postclypeal 4, upper frontal 5, lower frontal 6, and preantennal 7 small, multiple. Lateral spine of maxilla strongly serrate on one side. Inner preclypeal spines much shorter than the distance between their bases. Prothoracic hairs: 1 long, usually 3- or 4-branched; 2 short, single; 3 short, usually multiple; 4 short, multiple; 5 long, 3- or 4-branched; 6 long, single; 7 long, double or triple. Lateral abdominal hair 6 single on segments I to VI. Comb of eighth segment of 6 to 9 scales in a single row; individual scale short and fringed apically with subequal spines. Pentad hair 3 of eighth abdominal segment long, 3- or 4-branched, weakly barbed. Siphon short, strongly attenuated, and heavily sclerotized beyond middle; attenuated part bearing sawlike projections dorsally and stout hooks apically; a long stout recurved dorsal spine, a single stout dorsolateral hair, and a small multiple siphonal tuft arising before the heavily sclerotized part; pecten absent. Anal segment much longer than wide, ringed by the saddle; saddle spiculate, the spicules are coarser toward the dorsoapical margin; lateral hair represented by a small multiple tuft, inserted well before the apical margin of the saddle; dorsal brush bilaterally consisting of long multiple lower and upper caudal tufts; ventral brush well developed, with four (occasionally five) precratal tufts piercing the saddle; gills shorter than the saddle, pointed.

DISTRIBUTION. Known to occur in the southeastern United States, the Antilles, Mexico, Panama, and northern South America (132, 146, 554, 558). *United States:* Florida (146, 516, 554, 558); Georgia (516).

BIONOMICS. The larvae and pupae attach themselves to the roots of water lettuce *(Pistia)* and remain attached until the adult is ready to emerge. In an area where larval populations of *Mansonia* were heavy at Boca Raton, Florida, during May, 1945, the senior author ob-

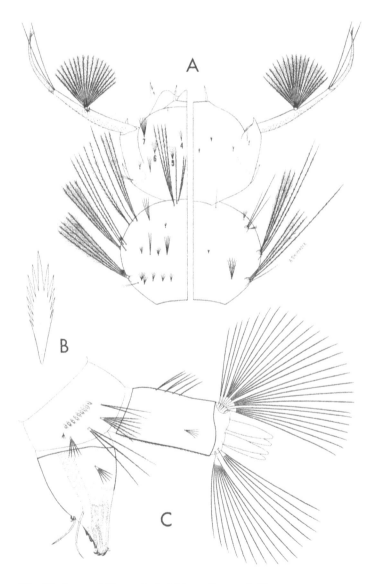

Fig. 84. Larva of *Mansonia indubitans* Dyar and Shannon. *A,* Head and thorax (dorsal and ventral). *B,* Comb scale. *C,* Terminal segments.

served that this species far outnumbered *M. titillans*; however, both species were collected from the same water-lettuce plants. The distribution and relative abundance of *M. indubitans* in relation to populations of *M. titillans* are not generally known at this time, since the procedure for separating the two species is a tedious one.

The females have been taken on several occasions, biting man and in animal-baited traps. Both the males and females are readily attracted to artificial light.

Mansonia (Mansonia) titillans (Walker)

Culex titillans Walker, 1848, List Dipt. Brit. Mus., 1:5.

ADULT FEMALE (pl. 30). Medium-sized species. *Head:* Proboscis dark, speckled with pale scales, and with a narrow white ring near apical third; palpi a little more than one-third as long as the proboscis (segment 4 twice as long as segment 3), dark-scaled, speckled with white, with small terminal segment white. Occiput with pale lanceolate scales and numerous dark-brown or black erect forked scales. Tori brown, darker on inner surface, a few grayish-white scales on dorsal and inner surfaces. *Thorax:* Scutum with integument mottled chocolate and dark brown; scutum clothed with dark-brown lanceolate scales intermixed with light golden-brown lanceolate scales; the prescutellar space surrounded by golden-brown scales. Scutellum with light golden scales and dark-brown setae on the lobes. Pleura with lightly scaled areas of grayish-white scales. Postspiracular bristles present. *Abdomen:* First tergite with median area pale-scaled; remaining tergites predominantly dark-scaled, with scattered yellow scales laterally, and with few to many yellow and white scales apically. Sternites with intermixed white, yellow, and brown scales. Eighth segment blunt, largely retracted within the seventh; eighth tergite (fig. 85, *B*) with many short stout spines in a curved posterior row and an irregular anterior row; the posterior row with a clump of six to nine closely set spines medially. Apical margin of the seventh tergite with a close-set row of short stout pointed spines. *Legs:* Femora and tibiae dark-brown-scaled, speckled with pale scales;

posterior surface of middle and hind femora predominantly pale-scaled. Tarsal segments 1 to 4 of front and middle legs each with a narrow basal white band; all segments of hind tarsus with white basal bands. *Wing:* Length about 4.0 mm. Scales mixed brown and white, very broad.

ADULT MALE. Coloration similar to that of the female. *Terminalia* (fig. 85). Ninth tergite (IX-T) with lobes indistinct. Tenth sternite (X-S) sclerotized, dorsal arm curved dorsally. Phallosome (Ph) large, open ventrally, closed dorsally near apex, broadly rounded at apex. Claspette absent. Basistyle (Bs) nearly three times as long as mid-width, clothed with scales and long and short setae, a group of strong setae at apex and several very long strong setae near base; basal lobe (B-L) long and slender, extending to apical fourth of basi-

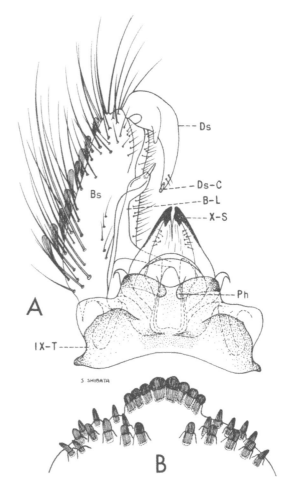

Fig. 85. *Mansonia titillans* (Walker). *A,* Male terminalia. *B,* Eighth abdominal tergite of female.

style, basal two-thirds of lobe narrow and stemlike, apical third swollen and bearing a short stout bluntly pointed terminal spine; apical lobe absent. Dististyle (Ds) nearly half as long as basistyle, broad on basal part, evenly curved and tapered, not convoluted, bearing a slender inner branch arising a little before middle; claw (Ds-C) short, stout.

LARVA (fig. 86). Head much broader than long. Antenna about twice as long as the head, whiplike, sparsely spined on basal half;

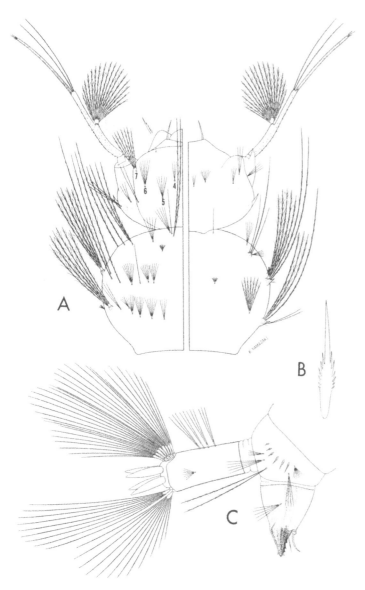

A

B

C

Fig. 86. Larva of *Mansonia titillans* (Walker). *A*, Head and thorax (dorsal and ventral). *B*, Comb scale. *C*, Terminal segments.

antennal tuft large, multiple, barbed, arising from a notch on basal third of shaft; a pair of long stout setae arising near middle of shaft and extending to tip; apex of antenna with a short spine, a seta, and a membranous papilla. Head hairs: Postclypeal 4, upper frontal 5, and lower frontal 6 small, multiple; preantennal 7 multiple, barbed. Lateral spine of maxilla strongly serrate on one side. Inner preclypeal spines stout, separated by more than the length of one spine. Prothoracic hairs: 1 long, usually double or triple; 2 short, single; 3 short, multiple; 4 short, double or multiple; 5 long, 2- to 4-branched; 6 long, single; 7 long, single or double. Lateral abdominal hair 6 single on segments I to VI. Comb of eighth segment of 6 to 8 scales in a single row, the scales becoming progressively much smaller ventrally; the dorsal scales long, spinelike, each with small lateral spinules on basal part. Pentad hair 3 on eighth abdominal segment long, double or triple, barbed. Siphon short, strongly attenuated and heavily sclerotized beyond middle; the attenuated part bearing sawlike projections dorsally and stout hooks apically; a long stout recurved dorsal spine, a single stout dorsolateral hair, and a multiple siphonal tuft arising before the heavily sclerotized part; pecten absent. Anal segment much longer than wide, ringed by the saddle, saddle spiculate; lateral hair represented by a small multiple tuft, inserted well before the apical margin of the saddle; dorsal brush bilaterally consisting of a long multiple lower caudal tuft and a moderately long multiple upper caudal tuft; ventral brush large, with about four precratal tufts piercing the saddle; gills shorter than the saddle, pointed.

DISTRIBUTION. Southern United States, the Antilles, Mexico, Central America, and South America. *United States:* Florida (146, 516, 554, 558); Texas (248, 463, 558).

BIONOMICS. After hatching, the larvae attach themselves to the submerged roots of aquatic plants from which they obtain oxygen. The pupae also remain attached to the roots of the plants until time for emergence of the adults. Water lettuce *(Pistia)* is claimed to be the principal host plant for the larvae and pupae of the species; however, Eads and Menzies (248) found the larvae constantly associated with water hyacinth *(Eichornia crassipes)* in the lower Rio Grande Valley in Texas where the species constitutes a serious problem. Pratt (558) says he has never found the species associated with *Pistia* in Florida and Jamaica. He collected the pupae from the roots of a *Pontederia*-like plant in Jamaica. The senior author has collected the larvae in small numbers from *Pistia* at Boca Raton, Florida, but has observed that *Pistia* is scarce in many areas in Panama where the species is often extremely abundant; therefore one must conclude that it is occupying some other host plant.

The females are troublesome out-of-doors biters and are known to fly several miles from marshes, ponds, and lakes where their immature stages occur. Both males and females are easily taken in light traps, and the females are readily attracted to animal-baited traps. The eggs and egg-laying habits of this species are described by Dyar (185).

MEDICAL IMPORTANCE. The virus causing Venezuelan equine encephalitis has been recovered from wild-caught *Mansonia titillans* in Trinidad (312), and it is believed that the species may have been an important vector of this disease during an epidemic in Trinidad in 1942–1943. According to Belding (47), this species is known to be a vector of filariasis.

Mansonia (Coquillettidia) perturbans (Walker)

Culex perturbans Walker, 1856, Ins. Saund., Dipt., p. 428.

Culex testaceus Van der Wulp, 1867, Tijds. voor Ent., *10*:128.

Culex ochropus Dyar and Knab, 1907, Jour. N.Y. Ent. Soc., *15*:100.

ADULT FEMALE (pl. 31). Moderately large species. *Head:* Proboscis dark, sprinkled with white scales basally and with a broad median ring of pale scales; palpi about one-fifth as long as the proboscis, dark-scaled, lightly speckled with pale scales. Occiput with pale-golden lanceolate scales and dark erect forked scales, a few pale forked scales on anterior part. Tori light brown on outer surface, darker and with a patch of grayish-white scales on inner surface. *Thorax:* Integument of scutum mottled dark brown and black; scutum clothed

with dark-brown lanceolate scales intermixed with pale-golden lanceolate scales; the golden scales are more numerous anteriorly, laterally, and on the prescutellar space. Scutellum with pale-golden scales and brown setae on the lobes. Pleura with patches of grayish-white scales. Spiracular and postspiracular bristles absent. *Abdomen:* First tergite dark-scaled; remaining tergites dark-scaled, with white or pale-yellow basolateral patches and occasionally with narrow basal segmental bands of pale scales. Venter with intermixed dark and pale scales, the pale scales more numerous on the basal part of the sternites. Eighth segment bluntly rounded, not largely retracted within the seventh; eighth tergite without short stout spines. *Legs:* Femora dark, speckled with pale scales, the apices almost entirely dark-scaled; hind femur with a narrow subapical, more or less distinct ring of pale scales; posterior surface of middle and hind femora predominantly pale-scaled except near apices. Front and middle tibiae dark-scaled, speckled with white, narrowly ringed with white scales at apices; hind tibia dark-scaled, speckled with white, ringed with white scales at outer third and at apex. First tarsal segment of all legs with a narrow white ring basally and a broader white ring a little beyond middle; remaining tarsal segments each with basal half white, apical half dark. *Wing:* Length about 4.0 mm. Scales broad, mixed dark and white, the dark scales predominating. .

ADULT MALE. Coloration similar to that of the female. *Terminalia* (fig. 87). Ninth tergite (IX-T) with the lobes (IXT-L) about as long as wide, sclerotized, each bearing several setae. Tenth sternite (X-S) heavily sclerotized, armed with blunt teeth at apex; dorsal arm (XS-DA) large, curved dorsad to a position just beneath lobe of ninth tergite. Phallosome (Ph) large, sclerotized, about two-thirds as broad as long, open dorsally and ventrally, constricted at middle, bluntly pointed apically; each plate with a longitudinal row of short, dark teeth directed ventrally. Claspette absent. Basistyle (Bs) about one and one half to two times as long as mid-width, clothed with broad scales and long and short setae; basal lobe (B-L) flat, triangular, fused to basistyle, bearing a thick, blunt, dark rod

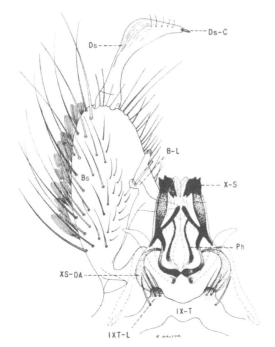

Fig. 87. Male terminalia of *Mansonia perturbans* (Walker).

and a smaller stout spine at apex; apical lobe absent. Dististyle (Ds) curved, about two-thirds as long as basistyle, enlarged at base, constricted and slender before middle, rounded and expanded on outer surface beyond middle, then tapered to a point; claw (Ds-C) short, stout.

LARVA (fig. 88). Head much broader than long. Antenna more than twice as long as the head, whiplike, sparsely spiculate basally; antennal tuft multiple, barbed, arising from a notch on basal third of shaft; a pair of short setae inserted at middle of shaft; a short spine, a seta, and a membranous papilla at tip. Head hairs: Postclypeal 4 small, multiple; upper frontal 5 multiple, smooth, much smaller than hair 6; lower frontal 6 large, multiple, barbed; preantennal 7 large, multiple, barbed. Lateral spine of maxilla smooth. Inner preclypeal spines long, separated by a little more than the length of one spine. Prothoracic hairs: 1–2 long, single; 3 long, multiple; 4 short, multiple; 5 long, single; 6 medium, single; 7 medium, double or triple. Lateral abdominal hair 6 single on segments I to VI. Comb of eighth segment of about 8 to 15 thorn-shaped scales in a single irregular row; individual

scale fringed on basal half with short spinules. Pentad hair 3 of eighth abdominal segment long, double, barbed. Siphon short, strongly attenuated and heavily sclerotized beyond middle; attenuated part of siphon bearing sawlike projections dorsally and stout hooks apically; arising before the heavily sclerotized part of the siphon are a long stout recurved dorsal spine, a single stout dorsolateral hair, and a multiple siphonal tuft; pecten absent. Anal segment much longer than wide, ringed by the saddle; lateral hair represented by a short multiple tuft, inserted well before the apical margin of the saddle; dorsal brush bilaterally consisting of a multiple lower caudal tuft and a shorter multiple upper caudal tuft; ventral brush large, usually restricted to the barred area, but often with one or two minute precratal tufts piercing the saddle; gills shorter than the saddle, pointed.

DISTRIBUTION. Occurs throughout most of the United States and southern Canada. It is also found in Mexico (481). *United States:* Alabama (135, 419); Arkansas (88, 213); California (287); Colorado (343); Connecticut (213, 494); Delaware (120, 167); District of Columbia (213); Florida and Georgia (135, 213); Idaho (672); Illinois (615); Indiana (152, 213); Iowa (213, 625); Kansas (14, 530); Kentucky (134, 570); Louisiana (213, 759); Maine (45, 270); Maryland (64, 161); Massachusetts (213, 270); Michigan (391, 529); Minnesota (533); Mississippi (135, 546); Missouri (213, 324); Montana (213); Nebraska (680); New Hampshire (66, 270); New Jersey (213, 350); New York (31); North Carolina (135, 419); North Dakota (553); Ohio (213, 482); Oklahoma (628); Oregon (672); Pennsylvania (298, 756); Rhode Island (35, 270); South Carolina (135, 742); South Dakota (14); Tennessee (135, 658); Texas (248, 463); Utah (575); Vermont (404, 648); Virginia (176, 178); Washington (67, 672); Wisconsin (171); Wyoming (578). *Canada:* British Columbia (352, 708); Manitoba (212, 466); Nova Scotia (708); Ontario (213, 409); Prince Edward Island (708); Quebec (536, 708); Saskatchewan (598).

BIONOMICS. The eggs are laid on the surface

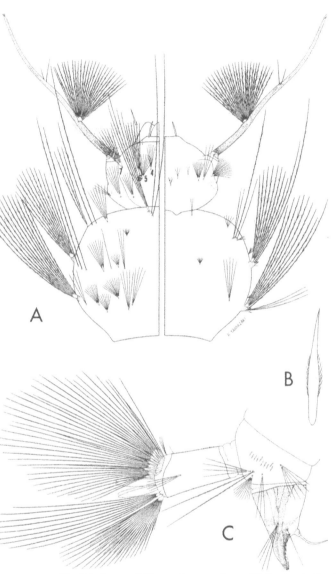

Fig. 88. Larva of *Mansonia perturbans* (Walker). *A*, Head and thorax (dorsal and ventral). *B*, Comb scale. *C*, Terminal segments.

of water in areas of heavy emergent vegetation. After hatching, the small larvae attach themselves with the modified siphon to the roots or submerged stems of plants where they remain throughout development. The pupa also attaches itself to plants by means of the modified respiratory trumpets and remains there until the adult is ready to emerge. The winter is passed as immature or mature larvae, and the adults emerge in the spring and summer.

The larvae are difficult to collect, as they quickly drop off the host plant whenever they are disturbed. McNeel (469) recommends the following procedure for collecting the larvae: Remove the host plants from a small area and lay them aside; scoop up the muck and debris with a large bucket and pour through a strainer. Transfer the material in small quantities from the strainer to clear water in a white porcelain pan and examine carefully for larvae.

The females bite principally at night, apparently being most active during the early part of the night. They occasionally attack man during daylight hours in shady places when their haunts are invaded. The adults are strong fliers and are important pests in communities near shallow lakes which are partly overgrown with emergent aquatic vegetation. The adults are readily attracted to light traps.

MEDICAL IMPORTANCE. Howitt *et al.* (383) recovered a filterable virus, immunologically and antigenically identical with that of eastern equine encephalitis virus, from wild-caught specimens of *M. perturbans* in Georgia.

Genus PSOROPHORA Robineau-Desvoidy

Psorophora Robineau-Desvoidy, 1827, Mem. Soc. d'Hist. Nat., *3*:412.

Janthinosoma Lynch Arribálzaga, 1891. Rev. Mus. de La Plata, *1*:374.

Conchyliastes Howard, 1902, Mosquitoes, p. 155.

Grabhamia Theobald, 1903, Mon. Culic., *3*:243.

Feltidia Dyar, 1905, Proc. Ent. Soc. Wash., *8*;45.

Ceratocystia Dyar and Knab, 1906, Jour. N.Y. Ent. Soc., *14*:178.

Lepidosia Coquillett, 1906, Science, n.s., *23*:314.

ADULT. Spiracular and postspiracular bristles present. Postnotum without setae. Squama fringed. Abdomen of female tapered; eighth segment retractile; cerci long, prominent. *Male terminalia.* Tenth sternite with few or no apical teeth. Phallosome not completely divided, variable in form in the different subgenera. Claspette and dististyle diversified in form. Basal and apical lobes of basistyle absent.

PUPA. Respiratory trumpet variable, long, tubular, slender in some species, moderate and broadly funnel-shaped in others. Abdomen usually quite stout. Paddles large, broadly rounded, with one or two small terminal setae.

LARVA. Anal segment completely ringed by the saddle, the saddle pierced on the mid-ventral line for nearly its entire length by tufts of the ventral brush. Siphon with a single pair of siphonal tufts, never inserted basally, reduced in size in some species. Comb of eighth segment present. Subgenus *Psorophora* has the mouth brushes prehensile, head broad and subquadrate, antennae small, siphon usually rather long, and the pecten teeth produced into hairs. Subgenera *Janthinosoma* and *Grabhamia* have the head and mouth brushes normal, antennae long, siphon moderate or short and usually inflated, pecten teeth broad and normal.

EGGS. Similar to those of *Aedes* but with surface finely spinose. The eggs are laid singly on the ground in depressions where water collects after rains or overflow from streams. In general the eggs are able to withstand long periods of drying.

BIONOMICS. The larvae develop in temporary rain-filled and flood-water pools. The develop-ment of the larvae is usually rapid. The larvae of the subgenus *Psorophora* are predacious on other mosquito larvae which occur in the same pools. The females are vicious biters.

CLASSIFICATION. Three subgenera, *Psorophora*, *Janthinosoma*, and *Grabhamia*, are recognized. All three subgenera are represented in this region. Several female specimens of *Psorophora* (*Janthinosoma*) *mexicana* (Bellardi), a species known to occur in Oaxaca, Mexico, were captured near the Port of Brownsville and the International Airport near Brownsville, Texas, and reported by Joyce (407). This neotropical species is believed to have been introduced to Texas either by ship or by aircraft and is not included in this study.

DISTRIBUTION. Widespread throughout the tropical and temperate regions of North and South America and near-by islands.

KEYS TO THE SPECIES

ADULT FEMALES

1. Wing scales mixed dark and white; hind femur with a more or less distinct narrow subapical ring of white scales (subgenus *Grabhamia*) 2
    Wing scales all dark or with only a few inconspicuous white scales on costa and subcosta; hind femur without a subapical ring of white scales....................................5

2 (1). Segment 1 of hind tarsus largely pale-scaled or with a white ring at middle................3
    Segment 1 of hind tarsus entirely dark-scaled except for a narrow basal pale ring................*pygmaea* (Theob.), p. 133

3 (2). Segment 1 of hind tarsus with white-scaled rings at base and at middle; wings speckled with brown and white scales in no definite pattern
    confinnis (L. Arr.), p. 129
    Segment 1 of hind tarsus largely pale-scaled; wings with definite areas of white and dark scales 4

4 (3). Wing fringe with alternating groups of dark and pale scales; vein 6 white-scaled apically................*signipennis* (Coq.), p. 134
    Wing fringe uniformly dark; vein 6 dark scaled apically......*discolor* (Coq.), p. 131

5 (1). Hind legs, including apical part of femora, with long erect scales, very shaggy; fifth segment of hind tarsus never entirely white; very large species (subgenus *Psorophora*) 6

Hind legs not particularly shaggy, apices of femora without erect scales (when tibiae are somewhat shaggy the fifth segment of hind tarsus is entirely white); medium-sized species (subgenus *Janthinosoma*)..7

6 (5). Scutum with a narrow median longitudinal stripe of golden scales; proboscis yellow-scaled on distal half, dark at tip
                    ciliata (Fabr.), p. 114
Scutum without a median longitudinal stripe of golden scales; proboscis entirely dark-scaled....................*howardii* Coq., p. 116

7 (5). Hind tarsi entirely dark-scaled; abdominal segments with apical submedian triangular patches of golden-yellow scales
                    cyanescens (Coq.), p. 118
Hind tarsi with white on apical segments; abdominal segments with pale-scaling restricted to apicolateral corners or the apical margin....................................8

8 (7). Segment 4 of hind tarsus white-scaled, at least on one side, segment 5 dark...........9
Segments 4 and 5 and often tip of segment 3 of hind tarsus white-scaled................10

9 (8). Scutum with a broad longitudinal median stripe of dark scales, yellowish-white scales laterally........*varipes* (Coq.), p. 127
Scutum clothed with yellowish-white scales, without a median longitudinal stripe of dark scales..*johnstonii* (Grabham), p. 124

10 (8). Scutum clothed with mixed dark-brown and golden-yellow scales in no definite pattern.........................*ferox* (Humb.), p. 120
Scutum with a broad median longitudinal stripe of dark bronze-brown scales, pale-yellow or grayish-white scales laterally..11

11(10). Pale knee spots present; palpus less than one-third as long as the proboscis
                    horrida (D. and K.), p. 122
Pale knee spots absent; palpus a little more than one-third as long as the proboscis
                    longipalpis Roth, p. 125

MALE TERMINALIA

1.   Phallosome with a pair of dorsal submedian longitudinal toothed ridges and a pair of broad lateral thornlike projections on apical half 2
Phallosome cylindrical or conical, without teeth or broad lateral thornlike projections 3

2 (1). Dististyle stout, with a very large lateral hatchet-shaped lobe directed mesad
                    howardii Coq., p. 117
Dististyle slender, with a small dorsal angular projection at distal fifth
                    ciliata (Fabr.), p. 115

3 (1). Phallosome conical, with a lateral flange on apical half and a basally divaricated carina on dorsal surface; crown of clasp-

ette with about eight or nine long flattened filaments and several feathered setae
                    cyanescens (Coq.), p. 118
Phallosome conical or cylindrical, without a lateral flange and dorsal divaricated carina; crown of claspette not as above......4

4 (3). Dististyle broad and square or truncate at apex; claw of dististyle inserted before apex.........................*varipes* (Coq.), p. 128
Dististyle not square or truncate at apex; claw of dististyle inserted at apex...........5

5 (4). Basistyle large, subspherical, about one and a half times as long as wide; dististyle constricted basally, strongly swollen, and rounded apically..*longipalpis* Roth, p. 126
Basistyle at least twice as long as broad, cylindrical; dististyle tapered apically to a narrow tip...6

6 (5). Claspette slender, about three-fourths as long as basistyle, with two contorted leaflike filaments, a short blunt filament, and numerous feathered setae on inner margin of apical fourth 7
Claspette stem slender, expanded into a broad crown apically, not extending beyond middle of basistyle; crown with five to seven narrow pointed blades or setae..8

7 (6). Basistyle three to three and a half times as long as broad; phallosome pointed at apex.........................*ferox* (Humb.), p. 120
                    horrida (D. and K.), p. 122
Basistyle two and a half times as long as broad; phallosome rounded at apex
                    johnstonii (Grabham), p. 124

8 (6). Dististyle expanded beyond middle, broadest at apical third..*pygmaea* (Theob.), p. 133
Dististyle broadest at middle........................9

9 (8). Claspette with about five or six setalike filaments on crown
                    signipennis (Coq.), p. 135
Claspette with five to seven long pointed apically feathered blades and a single apically feathered seta............................10

10 (9). Inner ventral margin of each plate of phallosome with a distinct triangular projection directed ventrolaterally; claspette with five or six blades and a single apically feathered seta........*discolor* (Coq.), p. 131
Inner ventral margin of each plate of phallosome with only a narrow rounded part directed ventrad, lacking a distinct triangular projection; claspette with six or seven blades and a single apically feathered seta..........*confinnis* (L. Arr.), p. 130

LARVAE (FOURTH INSTAR)

1.   Pecten teeth numerous (about 18 or more), each terminating in a hairlike filament; siphonal tuft represented by a single long hair 2
Pecten teeth few (less than 10), not pro-

longed into hairlike filaments; siphonal tuft multiple, large, small, or obsolete......3

2 (1). Lateral hair of anal segment 3- or 4-branched near base...............*ciliata* (Fabr.), p. 115
Lateral hair of anal segment single or forked beyond middle........*howardii* Coq., p. 117

3 (1). Siphonal tuft large, multiple, as long as the siphon; siphon small, not inflated; antennae inflated......*discolor* (Coq.), p. 132
Siphonal tuft small or obsolete, multiple; siphon large, more or less inflated medially; antennae not inflated...................4

4 (3). Upper frontal and lower frontal head hairs 5 and 6 multiple..*confinnis* (L. Arr.), p. 130
Upper frontal head hair 5 single or double (rarely triple), lower frontal 6 single, double, or triple.........................5

5 (4). Upper frontal and lower frontal hairs 5 and 6 with one or more hairs single................6
Upper frontal head hair 5 double, lower frontal 6 double or triple...........................8

6 (5). Hair of lateral valve of siphon more than half as long as apical width of siphon......7
Hair of lateral valve of siphon less than half as long as apical width of siphon
pygmaea (Theob.), p. 134

7 (6). Antennal and preantennal tufts strongly multiple (antennal tuft usually 8- to 15-branched, preantennal tuft about 6- to 8-branched), conspicuously barbed
signipennis (Coq.), p. 135
Antenna and preantennal tufts about 2- to 4-branched, occasionally rebranched toward tip, sparsely barbed
cyanescens (Coq.), p. 119

8 (5). Antennae distinctly longer than the median length of the head....................................9
Antennae about as long or rarely slightly longer than the median length of the head 10

9 (8). Lateral abdominal hair 6 single or double on segments IV–VI; branches of upper frontal head hair 5 and lower frontal 6 nearly equal.......................*ferox* (Humb.), p. 120
Lateral abdominal hair 6 multiple on segments IV–VI; branches of upper frontal head hair 5 and lower frontal 6 not equal, one branch of each being shorter and weaker than the other
longipalpis Roth, p. 126

10 (8). Siphon only slightly inflated
varipes (Coq.), p. 128
Siphon strongly inflated............................11

11(10). Lower frontal head hair 6 reaching near base of antenna; siphonal index about 3.5 to 4.0.............*horrida* (D. and K.), p. 122
Lower frontal head hair 6 reaching beyond base of antenna; siphonal index about 2.5 to 3.0.........*johnstonii* (Grabham), p. 124

Psorophora (Psorophora) ciliata
(Fabricius)

Culex ciliata Fabricius, 1794, Ent. Syst., 4:401.
Culex cyanopennis Humboldt, 1820, Voy. Reg. Equin., 7:119.
Culex molestus Wiedemann, 1821, Dipt. Exot., p. 7.
Psorophora boscii Robineau-Desvoidy, 1827, Mem. Soc. Nat. Hist. Paris, 3:413.
Culex conterrens Walker, 1856, Ins. Saund., Dipt., p. 427.
Culex perterrens Walker, 1856, Ins. Saund., Dipt., p. 431.
Psorophora holmbergii Theobald, 1901, Mon. Culic., 1:264.
Psorophora lynchi Brethes, 1916, An. Mus. Nac. Buenos Aires, 28:204.
Psorophora ctites Dyar, 1918, Ins. Ins. Mens., 6:126.

ADULT FEMALE (pl. 32). Very large species. *Head:* Proboscis long, scales on basal half long, dark brown, suberect, yellow and speckled with dark beyond middle, tip dark. Palpi about one-third as long as the proboscis, clothed with dark-brown suberect scales, those at tip black. Occiput with broad appressed grayish-white scales dorsally and with a narrow median bare stripe, numerous brown setae anteriorly and pale erect forked scales posteriorly; lateral region of occiput with brown setae but without scales. Tori brown, with a few small dark setae on inner surface. *Thorax:* Integument of scutum dark brown. Scutum marked as follows: A narrow median stripe of pale-golden lanceolate scales extending to the scutellum; a narrow nude submedian stripe on either side extending from the anterior margin of the scutum to the prescutellar space; a narrow stripe of narrow brown scales and dark setae extending from anterior third of scutum posteriorly along the sides of the prescutellar space, followed by a broad nude area on either side; sides of scutum with broad appressed pale-yellow to white scales; prescutellar space with white scales on either side of the median gold stripe. Scutellum with grayish-white and pale-golden scales and light-brown setae on the lobes. Pleura with patches of grayish-white appressed scales. *Abdomen:* First tergite with a median patch of grayish-white scales; remaining tergites pale yellow to brown-scaled, the pale scales predominating dorsally, the brown laterally. Venter primarily white-scaled. *Legs:* Femora yellow-scaled, speckled with dark on basal two-thirds to three-fourths, apical part of each densely clothed with long dark erect scales. Tibiae each with long dark erect scales except for a narrow basal ring of yellow appressed scales. Front and middle tarsi with appressed scales; segments 1 to 3 pale basally, dark apically; segments 4 and 5 usually all dark. Each segment of hind tarsus with a basal ring of pale appressed scales; scales beyond basal ring long, dark, erect on segments 1 and 2, dark and suberect or recumbent on segments 3 to 5. *Wing:* Length about 6.0 to 6.5 mm. Scales narrow, brown; a few inconspicuous pale scales distributed on costa, subcosta, and vein 1.

ADULT MALE. Coloration similar to that of the female. *Terminalia* (fig. 89). Lobes of ninth tergite (IXT-L) broadly rounded, each bearing many long setae. Tenth sternite (X-S) heavily sclerotized and toothed apically. Phallosome (Ph) sclerotized, conical, open ventrally, closed dorsally on basal half; apical half of each plate with a broad lateral triangular projection and a dorsal longitudinal toothed ridge. Claspette (Cl) with a slender stem and an expanded crown clothed with

numerous setae; a broad glabrous sickle-shaped filament inserted at apex on outer angle, and a strong thumblike projection directed mesad. Basistyle (Bs) rounded apically, about two and a half times as long as mid-width, clothed with scales and long and short setae; basal and apical lobes absent. Dististyle (Ds) nearly two-thirds as long as basistyle, curved, with several short stout setae on distal half or third; a dorsal angular projection at apical fifth; claw (Ds-C) short, peglike.

LARVA (fig. 90). Large larva. Head quadrate and concave anteriorly, broader than long. Antenna about one-third as long as the head, inserted at distal third of head, sparsely spiculate; antennal hair short, single, inserted at distal third or fourth of antenna. Mouth brushes composed of stout prehensile hairs, each hooked at tip and with a row of comblike teeth along the sides. Head hairs: Postclypeal 4 long, single; upper frontal 5 very short, branched distally; lower frontal 6 single, branched distally, inserted near anterior margin of clypeus; preantennal 7 single, branched distally. Comb of eighth segment with about 12 to 16 scales in a single curved row; individual scale thorn-shaped, with the large basal spines about one-fourth or one-third as long as the median spine. Siphonal index 3.0 to 4.0; pecten of numerous teeth on basal half of siphon; individual tooth long, hairlike, with one or more short basal teeth; siphonal tuft represented by a long single hair inserted beyond pecten. Anal segment ringed by the saddle; lateral hair represented by a tuft, 3- or 4-branched from near base, shorter than the saddle; dorsal brush bilaterally consisting of a long lower caudal hair and a short multiple upper caudal tuft; ventral brush short, extending almost the entire length of the anal segment, and piercing the saddle; gills about three times as long as the saddle or longer, sharply pointed.

DISTRIBUTION. Southern Canada; eastern United States west to South Dakota, Nebraska, and Texas; Mexico, Central and South America. It has also been found in Cuba (543). *United States:* Alabama (135, 419); Arkansas (88, 121); Connecticut (270, 494); Delaware (120, 167); District of Columbia (213, 318); Florida and Georgia (135, 213); Illinois (615); Indiana (152, 345); Iowa (213, 625);

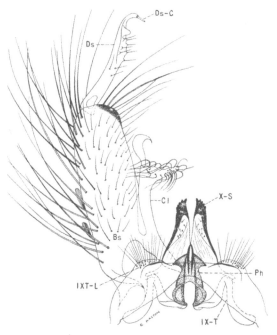

Fig. 89. Male terminalia of *Psorophora ciliata* (Fabricius).

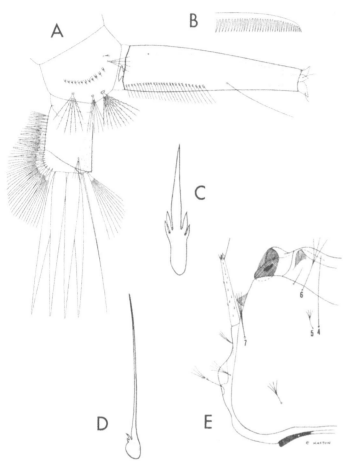

Fig. 90. Larva of *Psorophora ciliata* (Fabricius). *A*, Terminal segments. *B*, Mouth brush. *C*, Comb scale. *D*, Pecten spine. *E*, Head.

Kansas (213, 360); Kentucky (134, 570); Louisiana (213, 759); Maryland (64, 213); Massachusetts (213, 270); Michigan (391, 529); Mississippi (135, 546); Missouri (6, 324); Nebraska (680); New Hampshire (66); New Jersey (213, 350); New York (31); North Carolina (135, 213); Ohio (731a); Oklahoma (213, 628); Pennsylvania (298, 756); Rhode Island (35, 270); South Carolina (135, 742); South Dakota (14); Tennessee (135, 658); Texas (213, 463); Virginia (176, 178); West Virginia (213); Wisconsin (171). *Canada:* Ontario (708); Quebec (306, 708).

BIONOMICS. According to Schwardt (647) the eggs of *Psorophora ciliata* are deposited in small depressions or cracks in the soil and probably do not hatch until after a winter in hibernation. The larvae occur principally in unshaded temporary rain-filled pools where they develop rapidly, usually requiring 4 to 6 days. The pupal stage is usually complete in 2 days. The larvae are predacious, commonly feeding on other mosquito larvae with which they may be associated in their aquatic habitats, *P. confinnis* for example. The larva hangs from the surface of the water nearly in a vertical position. The larvae may be found from March to October in the southern United States and from May to September in the north. The species is common in the rice fields of Arkansas and Louisiana where conditions are generally favorable for its development throughout most of the summer; however, it is seldom abundant anywhere. The females are persistent biters, attacking any time during the day when their haunts are invaded and inflicting a painful injury.

Psorophora (Psorophora) howardii
Coquillett

Psorophora howardii Coquillett, 1901, Can. Ent., *33*:258.

Psorophora virescens Dyar and Knab, 1906, Proc. Biol. Soc. Wash., *19*:133.

ADULT FEMALE (pl. 33). Very large species. *Head:* Proboscis long, brown, darker apically, brown setae on all aspects. Palpi nearly half as long as the proboscis, brown, with setae on all aspects. Occiput sparsely clothed dorsally with broad flat grayish-white scales except for a narrow median bare stripe; lateral region bare except for scattered broad white scales. Tori brown, a few small dark setae on inner surface. *Thorax:* Integument of scutum shiny, dark brown to black; scutum with a narrow median stripe of narrow dark-bronze scales and black setae, a narrow submedian stripe of similar scales, and a large lateral area of broad appressed white scales on either side; a nude stripe on anterior half of the scutum between the median and submedian stripes; a broader nude area present on either side of the white-scaled prescutellar space. Scutellum with broad white scales and brown setae on the lobes. Pleura clothed with broad grayish-white scales. *Abdomen:* First tergite with a broad median patch of white scales; remaining tergites blue-black scaled dorsally, each with white scales laterally and apically. Dark scales of tergites II–V with purple reflection, VI with blue to green reflection, VII with greenish-golden reflection. Venter white-scaled. *Legs:* Femora yellow-scaled, speckled with dark, apical part with long dark erect or suberect scales. Tibiae clothed with yellow and purple recumbent scales, those of apices dark, suberect, particularly on hind legs. Tarsi primarily dark-scaled, blended with yellow and with purplish reflection. Second segment and often first segment of each tarsus with a narrow basal pale ring, segments 3, 4, and 5 dark. *Wing:* Length usually about 6.0 to 6.5 mm. Scales narrow, dark brown.

ADULT MALE. Coloration similar to that of the female. *Terminalia* (fig. 91). Ninth tergite (IX-T) with band narrow, sclerotized; lobes (IXT-L) prominent, each bearing several long setae. Tenth sternite (X-S) heavily sclerotized and toothed apically. Phallosome (Ph) long, conical, open ventrally, closed dorsally on basal half; apical third of each plate with a dorsal longitudinal toothed ridge and a broad lateral triangular projection. Claspette (Cl) consisting of a short broad stem fused with the basistyle, and a large subspherical crown clothed with short curved setae and bearing a large erect subapical filament on outer margin.

Basistyle (Bs) about twice as long as mid-width, clothed with sparse scales and long and short setae; basal and apical lobes absent. Dististyle (Ds) clothed with long fine setae on ventral aspect, a large lateral hatchet-shaped lobe projecting mesad, followed by a long slender subapical projection, also directed mesad; claw (Ds-C) inserted slightly before apex.

LARVA (fig. 92). Large larva. Head quadrate and concave anteriorly, broader than long. Antenna about one-third as long as the head, inserted at distal third of head, sparsely spiculate; antennal tuft represented by a single hair inserted at outer fourth or fifth of shaft. Mouth brushes composed of stout prehensile hairs, each hooked at tip and with a row of comblike teeth along the side. Head hairs: Postclypeal 4 long, single; upper frontal 5 very short, branched distally; lower frontal 6 single, branched distally, inserted near anterior margin of clypeus; preantennal 7 single, 3- or 4-branched distally. Comb of eighth segment with about 15 to 18 scales in a single curved row; individual scale thorn-shaped, with the large basal spines about one-third as long as the median spine. Siphonal index about 4.0; pecten of numerous teeth on basal half of siphon; individual tooth long, hairlike, with one or more short basal teeth; siphonal tuft represented by a single long hair inserted beyond pecten. Anal segment ringed by the saddle; lateral hair single or branched beyond middle, shorter than the saddle; dorsal brush bilaterally consisting of a long lower caudal hair and a short multiple upper caudal tuft; ventral brush extending almost the entire length of the anal segment and piercing the saddle; gills about three to four times as long as the saddle, pointed.

DISTRIBUTION. Southeastern United States, Mexico, the West Indies, and Central America. *United States:* Alabama (135, 514); Arkansas (88, 121); District of Columbia (213, 318); Florida (213, 419); Georgia (135, 608); Illinois (615); Indiana (152, 345); Kansas (530, 531); Kentucky (570): Louisiana (419, 759); Maryland (213); Mississippi (135, 546); Missouri (6, 324); Nebraska (14); North Carolina (135, 646); Oklahoma (628); South Carolina (135, 742); Tennessee (135,

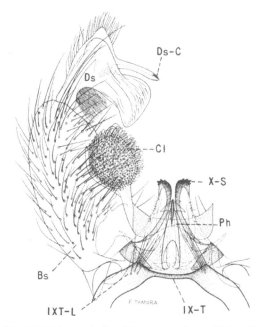

Fig. 91. Male terminalia of *Psorophora howardii* Coquillett.

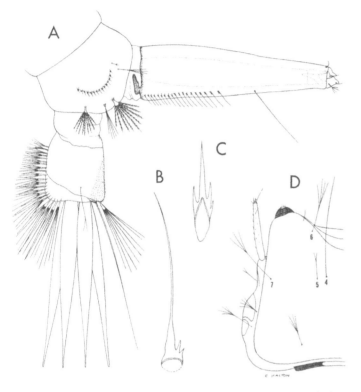

Fig. 92. Larva of *Psorophora howardii* Coquillett. *A*, Terminal segments. *B*, Pecten tooth. *C*, Comb scale. *D*, Head.

658); Texas (463, 634); Virginia (176, 178).

BIONOMICS. The larvae occur mostly in unshaded or partly shaded temporary rain-filled pools and are predacious, feeding on the larvae of other mosquitoes with which they are associated. The larvae occur from March to October in the southern United States and from about May to September in the north. The females are persistent biters, rivaling *Psorophora ciliata* in this respect. The adults are usually found near their larval habitats and will attack any time during the day when their haunts are invaded.

Psorophora (Janthinosoma) cyanescens
(Coquillett)

Culex cyanescens Coquillett, 1902, Jour. N.Y. Ent. Soc., *10*:137.

Psorophora (Janthinosoma) purpurascens Edwards, 1922, Bull. Ent. Res., *13*:77.

Psorophora (Janthinosoma) tovari Evans, 1922, Ann. Trop. Med., *16*:218.

Psorophora dyari Petrocchi, 1927, Rev. del Inst. Bact., *4*:725.

ADULT FEMALE. (pl. 34). Medium-sized to rather large species. *Head:* Proboscis long, dark, with purple reflection; palpi short, dark, with purple reflection. Occiput with ovate curved white to pale-golden scales and numerous pale and a few dark erect forked scales dorsally; a small blue-black-scaled patch, followed by broad yellowish scales laterally. Tori dark, with grayish-white to pale-golden scales on inner surface. *Thorax:* Integument of scutum black; scales of the median region intermixed lanceolate and ovate, mostly pale yellow to gold or golden brown, those of the lateral areas and prescutellar space ovate, gold to pale yellow or white. Scutellum with pale-yellow ovate scales and brown setae on the lobes. Pleura with grayish-white ovate scales. *Abdomen:* First tergite largely white-scaled; tergites II–VI black-scaled, with blue or purplish reflection, and with apical submedian triangular patches of golden-yellow scales; these apical patches narrow laterally, broadened toward the middorsal line, those of the anterior segments usually joined medially, tergite VII entirely dark. Sternites II–VI whitish-yellow to yellow-scaled, VII dark. *Legs:* Femora yellow-scaled, speckled with dark, apices dark with purplish reflection. Tibiae and tarsi entirely dark, with blue to purple reflection, scales appressed or recumbent. *Wing:* Length about 4.0 to 4.3 mm. Scales narrow, dark brown.

ADULT MALE. Coloration similar to that of the female. *Terminalia* (fig. 93). Ninth tergite (IX-T) with lobes (IXT-L) broadly rounded, each bearing several scattered setae. Tenth sternite (X-S) heavily sclerotized and toothed apically. Phallosome (Ph) sclerotized, conical, broad at base, rounded at apex, closed dorsally, open ventrally; apical half with a lateral flange and a middorsal longitudinal basally divaricated carina. Claspette (Cl) with stem slender, fused to the basistyle, expanded apically into a broad triangular crown which bears about nine elongated pointed flattened filaments and several feathered setae. Basistyle (Bs) about three times as long as mid-width, clothed with scales and long and short setae; basal and apical lobes absent. Dististyle (Ds) reticulated, about three-fifths as long as basistyle, expanded near middle, tapered and curved apically; the expanded inner margin is pubescent and bears

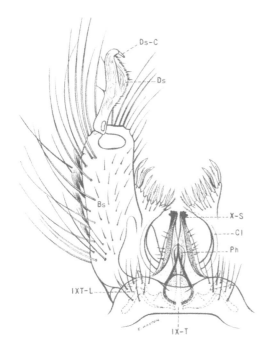

Fig. 93. Male terminalia of *Psorophora cyanescens* (Coquillett).

saddle or longer, the dorsal pair longer than the ventral pair, pointed.

DISTRIBUTION. Southeastern United States, Mexico, Central and South America. *United States:* Alabama (135, 419); Arkansas (88, 121); Florida (136, 516); Georgia (135, 420); Illinois (615); Indiana (152, 345); Kansas (213, 360); Kentucky (134, 570); Louisiana (213, 759); Mississippi (135, 546); Missouri (6, 324); Nebraska (14); New Mexico (471); North Carolina (516, 646); Ohio (731a); Oklahoma (628); South Carolina (135, 742); Tennessee (135, 658); Texas (213, 463); Virginia (176).

BIONOMICS. The eggs of *P. cyanescens* are deposited on the surface of the soil or in small cracks or holes at the bottom of dried pools. Observations made by Schwardt (647) indicate that hatching never occurs during the same year the eggs are laid. The larvae develop in rain-filled pools, in open or partly shaded

four or five small setae; claw (Ds-C) short, blunt.

LARVA (fig. 94). Antenna stout, about two-thirds as long as the head, coarsely spiculate; antennal tuft short, 2- to 4-branched, sparsely barbed, inserted slightly beyond middle of shaft. Head hairs: Postclypeal 4 small, branched distally; upper frontal 5 and lower frontal 6 long, single; preantennal 7 double or triple (rarely single on one side). Comb of eighth segment with 4 (rarely 5) scales on the posterior margin of a weakly sclerotized area; individual scale thorn-shaped, with the large basal spines about half as long as the median spine. Siphonal index 2.5 to 3.0; siphon strongly inflated; pecten of 3 to 5 strong teeth on basal third of siphon; siphonal tuft minute, multiple, inserted laterally beyond middle of siphon; hair of lateral valve longer than apical diameter of siphon. Anal segment ringed by the saddle; lateral hair represented by a minute multiple tuft; dorsal brush bilaterally consisting of a rather short lower caudal hair and a shorter multiple upper caudal tuft; ventral brush extending almost the entire length of the anal segment and piercing the saddle; gills three times as long as the

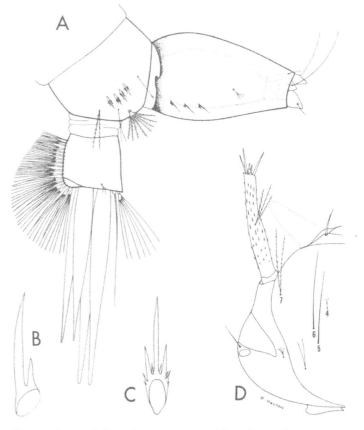

Fig. 94. Larva of *Psorophora cyanescens* (Coquillett). *A,* Terminal segments. *B,* Pecten tooth. *C,* Comb scale. *D,* Head.

areas. The larval period is unusually short, 3 to 4 days. The pupal period lasts 24 hours or a little longer. Schwardt (647) reared approximately 200 individuals from egg to adult in an average of 4.6 days. The males generally emerge before the females and rest on the vegetation at the site of the pool in which the immature stages developed.

The population of this species is frequently heavy after heavy rains in July and August in the south central states, particularly Alabama, Arkansas, Mississippi, and Louisiana. The females are persistent biters and will engorge themselves with blood until they are barely able to fly. The adults are often so numerous following summer rains that they seriously annoy livestock and interfere with farm work and sports in some areas. The adults are more commonly encountered in or near thickets or woodlands, but on cloudy days they will come out into the open in large numbers to bite.

Psorophora (Janthinosoma) ferox
(Humboldt)

Culex ferox Humboldt, 1820, Voy. Reg. Equin., 7:119.

Culex posticatus Wiedemann, 1821, Dipt. Exot., *1*:43.

Culex musicus Say (not Leach), 1827, Jour. Acad. Nat. Sci. Phila., 6:149.

Janthinosoma echinata Grabham, 1906, Can. Ent., *38*:311.

Janthinosoma vanhalli Dyar and Knab, 1906, Proc. Biol. Soc. Wash., *19*:134.

Janthinosoma terminalis Coquillett, 1906, U.S. Bur. Ent. Tech. Ser., *11*:8.

Janthinosoma sayi Dyar and Knab, 1906, Jour. N.Y. Ent. Soc., *14*:181.

Janthinosoma sayi Theobald, 1907, Mon. Culic., *4*:155.

Janthinosoma coquilletti Theobald, 1907, Mon. Culic., *4*:153.

ADULT FEMALE (pl. 35). Medium-sized species. *Head:* Proboscis long, dark-scaled, palpi short, dark. Occiput clothed dorsally with broad curved whitish-yellow to golden-yellow scales, paler anteriorly, with broad appressed yellowish scales laterally; yellow erect forked scales numerous on posterior half of occiput. Tori light brown on outer surface, inner surface dark brown. *Thorax:* Scutum with integument black; scutum clothed with rather broad dark-

brown and golden-yellow or yellowish-white scales in no definite pattern, the dark scales more abundant. Scutellum with rather broad yellowish-white scales and dark-brown setae on the lobes. Pleura rather densely clothed with grayish-white appressed scales. *Abdomen:* First tergite with a median patch of dark purplish scales; remaining tergites dark-scaled with purplish reflection on the dorsum, with prominent apicolateral triangular patches of whitish-yellow to golden-yellow scales. Sternites yellow-scaled on segments II–VI, mainly dark on segment VII. *Legs:* Femora dark, pale on posterior surface; knee spots present. Tibiae and tarsi of front and middle legs dark-scaled; segments 4 and 5 of hind tarsus, and often the apex of 3, white-scaled. Scales suberect, appearing rather shaggy and with purple reflection on apical part of hind tibia and on segments 1 and 2 of hind tarsus. *Wing:* Length about 3.7 to 4.0 mm. Scales dark, narrow.

ADULT MALE. Coloration similar to that of the female. *Terminalia* (fig. 95). Lobes of ninth tergite (IXT-L) broadly rounded, each bearing many setae. Tenth sternite (X-S) heavily sclerotized and toothed apically. Phallosome (Ph) sclerotized, conical, open ventrally, closed dorsally. Claspette (Cl) slender, about three-fourths as long as basistyle, expanded at apical fourth, then tapered, bearing numerous slender feathered filaments on inner margin at apical fourth; a large leaflike contorted pointed filament at apex, a similar filament and a shorter straight blunt filament near apex. Basistyle (Bs) about three to three and a half times as long as mid-width, clothed with scales and numerous long and short setae. Dististyle (Ds) reticulated, about three-fourths as long as basistyle, slender basally, strongly inflated medially, broadest at apical third, tapered distally, pubescent on broad inner margin; claw (Ds-C) short, blunt.

LARVA (fig. 96). Antenna much longer than the head, spiculate; antennal tuft multiple, barbed, inserted near middle of shaft. Head hairs: Postclypeal 4, small, branched; upper frontal 5 and lower frontal 6 long, usually double (hair 6 occasionally triple), barbed, about equal in length; preantennal 7 multiple, barbed. Lateral abdominal hair 6 multiple on segments I and II, double or triple on III, and

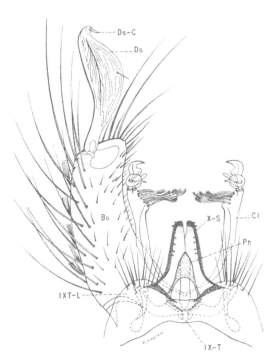

Fig. 95. Male terminalia of *Psorophora ferox* (Humboldt).

tucky (134, 570); Louisiana (213, 759); Massachusetts (270); Michigan (391); Minnesota (604); Mississippi (135, 546); Missouri (6, 324); Nebraska (530, 680); New Hampshire (66); New Jersey (350); New York (31); North Carolina (135, 213); Ohio (731a); Oklahoma (628); Pennsylvania (298, 756); South Carolina (135, 742); South Dakota (14); Tennessee (135, 658); Texas (463, 634); Virginia (176, 178); Wisconsin (171, 213). *Canada:* Ontario (212, 708).

BIONOMICS. The larvae occur in temporary rain-filled pools, particularly in or near thickets, in overflow pools along streams, and occasionally in potholes in stream beds after summer rains. They develop rapidly. The larvae occur from March to November in the southern United States and from early May to September in the north. The females are persistent and

single or double on IV–VI. Comb of eighth segment with about 6 to 8 scales in a curved row on the posterior margin of a weakly sclerotized area; individual scale thorn-shaped, with the large basal spines about one-third as long as the median spine. Siphonal index about 4.0; siphon strongly inflated; pecten of 3 to 5 widely spaced teeth on basal fourth of siphon; siphonal tuft minute, multiple, inserted laterally beyond middle of siphon. Anal segment ringed by the saddle; lateral hair represented by a small usually multiple tuft; dorsal brush bilaterally consisting of a long lower caudal hair and a short multiple upper caudal tuft; ventral brush extending almost the entire length of the anal segment and piercing the saddle; gills much longer than the saddle, each gradually tapering to a sharp point.

DISTRIBUTION. Southeastern Canada, eastern United States, Mexico, the West Indies, Central and South America. *United States:* Alabama (135, 419); Arkansas (88, 121); Connecticut (494); Delaware (120, 167); District of Columbia (213, 318); Florida and Georgia (135, 419); Illinois (615); Indiana (152, 345); Iowa (625, 626); Kansas (14, 530); Ken-

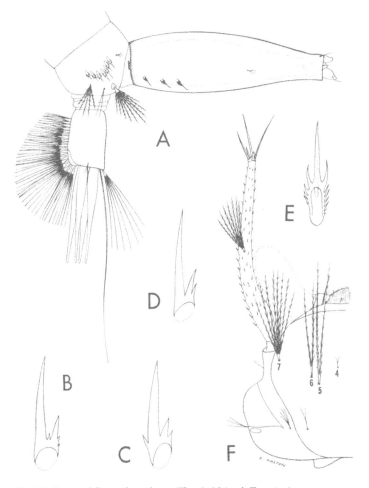

Fig. 96. Larva of *Psorophora ferox* (Humboldt). *A*, Terminal segments. *B*, *C*, and *D*, Pecten teeth. *E*, Comb scale. *F*, Head.

painful biters, even attacking in the open on cloudy days.

MEDICAL IMPORTANCE. Bates (40) has shown that *Psorophora ferox* frequently carries the eggs of *Dermatobia* in eastern Colombia. The senior author has also found this mosquito infested with *Dermatobia* eggs in Panama.

Psorophora (Janthinosoma) horrida
(Dyar and Knab)

Aedes horridus Dyar and Knab, 1908 (in part), Proc. U.S. Nat. Mus., *35*:56.

Psorophora horridus, Howard, Dyar, and Knab, 1917 (in part), Carnegie Inst. Wash. Pub., *159,* 2:561.

Psorophora (Janthinosoma) horrida, Roth, 1945, Proc. Ent. Soc. Wash., *47*:1.

ADULT FEMALE (pl. 36). Medium-sized species. *Head:* Proboscis long, dark-scaled; palpi dark, less than one-third as long as the proboscis, the fourth segment usually curved and about equal in length to the first three combined. Occiput with a large patch of broad appressed yellowish-white scales dorsally; a small dorsolateral patch of broad appressed purplish scales, followed by a lateral patch of broad appressed pale-yellow scales; pale erect forked scales on central part. Tori brown, inner surface darker and clothed with grayish-white scales. *Thorax:* Integument of scutum black; scutum with a broad median stripe of narrow dark bronze-brown scales; this stripe bordered laterally with broad grayish-white to pale-yellow scales; prescutellar space with broad pale scales. Scutellum with broad appressed grayish-white scales and dark setae on the lobes. Pleura with dense areas of broad grayish-white appressed scales. *Abdomen:* First tergite largely pale-scaled; remaining tergites dark purplish on the dorsum, with apicolateral patches of pale-yellow scales, often absent on VII. Sternites yellowish-scaled, except for the dark-scaled bases of IV to VI, and VII. *Legs:* Front and middle femora dark-purplish-scaled, the posterior surface pale; hind femur with the basal half to two-thirds pale on all aspects, dark apically; knee spots present. Front and middle legs with tibiae and tarsi entirely dark-scaled. Segments 4 and 5 of hind tarsus, and occasionally the apex of 3, white-scaled, other segments dark (segment 4 rarely with a few dark scales). Scales somewhat suberect, appearing rather shaggy on apical part of hind tibia and segments 1 and 2 of hind tarsus. *Wing:* Length about 3.8 to 4.0 mm. Scales narrow, dark.

ADULT MALE. Coloration similar to that of the female. Proboscis slender, not swollen apically. Palpi with the last two segments much stouter than the preceding segments. Antenna reaching apex of the third palpal segment or only slightly beyond; the last two antennal segments comprising less than half the length of the entire antenna. *Terminalia* (fig. 97). The teminalia is very similar to that of *Psorophora ferox.* Lobes of ninth tergite (IXT-L) widely separated, broadly rounded, each bearing many slender setae. Tenth sternite (X-S) heavily sclerotized and toothed apically. Phallosome (Ph) sclerotized, short and conical, open ventrally, closed dorsally. Claspette (Cl) slender, about three-fourths as long as the basistyle, expanded subapically, bearing numerous feathered filaments on inner margin before apex, a large leaflike contorted filament at apex, a similar filament and a shorter straight filament near apex. Basistyle (Bs) about three to three and a half times as long as mid-width, clothed with scales and long and short setae. Dististyle (Ds) reticulated, about three-fourths as long as the basistyle, slender near base, broad at apical third, and tapered distally; claw (Ds-C) short, blunt.

LARVA (fig. 98). Antenna about as long as the head, spiculate; antennal tuft multiple, barbed, inserted near middle of shaft. Head hairs: Postclypeal 4 small, multiple; upper frontal 5, short, double, barbed; lower frontal 6 short, triple (sometimes double), barbed, reaching near base of antenna (hairs 5 and 6 nearly equal in length); preantennal 7 multiple, barbed. Lateral abdominal hair 6 multiple and longer on segments I and II, 3- to 4-branched on III and usually double or triple and difficult to see on IV–VII. Comb of eighth segment with about 6 to 9 scales in a curved row; individual scale thorn-shaped, with the subapical spines about one-third as long as the median spine. Siphonal index about 3.5 to 4:0: siphon strongly inflated; pecten of about 3 to 5 widely spaced teeth on basal fourth of siphon; siphonal

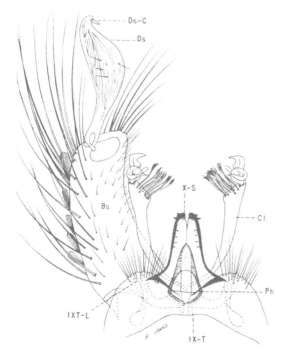

Fig. 97. Male terminalia of *Psorophora horrida* (Dyar and Knab).

BIONOMICS. The larvae occur in temporary shaded pools following heavy and prolonged summer rains. The eggs, which are deposited by the females in moist or dry depressions or scattered over the surface of the soil, generally require ripening for a few days before hatching takes place. They hatch shortly after flooding, and the larvae develop rapidly. Adults of a given generation emerge almost simultaneously. The females rest on vegetation near the breeding site and readily attack whenever their haunts are invaded, becoming serious pests in some localities.

Lt. C. B. Eaton was able to obtain eggs from wild-caught females on the Fort Benning Reservation and hatched and reared them in the laboratory. Females in captivity took from four to nine blood meals during their period of confinement.

tuft minute, multiple, about as long as apical pecten tooth, inserted laterally at apical third of siphon; hair of dorsal valve shorter than apical diameter of siphon. Anal segment completely ringed by the saddle; lateral hair small, usually 6- to 8-branched near base; dorsal brush bilaterally consisting of a long lower caudal hair and a short multiple upper caudal tuft; ventral brush extending almost the entire length of the anal segment and piercing the saddle; gills usually two to three times as long as the saddle, pointed.

DISTRIBUTION. Southeastern United States, north to Ohio and Pennsylvania and west to Nebraska and Texas. *United States:* Alabama (135, 516); Arkansas and District of Columbia (618); Florida (516); Georgia (135, 618); Illinois (615); Indiana (133, 618); Iowa, Kansas and Kentucky (618); Louisiana (618, 759); Maryland (618); Mississippi (135, 516); Missouri (618); Nebraska (14); North Carolina (135, 618); Ohio and Oklahoma (618); Pennsylvania (756); South Carolina (135, 742); Tennessee (516, 618); Texas (618, 634); Virginia (618).

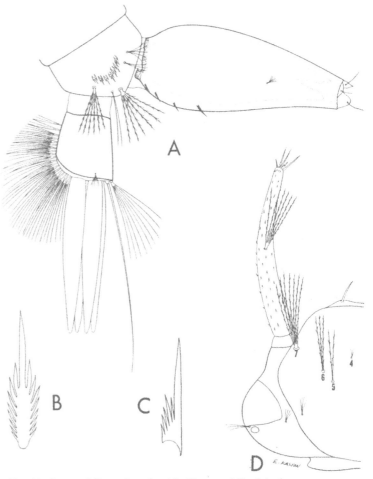

Fig. 98. Larva of *Psorophora horrida* (Dyar and Knab.) *A*, Terminal segments. *B*, Comb scale. *C*, Pecten tooth. *D*, Head.

124

Psorophora (Janthinosoma) johnstonii
(Grabham)

Janthinosoma johnstonii Grabham, 1905, Can. Ent.,
37:410.

Janthinosoma coffini Dyar and Knab, 1906, Proc.
Biol. Soc. Wash., *19*:134.

Psorophora coffini, Pratt, 1946, Proc. Ent. Soc.
Wash., *48*:209.

Psorophora (Janthinosoma) johnstonii, Thurman
et al., 1951, Ann. Ent. Soc. Amer., *44*:144.

ADULT FEMALE (pl. 37). Small to medium-sized species. *Head:* Proboscis long, dark-scaled with purplish reflection; palpi short, dark with purplish reflection. Occiput with broad recumbent yellowish-white scales and pale erect forked scales dorsally; a large patch of broad dingy scales laterally. Tori with integument dark brown, a few grayish-white scales on inner surface. *Thorax:* Integument of scutum dark; scutum clothed with scattered dull yellowish-white recumbent scales in no definite pattern. Scutellum with grayish-white scales and dark setae on the lobes. Pleura with dense areas of grayish-white appressed scales. *Abdomen:* First tergite largely white-scaled; remaining tergites dark with purplish reflection, with apicolateral triangular patches of white scales. Sternites yellowish-scaled, except for bases of V and VI and all of VII which are dark-scaled. *Legs:* Femora pale-scaled basally, dark purplish on apical third to one-half. White knee spots present on some specimens. Tibiae dark with purplish reflection. Tarsi dark with purplish reflection; basal four-fifths of segment 4 of hind tarsus usually white-scaled, but may vary from all white to all dark. *Wing:* Length about 3.0 to 3.5 mm. Scales narrow, dark.

ADULT MALE. Coloration similar to that of the female. *Terminalia* (fig. 99). Lobes of ninth tergite (IXT-L) broadly rounded, each bearing many setae. Tenth sternite (X-S) heavily sclerotized, bearing 2 to 4 teeth near apex. Phallosome (Ph) heavily sclerotized, conical, broad basally, somewhat narrowed medially, rounded apically, open ventrally, closed dorsally. Claspette (Cl) slender, about three-fourths as long as basistyle, expanded on apical third, then tapered to tip; inner margin of expanded area bearing numerous slender, feathered filaments; a large leaflike contorted pointed filament at apex, a similar one and a shorter

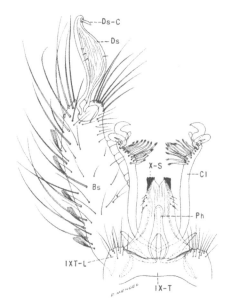

Fig. 99. Male terminalia of *Psorophora johnstonii*
(Grabham).

straight blunt filament subapically. Basistyle (Bs) about two and a half times as long as midwidth, curved, sides nearly parallel, clothed with scales and numerous long and short setae, the setae being longer near apex; basal and apical lobes absent. Dististyle (Ds) about three-fourths as long as basistyle, reticulated, slender basally, strongly inflated medially, broadest at apical third, tapered distally; claw (Ds) short, blunt.

LARVA (fig. 100). Antenna about as long as the head, spiculate; antennal tuft multiple, barbed, inserted near middle of shaft. Head hairs: Postclypeal 4 small, branched beyond base; upper frontal 5 long, double, barbed; lower frontal 6 long, double or triple, reaching beyond base of antenna; preantennal 7 multiple, barbed. Lateral abdominal hair 6 usually multiple on segments III–VI, single on VII. Comb of eighth segment with 5 to 7 thorn-shaped scales in a curved row; individual scale with a strong central spine, one or two smaller lateral spinules and three to five much smaller spinules. Siphonal index about 2.5 to 3.0; siphon strongly inflated; pecten of 3 to 8 widely spaced teeth on basal third of siphon. Anal segment completely ringed by the saddle; lateral hair small, multiple; dorsal brush bilaterally consisting of a long lower caudal hair and a shorter multiple upper caudal tuft; ventral brush extending almost the entire length of the anal segment and piercing the saddle; gills about two to four times as long as the saddle, tapered.

DISTRIBUTION. Florida and the Greater Antilles. The first record of this species in Florida was from Cudjoe Key and consisted of biting collections of females and a male captured in a light trap (562). The species has since been found on many of the Florida Keys listed by Thurman *et al.* (689).

BIONOMICS. The larvae occur in shallow, temporary, shaded, rain-filled depressions. Thurman *et al.* (689) report the finding of first instar larvae 12 to 18 hours after a heavy rain, and field observations indicate that about 6 days were required for the emergence of adults. Females were collected as far as one mile from their aquatic habitats. The females are said to be nervous and difficult to capture until actual feeding begins; however, they are claimed to be vicious biters and will pursue a victim. Biting occurs any time during the day both in shade and sunlight, on clear or cloudy days. A biting rate of approximately one hundred per minute was recorded by the above writers.

The swarming of the males is described by Thurman *et al.* (689) briefly as follows: Mating swarms were observed on Big Pine Key on August 5, 1948, at 4:00 P.M. Thirty-five to 50 males were seen hovering in a circular mass 2 to 3 feet in diameter about 3 to 12 inches above the water. Unengorged females were seen to fly into the swarm. Swarming lasted about one-half hour.

Psorophora (Janthinosoma) longipalpis Roth

Aedes horridus Dyar and Knab, 1908 (in part), Proc. U.S. Nat. Mus., *35*:56.

Psorophora horridus, Howard, Dyar, and Knab, 1917 (in part), Carnegie Inst. Wash. Pub., *159*, 2:561.

Psorophora (Janthinosoma) horridus, Matheson, 1934, Proc. Ent. Soc. Wash., *36*:41 (describes male).

Psorophora (Janthinosoma) horrida, Rozeboom, 1939, Jour. Parasit.. 25:145 (describes larva).

Psorophora (Janthinosoma) longipalpis Roth, 1945, Proc. Ent. Soc. Wash., 47:1.

ADULT FEMALE (pl. 38). Medium-sized species. *Head:* Proboscis long, dark-scaled with violet reflection; palpi dark with violet reflection, usually a little more than one-third as long as the proboscis, with the fourth segment straight, more than one and one-half times as long as the other segments combined. Occiput with a large median patch of broad recumbent white scales, with a lateral patch of broad appressed violet scales margined with broad appressed white scales above and broad yellowish-white scales below; erect forked scales on central part of occiput pale. Tori dark brown to black, with a patch of pale scales on inner surface. *Thorax:* Integument of scutum black; scutum with a broad median stripe of narrow

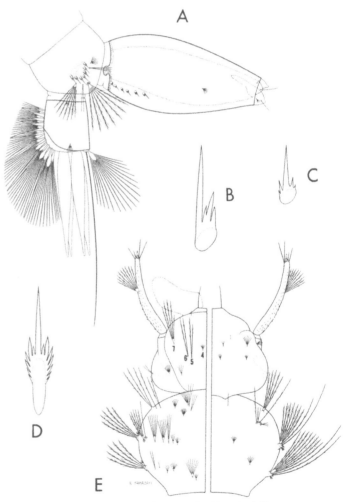

Fig. 100. Larva of *Psorophora johnstonii* (Grabham), *A*, Terminal segments. *B*, and *C*, Pecten teeth. *D*, Comb Scale. *E*, Head and thorax (dorsal and ventral).

dark bronze-brown scales, bounded laterally with broad white to yellowish scales; narrow pale scales intermixed with the broad ones on the posterior third of the scutum; prescutellar space margined laterally with broad pale scales. Scutellum with broad pale scales and dark setae on the lobes. Pleura with dense areas of appressed white scales. *Abdomen:* First tergite with yellowish-white scales; remaining tergites violet-scaled dorsally, with small apical patches of yellowish scales laterally on IV to VI (sometimes VII). Venter yellow-scaled, except for dark scales of segment VII, on base of VI, and occasionally on base of V. *Legs:* Front and middle femora dark, posterior surface pale; hind femur with basal two-thirds pale on all aspects. Tibiae and tarsi of front and middle legs entirely dark-scaled with violet reflection; segments 4 and 5 of hind tarsus, and sometimes the apex of segment 3, white; other segments dark (segment 4 rarely with a few dark scales); scales of hind tibia and of segments 1 to 3 of hind tarsus suberect. *Wing:* Length about 4.3 mm. Scales narrow, dark.

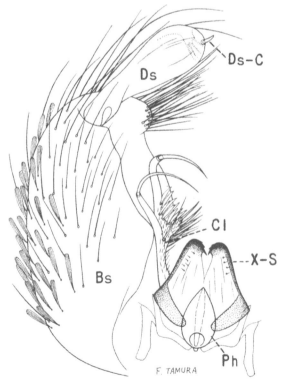

Fig. 101. Male terminalia of *Psorophora longipalpis* Roth.

ADULT MALE. Coloration similar to that of the female. Proboscis slender, swollen apically. Palpi long, the last two segments not much stouter than preceding segments. The last two segments of the antenna as long or longer than the preceding segments combined. *Terminalia* (fig. 101). Ninth tergite (IX-T) greatly modified, with lateral parts separated by a chitinized median structure, which is narrow basally, broadened beyond middle into a pair of elongate lateral projections, somewhat constricted subapically, and truncate at tip, often with a median terminal tooth. Tenth sternite (X-S) heavily sclerotized beyond middle. Phallosome (Ph) conical, bluntly pointed at apex, open ventrally, closed dorsally. Claspette (Cl) slender, unequally bifurcate near base; the shorter branch armed with a strong seta arising from a tubercle near base and a long slender apical spine; the longer branch bearing two spinelike setae arising from a short stout projection at basal third and a long slender apical spine. Basistyle (Bs) about one and a half times as long as mid-width, hemispherical in shape, clothed with numerous scales on basal half and long and short setae on apical half, a dense patch of setae at apex; basal and apical lobes absent. Dististyle (Ds) about half as long as the basistyle, the basal half constricted, the apical half strongly swollen and rounded at tip; a minute subventral lobe bearing a few small setae present at apical fourth near outer margin; claw (Ds-C) short, peglike.

LARVA (fig. 102). Antenna longer than the head, spiculate, darker on distal half; antennal tuft multiple, barbed, inserted a little beyond middle of shaft, not reaching tip. Head hairs: Postclypeal 4 small, multiple; upper frontal 5 and lower frontal 6 double, barbed, extending to or beyond the preclypeus (one branch of hairs 5 and 6 a little shorter and weaker than the other); preantennal 7 multiple, barbed, not reaching insertion of antennal tuft. Lateral abdominal hair 6 multiple on segments I–VII, longer on segments I and II. Comb of eighth segment with about 7 (rarely 5 or 6) scales in a curved row; individual scale with the subapical spines about one-third as long as the large median spine. Siphonal index about 4.0; siphon strongly inflated; pecten of 3 or 4 widely spaced teeth on basal fourth of siphon; siphonal

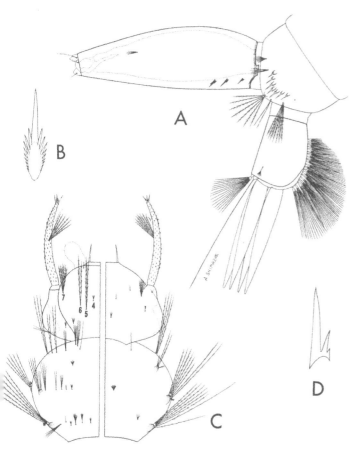

Fig. 102. Larva of *Psorophora longipalpis* Roth. *A*, Terminal segments. *B*, Comb scale. *C*, Head and thorax (dorsal and ventral). *D*, Pecten tooth.

during August, 1938, in a park at Tulsa, Oklahoma. The larvae were found in heavily shaded temporary pools following a heavy rain and were associated with the immature stages of *P. confinnis, P. signipennis, P. cyanescens, P. ferox, Aedes vexans,* and *A. trivittatus.* He further states that *P. longipalpis* was never found in water exposed to full sunlight. The males can be recognized even while resting, by their bulbous terminalia. The adults have been observed resting in shaded underbrush near the aquatic habitats of the immature stages.

Psorophora (Janthinosoma) varipes
(Coquillett)

Conchiliastes varipes Coquillett, 1904, Can. Ent., *36*:10.

Janthinosoma albigenu Peryassú, 1908, Os Culic. do Brazil, p. 155.

Psorophora discrucians Howard, Dyar, and Knab (not Walker), 1917, Carnegie Inst. Wash. Pub., *159*, 4:569.

Psorophora bruchi Petrocchi, 1927, Rev. del Inst. Bact. Buenos Aires, *4*:728.

ADULT FEMALE (pl. 39). Medium-sized species. *Head:* Proboscis dark-scaled; palpi short, dark. Occiput with broad appressed dingy-white, cream-colored, or yellow scales dorsally, pale erect forked scales on median area, with broad appressed dark and purplish scales laterally. Tori dark brown, with bluish-white scales on inner surface. *Thorax:* Integument of scutum dark brown or black; scutum with a broad median stripe of dark brown, rather broad lanceolate scales; this stripe margined laterally with broad yellowish-white scales; prescutellar space with pale scales. Scutellum with dingy-white scales and dark-brown setae on the lobes. Pleura with extensive areas of dingy-white appressed scales. *Abdomen:* First tergite primarily dingy-white-scaled; remaining tergites dark-scaled with purplish reflection, except for whitish or pale-yellow scales on apicolateral angles. Sternites mostly white scaled on segments II–VI, dark on VII. *Legs:* Front and middle femora dark on anterior surface and subapically, pale on basal two-thirds of posterior surface, hind femur pale on basal two-thirds, dark apically. Small white knee spots present. Tibiae and tarsi of front and

tuft minute, multiple, about as long as the apical pecten tooth, inserted laterally at apical third of siphon; hair of dorsal valve as long as the apical width of the siphon. Anal segment completely ringed by the saddle; lateral hair small, usually with many branches beyond middle; dorsal brush bilaterally consisting of a long lower caudal hair and a short multiple upper caudal tuft; ventral brush extending almost the entire length of the anal segment and piercing the saddle; gills about two to three times as long as the saddle, tapered.

DISTRIBUTION. Known only from the midwestern United States. *United States:* Arkansas (618); Kansas (14, 618); Louisiana (759); Missouri (14, 618); Oklahoma (618, 627); South Dakota (618); Texas (618, 634).

BIONOMICS. Rozeboom (627) describes the habits of the larvae of this species as observed

middle legs entirely dark-scaled; fourth tarsal segment of hind leg white at least on one side, other segments dark. *Wing:* Length about 3.0 to 3.5 mm. Scales dark, narrow.

ADULT MALE. Coloration similar to that of the female *Terminalia* (fig. 103). Lobes of ninth tergite (IXT-L) broadly rounded, bearing many scattered setae. Tenth sternite (X-S) heavily sclerotized and toothed apically. Phallosome (Ph) sclerotized, broad at base and pointed apically, open ventrally, closed dorsally. Claspette (Cl) flattened dorsoventrally, about three-fourths to four-fifths as long as the basistyle; inner surface with a strong down-curved clawlike appendage at tip; two long smooth setae on inner margin near middle, followed by a few smaller setae, some of which are feathered. Basistyle (Bs) nearly two and a half times as long as mid-width, clothed with scales and long and short setae; basal and apical lobes absent. Dististyle (Ds) about two-thirds as long as the basistyle, narrow near base, expanded beyond basal third, constricted beyond claw, square or slightly truncate at tip, surface with many reticulations and small papillated hairs; claw (Ds-C) short, peglike, inserted on inner margin of dististyle just before apical constriction.

LARVA (fig. 104). Antenna about as long as the head, spiculate; antennal tuft long, multiple, barbed, inserted near middle of shaft. Head hairs: Postclypeal 4 small, multiple; upper frontal 5 long, usually double, sparsely barbed; lower frontal 6 long, double or triple, sparsely barbed; preantennal 7 multiple, barbed. Lateral abdominal hair 6 multiple on segments I and II, variable on III–VI. Comb of eighth segment with about 7 (sometimes 6) scales in a V-shaped row, the median scales larger than the outer ones; individual scale thorn-shaped, with the subapical spines one-fifth to one-fourth as long as the median spine. Siphonal index about 4.0; siphon slightly inflated; pecten of 3 or 4 widely spaced teeth on basal third of siphon; siphonal tuft small, 3- to 6-branched, inserted beyond middle of siphon. Anal segment completely ringed by the saddle; lateral hair small, single or branched distally; dorsal brush bilaterally consisting of a long lower caudal hair and a shorter multiple upper caudal tuft; ventral brush extending

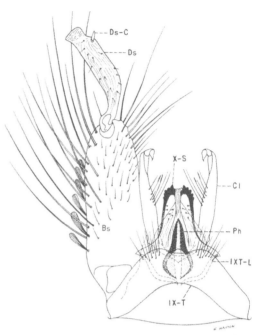

Fig. 103. Male terminalia of *Psorophora varipes* (Coquillett).

almost the entire length of the anal segment and piercing the saddle; gills about two to three times as long as the saddle, pointed.

DISTRIBUTION. Southern United States north to New York and west to Texas, Mexico, Central and South America. *United States:* Alabama (135, 419); Arkansas (88, 121); Florida (419); Georgia (135, 516); Illinois (615); Indiana (152, 345); Kentucky (570); Louisiana (213, 759); Mississippi (135, 546); Missouri (6, 213); New York (31); North Carolina (135, 646); Ohio (731a); Oklahoma (213, 628); South Carolina (135, 742); Tennessee (419, 658); Texas (532); Virginia (176).

BIONOMICS. The larvae develop in temporary floodwater pools. Buren (111) describes collections of the larvae taken during 1944 and 1945 from overflow areas and rain pools in a dense woodland near Alexandria, Louisiana. He states that the larvae were usually found in floating mats of debris consisting of dead twigs, leaves, and the like. The adults of this species have been described by Horsfall (375) and Carpenter (121) as occurring in swarms in woodlands along creeks in Arkansas following flooding in the spring. The adults are known to occur from April to September in the southern United States. The females are persistent biters, attacking any time during the day in woodlands.

Psorophora (Grabhamia) confinnis
(Lynch Arribálzaga)

Taeniorhynchus confinnis Lynch Arribálzaga, 1891, Rev. Mus. de la Plata, *2*:149.

Culex jamaicensis Theobald, 1901, Mon. Culic., *1*:345.

Janthinosoma texanum Dyar and Knab, 1906, Proc. Biol. Soc. Wash., *19*:135.

Janthinosoma floridense Dyar and Knab, 1906, Proc. Biol. Soc. Wash., *19*:135.

Janthinosoma toltecum Dyar and Knab. 1906, Proc. Biol. Soc. Wash., *19*:135.

Janthinosoma columbiae Dyar and Knab, 1906, Proc. Biol. Soc. Wash., *19*:135.

Taeniorhynchus walsinghami Theobald, 1907, Mon. Culic., *4*:484.

Culex scutipunctatus Lutz and Neiva, 1911, Mem. Inst. Osw. Cruz, *3*:298.

Psorophora funiculus Dyar, 1920, Ins. Ins. Mens., *8*:141.

Psorophora (Grabhamia) confinnis, Aitken, 1940, Rev. de Ent., *11*:677.

ADULT FEMALE (pl. 40). Medium-sized to rather large species. *Head:* Proboscis dark-scaled, except for a very wide yellowish-white median band; palpi short, dark, apical half of segment 4 white. Occiput clothed dorsally with narrow white to pale-violet scales and numerous black erect forked scales; a dorsolateral patch of broad flat dark scales followed by broad dingy-white to light-brown scales laterally. Tori light brown, inner surface darker and clothed with grayish-white scales. *Thorax:* Integument of scutum dull black; scutum clothed with fine narrow bronze-brown to blackish scales, except for lavender-tinted narrow white scales on the prescutellar space, the anterolateral angle of scutum, a streak on scutal angle, a patch above wing base, and a small submedian spot near middle of scutum. Scutellum with long narrow whitish scales and dark setae on the lobes. Pleura with patches of broad appressed whitish scales. *Abdomen:* First tergite with a median patch of grayish-white scales; remaining tergites dark, with white to pale-yellow-scaled apical markings,

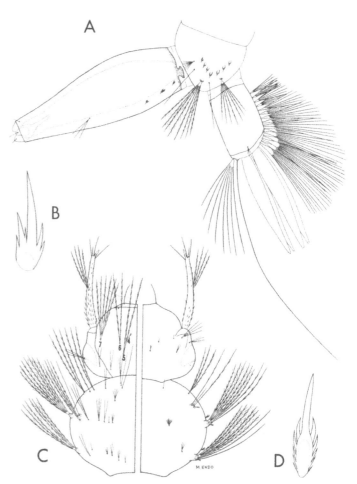

Fig. 104. Larva of *Psorophora varipes* (Coquillett). *A*, Terminal segments. *B*, Pecten tooth. *C*, Head and thorax (dorsal and ventral). *D*, Comb scale.

triangular in shape on II and III, divided into paired submedian patches on IV–VII. Venter with intermixed dark and pale scales. *Legs:* Femora dark-brown to black-scaled, liberally speckled with white scales, posterior surface largely pale-scaled; each femur with a narrow subapical white-scaled ring; knee spots white. Tibiae black, with numerous small white-scaled spots on outer surface. Hind tarsi with a broad basal white ring on each segment, first segment with a median white ring as well; front and middle tarsi similarly marked, but with white rings reduced or lacking on segment 4, absent on 5. *Wing:* Length about 4.0 to 4.5 mm. Scales rather broad, speckled dark brown and white, the white scales in no definite pattern; fringe entirely dark.

ADULT MALE. Coloration similar to that of female, but with the band on proboscis narrower. *Terminalia* (fig. 105). Lobes of ninth tergite (IXT-L) broadly rounded, each bearing many scattered setae. Tenth sternite (X-S) heavily sclerotized and toothed apically. Phallosome (Ph) long, bluntly pointed at apex, open ventrally, closed dorsally; middle of each plate with a narrow rounded part of its inner ventral margin turned ventrad (not a distinct triangular projection as in *P. discolor*). Claspette (Cl) with stem curved, expanded apically forming a triangular crown which is attached to basal third of basistyle; apical part free and bearing six or seven long pointed apically feathered blades and a single apically feathered seta. Basistyle (Bs) about two and one-half times as long as mid-width, clothed with scales and long and short setae; basal and apical lobes absent. Dististyle (Ds) reticulated, about three-fifths as long as basistyle, expanded medially, tapered distally, curved toward tip; claw (Ds-C) strong, dark.

LARVA (fig. 106). Antenna shorter than the head, spiculate; antennal tuft long, multiple, barbed, inserted near middle of shaft. Head hairs: Postclypeal 4 small, multiple; upper frontal 5, lower frontal 6, and preantennal 7 multiple, strongly barbed. Lateral abdominal hair 6 multiple on segments I and II, double

on III–V, single on VI. Comb of eighth segment with 6 (rarely 5) scales on the posterior margin of a weakly sclerotized area; individual scale thorn-shaped, with the large basal spines about one-third as long as the median spine. Siphonal index about 3.0; siphon slightly inflated; pecten of 3 to 6 widely spaced teeth, not reaching middle of siphon; siphonal tuft small, multiple, inserted at outer third of siphon. Anal segment ringed by the saddle; lateral hair small, double or triple; dorsal brush bilaterally consisting of a long lower caudal hair and a short multiple upper caudal tuft; ventral brush extending almost the entire length of the anal segment and piercing the saddle; gills about two to three times as long as the saddle, pointed.

DISTRIBUTION. United States, Mexico, the West Indies, Central and South America. *United States:* Alabama (135, 419); Arizona (9); Arkansas (88, 121); California (9, 287); Colorado (531); Delaware (120, 167); District of Columbia (213, 318); Florida and Georgia (135, 419); Illinois (615); Indiana (152, 345); Iowa (625, 626); Kansas (14, 360); Kentucky (134, 570); Louisiana (213, 759); Maryland (64, 161); Massachusetts (701); Mississippi (135, 546); Missouri (6, 324); Nebraska (680); New Jersey (213, 350); New Mexico (273); New York (31); North Carolina (135, 213); Ohio (731a); Oklahoma (368, 628); Pennsylvania (298, 756); South Carolina (135, 742); South Dakota (14); Tennessee (135, 658); Texas (249, 463); Virginia (176, 178); West Virginia (213).

BIONOMICS. The females lay their eggs on damp soil in depressions subject to flooding by rainfall or overflow from streams and irrigation canals. The winter is passed in the egg stage. Horsfall (377) found that areas covered by rice provided good oviposition sites for *Psorophora confinnis*, whereas areas bare of vegetation were of little importance. He found that rank low-growing vegetation furnished ideal oviposition sites. The larval period is relatively short, usually requiring 4 to 10 days. The males, on the average, emerge several hours before the females and remain on the vegetation at the breeding site until the females emerge.

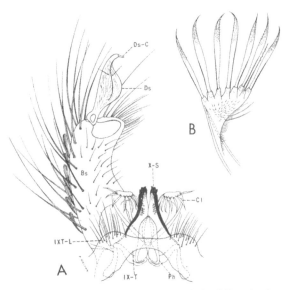

Fig. 105. *Psorophora confinnis* (Lynch Arribálzaga). *A,* Male terminalia. *B,* Crown of claspette.

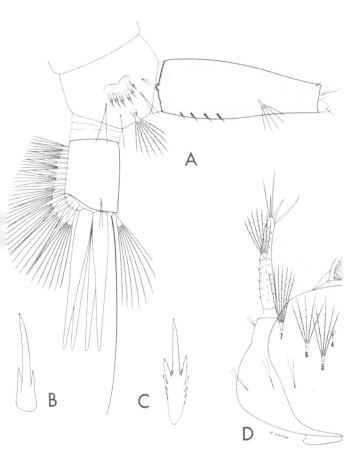

Fig. 106. Larva of *Psorophora confinnis* (Lynch Arribál-zaga). *A*, Terminal segments. *B*, Pecten tooth. *C*, Comb scale. *D*, Head.

The adults are readily attracted to light traps. Horsfall (378) found that the flight range of *P. confinnis* may be as great as 9 miles. Studies conducted by Whitehead (750) in Arkansas indicate that under the conditions present at that time cattle constituted the chief source of blood for this species, with horses, mules, and hogs of intermediate attractiveness. Birds were rarely fed upon. The females are persistent biters, attacking any time during the day or night. In the Florida Everglades and the rice fields of Arkansas where the species probably reaches its greatest abundance, livestock are occasionally killed. Houses are well screened, and people are inclined to stay indoors at night during the mosquito season. Because of its abundance in rice fields, it is commonly referred to as the dark rice-field mosquito.

Psorophora (Grabhamia) discolor
(Coquillett)

Culex discolor Coquillett, 1903, Can. Ent., *35*:256.

ADULT FEMALE (pl. 41). Medium-sized species. *Head:* Proboscis dark-scaled, except for a very wide median pale-yellow band; palpi short, dark, the apices pale. Occiput with narrow light-golden scales and dark erect forked scales dorsally, the forked scales of the central region pale, with broad flat appressed yellowish scales laterally. Tori brown, with a few narrow light-golden scales on inner surface. *Thorax:* Integument of scutum dark brown, scutum clothed with fine narrow pale-yellow to golden-brown scales. Scutellum with pale-golden scales and golden setae on the lobes. Pleura with patches of broad appressed grayish-white scales. *Abdomen:* First tergite grayish-white scaled on median area; remaining tergites almost entirely covered with grayish-white to pale-yellow scales, more or less speckled with dark scales; the dark scales usually more numerous basolaterally. Venter pale-scaled, frequently speckled with dark scales. *Legs:* Femora clothed with intermixed dark-brown and pale scales, the posterior surface largely pale; each femur with a narrow, often indefinite, subapical pale-scaled ring; knee spots present. Tibiae pale, sparsely speckled with dark scales, the apices usually dark, particularly on hind tibiae. Hind tarsus with segment 1 mostly pale, usually speckled with dark scales, dark at apex; segments 2 to 5 variable, usually with basal half pale, apical half dark. Front and middle tarsi similarly marked, but with basal pale ring lacking or reduced on segment 4, usually absent on 5. *Wing:* Length 3.0 to 3.7 mm. Scales rather broad, dark brown and pale, arranged more or less in definite patterns. Costa with pale scales at base and at point of union with subcosta, the remainder speckled with dark scales. Vein 6 with basal two-thirds pale, apical third dark. Remaining veins with intermixed dark and pale scales and with definite areas of all-dark and all-pale scales. Fringe uniformly rather dark.

ADULT MALE. Coloration similar to that of the female, but with band on proboscis narrower. *Terminalia* (fig. 107). Lobes of ninth

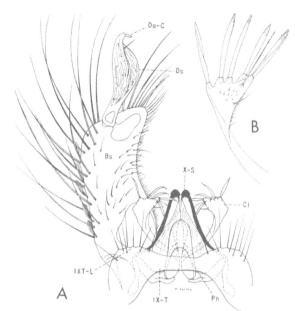

Fig. 107. *Psorophora discolor* (Coquillett). *A*, Male terminalia. *B*, Crown of claspette.

segment with six large scales on the posterior margin of a weakly sclerotized area; individual scale thorn-shaped, with the large lateral spines a little more than one-third as long as the median spine. Siphonal index about 3.0; siphon small, not inflated; pecten of about 6 to 8 long teeth on basal half of siphon; siphonal tuft long, multiple, barbed, inserted slightly beyond middle of siphon; dorsal preapical spine strong, curved, about as long as apical diameter of siphon. Anal segment ringed by the saddle, a row of spicules on the dorsoapical margin of the saddle; lateral hair single, about twice as long as the saddle; dorsal brush bilaterally consisting of a long lower caudal hair and a shorter 3- or 4-branched upper caudal tuft; ventral brush sparse, with about three or four tufts piercing the saddle; gills four times

tergite (IXT-L) broadly rounded, each bearing several scattered setae. Tenth sternite (X-S) heavily sclerotized apically. Phallosome (Ph) rounded at apex, open ventrally, closed dorsally; each plate with a distinct median triangular projection turned ventrolaterally from its inner ventral margin. Claspette (Cl) with stem expanded apically into a triangular crown attached to basal third of basistyle; apical part of crown free and bearing five or six long pointed apically feathered blades and a single apically feathered seta. Basistyle (Bs) about two and one-half times as long as mid-width, clothed with scales and long and short setae; basal and apical lobes absent. Dististyle (Ds) reticulated, about three-fifths as long as basistyle, expanded medially, tapered toward tip; claw (Ds-C) short, blunt.

LARVA (fig. 108). Antenna longer than the head, inflated, sinuate, spiculate; antennal tuft large, multiple, barbed; a pair of long dark spines arising from a projection at distal third of the shaft. Head hairs: Postclypeal 4 small, branched, inserted posterior to hair 5; upper frontal 5 and lower frontal 6 long, single, lightly barbed; preantennal 7 long, double, lightly barbed. Lateral abdominal hair 6 multiple on segments I and II, usually double on segments III–V, single on VI. Comb of eighth

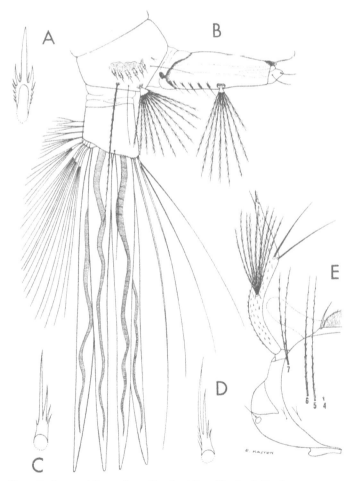

Fig. 108. Larva of *Psorophora discolor* (Coquillett). *A*, Comb scale. *B*, Terminal segments. *C*, and *D*, Pecten teeth. *E*. Head.

as long as the saddle or longer, each conspic-uously tracheate.

DISTRIBUTION. Mexico (481) and the south-ern United States, north to New Jersey and west to Nebraska and New Mexico. *United States:* Alabama (135, 419); Arkansas (88, 121); Delaware (167); District of Columbia (213, 318); Florida and Georgia (135, 419); Illinois (615); Iowa (531); Kansas (14, 530); Kentucky (134, 570); Louisiana (213, 759); Maryland (161); Mississippi (135, 546); Missouri (6, 324); Nebraska (14, 530); New Jersey (213, 350); New Mexico (273); North Carolina (135, 646); Ohio (731a); Oklahoma (628); South Carolina (135, 742); Tennessee (135, 658); Texas (249, 463); Virginia (213).

BIONOMICS. The larvae develop in temporary rain-filled pools, in overflow pools along streams, and in rice fields. The species appears to reach its greatest abundance in Arkansas, Oklahoma, and Texas. Horsfall (378) states that this species constituted 36 per cent of the total moquitoes collected in light traps oper-ated in the Arkansas rice fields from 1939 to 1941. Horsfall found that *Psorophora discolor* requires longer for development in the aquatic stages than *P. confinnis*, usually about 10 days to 2 weeks, and that the larvae do not suspend themselves at the surface film until just before pupation. This may explain the infrequency of larval collections of this mosquito.

The females are troublesome biters when abundant, particularly at night. Whitehead (750) conducted studies in the rice fields of Arkansas which indicate that, under conditions prevailing at the time of the study, cattle pro-vided the principal source of blood meals for the species, with horses, mules, and hogs of intermediate importance. Birds were rarely fed upon. The adults are readily attracted to light traps.

Psorophora (Grabhamia) pygmaea
(Theobald)

Grabhamia pygmaea Theobald, 1903, Mon. Culic., 3:245.
Culex nanus Coquillett, 1903, Can. Ent., 35:256.
Taeniorhynchus antiguae Giles, 1904, Jour. Trop. Med., 7:384.

ADULT FEMALE (pl. 42). Rather small spe-cies. *Head:* Proboscis dark-scaled, with a broad median whitish band; palpi short, dark, a few scattered white scales present. Occiput with narrow pale-yellow scales and numerous erect forked scales dorsally; the forked scales of the median area usually pale, others dark; occiput laterally with broad flat whitish scales surrounding a small black-scaled patch near eye margin. *Thorax:* Integument of scutum dark brown to black; scutum clothed with fine narrow pale-yellow to golden-brown scales, a little paler on anterior and lateral margins and on the prescutellar space. Scutellum with pale-yellow or golden scales and brown setae on the lobes. Pleura with patches of grayish-white appressed scales. *Abdomen:* First tergite with a median patch of white scales; remaining tergites dark basally, with broad apical white bands; tergites VI and VII with intermixed white and dark scales. Venter whitish-scaled, speckled or spotted with dark scales. *Legs:* Femora dark, speckled with pale scales, pos-terior surface pale; each femur with a narrow subapical more or less distinct white-scaled band; knee spots white. Tibiae dark, speckled with white. Each segment of hind tarsus with a narrow white-scaled basal ring; segments 1 to 3 of front and middle tarsi each with a very narrow white-scaled basal ring, reduced or absent on 4 and 5. *Wing:* Length about 3.0 mm. Scales rather broad, intermixed white and dark brown to black, the dark scales pre-dominating; fringe entirely dark.

ADULT MALE. Coloration similar to that of the female, but with median band of proboscis and the apical white bands of the terminal abdominal segments narrower. *Terminalia* (fig. 109). Lobes of ninth tergite (IXT-L) broadly rounded, bearing many scattered setae. Tenth sternite (X-S) heavily sclerotized beyond middle. Phallosome (Ph) bluntly con-ical, open ventrally, closed dorsally. Claspette (Cl) slender, triangularly expanded at apex; the expanded part weakly attached to basistyle, bearing a row of four or five long slender fila-ments on crown. Basistyle (Bs) a little more than twice as long as mid-width, clothed with sparse scales and numerous long and short setae; basal and apical lobes absent. Dististyle (Ds) reticulated, nearly two-thirds as long as

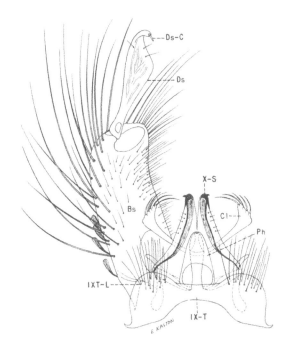

Fig. 109. Male terminalia of *Psorophora pygmaea*
(Theobald).

basistyle, broadly expanded at apical third
and tapered to a slender tip, bearing a few
small setae on distal part; claw (Ds-C) short,
peglike.

LARVA (fig. 110). Antenna much shorter
than the head, spiculate; antennal tuft short,
multiple, inserted a little before middle of
shaft. Head hairs: Postclypeal 4 small,
branched; upper frontal 5 and lower frontal
6 long, single, weakly and sparsely barbed;
preantennal 7 multiple, weakly and sparsely
barbed. Comb of eighth segment with about 6
scales in a curved row on posterior margin of
a weakly sclerotized area; individual scale
with the median spine about twice as long as
the stout subapical spines. Siphonal index
about 3.0; siphon somewhat inflated; pecten of
about 4 to 6 teeth on basal third of siphon;
siphonal tuft small, multiple, inserted beyond
middle of siphon; hair of lateral valve short,
less than half as long as the apical diameter
of the siphon. Anal segment completely ringed
by the saddle, the saddle spiculate on the
apical part; lateral hair multiple, much shorter
than the saddle; dorsal brush bilaterally consisting of a long lower caudal hair and a
shorter multiple upper caudal tuft; ventral

brush extending almost the entire length of the
anal segment and piercing the saddle; gills
usually about twice as long as the saddle,
pointed.

DISTRIBUTION. This mosquito occurs in the
Bahamas and West Indies but has been taken
at Key West, Florida, August, 1901, by A.
Busck and April 1–3, 1903, by E. A. Schwartz
(382). Additional collections were obtained
at Key West, Florida, by Sargent Mead, June
13 and October 8, 1924, and at Fisher's Island,
Miami Beach, Florida, during June and July,
1946 (112).

BIONOMICS. The larvae occur in temporary
rain pools and develop rapidly. According to
Howard, Dyar, and Knab (382), Dr. Grabham
obtained eggs from captured gravid females.
The larvae hatched in 2 days and were fully
grown in 8 days. The females are readily
attracted to light traps.

Psorophora (Grabhamia) signipennis
(Coquillett)

Taeniorhynchus signipennis Coquillett, 1904, Proc.
Ent. Soc. Wash., 6:167.

ADULT FEMALE (pl. 43). Medium-sized species. *Head:* Proboscis dark-scaled basally and
apically, with a very wide whitish-yellow median band; palpi short, dark, speckled with a
few pale scales. Occiput with narrow pale-
golden scales and erect forked scales dorsally;
the forked scales of central region pale, others
dark; occiput laterally with a patch of broad
appressed yellowish scales, either speckled
with dark scales or partly enclosing a small
dark-scaled patch behind the eye. Tori light
brown, densely clothed with narrow yellowish
scales on inner and dorsal surfaces. *Thorax:*
Integument of scutum dark brown; scutum
clothed with fine narrow golden-brown scales,
becoming pale yellow on sides and on prescutellar space. Scutellum with narrow pale-yellow scales and brown setae on the lobes. Pleura
with patches of broad appressed yellowish-
white scales. *Abdomen:* First tergite with median area white-scaled; remaining tergites primarily white-scaled, more or less speckled or
spotted with dark scales. Venter pale-scaled,
speckled, or spotted with dark scales. *Legs:*

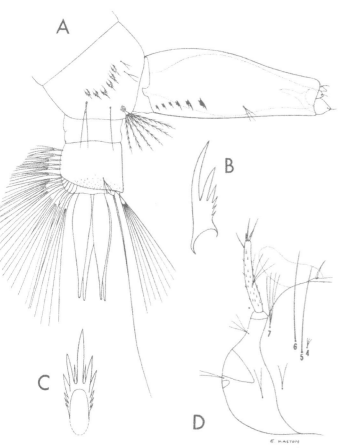

Fig. 110. Larva of *Psorophora pygmaea* (Theobald). *A*, Terminal segments. *B*, Pecten tooth. *C*, Comb scale. *D*, Head.

Femora dark, liberally speckled with pale scales; largely pale on posterior surface; each with a narrow, often indefinite, subapical whitish-scaled ring; knee spots present. Tibiae dark, speckled with pale scales. Hind tarsus with segment 1 predominantly pale-scaled, sprinkled with dark scales, dark-ringed sub-basally and apically; segments 2 to 5 pale-scaled basally, dark-scaled apically; front and middle tarsi similarly marked, but with the pale scaling reduced on segment 4, absent on 5. *Wing:* Length about 3.2 to 4.0 mm. Scales rather broad, intermixed white and dark brown to black. Apical half of costa with two dark-scaled spots separated by an area of white scales; vein 6 with intermixed dark and white scales on basal half to two-thirds, a dark-scaled patch at distal two-thirds, distal part white-scaled. Fringe scales arranged in alternating dark and pale patches.

ADULT MALE. Coloration similar to that of the female. *Terminalia* (fig. 111). Lobes of ninth tergite (IXT-L) broadly rounded, each bearing several setae. Tenth sternite (X-S) heavily sclerotized apically. Phallosome (Ph) rounded apically, open ventrally, closed dorsally. Claspette (Cl) slender, curved and expanded apically; apical part attached to basistyle and bearing five setalike filaments. Basistyle (Bs) about twice as long as mid-width, square or slightly truncate at apex, clothed with sparse scales and large and small setae; basal and apical lobes absent. Dististyle (Ds) a little more than one-half as long as basistyle, reticulated, somewhat expanded medially, slender and curved near tip; claw (Ds-C) short, peg-like.

LARVA (fig. 112). Antenna a little shorter than the head, spiculate; antennal tuft about 8- to 15-branched, barbed, inserted near middle of shaft. Head hairs: Postclypeal 4 small, multiple; upper frontal 5 and lower frontal 6 long, single to triple (one or more usually single), barbed; preantennal 7 about 6- to 8-branched, barbed. Comb of eighth segment with 4 to 8 scales in a curved row on the posterior mar-

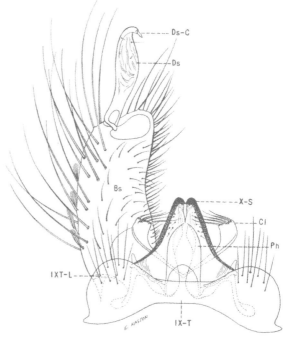

Fig. 111. Male terminalia of *Psorophora signipennis* (Coquillett).

gin of a weakly sclerotized area; individual scale thorn-shaped, with a long central tooth and much shorter basal teeth. Siphonal index about 3.0; siphon moderately inflated; pecten usually of 4 to 6 progressively longer teeth on basal third of siphon; siphonal tuft minute, multiple, inserted near apical third of siphon; hair of lateral valve more than half as long as apical diameter of siphon. Anal segment completely ringed by the saddle; lateral hair small, much shorter than the saddle, double or multiple; dorsal brush bilaterally consisting of a long lower caudal hair and a shorter multiple upper caudal tuft; ventral brush extending almost the entire length of the anal segment and piercing the saddle; gills about two to three times as long as the saddle.

DISTRIBUTION. Central United States, south to Mexico. *United States:* Arizona (213); Arkansas (88, 121); Colorado (343); Iowa (625, 626); Kansas (14, 360); Kentucky (570); Missouri (324, 530); Montana (213, 475); Nebraska (14, 680); New Mexico (213, 273); North Dakota (553); Oklahoma (628); South Dakota (14, 530); Tennessee (658); Texas (249, 463); Wyoming (578).

BIONOMICS. The larvae occur in temporary ground pools, roadside pools, and occasionally in recently filled depressions in dried-up stream beds. According to Mail (475) the species is well adapted to transient pools on the arid plains where it may pass from the egg to

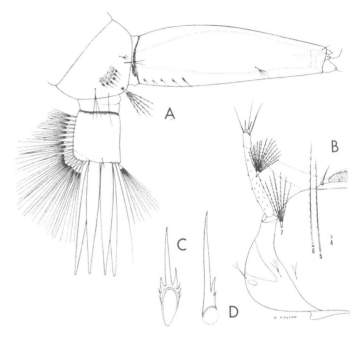

Fig. 112. Larva of *Psorophora signipennis* (Coquillett). *A,* Terminal segments. *B,* Head. *C,* Comb scale. *D,* Pecten tooth.

the adult stage in 5 days under favorable conditions. Rozeboom (628) states that it is one of the commonest mosquitoes in Oklahoma, but in spite of the abundance of the larvae and of the adults in light traps, the females were rarely observed biting. Tate and Gates (680) found the species to be quite annoying at times in Nebraska, and Rowe (625) made similar observations in Iowa.

Genus AEDES Meigen[8]

Aedes Meigen, 1818, Syst. Beschr. Zweifl. Ins., *1*:13.

ADULT. Palpi of male rarely longer than the proboscis, sometimes rather short; palpi of female rarely more than one-fourth as long as the proboscis. Ornamentation of the scales on the dorsal region of the occiput varying with the species; erect forked scales numerous. Thorax with variable ornamentation; scutal bristles well developed. Posterior pronotal bristles usually four to six in a row, sometimes more. Spiracular bristles absent. Postspiracular bristles present. Prealar, sternopleural, and upper mesepimeral bristles rather numerous. Lower mesepimeral bristles present or absent. Pleura usually well scaled. Abdomen of female with tip more or less pointed, the eighth segment retractile. Pulvilli absent or hairlike. Tarsal claws of front and middle legs of male unequal, each with one tooth. Tarsal claws of the female usually toothed. Wing with squama fringed; fork of vein 2 much longer than its petiole; wing membrane with distinct microtrichia. *Male terminalia* (fig. 8). The structures vary greatly in the different subgenera and provide useful characters for classification. The extent of variation in the species of this region is noted in the individual descriptions and illustrations.

PUPA. Respiratory trumpets usually short, somewhat dilated at apices. Dendritic tufts on first abdominal segment well developed. Paddles with margin usually serrate or fringed and with a single or branched apical hair of moderate length.

LARVA. Head hairs vary greatly in position and branching. Abdomen without dorsal chitinous plates except on anal segment. Comb scales variable in shape and number. Siphon usually short, rarely more than four times as long as its basal diameter. Pecten well developed. Siphon with a single pair of siphonal tufts, never inserted near base (rarely with accessory dorsal hairs or tufts). Anal segment always with a barred area. Ventral brush of anal segment usually well developed, rarely reduced. Dorsal brush of anal segment with the lower paired caudal hairs long and simple; upper paired caudal tufts usually shorter and multiple.

EGGS. Spindle shaped or elliptical, usually with fine hexagonal sculpturing on exochorion but without spiny processes. In general, the eggs are able to withstand long periods of drying. The eggs of most species are laid on damp soil in depressions subject to filling by rains, melting snow, or floodwater, although a few species utilize more restricted habitats such as rot cavities in trees and artificial containers.

BIONOMICS. The members of this genus are frequently of considerable economic importance, since the majority are vicious biters and several are vectors of diseases of man and animals.

CLASSIFICATION. The form of the male palpi and the characters of the male terminalia are the chief means of separating the subgenera. The following subgenera are recognized for this area and salient characters are given for each.

Subgenus *Ochlerotatus* Lynch Arribálzaga. Palpi of the male usually longer than the proboscis, sometimes of about equal length or a little shorter. Male terminalia with phallosome simple, smooth, not divided into lateral plates. Claspettes usually well developed with columnar stem and flattened filament (filament rarely setiform). Basistyle with a distinct basal lobe and usually with a more or less developed apical lobe.

Many of the dark-legged *Aedes*, particularly in Edwards' (251) Group G, have variable color characters, especially in the markings of the scutum of the adult female, and are often difficult to identify with certainty in this stage. Studies of the *punctor* subgroup by Knight (430), based on larval-associated specimens, show that *Aedes aboriginis* Dyar, *A. hexodontus* Dyar, *A. abserratus* (Felt and Young), *A. punctor* (Kirby), and *A. punctodes* Dyar are extremely variable in scutal markings in the female. At present the identification of these species should be based, for the most part, on larvae or larval-associated specimens, particularly where their range overlaps.

Subgenus *Finlaya* Theobald. Palpi of the male usually a little shorter than the proboscis. Male terminalia with phallosome simple and undivided. Claspettes well developed, with

---

[8]Consult Dyar (1928) or Edwards (1932) for synonyms.

moderate stem and with filament usually long and slightly flattened. Basistyle with the basal lobe, at most, weakly developed; anal lobe absent.

Subgenus *Stegomyia* Theobald. Palpi of male as long or nearly as long as the proboscis. Male terminalia with the phallosome divided into two lateral plates, each plate bearing numerous teeth distally. Claspettes absent. Basistyle with the basal lobe present, apical lobe absent.

Subgenus *Aedimorphus* Theobald. Palpi of male about as long as the proboscis. Male terminalia with the phallosome divided into two lateral plates, each plate bearing teeth distally. Claspette filament absent. Claw of dististyle subterminal or well removed from tip. Dististyle often highly modified.

Subgenus *Aedes* Meigen. Palpi very short in both sexes. Male terminalia with the phallosome divided into a pair of plates or rods. Claspette filament absent. Basistyle short and stout. Dististyle varies in the different species; terminal claw absent.

KEYS TO THE SPECIES

ADULT FEMALES[9]

1.   Tarsal segments, at least on hind legs, with white rings 2
Tarsal segments without white rings......26

2 (1). Tarsal segments white-ringed on basal part of segment only 3
Tarsal segments white-ringed both apically and basally (at least on some of the segments) 20

3 (2). Proboscis with a white ring near middle....4
Proboscis lacking a white ring near middle 7

4 (3) Abdomen with a white to yellow dorsal median longitudinal stripe or dorsal median longitudinal row of disconnected spots; wing scales either intermixed dark brown and white or entirely dark......................5
Abdomen with transverse basal bands of white scales, but lacking a pale median longitudinal stripe; wing scales entirely dark.......*taeniorhynchus* (Wied.), p. 238

5 (4). Wing entirely dark-scaled; first segment of

hind tarsus without a pale median ring
mitchellae (Dyar), p. 204
Wing with intermixed dark-brown and white scales; first segment of hind tarsus with or without a pale median ring......6

6 (5). Lateral pale markings of abdomen white, the dorsal pale markings yellowish; first segment of hind tarsus with a definite yellow median ring; last segment of hind tarsus mostly white
sollicitans (Walker), p. 227
Lateral and dorsal pale markings of abdomen both yellowish; first segment of hind tarsus variably marked (if a median pale ring is present, it is white, not yellow); last segment of hind tarsus white basally, dark apically, rarely all white
(in part) *nigromaculis* (Ludlow), p. 208

7 (3). Scutum with a conspicuous lyre-shaped marking of silver-white scales against a dark background. *aegypti* (Linn.), p. 261
Scutum without such markings..................8

8 (7). Hind femora entirely pale on all aspects of basal half......*zoosophus* D. and K., p. 259
Hind femora with basal half of anterior surface all dark or with intermixed dark and pale scales 9

9 (8). Basal white rings of tarsal segments broad, especially on hind legs.........................10
Basal white rings of tarsal segments narrow 19

10 (9). Wing scales very large, broad, triangular-shaped, with dark and pale scales rather evenly intermixed.........................11
Wing scales moderate in size, narrow, ligulate, with dark and white scales unevenly distributed or with pale scales absent..12

11 (10). Proboscis with many pale scales intermixed to near apex; scutum with a broad median brown stripe; sides of scutum with mottled areas of brown and yellowish-white scales, the yellowish-white scales predominating..*squamiger* (Coq.), p. 231
Proboscis with a few scattered pale scales on basal half; scutum with a broad median dark brown stripe becoming golden-brown anteriorly; sides of scutum white
grossbecki D. and K., p. 187

12 (10). Palpi all dark; abdominal tergites basally, laterally, and medially yellowish-scaled, nearly surrounding large patches of dark scales, the median pale scales forming a longitudinal stripe
(in part) *nigromaculis* (Ludlow), p. 208
Palpi with some pale scales; pale scales of abdominal tergites never forming a distinct and complete median longitudinal line 13

13 (12). Abdominal tergites clothed with yellow scales, without basal bands
flavescens (Müller), p. 183

---

[9]The specimens of *Aedes cacothius* Dyar which we saw were generally brushed and unsatisfactory for keying. Specimens of *Aedes rempeli* Vockeroth were not available to us for keying.

Abdominal tergites dark-scaled, with pale basal bands (dark-scaled areas with few to many pale scales in some species)..14

14 (13). Tarsal claws with main tooth abruptly bent near the base of the lateral tooth, main tooth and lateral tooth parallel with each other (fig. 143, *B*)
excrucians (Walker), p. 178
Tarsal claws with the main tooth bent beyond the base of the lateral tooth..........15

15 (14). Tarsal claws with lateral tooth short and blunt, less than one-half as long as the main tooth, the main tooth bent before tip (fig. 183, *B*)
riparius D. and K., p. 222
Tarsal claws with lateral tooth long and slender, about one-half as long as the main tooth, the main tooth bent beyond the base of the lateral tooth (fig. 145, *B*) 16

16 (15). Lower mesepimeral bristles absent
(in part) *fitchii* (F. and Y.), p. 181
Lower mesepimeral bristles present..........17

17 (16). Tori without white scales on dorsal surface; palpi lacking hairs on basal half of apical segment at inner ventral surface
increpitus Dyar, p. 197
Tori with white scales on dorsal surface or with apical segment of palpi bearing many large hairs on inner ventral edge, or with both 18

18 (17). Lower mesepimeral bristles rarely more than two; tori with white scales on dorsal half....(in part) *fitchii* (F. and Y.), p. 181
Lower mesepimeral bristles usually three or more; tori with or without white scales on dorsal surface
stimulans (Walker), p. 236

19 (9). Lower mesepimeral bristles absent; seventh abdominal tergite mostly dark-scaled, with white scales at apex of segment when present..............*vexans* (Meigen), p. 263
Lower mesepimeral bristles present; seventh abdominal tergite entirely pale-scaled.................*cantator* (Coq.), p. 164

20 (2). Wing with dark and white scales intermixed 21
Wing with scales all dark or with some white scales on anterior veins........................23

21 (20). Wing with dark and white scales evenly intermixed 22
Wing with dark and white scales not evenly intermixed, vein 3 with more dark scales than veins 2 and 4
dorsalis (Meigen), p. 174

22 (21). Wing with white scales predominating
campestris D. and K., p. 159
Wing with dark scales usually predominating..................*melanimon* Dyar, p. 202

23 (20). Wing with a patch of white scales at base of costa 24
Wing with base of costa dark-scaled......25

24 (23). Palpi with white bands; scutellum with broad white scales
varipalpus (Coq.), p. 257
Palpi entirely dark; scutellum with narrow yellowish scales
atropalpus (Coq.), p. 253

25 (23). Hind tarsi with broad white rings present basally and apically on all segments except the last, which is entirely white; scutum clothed dorsally with golden-brown scales
canadensis canadensis (Theob.), p. 161
Hind tarsi with narrow white rings present basally and apically on segments 1 and 2 and basally on segment 3, the remaining segments entirely dark; scutum clothed dorsally with blackish-brown scales and a very narrow median longitudinal line of golden-brown scales
canadensis mathesoni Middlekauff p. 163

26 (1). Postspiracular bristles absent
purpureipes Aitken, p. 220
Postspiracular bristles present..............27

27 (26). Integument of scutum with a pair of dark-brown or black posterolateral spots......28
Integument of scutum lacking posterolateral dark-brown or black spots.....................29

28 (27). Pleuron of thorax with a dark-brown spot beneath the anterior spiracle; abdominal tergites yellow-scaled basally and laterally, dark-scaled apically
fulvus pallens Ross, p. 185
Pleuron of thorax without a dark spot; abdomen almost entirely yellow-scaled
bimaculatus (Coq.), p. 157

29 (27). Scutum with a broad median longitudinal stripe or patch of silver-white or pale-yellow scales or with the sides and anterior margin clothed with silver-white scales 30
Scutum not marked with silver-white scales 33

30 (29). Scutum with a broad median stripe of dark-brown scales, this stripe broader posteriorly, sides and anterior margin with silver-white scales
triseriatus (Say), p. 255
Scutum with a broad median stripe or patch of silver-white scales or pale-yellow scales 31

31 (30). Median longitudinal stripe of scutum extending from the anterior margin to a little beyond middle, much broader than the dark-scaled area on either side
infirmatus D. and K., p. 199;
scapularis (Rond.), p. 224

Median longitudinal stripe extending the full length of the scutum, usually narrower than the dark-scaled area on either side (variable in *A. dupreei*)..............32

32 (31). Small species (wing length about 2.5 mm); occiput with a narrow median stripe of narrow white scales bounded submedially by a patch of broad white scales
dupreei (Coq.), p. 177

Medium-sized species (wing length about 3.5 mm); occiput dorsally with a median stripe of narrow white scales bounded submedially by a patch of broad dark scales...........*atlanticus* D. and K., p. 151; *tormentor* D. and K., p. 243

33 (29). Scutum with a pair of broad submedian white or yellowish-white stripes separated by a brown stripe of about the same width...............*trivittatus* (Coq.), p. 249

Scutum not marked with two broad submedian white or yellowish-white stripes 34

34 (33). Abdominal tergites with median basal patches of white scales
thelcter Dyar, p. 240

Abdominal tergites with complete or partial basal pale bands, without pale bands or almost completely pale-scaled................35

35 (34). Wing scales distinctly bicolored..............36

Wing scales entirely dark or with only white-scaled patches at bases of the veins or with scattered pale scales on the anterior veins 38

36 (35). Wing with dark and pale scales intermixed, the dark predominating; lower mesepimeral bristles usually present
niphadopsis D. and K., p. 210

Wing veins alternating black- and white-scaled; lower mesepimeral bristles absent 37

37 (36). Abdomen with a dorsal median longitudinal stripe of pale scales or almost entirely pale-scaled; scales on dorsal half of posterior pronotum brown
spencerii (Theob.), p. 229

Abdomen lacking a dorsal median longitudinal stripe of pale scales; scales on dorsal half of posterior pronotum white or whitish-brown
idahoensis (Theob.), p. 191

38 (35). Scutum lacking contrasting lines or stripes 39

Scutum with contrasting lines or stripes..45

39 (38). Lower mesepimeral bristles present..........40

Lower mesepimeral bristles absent..........44

40 (39). Scutum with many long black or brownish-black setae, thus giving the insect a hairy appearance 41

Scutum moderately clothed with normal setae 42

41 (40). Sternopleuron with scales extending to anterior angle; scutellum with golden-brown scales........*nigripes* (Zett.), p. 206

Sternopleuron with scales not reaching anterior angle; scutellum with yellowish-white scales......*impiger* (Walker), p. 193

42 (40). Mesepimeron bare on lower one-fourth to one-third...............*intrudens* Dyar, p. 200

Mesepimeron with scales reaching to lower margin 43

43 (42). Wing with scattered to numerous whitish scales along costa, subcosta, and vein 1
cataphylla Dyar, p. 166

Wing scales dark except for a patch of white scales at base of costa
(in part) *hexodontus* Dyar, p. 188; *punctodes* Dyar, p. 216; (in part) *punctor* (Kirby), p. 217

44 (39). Coxa of front leg with a central patch of brown scales on anterior margin; occiput dorsally with a submedian patch of broad dark appressed scales
cinereus Meigen, p. 266

Coxa of front leg clothed with white scales on anterior surface; occiput lacking a submedian patch of broad dark appressed scales on dorsal surface
ventrovittis Dyar, p. 251

45 (38). Lower mesepimeral bristles present..........46

Lower mesepimeral bristles absent..........54

46 (45). Hypostigial spot of few to many white scales present 47

Hypostigial spot of scales absent..............50

47 (46). Proboscis with scattered pale scales to near tip...............*bicristatus* T. and W., p. 155

Proboscis entirely dark............................48

48 (47). Scutum with paired bare submedian stripes separated by a pair of narrow stripes of light-brown scales with a narrow bare faint line between them
pullatus (Coq.), p. 214

Scutum with a broad median stripe or narrowly divided stripes of brown scales....49

49 (48). Sternopleuron with scales extending about halfway to anterior angle
implicatus Vockeroth, p. 195

Sternopleuron with scales extending to anterior angle........*trichurus* (Dyar), p. 247

50 (46). Abdominal tergites with narrow basal pale bands on less than half the segments when present
(in part) *diantaeus* H., D., and K., p. 172

Abdominal tergites with basal white bands on more than half the segments............51

51 (50). Proboscis with yellowish-gray scales on ventral surface..*schizopinax* Dyar, p. 226

Proboscis with brown scales on ventral surface 52

52 (51). Wing usually with a patch of white scales at base of costa[10].........................53
Wing usually lacking a patch of white scales at base of costa[11] *aboriginis* Dyar, p. 147;
abserratus (F. and Y.), p. 149;
(in part) *punctor* (Kirby), p. 217

53 (52). Scutum clothed with yellow or yellowish-white scales and a pair of rather well-defined submedian dark-brown stripes or with a varying pattern of yellowish-white to yellow and dark-brown scales; supra-alar bristles generally dark brown or black...........*communis* (De G.), p. 168;
pionips Dyar, p. 212
Scutum clothed with yellow scales (occasionally yellowish white or yellowish brown) and a broad median or narrowly divided stripe of dark-brown scales; supra-alar bristles generally yellow to dark brown
(in part) *hexodontus* Dyar, p. 188

54 (45). Abdomen with basal white bands on more than half of the segments.....................55
Abdomen without basal white bands or with narrow bands on less than half of the segments 57

55 (54). Hypostigial spot of scales present
muelleri Dyar, p. 206
Hypostigial spot of scales absent..............56

56 (55). Pleuron with medium-sized patches of pale scales; sternopleuron with a narrow line of scales reaching near the anterior angle and with the scales narrowly separated from the patch of pale scales on the pre-alar area..........*sticticus* (Meigen), p. 233
Pleuron with small patches of pale scales; sternopleuron with scales reaching about halfway to anterior angle and with scales well separated from patch on prealar area
tortilis (Theob.), p. 245

57 (54). Scutum with a broad median dark stripe, much broader posteriorly.....................58
Scutum with two narrow brown stripes separated by a narrow line of pale-yellow scales, brown stripes sometimes fused, thus forming one broad brown stripe but not distinctly broader posteriorly........59

58 (57). Venter of abdomen clothed with grayish-white scales; scutum with median dark stripe gradually broadened posteriorly
aurifer (Coq.), p. 153
Venter of abdomen with some of the segments brown-scaled apically; scutum with median dark stripe abruptly broadened posteriorly......*thibaulti* D. and K., p. 242

---

[10]Costa occasionally entirely dark in *A. communis, A. hexodontus,* and *A. pionips.*

[11]Costa occasionally with a white patch of scales in *A. aboriginis;* rarely a few white scales at base of costa in *A. punctor.*

59 (57). Occiput clothed dorsally with narrow curved pale-yellow scales; sternopleuron with about twelve to twenty setae
(in part) *diantaeus* H., D., and K., p. 172
Occiput dorsally with narrow curved pale-yellow scales on median area and with subdorsal patches of dark-brown appressed scales; sternopleuron usually with five or six, not more than ten, setae
decticus H., D., and K., p. 170

MALE TERMINALIA[12]

1. Dististyle furcate near base, inserted well before apex of the basistyle
cinereus Meigen, p. 266
Dististyle not furcate, inserted at apex of the basistyle 2

2 (1). Claspettes absent; basistyle very short (about one and one-half times as long as wide)....................*aegypti* (Linn.), p. 261
Claspettes present; basistyle elongate (two and one-half to four times as long as wide) 3

3 (2). Claspette stem crowned with a dense tuft of setae, filament absent; claw of dististyle subapical in position, arising at apical fifth of the dististyle
vexans (Meigen), p. 264
Claspette with a distinct filament; claw of dististyle apical in position.....................4

4 (3). Claspette filament arising from a short lateral branch a little beyond the middle of the claspette stem
thibaulti D. and K., p. 242
Claspette filament arising at apex of the claspette stem 5

5 (4). Terminal claw of dististyle one-half as long as the dististyle 6
Terminal claw of dististyle never more than one-third as long as the dististyle..........7

6 (5). Basistyle with a small dense patch of long hairs arising on the inner face at apical third.................*triseriatus* (Say), p. 255
Basistyle without a small dense patch of long hairs on the inner surface at apical third..............*zoosophus* D. and K., p. 259

7 (5). Apical lobe of basistyle absent.................8
Apical lobe of basistyle present.................12

8 (7). Claspette filament sharply pointed, bearing a prominent sharp retrorse median projection on the convex side
taeniorhynchus (Wied.), p. 238
Claspette filament slender, curved, without a retrorse median projection on the convex side 9

9 (8). Basal lobe of basistyle large, elongate, bear-

---

[12]Male specimens of *Aedes muelleri* Dyar and *A. purpureipes* Aitken were not available to us for keying.

ing a dense basal clump of long rather strong apically curved setae and a secondary area of short setae extending to the apical fourth of the basistyle
varipalpus (Coq.), p. 257
Basal lobe of basistyle small, rounded, clothed with slender, nearly straight setae, without a secondary area of short setae.10

10 (9). Dististyle of nearly equal width throughout, glabrous except for short subapical setae; lobes of ninth tergite inconspicuous, without setae..........*atropalpus* (Coq.), p. 253
Dististyle distinctly broadened before the middle, pilose; lobes of ninth tergite small but distinct, armed with short spinelike setae 11

11 (10). Basal lobe of basistyle prominent and rounded apically
mitchellae (Dyar), p. 204
Basal lobe of basistyle only slightly elevated, not prominent
sollicitans (Walker), p. 228;
nigromaculis (Ludlow), p. 209

12 (7). Claspette stem stout, broad medially, constricted basally and apically, the median part much broader than the claspette filament.........*atlanticus* D. and K., p. 152
Claspette stem slender, never more than slightly broader than the claspette filament 13

13 (12). Basistyle with a dense brushlike tuft of long setae on ventral side near base of apical lobe 14
Basistyle lacking a dense brushlike tuft of long setae on ventral side near base of apical lobe 15

14 (13). Apical lobe of the basistyle elongate; claspette stem with a sharp inner angle near the distal fourth
diantaeus H., D., and K., p. 172
Apical lobe of the basistyle short and broadly rounded; claspette stem curved but lacking a sharp inner angle
decticus H., D., and K., p. 170

15 (13). Claspette filament very short and partly cone-shaped 16
Claspette filament rather long, ligulate, or bladelike 17

16 (15). Basal lobe of the basistyle bearing two or three heavy setae on the dorsal surface; sclerotized cap of the claspette filament has transverse ridges
trichurus (Dyar), p. 247
Basal lobe of the basistyle bearing seven or eight long setae on the dorsal surface; sclerotized cap of the claspette filament has fingerlike projections
bicristatus T. and W., p. 155

17 (15). Basal lobe of the basistyle bearing setae but lacking a distinctly enlarged spine..18

Basal lobe of the basistyle bearing one or more distinctly enlarged spines............23

18 (17). Apical lobe of the basistyle bearing numerous short flattened spatulate setae
canadensis canadensis (Theob.), p. 162;
canadensis mathesoni Middlekauff, p. 164
Apical lobe of the basistyle bearing slender hairlike or slightly broadened setae....19

19 (18). Claspette filament with a sharp angular or retrorse projection on the convex side near base or middle..............................20
Claspette filament without an angular or retrorse projection on convex side near base or middle.....................................21

20 (19.) Basal lobe of basistyle elongate, extending nearly to base of the apical lobe; claspette filament with an angular projection on basal third of convex side
excrucians (Walker), p. 179
Basal lobe of basistyle not extending more than halfway to base of the apical lobe; claspette filament with a sharp angular or slightly retrorse projection near middle of convex side......*increpitus* Dyar, p. 197

21 (19). Basal lobe of basistyle narrowly elongate, semidetached, directed distad; apical lobe bearing a few strong setae
bimaculatus (Coq.), p. 157
Basal lobe of basistyle short, conical, directed mesad; apical lobe bearing a few short setae 22

22 (21). Phallosome deeply incised apically, with the points long and turned inward
nigripes (Zett.), p. 207
Phallosome notched apically, the points short and not turned inward
rempeli Vockeroth, p. 221

23 (17). Basistyle with a conspicuous dense tuft of setae near the apex................................24
Basistyle without a conspicuous dense tuft of setae near the apex.........................25

24 (23). Claspette filament with a large median barblike retrorse projection arising from the convex side; basal lobe of the basistyle bearing a single long stout spine at apex.
aurifer (Coq.), p. 153
Claspette filament with a median angular expansion on the convex side, not barblike; basal lobe of the basistyle bearing two rather stout spines at apex and a large dorsal spine at base
intrudens Dyar, p. 201

25 (23). Basal lobe of the basistyle cylindrical, fingerlike, nearly four times as long as its basal width.....*tormentor* D. and K., p. 244
Basal lobe of the basistyle never more than twice as long as its basal width...........26

26 (25). Basal lobe of the basistyle bearing one or two stout spines in addition to the normal long stout dorsal spine............................27

Basal lobe of the basistyle without differentiated strong spines other than a long stout one (strong dorsal spine often followed by progressively weaker spines or setae) 29

27 (26). Basal lobe of the basistyle with two strong inner spines situated close together and a long strong outer spine; claspette stem sharply angled near middle
pullatus (Coq.), p. 215

Basal lobe of the basistyle with one short stout spine, a long curved spine, and a dense group of slender setae; claspette stem not sharply angled near middle....28

28 (27). Claspette filament roundly expanded near middle of the convex side; basal lobe of basistyle constricted basally, somewhat expanded apically
dorsalis (Meigen), p. 175

Claspette filament angularly expanded near middle of the convex side; basal lobe of basistyle conical. *melanimon* Dyar, p. 203

29 (26). Claspette filament with a spinelike retrorse projection or barb on the convex side forming an acute angle...........................30

Claspette filament ligulate, roundly expanded, or with an angular expansion on the convex side forming an obtuse angle 33

30 (29). Claspette filament with one or more small or minute accessory retrorse spines present on the inner margin of the large retrorse spine or between the large retrorse spine and the base of the filament 31

Claspette filament without accessory retrorse spines on the inner margin of the large retrorse spine, or between the large retrorse spine and the base of the filament 32

31 (30). Accessory retrorse spines on the claspette filament minute, often indistinct
trivittatus (Coq.), p. 249

Accessory retrorse spines on the claspette filament prominent and distinct
infirmatus D. and K., p. 199;
tortilis (Theob.), p. 245

32 (30). Retrorse projection on the convex side of the claspette filament long and slender
scapularis (Rond.), p. 224;
thelcter Dyar, p. 240

Retrorse projection on the convex side of the claspette filament short and triangular
squamiger (Coq.), p. 232

33 (29). Filament of the claspette ligulate, not expanded on the convex side, little if any broader at middle than the distal end of the claspette stem.............................34

Filament of claspette roundly or angularly expanded on the convex side, broader at

middle than the distal end of the claspette stem 35

34 (33). Apical lobe of the basistyle slender, triangular; basal lobe longer than broad
dupreei (Coq.), p. 177

Apical lobe of the basistyle broadly rounded; basal lobe shorter than broad
campestris D. and K., p. 159

35 (33). Enlarged spine of the basal lobe of the basistyle with tip broadened, flattened, and usually slightly retrorse
fulvus pallens Ross, p. 186

Enlarged spine of the basal lobe of the basistyle with tip sharply pointed and usually curved 36

36 (35). Basal lobe of the basistyle represented by a large rugose area raised basally, flattened distally, and reaching well beyond the middle of the basistyle
flavescens (Müller), p. 184

Basal lobe of the basistyle triangular, conical, or bluntly rounded, not reaching beyond basal two-fifths of the basistyle....37

37 (36). Claspette filament with a sharp angle or notch near base of the concave side....38

Claspette filament without a sharp angle or notch near base of the concave side....40

38 (37). Apical lobe of the basistyle clothed with normal rather straight setae; claspette filament less than twice as wide at basal one-third as the apical diameter of the claspette stem....*fitchii* (F. and Y.), p. 182

Apical lobe of the basistyle clothed with short broadened recurved setae; claspette filament expanded at basal one-third to about twice the apical diameter of the claspette stem 39

39 (38). Basal lobe of basistyle bearing a dense tuft of fine long setae adjacent to the strong dorsal spine....*sticticus* (Meigen), p. 234

Basal lobe of basistyle with fewer fine long setae adjacent to the strong dorsal spine
idahoensis (Theob.), p. 191;
spencerii (Theob.), p. 230;
ventrovittis[13] Dyar, p. 251

40 (37). Claspette filament with a sharp angular expansion in the form of an obtuse angle on the convex side.................................41

Claspette filament roundly expanded or evenly curved on the convex side........43

41 (40). Basal lobe of the basistyle long, narrow and conical or wedge-shaped
impiger (Walker), p. 194

Basal lobe of the basistyle short, conical, or rounded 42

42 (41). Basal lobe of the basistyle with an apical

---

[13]The sharp angle or notch near the base of the concave side is somewhat more prominent than in *A. idahoensis* and *A. spencerii*.

row of rather stout setae preceded by a large strong dorsal spine
implicatus Vockeroth, p. 195
Basal lobe of the basistyle with a large strong dorsal spine and short setae but without a row of rather stout setae
stimulans (Walker), p. 236

43 (40). Claspette filament narrow and necklike on the basal third, abruptly expanded on the convex side near middle.........................44
Claspette filament not narrow and necklike on the basal third, uniformly rounded on the convex side, or gradually expanded near the base or near the basal third....45

44 (43). Basal lobe of the basistyle long, conical, bearing small apical setae and a stout curved dorsal spine
grossbecki D. and K., p. 187
Basal lobe of the basistyle short, bluntly conical, bearing a stout dorsal spine, two or three long setae and many short setae
cantator (Coq.), p. 164

45 (43). Apical lobe of the basistyle bearing few to many straight setae on the inner surface 46
Apical lobe of the basistyle bearing many short broadened curved setae on the inner surface 47

46 (45). Apical lobe of the basistyle bearing only a few setae.............*cataphylla* Dyar, p. 166;
niphadopsis Dyar and Knab, p. 211
Apical lobe of the basistyle bearing many setae......*communis*[14] (De Geer), p. 168;
pionips Dyar, p. 212

47 (43). Claspette filament roundly expanded near base of the convex side
riparius D. and K., p. 222
Claspette filament uniformly curved on the convex side 48

48 (47). Claspette stem pilose to near apex
schizopinax Dyar, p. 226
Claspette stem glabrous on distal one-third to one-half 49

49 (48). Basal lobe of the basistyle large, bearing a small dorsobasal protuberance
aboriginis Dyar, p. 148;
hexodontus Dyar, p. 189;
punctor (Kirby), p. 218
Basal lobe of the basistyle medium-sized, bearing a rather large dorsobasal protuberance
aberratus (F. and Y.), p. 150;
punctodes Dyar, p. 216

---

[14]The claspette filament of *A. communis* appears to have a low wing when viewed from the side, whereas the claspette filament of *A. pionips* appears flat when viewed from this position.

LARVAE (FOURTH INSTAR)[15]

1. Anal segment completely ringed by the saddle 2
Anal segment not completely ringed by the saddle 20

2 (1). Dorsal brush of the anal segment represented by two long strong hairs on either side; head and body hairs coarse and of about equal diameter throughout
aberratus (F. and Y.), p. 150
Dorsal brush of the anal segment composed of a long single lower caudal hair and a shorter multiple upper caudal tuft of two or more hairs on either side; head and body hairs gradually tapered..................3

3 (2). Pecten with distal teeth detached................4
Pecten with all teeth rather evenly spaced..7

4 (3). Individual comb scale thorn-shaped, with minute spinules on the basal part..........5
Individual comb scale rounded apically and fringed with subequal spinules or with the subapical spinules at least half as long as the median spine........................6

5 (4). Siphonal tuft with the hairs less than half as long as the basal diameter of the siphon; dorsal preapical spine of the siphon more than half as long as the apical pecten tooth........*nigromaculis* (Ludlow), p. 209
Siphonal tuft with the hairs as long or longer than the basal diameter of the siphon; dorsal preapical spine of the siphon less than half as long as the apical pecten tooth..........*nigripes* (Zett.), p. 207

6 (4). Lower frontal head hair 6 double or triple; individual comb scale rounded apically and fringed with subequal spinules
fulvus pallens Ross, p. 186
Lower frontal head hair 6 single; individual comb scale with the subapical spinules about half as long and about one-fourth to one-third as broad as the median spine
thelcter Dyar, p. 240

7 (3.) Siphonal tuft inserted within the pecten..8
Siphonal tuft inserted beyond the pecten..9

8 (7). Comb of the eighth segment with about 9 to 12 scales in a single curved or irregular row................*tormentor* D. and K., p. 244
Comb of the eighth segment with 30 to 40 scales in a patch
bimaculatus (Coq.), p. 158

9 (7). Anal gills eight times as long as the saddle of the anal segment or longer, with conspicuous darkly pigmented tracheae; siphonal index 3.5 to 4.0
dupreei (Coq.), p. 177
Anal gills less than eight times as long as the saddle, lacking conspicuous darkly

---

[15]Larval specimens of *Aedes melanimon* Dyar were not available to us for keying. Specimens of *A. purpureipes* Aitken were not received in time to be inserted in the key.

pigmented tracheae; siphonal index rarely more than 3.0............................10

10 (9). Comb of the eighth segment with 4 to 9 scales in a single row............................11
Comb of the eighth segment usually with more than 9 scales in a single regular or irregular row or a patch..........................12

11 (10). Siphonal index about 2.0; lateral hair of the anal segment shorter than the saddle
atlanticus D. and K., p. 152
Siphonal index about 3.0; lateral hair of the anal segment as long or longer than the saddle................*hexodontus* Dyar, p. 190

12 (10). Individual comb scale rounded apically and fringed with subequal spinules...............13
Individual comb scale thorn-shaped, with the subapical spinules not more than two-thirds as long as the median spine......16

13 (12). Anal gills much shorter than the saddle of the anal segment, budlike and bluntly rounded; lateral abdominal hair 6 usually multiple on segments III–V
taeniorhynchus (Wied.), p. 239
Anal gills as long or longer than the saddle, each tapering to a point; lateral abdominal hair 6 single or double on segments III–V 14

14 (13). Thorax and abdomen usually densely spiculate, the spicules appearing short and dark................*scapularis* (Rond.), p. 225
Thorax and abdomen with skin smooth or sparsely spiculate 15

15 (14). Siphonal index about 4.0; dorsal preapical spine of siphon longer than the apical pecten tooth......*rempeli* Vockeroth p. 222
Siphonal index 2.5 to 3.0; dorsal preapical spine of siphon much shorter than the apical pecten tooth
tortilis (Theob.), p. 246

16 (12). Dorsal preapical spine of the siphon as long as the apical pecten tooth....................17
Dorsal preapical spine of the siphon much shorter than the apical pecten tooth......18

17 (16). Siphonal index 3.0 to 3.5; pecten not quite reaching middle of the siphon; upper frontal head hair 5 and lower frontal 6 well barbed......*mitchellae* (Dyar), p. 205
Siphonal index 2.0 to 2.5; pecten reaching middle of the siphon or slightly beyond; upper frontal head hair 5 and lower frontal 6 smooth or very finely barbed
sollicitans (Walker), p. 228

18 (16). Individual comb scale thorn-shaped, with small lateral spinules on basal part
punctor (Kirby), p. 218
Individual comb scale with subapical spinules one-half to two-thirds as long as the median spine 19

19 (18). Median spine of comb scale three or four

times as broad and about twice as long as the subapical spinules
infirmatus D. and K., p. 200
Median spine of comb scale about twice as broad and one and one-third times as long as the subapical spinules
trivittatus (Coq.), p. 250

20 (1). Pecten with one or more of the distal teeth detached 21
Pecten with all teeth rather evenly spaced.38

21 (20). Siphon with several prominent paired subdorsal hair tufts..*trichurus* (Dyar), p. 248
Siphon lacking subdorsal hair tufts (one small paired subdorsal tuft present in *A. bicristatus*) 22

22 (21). Siphon with a small multiple lateral tuft inserted before the siphonal tuft
bicristatus T. and W., p. 156
Siphon lacking a small lateral tuft before the siphonal tuft.................................23

23 (22). Siphonal tuft inserted within the pecten..24
Siphonal tuft inserted beyond the pecten (rarely near base of distal pecten tooth) 25

24 (23). Siphonal index rarely more than 2.0; individual comb scale rounded and fringed apically with subequal spinules
atropalpus (Coq.), p. 254
Siphonal index about 2.5 to 3.5; individual comb scale thorn-shaped, with short lateral spinules on basal part
cataphylla Dyar, p. 166

25 (23). Antenna as long or longer than the head..26
Antenna shorter than the head.................27

26 (25). Antennal tuft inserted near middle of the shaft; comb of the eighth segment with no more than 15 scales in an irregular row..........*diantaeus* H., D., and K., p. 173
Antennal tuft inserted beyond middle of the shaft; comb of the eighth segment with more than 15 scales in a patch
aurifer (Coq.), p. 154

27 (25). Scales of comb of the eighth segment arranged in a patch three or more rows deep 28
Scales of comb arranged in a single or irregular double row............................30

28 (27). Siphonal index about 5.0, siphon slender; lateral abdominal hair 6 usually double on segments I and II, single on III–VI
excrucians (Walker), p. 179
Siphonal index not more than 4.0, siphon stout; lateral abdominal hair 6 double on segments I–VI 29

29 (28). Anal gills budlike, much shorter than the saddle of the anal segment; pecten reaching beyond middle of the siphon
(in part) *campestris* D. and K., p. 160
Anal gills not budlike, varying from shorter

than the saddle to about twice as long;
pecten not reaching middle of the siphon
(in part) *flavescens* (Müller), p. 184

30 (27). Upper frontal head hair 5 with three or
more branches 31
Upper frontal head hair 5 single or double
(rarely triple on both sides) 33

31 (30). Upper frontal head hair 5, lower frontal 6,
and preantennal 7 inserted nearly in a
straight line........*cinereus* Meigen, p. 266
Upper frontal head hair 5, lower frontal 6,
and preantennal 7 not inserted in a
straight line 32

32 (31). Hairs of siphonal tuft about as long as the
basal diameter of the siphon; saddle of
the anal segment deeply incised on ventral margin............*intrudens* Dyar, p. 201
Hairs of siphonal tuft rarely more than
half as long as the basal diameter of the
siphon; saddle of the anal segment not
deeply incised on ventral margin
vexans (Meigen), p. 264

33 (30). Antenna two-thirds to three-fourths as long
as the head; upper frontal head hair 5,
lower frontal 6, and preantennal 7 coarse
and of about equal diameter throughout
decticus H., D., and K., p. 171
Antenna not more than half as long as the
head; upper frontal head hair 5, lower
frontal 6, and preantennal 7 gradually
tapered 34

34 (33). Thorax and abdomen spiculate.................35
Thorax and abdomen with skin smooth...36

35 (34). Individual comb scale with the median spine
long and gradually tapered from the basal
part; basal part with minute lateral
spinules..........*spencerii* (Theob.), p. 230
Individual comb scale with the median spine
of medium length and arising abruptly
from the basal part; basal part with
rather prominent lateral spinules
idahoensis (Theob.), p. 192

36 (34). Anal gills much shorter than the saddle of
the anal segment, budlike; pecten on
basal one-fourth to one-third of the
siphon........*niphadopsis* D. and K., p. 211
Anal gills as long or longer than the saddle;
pecten on basal half to three-fifths of the
siphon 37

37 (36). Upper frontal head hair 5 and lower frontal
6 usually single..*ventrovittis* Dyar, p. 252
Upper frontal head hair 5 and lower frontal
6 usually double
riparius D. and K., p. 223

38 (20). Antenna smooth; antennal tuft represented
by a single small hair............................39
Antenna spiculate; antennal tuft double or
multiple 42

39 (38). Upper frontal head hair 5, lower frontal 6,

and preantennal 7 single; individual
comb scale with a strong median spine
and several shorter stout thornlike lateral
spines.....................*aegypti* (Linn.), p. 262
Upper frontal head hair 5 and lower frontal
6 single, preantennal 7 double or multiple; individual comb scale long, gradually tapered, and evenly fringed with
short lateral spinules................................40

40 (39). Anal gills at least two and one-half times as
long as the saddle of the anal segment,
each very stout, sides parallel, bluntly
rounded, the dorsal and ventral pairs of
about equal length; ventral brush of the
anal segment sparsely developed, with
single hairs on the barred area
varipalpus (Coq.), p. 258
Anal gills rarely more than twice as long as
the saddle, tapered, and bluntly rounded,
the dorsal pair distinctly longer than the
ventral pair; ventral brush well developed, with multiple tufts on the barred
area 41

41 (40). Lateral hair of the anal segment inserted
near the center on posterior border of
the saddle, a light-colored depression on
either side near the ventral margin of the
saddle; larva light in color
zoosophus D. and K., p. 260
Lateral hair of anal segment inserted near
ventrolateral margin of the saddle, without a light-colored depression on the
saddle; larva dark in color
triseriatus (Say), p. 256

42 (38). Individual comb scale with the subapical
spinules weak, not more than half as long
as the strong median spine....................43
Individual comb scale rounded apically and
fringed with subequal spinules or with the
subapical spinules at least two-thirds as
long as the median spine........................49

43 (42). Siphon slender, index 4.0 to 5.0; apical
pecten tooth nearly as long as the apical
diameter of the siphon
fitchii (F. and Y.), p. 182
Siphon stout, index usually less than 4.0;
apical pecten tooth not more than half as
long as the apical diameter of the
siphon 44

44 (43). Comb of the eighth segment with 8 to 16
scales 45
Comb of the eighth segment with 18 or
more scales 46

45 (44). Saddle of the anal segment extending to
near mid-ventral line; anal gills usually
less than one and one-half times as long
as the saddle........*punctodes* Dyar, p. 217
Saddle of the anal segment extending halfway or a little more down the sides; anal

gills about two to four times as long as the saddle........*impiger* (Walker), p. 194

46 (44). Lateral hair of anal segment shorter than the saddle 47

Lateral hair of anal segment longer than the saddle 48

47 (46). Ventral brush of anal segment with four or more precratal tufts inserted to near the base of the segment
(in part) *flavescens* (Müller), p. 184

Ventral brush of anal segment with fewer than four precratal tufts, all inserted on the distal half of the segment
sticticus (Meigen), p. 234

48 (46). Upper frontal head hair 5 usually 4- to 6-branched..............*aboriginis* Dyar, p. 148

Upper frontal head hair 5 usually triple
schizopinax Dyar, p. 226

49 (42). Upper frontal head hair 5 usually with four or more branches, lower frontal 6 with three or more branches.........................50

Upper frontal head hair 5 single to triple (rarely 4-branched), lower frontal 6 single or double (rarely triple) 54

50 (49). Comb of the eighth segment with 70 or more scales in a patch........*pionips* Dyar, p. 213

Comb of the eighth segment usually with fewer than 60 scales in a patch.............51

51 (50). Inner preclypeal spines separated by a little more than the length of one spine, lightly pigmented; siphonal index 4.5 to 5.0
thibaulti D. and K., p. 242

Inner preclypeal spines not separated by more than the length of one spine, darkly pigmented; siphonal index 2.5 to 4.0..52

52 (51). Anal gills very short, budlike; siphonal index 2.5 to 3.0....*cantator* (Coq.), p. 165

Anal gills as long as the saddle of the anal segment or longer; siphonal index 3.0 to 4.0 53

53 (52). Prothoracic hair 1 usually double, hair 5 usually triple........*pullatus* (Coq.), p. 215

Prothoracic hair 1 single, hair 5 single or double
canadensis canadensis (Theob.), p. 162;
canadensis mathesoni Middlekauff, p. 164

54 (49). Lateral hair of the anal segment single or double and as long or longer than the saddle..............*squamiger* (Coq.), p. 232

Lateral hair of the anal segment single (rarely double) and shorter than the saddle 55

55 (54). Siphonal tuft inserted near distal two-fifths of the siphon (occasionally near middle in *A. dorsalis*); anal gills short and budlike (sometimes as long or longer than the saddle in *A. dorsalis*) 56

Siphonal tuft usually inserted before middle of the siphon; anal gills about as long

or longer than the saddle of the anal segment 57

56 (55). Upper frontal head hair 5 single (occasionally double); pecten not extending beyond middle of the siphon
dorsalis (Meigen), p. 175

Upper frontal head hair 5 double or triple; pecten extending to near distal two-fifths of the siphon
(in part) *campestris* D. and K., p. 160

57 (55). Spicules toward apex of the saddle of the anal segment weakly developed, less than twice as long as those toward base of the saddle and no longer than the diameter of the setal ring of the lateral hair
communis (De G.), p. 168

Spicules toward apex of the saddle well developed, more than twice as long as those toward base of the saddle and usually longer than the diameter of the setal ring of the lateral hair..............................58

58 (57). Individual comb scale with the median spine about one and one-half times as long as the large subapical spinules....59

Individual comb scale with the median spine a little longer than the large subapical spinules but less than one and one-half times as long........*increpitus* Dyar, p. 198

59 (58). Upper frontal head hair 5 single or double (occasionally triple); lower frontal 6 usually single (rarely double) 60

Upper frontal head hair 5 usually triple (rarely double or 4-branched); lower frontal 6 double (occasionally triple)
grossbecki D. and K., p. 188

60 (59). Upper frontal head hair 5 usually single (occasionally one double, rarely both double) ......*implicatus* Vockeroth, p. 196

Upper frontal head hair 5 usually double (occasionally one single, rarely both single)..........*stimulans* (Walker), p. 237

Aedes (Ochlerotatus) aboriginis Dyar

Aedes aboriginis Dyar, 1917, Ins. Ins. Mens., 5:99.

ADULT FEMALE (pl. 44). Medium-sized to rather large species. *Head:* Proboscis dark-scaled; palpi short, dark-scaled. Occiput with dark-yellow curved scales and pale erect forked scales dorsally, with broad cream-colored scales laterally. Tori light brown, dark brown on inner surface. *Thorax:* The scutum is marked similar to that of "type *A. punctor*" variety, "type *A. hexodontus*" variety, and *A. abserratus*. Integument of scutum dark brown to black; scutum clothed with narrow yellowish to light golden-brown lanceolate scales; a dark-brown median stripe on anterior three-fourths, posterior dark-brown half lines present (median stripe sometimes divided by a thin line or stripe of paler scales); prescutellar space with yellow scales. Posterior pronotum with narrow curved light golden-brown scales on dorsal three-fourths. becoming broader and yellowish or grayish white ventrally. Scutellum with narrow yellow scales and light golden-brown setae on the lobes. Pleura with extensive patches of grayish-white scales. Scales on sternopleuron extending to anterior angle, narrowly separate from patch on prealar area. Mesepimeron with scales extending to lower margin. Hypostigial spot of scales absent. Lower mesepimeral bristles one or two. *Abdomen:* First tergite with a median patch of white scales; remaining tergites each black with a narrow basal white band widening at the sides. Venter clothed with grayish-white scales. *Legs:* Front and hind femora dark with scattered pale scales; middle femur dark; all femora pale on posterior surface; pale knee spots present. Tibiae dark, sprinkled with white scales. Tarsi dark-scaled. *Wing:* Length about 4.2 to 5.0 mm. Scales narrow, dark.

ADULT MALE. Coloration similar to that of the female. *Terminalia* (fig. 113). Appears to be indistinguishable from *A. hexodontus* and *A. punctor,* but is separable from *A. abserratus* and *A. punctodes* on the shape of the basal lobe of the basistyle. Lobes of ninth tergite (IXT-L) about as long as broad, each bearing several short stout spines. Tenth sternite (X-S) heavily sclerotized and hooked apically. Phallosome (Ph) cylindrical, longer than broad, open ventrally, closed dorsally, rounded and notched at apex. Claspette stem (Cl-S) rather stout, pilose on basal half; claspette filament (Cl-F) shorter than the stem, sclerotized, roundly expanded before middle, tapered, and curved at tip. Basistyle (Bs) cylindrical, about three and a half times as long as mid-width, clothed with scales and long and short setae; basal lobe (B-L) large and triangular-shaped, slightly detached basally, surface clothed with short setae except for a basal row of longer setae preceded by a rather strong dorsal spine, dorsobasal protuberance small; apical lobe (A-L) broadly rounded, clothed with short flattened

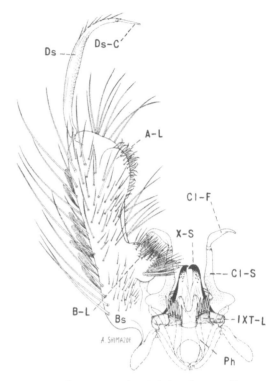

Fig. 113. Male terminalia of *Aedes aboriginis* Dyar.

curved setae. Dististyle (Ds) nearly two-thirds as long as the basistyle, broader medially, pilose, bearing small setae before apex; claw (Ds-C) slender, about one-fifth as long as the dististyle.

LARVA (fig. 114). Antenna less than half as long as the head, spiculate; antennal tuft multiple, sparsely barbed, inserted just before middle of shaft, reaching near tip. Head hairs: Postclypeal 4 small, single or branched distally; upper frontal 5, 3- to 5-branched, sparsely barbed; lower frontal 6, 2- to 4-branched, sparsely barbed, reaching preclypeus; preantennal 7 usually 6- to 11-branched, sparsely barbed, reaching insertion of antennal tuft. Prothoracic hairs: 1 long, double or triple; 2 long, single; 3 medium, usually double; 4 short, single or double; 5 long, usually double; 6 long, single; 7 long, usually triple. Lateral abdominal hair 6 usually double on segments III–VI. Comb of eighth segment with about 23 to 40 scales in a patch; individual scale with a strong apical spine and several shorter rather stout lateral spines. Siphonal index about 2.5; pecten of numerous evenly spaced teeth on basal two-fifths of siphon; siphonal tuft large, about 4- to 6-branched, sparsely barbed, inserted beyond the pecten; dorsal preapical spine little more than half as long as apical pecten tooth. Anal segment with the saddle reaching near mid-ventral line; lateral hair single, longer than the saddle; dorsal brush bilaterally consisting of a long lower caudal hair and a shorter multiple upper caudal tuft; ventral brush well developed, with one or two precratal tufts; gills about one to two and a half times as long as the saddle, pointed.

DISTRIBUTION. Northwestern United States, Canada, and Alaska. *United States:* Idaho and Oregon (672); Washington (67, 672). *Canada:* British Columbia (352, 430); Saskatchewan (708). *Alaska:* (213, 699).

BIONOMICS. The larvae are found in ditches and abandoned excavation pits and in early snow pools in deep woodlands, especially in mountainous areas in the north. The females bite either by day or by night in the forests. Hearle (351) states that the females are found in the greatest numbers about the middle of June in British Columbia, but are neither vicious nor persistent in their attacks. Dyar (188) observed the males swarming at Lake Cushman, Washington, in the forenoon where they occurred in groups of a few to fifty individuals on the dark side of the trunks of cedar trees in the forest, 6 to 10 feet above the ground.

Aedes (Ochlerotatus) abserratus
(Felt and Young)

Culex abserratus Felt and young, 1904, Science, n.s., *20*:312.

Aedes centrotus Howard, Dyar, and Knab, 1917, Carnegie Inst. Wash. Pub., *159, 4*:747.

Aedes dysanor Dyar, 1921, Ins. Ins. Mens., *9*:70.

Aedes punctor (of authors, in part, not Kirby).

Aedes implacabilis, Dyar, 1924 (not Walker), Ins. Ins. Mens., *12*:26.

Aedes (Ochlerotatus) implacabilis, Dyar, 1928 (not Walker), Carnegie Inst. Wash. Pub., *387*:186.

Aedes (Ochlerotatus) implacabilis, Matheson, 1944 (not Walker), Handbook Mosq. No. Amer., p. 170.

Aedes (Ochlerotatus) implacabilis, Gjullin, 1946 (not Walker), Proc. Ent. Soc. Wash., *48*:288

Aedes (Ochlerotatus) implacabilis, Knight, 1951 (not Walker), Ann. Ent. Soc. Amer., *44*:98.

Aedes (Ochlerotatus) implacabilis, Darsie, 1951 (not Walker), Cornell Agr. Exp. Sta. Mem., *304*:23.

Aedes abserratus, Vockeroth, 1954, Can. Ent., *86*:114.

Vockeroth (734a) has shown that the type of *Culex implacabilis* Walker is a specimen of *Aedes punctor* (Kirby) and that the name *A. abserratus* (Felt and Young) should be used for *Aedes implacabilis* of authors.

ADULT FEMALE (pl. 45). Medium-sized species. *Head:* Proboscis dark-scaled; palpi short, dark-scaled. Occiput with narrow curved yellowish-white scales and yellow erect forked scales dorsally, with broad appressed white scales laterally. Tori brown, inner surface darker and with a patch of grayish-white scales. *Thorax:* The scutum is marked similar to that

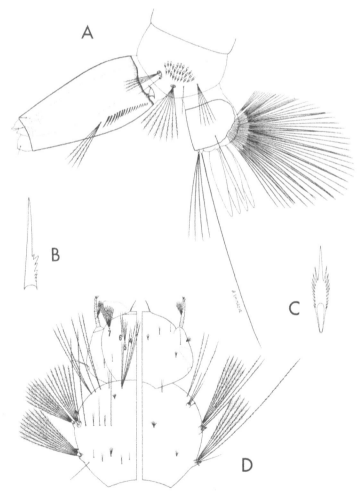

Fig. 114. Larva of *Aedes aboriginis* Dyar. *A,* Terminal segments. *B,* Pecten tooth. *C,* Comb scale. *D,* Head and thorax (dorsal and ventral).

of *A. aboriginis*, "type *A. hexodontus*" variety, and "type *A. punctor*" variety. However, the female is usually recognizable by the fine reddish-brown scales on the median area of the scutum. Integument of scutum dark brown; scutum clothed with yellowish-brown scales; a broad, more or less distinct, median longitudinal stripe of reddish-brown scales reaching from the anterior margin to the prescutellar space; pale yellowish-brown scales above the wings and on the prescutellar space. Posterior pronotum with narrow curved brown scales, becoming yellowish or grayish white ventrally. Scutellum with yellowish-brown scales and light-brown setae on the lobes. Pleura with patches of grayish-white scales. Scales on sternopleuron extending to anterior angle, narrowly separate from patch on prealar area. Mesepimeron with scales extending to lower margin. Hypostigial spot of scales absent. Lower mesepimeral bristles about one to three. *Abdomen:* First tergite with a median area of grayish-white scales; remaining tergites dark-scaled, each with a basal band of white scales. Venter pale-scaled, the apices of some of the segments black-scaled medially. *Legs:* Femora of front and hind legs dark-scaled, sprinkled with white; middle femur dark-scaled, with few or no scattered pale scales; all femora largely pale-scaled on posterior surface; knee spots present. Tibiae dark, posterior surface pale. Tarsi dark, first segment pale on posterior surface. *Wing:* Length about 4.0 to 4.7 mm. Scales narrow, dark brown.

ADULT MALE. Coloration similar to that of the female. *Terminalia* (fig. 115). Indistinguishable from *A. punctodes*, but the ranges of the two species are not known to overlap; separable from *A. aboriginis*, *A. hexodontus*, and *A. punctor* on the shape of the basal lobe of the basistyle. Lobes of ninth tergite (IXT-L) sclerotized, as long as broad, each bearing several strong spines. Tenth sternite (X-S) heavily sclerotized distally. Phallosome (Ph) long, cylindrical, rounded and notched at apex, open ventrally, closed dorsally. Claspette stem (Cl-S) short, pilose on basal half; claspette filament (Cl-F) nearly as long as the stem, weakly sclerotized, roundly expanded before middle, tapered and recurved at tip. Basistyle (Bs) about three times as long as mid-width, clothed with

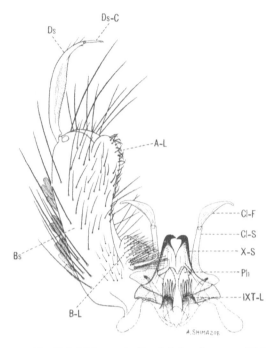

Fig. 115. Male terminalia of *Aedes abserratus* (Felt and Young).

scales and long and short setae; basal lobe (B-L) medium-sized, irregularly triangle-shaped, semidetached basally, surface clothed with short setae except for a basal row of long setae preceded by a weak dorsal spine, the dorsobasal angle of the lobe has a prominent cone-shaped protuberance; apical lobe (A-L) broadly rounded apically, bearing short, flattened curved setae. Dististyle (Ds) nearly two-thirds as long as the basistyle, slightly expanded medially, pilose, bearing a few subapical setae; claw (Ds-C) about one-sixth as long as the dististyle.

LARVA (fig. 116). The larva has coarse head and body hairs of rather uniform diameter for their entire length. Antenna shorter than the head, spiculate; antennal tuft multiple, inserted just before middle of shaft. Head hairs: Postclypeal 4 small, branched; upper frontal 5 single (occasionally double), stout, sparsely barbed; lower frontal 6 usually single, stout, sparsely barbed; preantennal 7 usually 2- to 4-branched, barbed. Prothoracic hairs: 1 long, single, sometimes double; 2 long, single; 3 long, single, occasionally double; 4 short, single, sometimes double or triple; 5 long, usually single; 6 long, single; 7 long, double or triple. Lateral abdominal hair 6 single on segments I–VI. Comb of eighth segment with about 5 to 7 scales in a curved row; individual scale long, thorn-shaped, with small lateral spinules on basal part. Siphonal index 3.5 to 4.0; pecten of about 11 to 21 teeth not reaching middle of siphon, the distal 1 to 3 teeth frequently detached; siphonal tuft usually 2- to 4-branched, inserted beyond pecten a little before middle of siphon, individual hairs longer than the basal diameter of the siphon; dorsal preapical spine no more than half as long as the apical pecten tooth. Anal segment ringed by the saddle; lateral hair single, a little longer than the saddle; dorsal brush bilaterally consisting of a single lower and a single upper caudal hair; ventral brush well developed, confined to the barred area; gills about one and a half to two and a half times as long as the saddle, tapered.

DISTRIBUTION. Southeastern Canada and the northeastern United States. *United States:* Connecticut (270, 494); Illinois (430, 615); Maine (45); Massachusetts (267, 270); Michigan (391); Minnesota (430); New Hampshire (430, 458); New Jersey (350); New York (31, 430); Ohio (731a); Pennsylvania (128); Rhode Island (270, 430); Vermont (430); Wisconsin (171, 430). *Canada:* Labrador (289, 430); Nova Scotia (289, 309); Ontario (289, 430); Prince Edward Island (710).

BIONOMICS. Only a single generation of this species occurs each year, in the late spring. The larvae are found in early cold pools. The senior author found the larvae of this species to be generally common in woodland pools in northern New Jersey during April and May, 1946. Ross (615) found the larvae in pools near the edge of bogs in northeastern Illinois. The females apparently do not bite readily during daylight hours, even on cloudy days, but are more active in the evening.

Aedes (Ochlerotatus) atlanticus
Dyar and Knab

Aedes atlanticus Dyar and Knab, 1906, Jour. N.Y. Ent. Soc., *14*:198.

ADULT FEMALE (pl. 46). The female appears to be indistinguishable from *A. tormentor.* Medium-sized species. *Head:* Proboscis dark-

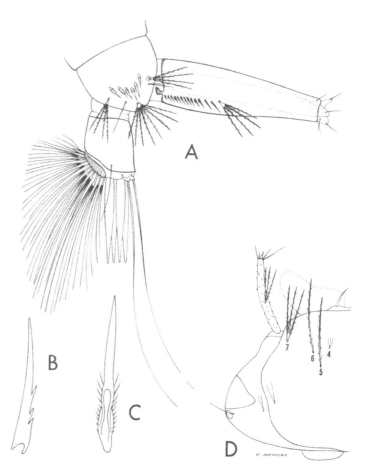

Fig. 116. Larva of *Aedes abserratus* (Felt and Young). *A*, Terminal segments. *B*, Pecten tooth. *C*, Comb scale. *D*, Head.

scaled; palpi short, dark. Occiput with a rather broad median stripe of white lanceolate scales extending forward between the eyes, and bounded on either side by a large submedian patch of broad dark scales; lateral region of occiput with broad appressed dingy-white scales; erect forked scales brown, pale on central part. Tori yellow, inner surface darker and bearing small brown setae. *Thorax:* Integument of scutum dark brown; scutum densely clothed with fine dark bronze-brown scales, except for a silver-white to pale-yellow-scaled narrow median stripe extending the entire length of the scutum and covering the middle lobe of the scutellum. Posterior pronotum with narrow curved dark-brown scales on the dorsal half. Pleura with small patches of broad flat grayish-white scales. Scales on sternopleuron reaching about halfway to anterior angle, well separated from patch on prealar area. Mesepimeron with a patch of scales on the upper half, lower half bare. Hypostigial spot of scales absent. Lower mesepimeral bristles absent. *Abdomen:* Tergites dark-scaled, with basal triangular patches of white scales laterally. Venter whitish-scaled, often speckled with dark scales. *Legs:* Legs dark-scaled except for pale posterior surface of femora. *Wing:* Length about 3.0 to 3.6 mm. Scales narrow, dark.

ADULT MALE. Coloration similar to that of the female, but usually has more pale scales on the occiput. *Terminalia* (fig. 117). Lobes of ninth tergite (IXT-L) nearly as long as broad, each bearing four to seven short spines. Tenth sternite (X-S) heavily sclerotized apically. Phallosome (Ph) stoutly conical, open ventrally, closed dorsally, rounded at apex. Claspette stem (Cl-S) large, stout, sinuous, broad medially, constricted basally and apically, pilose; claspette filament (Cl-F) less than half as long as the stem, narrower than the stem, striated, curved and tapered apically. Basistyle (Bs) about three times as long as mid-width, clothed with scales and long and short

setae; basal lobe (B-L) large, partly detached at base, expanded apically, bearing numerous rather short setae on apex and a large stout recurved dorsal spine near middle; apical lobe (A-L) long, triangular, bearing several short setae. Dististyle (Ds) approximately two-thirds as long as the basistyle, pilose; claw (Ds-C) slender, about one-fifth as long as the dististyle.

LARVA (fig. 118). Antenna a little less than half as long as the head, sparsely spiculate; antennal tuft small, multiple, barbed, inserted near middle of shaft, not reaching tip. Head hairs: Postclypeal 4 small, branched; upper frontal 5 and lower frontal 6 single, sparsely barbed; preantennal 7 usually 7- to 9-branched, sparsely barbed, reaching beyond insertion of the antennal tuft. Prothoracic hairs: 1 long, single; 2 short, single; 3 short, double or triple; 4 short, single; 5 long, single; 6 long, single or double; 7 long, usually 2- to 4-branched. Lateral abdominal hair 6 double on segments I and II, single on segments III–VI. Comb of eighth segment with 4 to 6 large scales in a single curved row; individual scale thorn-shaped, with minute lateral spinules on basal part. Siphonal index about 2.0; pecten of about 7 to 12 dark, evenly spaced teeth on basal two-fifths of siphon; siphonal tuft multiple, barbed, inserted beyond the pecten, about as long as basal diameter of the siphon; dorsal preapical spine much shorter than the apical pecten tooth. Anal segment ringed by a heavily sclerotized saddle; lateral hair single or double, shorter than the saddle; dorsal brush bilaterally consisting of a long lower caudal hair and a shorter multiple upper caudal tuft; ventral brush large, with about two precratal tufts; gills with the dorsal pair about three to four times as long as the saddle or a little longer, dorsal pair somewhat longer than the ventral pair, pointed.

DISTRIBUTION. Southeastern United States, north to New York and west to Kansas and Texas. *United States:* Alabama (135, 325); Arkansas (88, 375); Delaware (120, 167); District of Columbia (213); Florida (135, 419); Georgia (135, 213); Kansas (14); Louisiana (213, 759); Maryland (64, 161); Mississippi (135, 546); Missouri (14, 324); New Jersey (213, 350); New York (31);

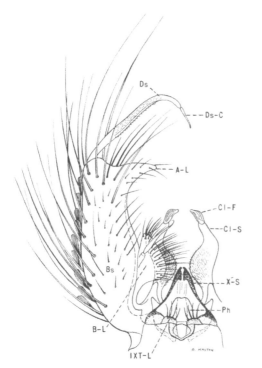

Fig. 117. Male terminalia of *Aedes atlanticus* Dyar and Knab.

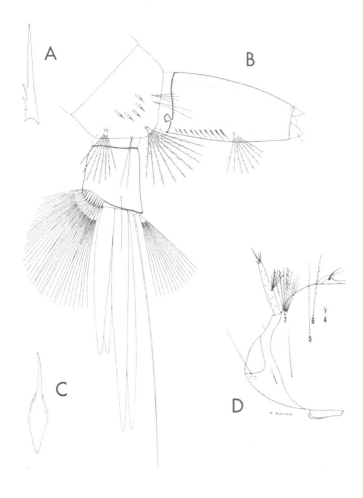

Fig. 118. Larva of *Aedes atlanticus* Dyar and Knab. *A*, Pecten tooth. *B*, Terminal segments. *C*. Comb scale. *D*, Head.

North Carolina (135, 516); Oklahoma (628); South Carolina (135, 742); Texas (463); Virgina (176, 178).

BIONOMICS. The larvae are found in temporary pools in open fields and in woodlands. Michener (510) found the larvae in Mississippi in clear shallow pools with grass and other types of vegetation. The larvae have been found from March to November in the southern states. The females are persistent biters and are often associated with *A. infirmatus* and other woodland species following summer rains. Michener (510) observed that the females bit readily and severely during daylight hours in light shade and even in open sunlight near Camp Shelby, Mississippi.

Aedes (Ochlerotatus) aurifer (Coquillett)

Culex aurifer Coquillett, 1903, Can. Ent., 35:255.

ADULT FEMALE (pl. 47). Medium-sized species. *Head:* Proboscis long, slender, dark scaled; palpi short, dark. Occiput with a large median patch of yellowish scales; a submedian patch of black scales on either side behind the eyes, with yellowish-white appressed scales laterally; erect forked scales dark. Tori dark brown, with minute setae on inner surface. *Thorax:* Integument of scutum dark brown; scutum with a broad median purplish-black stripe widening posteriorly, the sides and most of the prescutellar space pale golden. Posterior pronotum with curved brown scales dorsally, becoming pale golden ventrally. Scutellum with dark-brown setae on the lobes; scales on middle lobe pale golden. Pleura with medium-sized patches of broad flat grayish-white scales. Scales on sternopleuron reaching about halfway to anterior angle, separate from patch on prealar area. Mesepimeron with lower one-fourth to one-third bare. Hypostigial spot of scales absent. Lower mesepimeral bristles absent. *Abdomen:* First tergite dark, with a few pale scales intermixed; remaining tergites dark-scaled with small basolateral triangular patches of white scales. Venter clothed with grayish-white scales. *Legs:* Femora with dark and pale scales intermixed, darker toward apices, pale on posterior surface; pale knee spots small. Tibiae and tarsi dark with bronze-blue reflection. *Wing:* Length about 3.7 to 4.0 mm. Scales narrow, bluish black.

ADULT MALE. Coloration similar to that of the female. *Terminalia* (fig. 119). Lobes of ninth tergite (IXT-L) nearly as long as broad, each bearing several short stout spines. Tenth sternite (X-S) sclerotized apically. Phallosome (Ph) broadly conical, rounded apically, narrowly open ventrally, closed dorsally. Claspette stem (Cl-S) pilose, slender, bearing a short seta near apex; claspette filament (Cl-F) large, about as long as the stem or slightly longer, with a large sharp retrorse projection arising from the convex side, tapered apically to a curved point. Basistyle (Bs) more than twice as long as mid-width, bearing a large tuft of conspicuous setae at apex, clothed with scales and long and short setae; basal lobe (B-L) rep-

154

Tribe Culicini

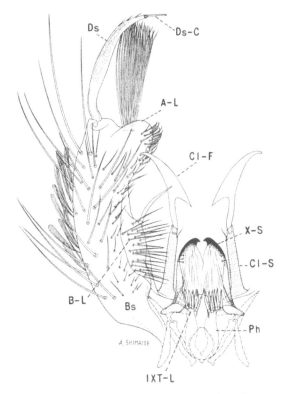

Fig. 119. Male terminalia of *Aedes aurifer* (Coquillett).

resented by a partly detached sclerotized plate on the mesial surface of the basistyle, bearing setae of medium length and a long stout spine at apex; apical lobe (A-L) prominent, bearing short straight and curved setae. Dististyle (Ds) pilose, bearing two or three small setae before apex; claw (Ds-C) slender, about one-fifth as long as the dististyle.

LARVA (fig. 120). Antenna about as long as the head, curved, spiculate, slender beyond insertion of antennal tuft, apical one-sixth very dark, terminal spines large; antennal tuft large, multiple, barbed, inserted beyond middle of shaft, reaching beyond tip. Head hairs: Postclypeal 4 small, multiple; upper frontal 5 and lower frontal 6 long, usually double (occasionally one of them triple); preantennal 7 large, multiple, barbed, reaching to or slightly beyond insertion of antennal tuft. Head hairs 5, 6, and 7 inserted in a straight line. Prothoracic hairs: 1 long, single; 2 short, single; 3 short, usually single; 4 short, single; 5 long, usually single; 6 long, single; 7 long, usually double. Lateral abdominal hair 6 multiple on segments I and II, usually double on III–V

and single on VI. Comb of eighth segment with about 20 to 30 scales in a patch; individual scale with the median spine much longer and broader than the subapical spinules. Siphonal index 3.5 to 4.0; pecten of about 12 to 20 teeth not quite reaching middle of siphon, with 1 to 3 distal teeth detached; siphonal tuft large, multiple, sparsely barbed, inserted beyond pecten, longer than the basal diameter of the siphon; dorsal preapical spine much shorter than the apical pecten tooth. Anal segment with saddle extending nearly to mid-ventral line; saddle spiculate on apical part; lateral hair single, a little shorter than the saddle; dorsal brush bilaterally consisting of a long lower caudal hair and a shorter multiple upper caudal tuft; ventral brush large, with about three or four precratal tufts; gills usually about as long as the saddle, pointed.

DISTRIBUTION. Northeastern and north central United States and southern Canada, west to Manitoba, Minnesota, and Iowa. *United States:* Connecticut (270, 494); Delaware (120, 167); Illinois (615); Iowa (625); Maine (45, 564); Maryland (161); Massachusetts (213, 270); Michigan (391, 529); Minnesota (533); New Hampshire (66); New Jersey (130, 350); New York (31); Ohio (731a); Rhode Island (35, 270); Vermont (649); Wisconsin (171). *Canada:* Manitoba (708); Ontario (304, 708); Quebec (305, 708).

BIONOMICS. This species overwinters in the egg stage, and there is a single generation a year. The larvae occur in temporary spring pools, especially those associated with cranberry bogs. Owen (533) found the larvae in roadside pools made by melting snow in Minnesota. Matheson (492) collected the larvae in great numbers from woodland pools in New York. Smith (662) found the larvae in larger bodies of water covering cranberry bogs, in reservoirs, and in near-by pools of considerable size. Smith further states that the larvae do not hug the edge of the pool as generally as some of the other species but prefer tufts of grass or vines several feet from the shore.

The senior author found the larvae of this species in a cranberry bog at Lakewood, New Jersey, in May, 1946, and captured females in a near-by woodland where they were biting

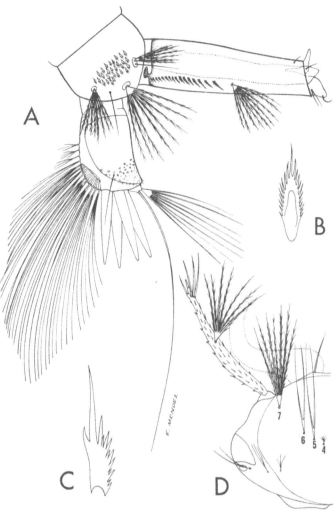

Fig. 120. Larva of *Aedes aurifer* (Coquillett). *A*, Terminal segments. *B*, Comb scale. *C*, Pecten tooth. *D*, Head.

at midmorning. On this occasion it was noticed that the larvae of *A. aurifer* were generally found several feet from the shore line although the larvae of *A. canadensis* were abundant at the edge of the pool. On other occasions the females have been observed biting freely in the shade during the day and in the evening. Smith (662) observed that it was one of the most bloodthirsty species in New Jersey, usually found near its larval habitat, but that it does fly some distance at night and does not enter houses.

Aedes (Ochlerotatus) bicristatus
Thurman and Winkler

Aedes bicristatus Thurman and Winkler, 1950, Proc. Ent. Soc. Wash., **52**:237.

ADULT FEMALE (pl. 48). Medium-sized species. *Head:* Proboscis dark-scaled, sprinkled with white; palpi short, dark-scaled, sprinkled with white. Occiput clothed with grayish-white scales; dark erect forked scales numerous on median posterior area. Tori dark, with white scales on inner and dorsal surfaces. *Thorax:* Integument of scutum dark brown to black; scutum with four, more or less distinct, narrow submedian stripes of dark-brown scales separated by narrow lines of light-golden to white scales; the outer of these stripes extend to the scutellum; patches of dark scales on either side at the scutal angle and on the posterior half of the scutum; yellowish-white scales laterally and on the prescutellar space. Posterior pronotum clothed with broad yellowish-white scales, with few to many brown scales on dorsal half. Scutellum with white scales and black setae on the lobes. Pleura with poorly defined patches of broad yellowish-white or cream-white scales. Scales on sternopleuron reaching anterior angle, narrowly separate from patch on prealar area. Mesepimeron with scales reaching near lower margin. Hypostigial spot of many scales present. Lower mesepimeral bristles about five to nine. *Abdomen:* First tergite with a large median patch of grayish-white scales; remaining tergites dark bronze-brown, with broad basal bands of grayish-white scales, a few scattered white scales apically on segments VI and VII. Venter predominantly white-scaled. *Legs:* Femora dark brown, sprinkled with white, posterior surface pale, dark preapically; knee spots white. Tibiae dark brown, sprinkled with white. Tarsi dark, sprinkled with pale scales on segments 1 to 3, segments 4 and 5 entirely dark. *Wing:* Length 3.5 to 4.0 mm. Scales narrow, brown, except for patches of white scales at bases of veins.

ADULT MALE. Coloration similar to that of the female. *Terminalia* (fig. 121). Lobes of ninth tergite (IXT-L) nearly as long as wide, each bearing several short spines. Tenth sternite (X-S) sclerotized apically. Phallosome (Ph) long, cylindrical, open ventrally, closed dorsally, fringed apically. Claspette stem (Cl-S) long, curved, pilose, bearing two small setae near apex; claspette filament (Cl-F) short, resembling a partial cone open on inner aspect and with a sclerotized cap bearing short

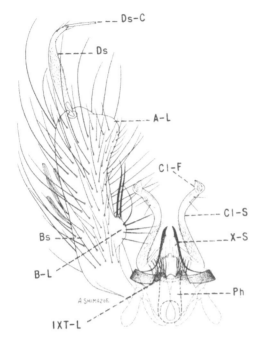

Fig. 121. Male terminalia of *Aedes bicristatus* Thurman
and Winkler.

projections. Basistyle (Bs) about three to three
and a half times as long as mid-width, bearing
a few scales and long and short setae; basal lobe
(B-L) rounded, dorsal surface bearing about
seven or eight long setae, the lobe tapered to a
ventral point which has a tuft of setae; apical
lobe (A-L) prominent, broadly rounded apically, bearing short setae. Dististyle (Ds) less
than two-thirds as long as the basistyle, pilose,
bearing small setae before apex; claw (Ds-C)
about one-fifth as long as the dististyle.

LARVA. (fig. 122). Antenna about half as
long as the head, spiculate; antennal tuft multiple, inserted near middle of shaft, not reaching tip. Head hairs: Postclypeal 4 small double
or triple; upper frontal 5 double (occasionally
one or both double), sparsely barbed; lower
frontal 6 double, rarely triple, sparsely barbed;
preantennal 7 usually 4- to 8-branched,
barbed, reaching insertion of antennal tuft. Prothoracic hairs: 1 long, single or double; 2 medium, single; 3 short, single or double; 4 medium, single; 5 long, 2- to 4-branched; 6 long,
single; 7 long, 2- to 4-branched. Lateral abdominal hair 6 usually multiple on segments I and
II, usually double on III–VI. Comb of eighth
segment with 4 or 5 (rarely 6) long scales in a

curved row; individual scale thorn-shaped,
with minute lateral spinules on basal part.
Siphonal index 3.0 to 3.5; pecten of about 14
to 22 teeth on basal half of siphon, with 1 to 3
distal teeth detached; siphonal tuft large, about
7- to 10-branched, barbed, inserted either within or beyond the pecten, longer than the basal
diameter of the siphon; a small multiple lateral
tuft inserted before the siphonal tuft, usually
within the pecten; a small single or double subdorsal tuft inserted at apical fifth of siphon;
dorsal preapical spine about as long as apical
pecten tooth. Anal segment with the saddle
reaching about three-fourths to four-fifths
down the sides, the saddle is deeply incised ventrally at apical third; lateral hair single, a little
shorter than the saddle; dorsal brush bilaterally consisting of a long lower caudal hair and a
shorter multiple upper caudal tuft; ventral
brush well developed, with two to four precratal tufts; gills varying from about two-thirds
to about one and a half times as long as the
saddle, the dorsal pair a little longer than the
ventral pair, each bluntly rounded.

DISTRIBUTION. Known only from California.
This species was described from specimens collected near Kelseyville, Lake County, California, in the Coast Range. Larval and pupal collections were also taken by Thurman and Winkler (691) at several additional sites in the same
general area in Lake County. Dr. Richard M.
Bohart has collected female specimens of *A.
bicristatus* in Sonoma and San Mateo counties.

BIONOMICS. The larvae of *A. bicristatus* were
found by Thurman and Winkler (691) during
late winter and early spring in flooded meadows and roadside ditches. The senior author
collected the larvae of this species during late
February, 1953, from several large meadow
pools and one large woodland pool in Lake
County. On the morning of February 23 two of
the meadow pools were frozen over with a continuous sheet of ice, and many larvae and pupae had been trapped and frozen in the ice.
Specimens were collected by breaking and collecting pieces of ice containing larvae and pupae. On thawing of the ice the larvae and pupae
became active and continued their development. Thurman and Winkler took two females
of *A. bicristatus* biting in the shade of Ponder-
osa pines near one of the larval habitats.

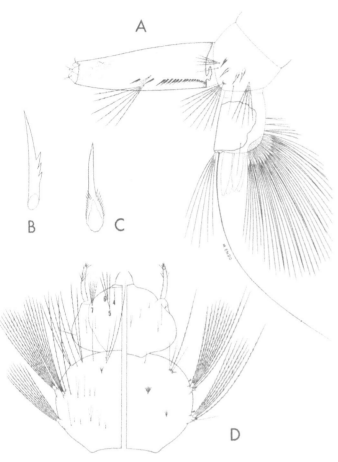

Fig. 122. Larva of *Aedes bicristatus* Thurman and Winkler. *A*, Terminal segments. *B*, Pecten tooth. *C*, Comb scale. *D*, Head and thorax (dorsal and ventral).

Aedes (Ochlerotatus) bimaculatus
(Coquillett)

Culex bimaculatus Coquillett, 1902, Proc. U.S. Nat. Mus., *25*:84.

Aedes bimaculatus, Howard, Dyar, and Knab, 1917 (in part), Carnegie Inst. Wash. Pub., *159*, 4:622.

Aedes fulvus, Dyar, 1928 (in part), Carnegie Inst. Wash. Pub., *387*:154.

Aedes rozeboomi Vargas, 1941, Graceta Medica de Mexico, *69*:393.

Aedes bimaculatus, Ross, 1943, Proc. Ent. Soc. Wash., *45*:143.

ADULT FEMALE. (pl. 49). A medium-sized to rather large bright orange-yellow mosquito. *Head:* Proboscis yellow-scaled, with an increasing mixture of dark scales on apical half, entirely black at apex; palpi about one-fifth as long as the proboscis, yellow-scaled, dark at apices. Occiput clothed dorsally with narrow golden scales, yellow setae and yellow erect forked scales, with broad flat yellow scales laterally. Tori yellow, with a few minute dark-brown hairs on inner surface. *Thorax:* Integument of scutum yellow except for a pair of large dark-brown to black posterolateral spots, separated by the width of the prescutellar space; scutum with narrow yellow and dark-brown scales; the dark spots clothed with black scales. Scutellum with narrow black scales and light-brown setae on the lobes. Pleura shiny yellow, with small patches of closely appressed silvery scales on prealar area and on upper half of the mesepimeron. Hypostigial spot of scales absent. Lower mesepimeral bristles absent. *Abdomen:* Tergites golden-scaled, with black scales on basal tergite. Venter golden-scaled. *Legs:* Femora yellow-scaled, black at apices. Tibiae of front and middle legs largely dark-scaled on anterior surface, apices gradually darkened; hind tibiae narrowly dark-scaled at base and broadly dark at apex. Front tarsi dark; middle tarsi dark except on basal part of segment 1; hind tarsi dark except on basal two-thirds of segment 1. *Wing:* Length about 4.5 to 5.0 mm. Scales dark brown, except near base of costa and subcosta where they are yellow.

ADULT MALE. Coloration similar to that of the female. *Terminalia* (fig. 123). Lobes of ninth tergite (IXT-L) about as long as broad, each bearing several strong spines. Tenth sternite (X-S) sclerotized apically. Phallosome (Ph) stout, conical, open ventrally, closed dorsally, notched at apex. Claspette stem (Cl-S) slender, pilose, reaching a little beyond middle of basal lobe; claspette filament (Cl-F) about two-thirds as long as the stem, rather slender, gradually tapered beyond basal third, tip slightly recurved. Basistyle (Bs) stout, about three times as long as mid-width, clothed with scales and long and short setae; basal lobe (B-L) partly detached, about half as long as the basistyle, dark, densely clothed with long fine slender setae; apical lobe (A-L) prominent, rounded, bearing a few rather strong setae. Dististyle (Ds) nearly as long as the basistyle, broader medially, pilose, bearing sev-

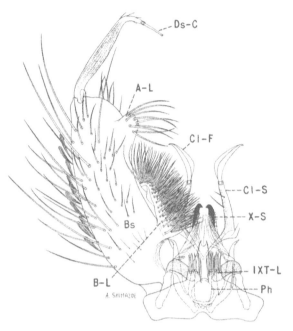

Fig. 123. Male terminalia of *Aedes bimaculatus* (Coquillett).

brush large, heavy, restricted to the barred area; gills usually about twice as long as the saddle, pointed.

DISTRIBUTION. Texas, Mexico, and El Salvador (613).

BIONOMICS. Larvae were found by W. C. Reeves in a roadside ditch near San Benito, Texas in May, 1942. Reeves observed that the larvae were nearly transparent except for the head and the sixth and seventh abdominal segments which were dark. Ross (613) found recently emerged adults resting on the grass and collected pupae from a near-by large, clear, semipermanent roadside pool near Brownsville, Texas, in September, 1942. Since no larvae were found in this collection, Ross believes that a noncontinuous breeding explains the infrequency with which the species is encountered.

eral small setae before apex; claw (Ds-C) slender, about one-fifth as long as the dististyle.

LARVA. (fig. 124). Antenna less than one-third as long as the head, sparsely spiculate; antennal tuft small, multiple, inserted near middle of shaft. Head hairs: Postclypeal 4 single or branched beyond base; upper frontal 5 single, sparsely barbed; lower frontal 6 usually double, sparsely barbed; preantennal 7 multiple, barbed. Prothoracic hairs: 1 long, single; 2 medium, single or double; 3 short, usually double; 4 short, usually triple; 5–6 long, single; 7 long, usually triple. Lateral abdominal hair 6 double on segments I and II, usually single on III–VI. Comb of eighth segment with about 30 to 40 scales in a patch; individual scale broadly rounded apically and fringed with subequal spinules. Siphonal index about 2.0; pecten of about 12 to 15 rather evenly spaced teeth on basal two-fifths or half of siphon; siphonal tuft large, multiple, heavily barbed, inserted within the pecten; dorsal preapical spine much shorter than the apical pecten tooth. Anal segment completely ringed by the saddle; lateral hair single, nearly as long as the saddle; dorsal brush bilaterally consisting of a long lower caudal hair and a shorter multiple brushlike upper caudal tuft; ventral

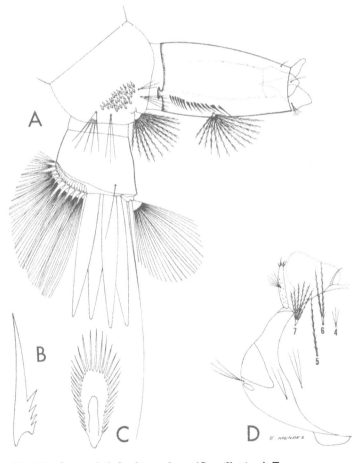

Fig. 124. Larva of *Aedes bimaculatus* (Coquillett). *A*, Terminal segments. *B*, Pecten tooth. *C*, Comb Scale. *D*, Head.

Aedes (Ochlerotatus) cacothius Dyar

Aedes cacothius Dyar, 1923, Ins. Ins. Mens., *11*:44.

ADULT FEMALE. Medium-sized species. *Head:* Proboscis long, dark; palpi short, dark. Occiput with white or yellowish-white lanceolate scales and white and brown erect forked scales dorsally; a patch of broad grayish-white scales surrounding a small patch of dark scales laterally. Tori black, with a few white scales on inner surface. *Thorax:* Integument of scutum black; scutum clothed with reddish-brown lanceolate scales; narrow yellowish-white scales on anterior and lateral margins and on the prescutellar space. Scutellum with narrow yellowish-white scales and light-brown setae. Pleura rather densely clothed with white appressed scales. Lower mesepimeral bristles absent. *Abdomen:* First tergite with a small median patch of brown scales, with a few pale scales intermixed; remaining tergites bronze-brown, with basal bands of white scales. Venter white-scaled, with brown scales apically on posterior sternites. *Legs:* Legs dark-scaled, except for pale posterior surface of femora, tibiae and first tarsal segment, and pale knee spots. *Wing:* Scales narrow, dark brown; a patch of pale scales on base of costa.

ADULT MALE. Unknown.

LARVA. Unknown.

DISTRIBUTION. Known only from Wyoming. This species is represented by three female specimens taken at Shoshone Point, Yellowstone National Park, Wyoming, by H. G. Dyar, June 27, 1922, at an altitude of 8,200 feet, in bright sunlight along a road on an exposed hillside.

BIONOMICS. Biting specimens were taken by Dyar near noon. Nothing more is known of the habits of this species.

Aedes (Ochlerotatus) campestris
Dyar and Knab

Aedes campestris Dyar and Knab, 1907, Jour. N.Y. Ent. Soc., *15*:213.

Aedes callithotrys Dyar, 1920, Ins. Ins. Mens., *8*:16.

ADULT FEMALE. (pl. 50). Rather large species. *Head:* Proboscis black, with pale scales intermixed on basal half to two-thirds; palpi short, black, speckled with white scales. Occiput with narrow yellowish-white scales and pale erect forked scales on median area, a small submedian patch of narrow golden scales and dark erect forked scales; lateral region with broad appressed white scales. Tori brown, with white scales on dorsal and inner surfaces. *Thorax:* Integument of scutum dark brown; scutum with a broad median stripe of narrow brown scales; lateral margins bronze-brown; remainder of scutum clothed with narrow yellowish-white scales. Posterior pronotum with narrow curved brown scales dorsally, with narrow curved yellowish-white scales ventrally. Scutellum with narrow yellowish-white scales and golden setae on the lobes. Pleura with large poorly defined patches of narrow curved yellowish-white scales. Scales on sternopleuron extending to anterior angle, continuous with patch on the prealar area. Mesepimeron with scales extending to near lower margin. Hypostigial spot of many scales present. Lower mesepimeral bristles about two to seven. *Abdomen:* First tergite with a large median patch of cream-white scales; remaining tergites each cream-white apically, basally, laterally, and medially, framing small irregular submedian dark patches on each tergite; posterior tergites usually all cream-white-scaled. Venter pale scaled. *Legs:* Femora and tibiae largely white-scaled, speckled with dark scales on anterior surface. First tarsal segments with white and black scales intermixed; hind tarsi with second to fourth segments dark, with apical and basal white bands, the fifth segment almost entirely white; middle tarsi with narrow apical and basal white bands on segments 2 and 3, segment 4 with a narrow basal white band, segment 5 almost entirely dark; front tarsi with narrow basal white bands on segments 2 and 3, 4 and 5 dark. *Wing:* Length about 4.3 to 5.0 mm. White scales predominating but with dark-brown scales uniformly intermixed.

ADULT MALE. Coloration similar to that of the female. *Terminalia.* (fig.125). Lobes of ninth tergite (IXT-L) about as long as broad, each bearing five to eight stout spines. Tenth sternite (X-S) heavily sclerotized apically. Phallosome (Ph) conical, longer than broad, open ventrally, closed dorsally, rounded and notched at apex. Claspette stem (Cl-S) pilose,

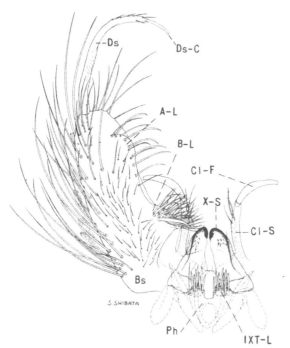

Fig. 125. Male terminalia of *Aedes campestris* Dyar and Knab.

somewhat curved, extending to distal margin of basal lobe, bearing two or three subapical setae; claspette filament (Cl-F) nearly as long as the stem, roundly expanded, sickle-shaped. Basistyle (Bs) nearly three times as long as broad, clothed with scales and long and short setae; basal lobe (B-L) conical or subquadrate, bearing many fine setae, a few coarser setae and a single large spine on the basal margin; apical lobe (A-L) low and broadly rounded, bearing rather long setae. Dististyle (Ds) about three-fifths as long as the basistyle, slightly broader medially, pilose, with a few small subapical setae; claw (Ds-C) slender, about one-fifth as long as the dististyle.

LARVA (fig. 126). Antenna shorter than the head, spiculate; antennal tuft multiple, barbed, inserted near middle of shaft. Head hairs: Postclypeal 4 small, multiple; upper frontal 5 double or triple (rarely 4-branched), weakly barbed; lower frontal 6 usually single (occasionally double), weakly barbed; preantennal 7 about 6- to 10-branched, barbed, reaching near insertion of antennal tuft. Prothoracic hairs: 1 long, usually double; 2 short, single; 3 short, single or double; 4 short, single; 5

long, usually triple (occasionally 2- or 4-branched); 6 long, single; 7 long, usually triple. Lateral abdominal hair 6 usually double on segments I–VI. Comb of eighth segment with about 19 to 33 scales in a triangular patch; individual scale broadly rounded and fringed with subequal spinules, the median one a little stronger. Siphonal index about 3.0; pecten of about 19 to 32 teeth on basal three-fifths of siphon, 1 to 4 distal teeth usually detached; siphonal tuft about 4- to 6-branched, inserted beyond pecten, about as long as the basal diameter of the siphon; dorsal preapical spine much shorter than the apical pecten tooth. Anal segment with the saddle extending about two-thirds down the sides, saddle spiculate on dorsoapical part; lateral hair single, shorter than the saddle; dorsal brush bilaterally consisting of a long lower caudal hair and a shorter multiple upper caudal tuft; ventral brush large, with about three or four precratal tufts; gills small, budlike, much shorter than the saddle.

DISTRIBUTION. The western semiarid plains of the United States, Canada, and Alaska. *United States:* Colorado (343); Idaho (344, 672); Iowa (625, 626); Michigan (391); Minnesota (533); Montana (475); Nebraska (14, 680); North Dakota (553); Oregon (672); South Dakota (14); Texas (463); Utah (213, 575); Washington (67, 672); Wisconsin (171); Wyoming (578). *Canada:* Alberta (598, 708); British Columbia (352, 708); Manitoba (466, 708); Ontario (289); Quebec (708); Saskatchewan (212, 598); Yukon (212, 708). *Alaska:* (213, 492).

BIONOMICS. The larvae develop in water in depressions filled by melting snow or rain. Rees (575) says he believes that the eggs overwinter before hatching and that hatching occurs whenever conditions are favorable during the following spring and summer. According to Rempel (598), in western Canada the species prefers pools rich in organic matter and a pH that is on the alkaline side. Twinn *et al.* (711) found *A. campestris* to be a rather late-appearing species in the vicinity of Churchill, Manitoba; the larvae were collected in late June in shallow open pools in marshy meadows near the Churchill River.

The females bite any time but are more

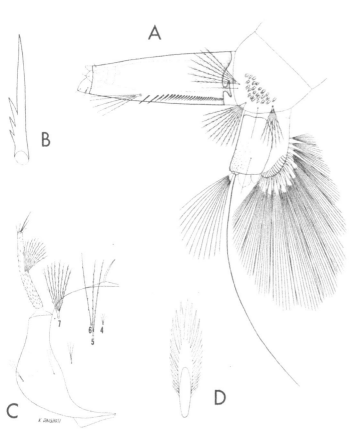

Fig. 126. Larva of *Aedes campestris* Dyar and Knab. *A*, Terminal segments. *B*, Pecten tooth. *C*, Head. *D*, Comb scale.

Aedes (Ochlerotatus) canadensis canadensis (Theobald)

Culex canadensis Theobald, 1901, Mon. Culic., 2:3.

Culex nivitarsis Coquillett, 1904, Proc. Ent. Soc. Wash., 6:168.

ADULT FEMALE (pl. 51). Medium-sized species. *Head:* Proboscis dark-scaled; palpi short, dark, with a few white scales at tip and at base of segment 4. Occiput with narrow yellowish-white scales in a broad median patch; scales of submedian area narrow, golden brown; lateral region with broad whitish scales, often surrounding a small dark-scaled patch; erect forked scales numerous on dorsal surface, those on central part pale. Tori dark brown, with a few small dark scales on inner surface. *Thorax:* Integument of scutum reddish brown; scutum clothed with narrow golden-brown scales, paler on anterior and lateral margins and on the prescutellar space; posterior half-stripes of pale scales are often present and extend anteriorly to middle of scutum. Posterior pronotum with narrow brown scales on dorsal half. Scutellum with narrow golden curved scales and dark-brown setae on the lobes. Pleura with small well-defined patches of appressed grayish-white scales. Scales on sternopleuron extending about halfway to anterior angle. Mesepimeron bare on lower one-third. Hypostigial spot of scales absent. Lower mesepimeral bristles absent. *Abdomen:* First tergite with a median patch of dark scales sprinkled with white; remaining tergites dark-scaled, with narrow basal white bands (lacking on some specimens) and prominent basolateral patches of white scales; the apical margin of the seventh segment, and often the sixth, white-scaled. Venter entirely white or with apices of segments dark. *Legs:* Femora dark, posterior surface pale; knee spots white. Tibiae dark, tipped with white basally and apically, posterior surface streaked with pale scales. Tarsal segments ringed with white both basally and apically; hind tarsi with rather broad white rings on segments 1 to 4, segment 5 entirely white; front and middle tarsi with white rings on segments 1 to 3 narrower than those on hind tarsi, white rings greatly reduced or entirely lacking on segments 4 and 5. *Wing:* Length about 3.2 to 4.0 mm. Scales narrow, dark.

annoying during the evening and early morning hours. It is typically a mosquito of the semi-arid regions. A ten-mile flight range has been recorded in Utah for this species. Rempel (598) says that the adults are on the wing about the end of June in Saskatchewan and that there appears to be only one brood. Rees (575) says that some of the adults survive the summer in Utah, a few having been collected early in the fall season. Twinn *et al.* (711) observed that the females were troublesome in a military camp at Churchill, Manitoba, during the latter part of July, especially toward evening and after sundown when they often invaded living quarters to bite. They were still common in the open country surrounding the camp in early August when observations ceased. Knab (382) obtained males during the daytime by beating bushes near an area where the females were biting.

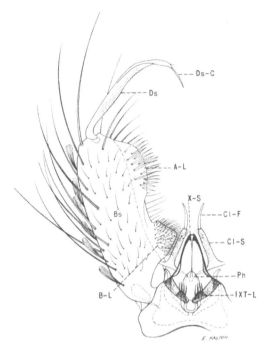

Fig. 127. Male terminalia of *Aedes canadensis canadensis*
(Theobald).

ADULT MALE. Coloration similar to that of
the female, but the dorsal abdominal bands are
much broader and the whitish scales on the
occiput are more prevalent. *Terminalia* (fig.
127). Lobes of ninth tergite (IXT-L) as long
as broad, each bearing a varying number of
about four to eight short irregularly placed
spines. Tenth sternite (X-S) heavily sclerotized
apically. Phallosome (Ph) cylindrical, nearly
twice as long as broad, rounded apically, open
ventrally, closed dorsally. Claspette stem (Cl-
S) slender, pilose, bearing two small setae
before apex; claspette filament (Cl-F) ligulate,
about two-thirds as long as the stem, uniformly
tapered. Basistyle (Bs) about three times as
long as mid-width, clothed with scales and long
and short setae; basal lobe (B-L) large, quadrate, or triangular, densely clothed with short
setae arising from small tubercles; apical lobe
(A-L) large, broadly rounded, clothed with
many short flattened spatulate setae. Dististyle
(Ds) about two-thirds as long as the basistyle,
expanded medially, pilose; claw (Ds-C) slender, about one-fifth as long as the dististyle.

LARVA (fig. 128). Antenna about half as long
as the head, spiculate; antennal tuft multiple,

inserted before middle of shaft, reaching near
tip. Head hairs: Postclypeal 4 small, multiple;
upper frontal 5 usually 4- to 9-branched,
barbed; lower frontal 6 usually 4- to 8-
branched, barbed; preantennal 7 usually 8- to
12-branched, barbed. Prothoracic hairs: 1
long, single; 2 medium to short, single; 3 short,
usually 1- to 3-branched; 4 short, single; 5
long, usually single; 6 long, single; 7 long,
usually triple. Lateral abdominal hair 6
usually double on segments I and II, double or
occasionally single on III–V, single on VI.
Comb of eighth segment with many scales in a
patch; individual scale pointed and fringed
with rather slender subequal spinules. Siphonal
index 3.0 to 4.0; pecten of about 13 to 24
evenly spaced teeth on basal two-fifths of
siphon; siphonal tuft usually 3- to 8-branched,
barbed, inserted beyond pecten, as long or a
little longer than the basal diameter of the
siphon. Anal segment with the saddle extending about two-thirds down the sides; lateral
hair single, a little shorter than the saddle;
dorsal brush bilaterally consisting of a long
lower caudal hair and a shorter multiple upper
caudal tuft; ventral brush large, with about two
precratal tufts; gills variable in length, about
one to one and seven-tenths times as long as the
saddle, tapered.

DISTRIBUTION. This species is widely distributed in the forested areas of the United States
and Canada. *United States:* Alabama (135,
420); Arkansas (88, 121); Colorado (343);
Connecticut (270, 494); Delaware (120,
167); District of Columbia (318); Florida
(135, 213); Georgia (135, 420); Idaho (344,
672); Illinois (615); Indiana (152, 345);
Iowa (530, 625); Kansas (14); Kentucky
(134, 570); Louisiana (419, 759); Maine
(45, 270); Maryland (64, 161); Massachusetts (267, 270); Michigan (391, 529); Minnesota (533); Mississippi (135, 546); Missouri (6, 324); Montana (213, 475); Nebraska
(530); New Hampshire (66); New Jersey
(350); New Mexico (273) New York (31);
North Carolina (135, 213); North Dakota
(553); Ohio (482); Oklahoma (628); Pennsylvania (298, 756); Rhode Island (35, 270);
South Carolina (135, 742); South Dakota
(14); Tennessee (135, 658); Texas (249,
532); Vermont (270); Virginia (176, 178);

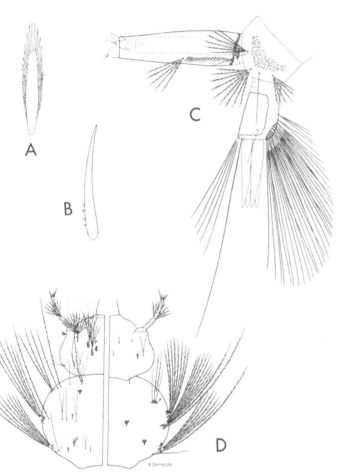

Fig. 128. Larva of *Aedes canadensis canadensis* (Theobald). *A*, Comb scale. *B*, Pecten tooth. *C*, Terminal segments. *D*, Head and thorax (dorsal and ventral).

during the summer and fall following the rains. It is not known whether these late season larvae are from overwintering eggs or from recently deposited eggs. Rudolphs (633) describes a pool where continuous breeding of *A. canadensis* occurred. Mail (475) believes that summer larvae are from overwintering eggs delayed in hatching. In the deeper pools the larvae invariably occupy the shallow margins, although the larvae of associated species often inhabit the deeper parts. The adults generally emerge in April, May, and early June.

Ross (615) describes the females as being persistent biters, attacking readily in shaded situations throughout most of the day. Mail (475) made similar observations on the biting habits of this mosquito in Montana. The senior author has observed that it is seldom a troublesome biter in the eastern part of its range, even in areas where large numbers have recently emerged. The females appear to be long-lived, persisting in small numbers in the woods until late summer.

Aedes (Ochlerotatus) canadensis mathesoni Middlekauff

Aedes mathesoni Middlekauff, 1944, Proc. Ent. Soc. Wash., *46*:42.

Aedes mathesoni, Rings and Hill, 1946, Proc. Ent. Soc. Wash., *48*:237.

Aedes canadensis mathesoni, Rings and Hill, 1948, Proc. Ent. Soc. Wash., *50*:41.

ADULT FEMALE. Similar to *A. canadensis canadensis*, but typical specimens differ as follows: There is a greater prevalence of black scales and more intense purplish-black coloration in *canadensis mathesoni* than in *canadensis canadensis*. The scutum of *mathesoni* appears blackish brown with characteristic areas of silvery or yellowish-white scales. The scutum has a narrow median line of small curved golden-brown scales bordered laterally by black scales. There is an elongate patch of yellowish-white scales on the anterior angles of the scutum and another patch laterally on either side of the middle. *A. canadensis canadensis* generally has golden-brown scales on the scutum and a greater amount of white scaling. The hind tarsi of *mathesoni* generally have narrow white rings present basally and apically

Washington (672); Wisconsin (171); Wyoming (578). *Canada:* Alberta (212, 598); British Columbia (212, 352); Labrador (289, 348); Manitoba (289, 466); New Brunswick (212, 708); Newfoundland (289); Northwest Territories (289); Nova Scotia (536, 708); Ontario (289, 708); Prince Edward Island (710); Quebec (708); Saskatchewan (212, 598); Yukon (289, 708).

BIONOMICS. The larvae develop in temporary or semipermanent shaded woodland pools containing fallen leaves, and to a lesser extent in pools in small stream beds and pools and ditches adjacent to wooded areas. The species overwinters in the egg stage, and the larvae hatch in large numbers in the late winter and spring. The larvae also occur in small numbers

on segments 1 and 2 and basally on segment 3, the remaining segments being entirely dark. Intergrades showing variations in scutal and tarsal patterns are illustrated by Rings and Hill (607).

ADULT MALE. Coloration similar to that of the female. *Terminalia.* The male terminalia appears to be almost identical with that of *A. canadensis canadensis* (fig. 127).

LARVA. Similar to *A. canadensis canadensis* (fig. 128) but displays a higher degree of branching of head hairs 5, 6, and 7. Specimens collected in the more southern parts of its geographical range usually show relatively more branching than those from the north.

DISTRIBUTION. A melanistic subspecies or geographical variation of *A. canadensis* known to occur in Florida (511), Georgia (606, 607), southern Alabama, and South Carolina (607). Intergradation in larval structures and in color patterns of the adults occur where the distribution of the two subspecies overlap.

BIONOMICS. The larvae were found in an abandoned foxhole at Camp Blanding, Florida, and in a similar habitat at Camp Gordon, Georgia (606). Little is known of the habits of the adults, but they are probably similar to those of *A. canadensis canadensis.*

Aedes (Ochlerotatus) cantator
(Coquillett)

Culex cantator, Coquillett, 1903, Can. Ent., *53*:255.

ADULT FEMALE (pl. 52). Medium-sized species. *Head:* Proboscis dark-scaled; palpi short, dark. Occiput with narrow yellowish-white to pale golden-brown scales and numerous dark erect forked scales on median area; lateral area with broad dingy-yellow scales, usually speckled with dark scales or surrounding a small dark patch. Tori brown, inner surface darker and bearing a few pale scales. *Thorax:* Integument of scutum reddish brown; scutum clothed with narrow golden-brown scales, those on anterior and lateral margins and prescutellar space paler. Scales of posterior pronotum light golden brown to dark brown. Scutellum with narrow golden scales and brown setae on the lobes. Pleura with medium-sized patches of broad appressed grayish-white scales. Scales on sternopleuron extending to

anterior angle in a narrow line, separate from patch on the prealar area. Mesepimeron bare on lower one-third. Lower mesepimeral bristles one to three, rarely none. *Abdomen:* First tergite with a median patch of pale scales; remaining tergites black with basal white bands, narrow medially, and widening laterally, the last two segments largely pale-scaled. Venter with yellowish-white scales. *Legs:* Femora and tibiae brown, with yellow scales intermixed, posterior surface yellow; knee spots pale. Tarsi dark, first segment often streaked on one side with pale scales; each segment of hind tarsus with a narrow basal white ring; front and middle tarsi with narrow basal white rings on segments 1 to 3, greatly reduced or lacking on 4 and 5. *Wing:* Length about 4.0 to 4.4 mm. Scales rather narrow, brown.

ADULT MALE. Coloration similar to that of the female. *Terminalia* (fig. 129). Lobes of

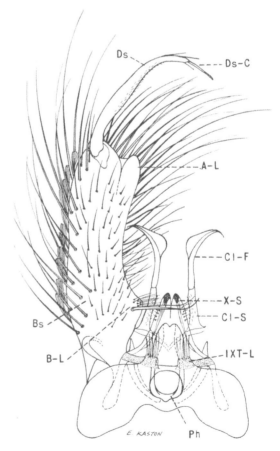

Fig. 129. Male terminalia of *Aedes cantator* (Coquillett).

ninth tergite (IXT-L) about as long as broad, each bearing four to eight spines. Tenth sternite (X-S) heavily sclerotized apically. Phallosome (Ph) cylindrical, much longer than broad, open ventrally, closed dorsally, notched at apex. Claspette stem (Cl-S) pilose to near apex, extending a little beyond basal lobe; claspette filament (Cl-F) longer than the stem, roundly expanded medially on convex side, curved and pointed at tip. Basistyle (Bs) about three and a half times as long as mid-width, clothed with scales and long and short setae, the setae are long and abundant on the inner ventral margin; basal lobe (B-L) short, bluntly rounded, bearing a strong dorsal spine and many short and two or three rather long setae apically; apical lobe (A-L) rounded, bearing a few short setae or nearly bare. Dististyle (Ds) nearly two-thirds as long as the basistyle, broader medially, pilose on inner surface, bearing three or four short setae before apex; claw (Ds-C) slender, about one-fourth as long as the dististyle.

LARVA (fig. 130). Antenna about half as long as the head, spiculate; antennal tuft multiple, finely barbed, inserted before middle of shaft, not reaching tip. Head hairs: Postclypeal 4 small, multiple; upper frontal 5, 4- to 10-branched, weakly barbed, reaching beyond preclypeus; lower frontal 6, 3- to 8-branched weakly barbed; preantennal 7 large, multiple, barbed, reaching beyond insertion of antennal tuft. Prothoracic hairs: 1 long, single; 2 medium, single; 3 short, double or triple; 4 medium, single; 5–6 long, usually single; 7 long, double or triple. Lateral abdominal hair 6 double, occasionally triple on segments I–V. Comb of eighth segment with many scales in a patch; individual scale fringed laterally with small spinules and apically with a few larger subequal spines, the median spine longer and stouter. Siphonal index 2.5 to 3.0; pecten of about 14 to 20 evenly spaced teeth on basal two-fifths of siphon; siphonal tuft about 4- to 9-branched, barbed, inserted beyond pecten, as long as basal diameter of siphon. Anal segment with the saddle reaching about two-thirds down the sides; lateral hair single, about as long as the saddle; dorsal brush bilaterally consisting of a long lower caudal hair and a much shorter multiple upper caudal tuft; ventral brush

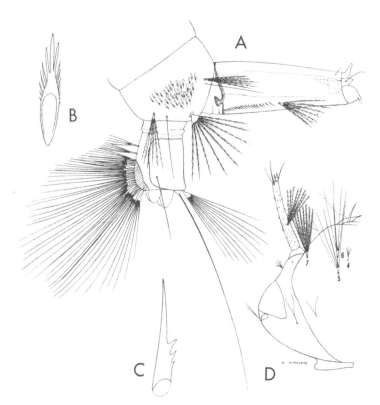

Fig. 130. Larva of *Aedes cantator* (Coquillett). *A*, Terminal segments. *B*, Comb scale. *C*, Pecten tooth. *D*, Head.

large, with about two precratal tufts; gills variable in length, but usually shorter than the saddle and budlike.

DISTRIBUTION. Along the Atlantic coast of the northeastern United States and southeastern Canada. It has been taken breeding in saline pools as far inland as Syracuse and Ithaca, New York. *United States:* Connecticut (270, 494); Delaware (120, 167); Maine (45, 270); Maryland (64, 161); Massachusetts (267, 270); New Hampshire (66, 270); New Jersey (213, 350); New York (31); Pennsylvania (298, 756); Rhode Island (35, 270); Virginia (178). *Canada:* New Brunswick, Nova Scotia, and Prince Edward Island (311, 708).

BIONOMICS. The larvae of *Aedes cantator* are found in coastal marshes, including both fresh and salt water, but less brackish water seems to be preferred. Fresh-water pools formed by rains or drainage from the uplands are preferable since larval production is nearly always much heavier on that part of the marsh adjacent to the uplands. In the New Jersey marshes, it is usually the dominant species in the spring with successive broods developing throughout the summer when weather conditions permit, but is usually less common than *A. sollicitans* during the summer months.

The females are persistent biters, attacking principally in the open, but also freely entering houses. The females bite any time during the day when their resting places among the vegetation are invaded, but they are essentially evening biters. They are somewhat less active biters than *A. sollicitans* during the daytime. Adults often migrate long distances from their larval habitats. According to Smith (662) the males accompany the females in flight even for several miles.

Aedes (Ochlerotatus) cataphylla Dyar

Aedes cataphylla Dyar, 1916, Ins. Ins. Mens., 4:86.
Aedes prodotes Dyar, 1917, Ins. Ins. Mens., 5:118.

ADULT FEMALE (pl. 53). Medium-sized species. *Head:* Proboscis dark brown, with a few pale scales intermixed on basal half; palpi short, dark, lightly speckled with gray scales. Occiput with narrow curved yellowish-white scales and yellowish and brown erect forked scales on median area, followed by a submedian patch of brown scales on either side, with broad appressed white scales laterally. Tori dark brown, with many white scales on inner and dorsal surfaces. *Thorax:* Integument of scutum black; scutum clothed with curved lanceolate scales, golden brown dorsally, those at the sides and on the prescutellar space whitish, colors frequently intermixed. Posterior pronotum with narrow curved yellowish-white to golden-brown scales. Scutellum with narrow whitish scales and brown setae on the lobes. Pleura with rather well-defined patches of broad flat grayish-white scales. Sternopleuron with scales extending to near the anterior angle and usually continuous with patch on prealar area. Mesepimeron with scales extending to near lower margin. Hypostigial spot of scales present or absent. Lower mesepimeral bristles two to seven. *Abdomen:* First tergite with a large median patch of white scales; remaining tergites black with basal white bands. Venter white-scaled. *Legs:* Femora dark brown,

sprinkled with white scales, posterior surface yellowish-white; knee spots small, white. Tibiae and tarsi dark with intermixed white scales, tarsi mostly dark on distal segments. *Wing:* Length about 3.7 to 4.5 mm. Scales narrow, dark brown; a prominent area of white scales at base of costa; scattered white scales also on costa, subcosta, and vein 1.

ADULT MALE. Coloration similar to that of the female. *Terminalia* (fig. 131). Lobes of ninth tergite (IXT-L) about as long as broad, each bearing several rather long spines. Tenth sternite (X-S) heavily sclerotized apically. Phallosome (Ph) cylindrical, tapered and notched apically, open ventrally, closed dorsally. Claspette stem (Cl-S) slender, curved, extending well beyond basal lobe, rather densely pilose on basal half; claspette filament (Cl-F) nearly three-fourths as long as the stem, broadly expanded at basal third, tapered apically to a recurved point. Basistyle (Bs) about three times as long as mid-width, clothed with long setae on inner ventral margin and with scales and long and short setae on outer aspect; basal lobe (B-L) bluntly conical, bearing many short setae apically and a stout dorsal spine followed by progressively weaker setae on basal part; apical lobe (A-L) thumb-shaped, bearing a few short straight setae, nearly bare. Dististyle (Ds) about three-fifths as long as the basistyle, curved, slightly expanded medially, finely pilose, bearing two to four small setae subapically; claw (Ds-C) slender, about one-fifth as long as the dististyle.

LARVA (fig. 132). Antenna short, less than half as long as the head, moderately spiculate; antennal tuft multiple, inserted near middle of shaft, not reaching tip. Head hairs: Postclypeal 4 small, branched; upper frontal 5 and lower frontal 6 long, single, smooth; preantennal 7 3- to 6-branched, barbed, reaching base of antennal tuft. Prothoracic hairs: 1 long, usually single, sometimes double; 2 medium, single; 3 short, single or double; 4 short, single; 5 long, double or triple; 6 long, single; 7 long, usually triple, sometimes double. Lateral abdominal hair 6 double or triple on segment I, usually double on II–V, single or double on VI. Comb of eighth segment with about 10 to 24 scales arranged in two irregular rows or a patch; individual scale thorn-shaped, with a

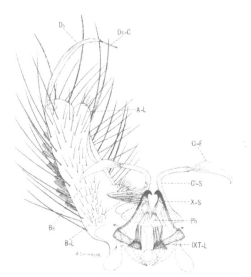

Fig. 131. Male terminalia of *Aedes cataphylla* Dyar.

long median spine and short lateral spinules. Siphonal index 3.0 to 3.5; pecten of about 11 to 20 evenly spaced teeth on basal fourth or third of siphon followed by 3 to 5 larger detached teeth reaching near apex; siphonal tuft usually 2· to 4-branched, inserted within the pecten at basal two-fifths of siphon, about as long as the basal diameter of the siphon; dorsal preapical spine much shorter than the apical pecten tooth. Anal segment with the saddle extending two-thirds to three-fourths down the sides; saddle spiculate on dorsoapical part; lateral hair single, shorter than the saddle; dorsal brush bilaterally consisting of a long lower caudal hair and a multiple upper caudal tuft; ventral brush large, with one or two precratal tufts; gills about one to one and four-fifths as long as the saddle, pointed.

DISTRIBUTION. The Sierra of California and the northern Rocky Mountain states, northward to Alaska. It also occurs in northern Europe and in Siberia. *United States:* California (287); Colorado (343); Idaho (344, 672); Montana (213, 475); Oregon (672); Utah (575); Washington (672); Wyoming (578). *Canada:* Alberta (212, 598); British Columbia (352, 598); Manitoba and Saskatchewan (598); Yukon (212, 289). *Alaska:* (213, 673).

BIONOMICS. The winter is passed in the egg stage, and there is a single generation a year.

The larvae develop in pools formed by melting snow and floodwaters. Dyar (237) claims that the species prefers grassy pools along river banks for breeding. It is most often found in open meadow pools in the Sierra; however, it does occur in pools in dense shade. Natvig (523) collected the larvae in pools, water-filled ditches, and flooded meadows in Norway. Wesenberg-Lund (749) says the species seems to prefer temporary ponds on plains, not overshadowed by trees, or ponds near the edges of forests in Denmark.

Dyar (199) observed the males at White Horse swarming some 10 feet in the air over the bare ground between willows or small pines just after sunset. Rees (575) says this is one of the most serious pest mosquitoes in Utah in wooded sections of mountains at elevations of more than 7,500 feet. The females bite both during the day and night. It is one of the earliest species on wing in many localities within its range, but the time of emergence depends somewhat on the elevation. Larvae occur from May to early July in Utah, and females have been taken as late as mid-August.

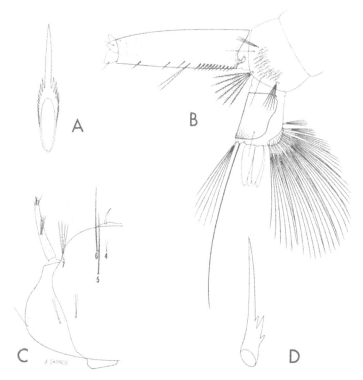

Fig. 132. Larva of *Aedes cataphylla* Dyar. *A*, Comb scale. *B*, Terminal segments. *C*, Head. *D*, Pecten tooth.

Aedes (Ochlerotatus) communis
(DeGeer)

Culex communis De Geer, 1776, Mem. des Ins.,
  6:316.
Culex nemorosus Meigen, 1818, Syst. Beschr. Zweifl.
  Ins., *1*:4.
Aedes obscurus Meigen, 1830, Abb. Zweifl. Ins. pl. 2,
  fig. 2.
Culex lazarensis Felt and Young, 1904, Science, n.s.,
  20:312.
Culex borealis Ludlow, 1911, Can. Ent., *43*:178.
Culicada nemorosa diplolineata Schneider, 1913,
  Verh. Nat. Ver. Bonn, *70*:37.
Aedes tahoensis Dyar, 1916, Ins. Ins. Mens., *4*:82.
Aedes altiusculus Dyar, 1917, Ins. Ins. Mens., *5*:100.
Aedes masamae Dyar, 1920, Ins. Ins. Mens., *8*:166.
Ochlerotatus palmeni Edwards, 1921, Ent. Tidskr.,
  p. 52.
Aedes prolixus Dyar, 1922, Ins. Ins. Mens., *10*:2.

ADULT FEMALE (pl. 54). Similar to *A. pionips* but differs in the scutal pattern and the shape of the tarsal claws, as described by Vockeroth (734). Medium-sized species. *Head:* Proboscis dark-scaled; palpi short, dark, lightly sprinkled with pale scales. Occiput with narrow curved yellowish scales in a broad median patch; erect forked scales dark, with a few yellowish ones intermixed; appressed broad yellowish-white scales laterally. Tori brown to black, with white scales on inner surface. *Thorax:* Integument of scutum dark; scutum clothed with a varying pattern of yellowish-white to yellow and dark-brown scales; the dark-brown scales are usually confined to a pair of submedian stripes and posterior half-stripes. (The line of pale scales separating the submedian dark stripes is usually broader than in *A. pionips*, and the scales on the scutum of *A. communis* are less regular and more loosely placed.) Posterior pronotum with rather narrow yellowish-white scales dorsally, becoming somewhat broader and paler ventrally. Scutellum with pale scales and dark setae on the lobes. Pleura with extensive, poorly defined patches of broad grayish-white scales. Scales on sternopleuron extending to anterior angle, narrowly separate from patch on prealar area. Mesepimeron with scales extending to near lower margin. Hypostigial spot of scales absent. Lower mesepimeral bristles two to six. *Abdomen:* First tergite with a median patch of white scales; remaining tergites each brownish black, with a broad basal band of cream-colored scales, widened laterally, particularly on posterior tergites. Venter mostly pale-scaled. *Legs:* Femora dark brown, sprinkled with pale scales, posterior surface yellowish white; knee spots white. Tibiae dark brown, sprinkled with pale scales. Tarsi dark brown, first segment more or less sprinkled with pale scales. The front tarsal claw is variable but is usually shorter than in *A. pionips* and bends just beyond the lateral claw (fig. 133, *B*). *Wing:* Length 4.5 to 5.0 mm. Scales dark, a small patch of pale scales usually present on base of costa and occasionally on base of vein 1.

ADULT MALE. Coloration similar to that of the female. Similar to *A. pionips* but has the palpi distinctly longer than the proboscis. *Terminalia* (fig. 133). Similar to *A. pionips*. Lobes of ninth tergite (IXT-L) longer than broad, each bearing a varying number of about four to eleven spines. Tenth sternite (X-S) heavily sclerotized beyond middle. Phallosome (Ph) conical, open dorsally, closed ventrally, notched at apex. Claspette stem (Cl-S) long, reaching well beyond distal margin of basal lobe, curved, pilose on basal half; claspette filament (Cl-F) shorter than the stem, blade-like, expanded near base, gradually tapered to a recurved tip, has a low wing running along the filament and appears grooved when viewed from one side. Basistyle (Bs) about three and a half to four times as long as mid-width, clothed with scales and long and short setae; basal lobe (B-L) prominent, conical or subquadrate, slightly detached at base, clothed with numerous setae, those on apical margin larger and stronger, a stout dorsal spine arising from near base; apical lobe (A-L) prominent, rounded, clothed with numerous setae of moderate length. Dististyle (Ds) about three-fifths as long as the basistyle, somewhat broader medially, pilose, bearing short setae before apex; claw (Ds-C) slender, about one-fifth as long as the dististyle.

LARVA (fig. 134). Antenna shorter than the head, spiculate; antennal tuft multiple, sparsely barbed, inserted before middle of shaft, not reaching tip. Head hairs: Postclypeal 4 small, branched; upper frontal 5 and lower frontal 6 long, single (occasionally double), extending beyond the preclypeus; preantennal 7 about 4- to 9-branched, weakly barbed. Prothoracic

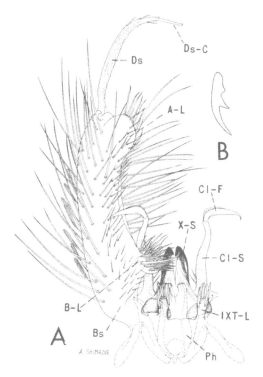

Fig. 133. *Aedes communis* (De Geer). *A*, Male terminalia.
B, Front tarsal claw of female.

hairs: 1 long, double or triple; 2 short, single; 3 short, single or double; 4 short, single; 5 long, double or triple; 6 long, single; 7 long, usually 3- or 4-branched. Lateral abdominal hair 6 usually double on segments I–V, single or double on VI. Comb of eighth segment with many scales in a large triangular patch; individual scale broadly rounded and fringed at apex with four to seven stout spines, nearly equal in size. Siphonal index 2.5 to 3.0; pecten of about 14 to 21 evenly spaced teeth not reaching middle of siphon; siphonal tuft 4- to 8-branched, barbed, inserted beyond pecten; dorsal preapical spine shorter than apical pecten tooth, inserted at least its length from apex of siphon. Anal segment with the saddle extending halfway to two-thirds down the sides; lateral hair single, shorter than the saddle; dorsal brush bilaterally consisting of a long lower caudal hair and a shorter multiple upper caudal tuft; ventral brush large, with two to four precratal tufts; gills about two to two and a half times as long as the saddle, pointed.

DISTRIBUTION. Occurs in the forested regions across the Northern United States, Canada,

Alaska, Siberia, and northern Europe. *United States:* California (287); Colorado (343); Maine (45, 564); Massachusetts (701); Michigan (391); Minnesota (533); Montana (475); New Hampshire (66, 213); New Jersey (130); New York (31); Oregon (672); Pennsylvania (128); Utah (575); Washington (67, 672); Wisconsin (171); Wyoming (578). *Canada:* Alberta (212, 708); British Columbia (352, 289); Labrador (289); Manitoba (289, 708); New Brunswick (708); Northwest Territories (289, 708); Nova Scotia (309, 708); Ontario (212, 289); Prince Edward Island (311, 708); Quebec (307, 289); Saskatchewan (310, 598); Yukon (289, 708). *Alaska:* (396, 673).

BIONOMICS. The larvae of *Aedes communis* are found in pools filled by melting snow in the spring. The senior author observed that the larvae were restricted to deep pools in both deciduous and evergreen forests in northern

Fig. 134. Larva of *Aedes communis* (De Geer). *A*, Terminal segments. *B*, Pecten tooth. *C*. Comb scale. *D*, Head.

New Jersey and in Pennsylvania and that, although the larvae of *A. canadensis* were numerous in the shallow margins, the larvae of *A. communis* were always confined to the deeper parts. Jenkins (396) found the larvae in bodies of water of all types and sizes in central Alaska, but they were most abundant in shallow depressions, either open or shaded. He also observed that the larvae were nearly always found in open water and did not hide in the vegetation. Twinn *et al.* (711) found the larvae of *A. communis* mostly in shallow snow-water pools in open forests near Churchill, Manitoba. Frohne (293) collected the larvae from brackish water habitats along the coastal marshes in Alaska. It is a single-brooded species, but a few adults may persist throughout most of the summer.

Mail (475) claims that it is one of the most common mosquitoes found at higher elevations, at 5,000 feet or more, in forested areas in Montana on both sides of the Rockies. Rees (575) found it to be widely distributed in the higher mountains of Utah where the adults emerge during June and early July. Although the females of this species cannot always be identified with certainty, Twinn *et al.* (711) believe that this species forms an important part of the immense numbers of mosquitoes found in the forests about Churchill, Manitoba, in late June and July.

The females are often serious pests in the northern forests where they may be encountered in swarms in the spring, biting mostly in the shade or after sundown. Jenkins (396) observed that the females were common pests in central Alaska throughout the summer, biting during the day in wooded areas and coming out in the open to attack during the evening. Natvig (523) states that in Norway the females attack by preference at night or in the shade or during cloudy weather by day. He further states that he has been attacked by this species in the afternoon even in bright sunshine in the mountains and that it is repeatedly found in houses and cowstables in Norway. Wesenberg-Lund (749) states that it is an exclusive forest mosquito in Denmark, inhabiting the deepest, darkest part of the forest where it will attack just as frequently at noon on a sunny day as in the evening.

Aedes (Ochlerotatus) decticus
Howard, Dyar, and Knab

Aedes decticus Howard, Dyar, and Knab, 1917, Carnegie Inst. Wash. Pub., *159, 4*:737.

Aedes pseudodiantaeus Smith, 1952, Bull. Brooklyn Ent. Soc., 47:19.

Dr. Alan Stone has examined specimens of *Aedes pseudodiantaeus* Smith in the United States National Museum and found them to be identical with *A. decticus;* hence *decticus* is the correct name of the species.

ADULT FEMALE (pl. 55). A rather small mosquito. *Head:* Proboscis black-scaled; palpi short, black-scaled. Occiput clothed dorsally with narrow curved pale-yellow scales and yellow to black erect forked scales; subdorsal patches of dark-brown appressed scales usually distinct; lateral region of occiput with broad appressed cream-colored scales. Tori deeply infuscated, with a few small dark setae on inner surface. *Thorax:* Integument of scutum dark; scutum clothed with pale-golden scales, except for a pair of narrow median longitudinal dark-brown stripes, usually separated by a narrow stripe of pale-yellow scales, but in some specimens they may be more or less fused, thus appearing as one broad stripe; posterior half-stripes of brown scales often distinct. Posterior pronotum with a few to many curved brown scales dorsally, becoming broader and grayish white ventrally. Scutellum with pale-golden scales and light-brown setae on the lobes. Pleura with rather well-defined patches of broad appressed grayish-white scales. Sternopleuron with scales extending a little more than half way to anterior angle, narrowly separate from patch on prealar area. Mesepimeron with lower one-third bare. Hypostigial spot of scales absent. Lower mesepimeral bristles absent. *Abdomen:* First tergite with a few grayish-white scales on median area; remaining tergites with purplish-brown scales; basolateral triangular patches of white scales present. Sternites with irregular-shaped broad basal bands of white scales, dark apically. *Legs:* Legs with purplish-brown scales; femora largely pale-scaled on posterior surface; small knee spots of pale scales present. *Wing:* Length about 3.0 to 3.5 mm. Scales narrow, brown.

ADULT MALE. Coloration similar to that of

the female, but with white scaling of abdomen reduced. *Terminalia* (fig. 135). Lobes of ninth tergite (IXT-L) small, each bearing four or five stout spines. Tenth sternite (X-S) slender, sclerotized and hooked at apex. Phallosome (Ph) cylindrical, narrowed and notched at apex, open ventrally, closed dorsally. Claspette stem (Cl-S) stout, curved sharply dorsad beyond middle, pilose; claspette filament (Cl-F) with base slender, broadly expanded beyond base, armed apically with two sharp recurved points. Basistyle (Bs) about three times as long as mid-width, slightly concave on outer surface, clothed with scales and long and short setae; the basistyle has on its ventral surface just beyond the middle a large brushlike tuft of fine setae directed mesocaudad and a small tuft of similar setae situated between the large brushlike tuft and the dististyle; basal lobe (B-L) large, reaching to middle of basistyle, bearing two stout curved apical spines arising from a common tubercle (inner spine larger), a long stout dorsal spine arising from a weak protuberance on basal part of lobe and scattered

small setae; apical lobe (A-L) inconspicuous, short, broadly rounded, clothed with short retrorse setae. Dististyle (Ds) about two-thirds as long as the basistyle, curved, bearing a few small subapical setae; claw (Ds-C) slender, about one-sixth as long as dististyle.

LARVA (fig. 136). The larva has coarse head and body hairs of rather uniform diameter for their entire length. Antenna about two-thirds to three-fourths as long as the head, slender, curved, spiculate; antennal tuft inserted just before middle of shaft, usually double, rarely single, and reaching to tip of shaft, sometimes 3- to 5-branched and not reaching tip. Head hairs: Postclypeal 4 small, usually branched; upper frontal 5 and lower frontal 6 long, single or double (hair 5 rarely triple); preantennal 7 long, usually double (occasionally 3- or 4-branched), sparsely and weakly barbed. Prothoracic hairs: 1 long, single; 2–4 minute; 5–7 long, single, rarely double or triple. Lateral abdominal hair 6 single on segments I–VII. Comb of eighth segment with 5 to 7 (usually 6) scales in a curved row; individual scale thorn-shaped and with lateral spinules which are stronger distally. Siphonal index 3.5 to 4.0; pecten of about 10 to 15 teeth reaching middle of siphon or slightly beyond, with 1 to 3 distal teeth detached; siphonal tuft 3- to 7-branched, inserted beyond pecten; dorsal preapical spine much shorter than the apical pecten tooth. Anal segment with the saddle extending almost to midventral line (sometimes touching but never completely fused), saddle spiculate on posterodorsal part; lateral hair single, shorter than the saddle; dorsal brush bilaterally consisting of a long lower caudal hair and a shorter upper caudal tuft of 2 to 5 hairs; ventral brush sparsely developed, with about one to three precratal tufts; gills about one and a half to two and a half times as long as the saddle, pointed.

DISTRIBUTION. Known from the northern United States, Canada, and Alaska. *United States:* Massachusetts, Michigan, New Hampshire, and New York (663). *Canada:* Labrador (663); Ontario (382). *Alaska:* (663).

BIONOMICS. Smith (663) found the larvae and pupae in a sphagnum bog swamp in Belchertown, Massachusetts, in considerable num-

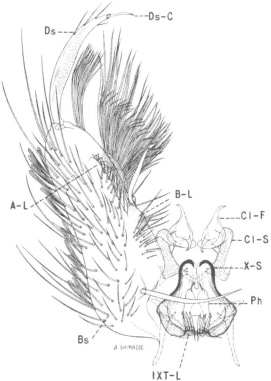

Fig. 135. Male terminalia of *Aedes decticus* Howard, Dyar, and Knab.

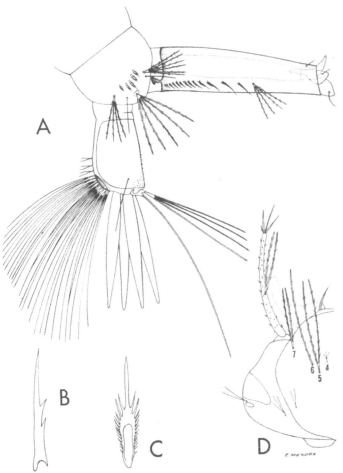

Fig. 136. Larva of *Aedes decticus* Howard, Dyar, and Knab. *A*, Terminal segments. *B*, Pecten tooth. *C*, Comb Scale. *D*, Head.

bers; they were associated with the immature stages of several other species of mosquitoes including *Aedes canadensis*, *A. excrucians*, *A. abserratus*, *A. cinereus*, and *Culiseta morsitans*. First instar larvae of *A. decticus* were found in the swamp as early as late March, and larvae and pupae were observed as late as May 15. The larvae are said to occur in the deeper and colder sphagnum pools. Smith also obtained the larvae from a woodland pool and a permanent bog in other localities in Massachusetts.

Many adult females were observed by Smith in the swamp in the early afternoon of May 29, 1951, which was a warm overcast day. They were attracted in numbers to the face, arms, and clothing. The bite is described as being scarcely noticeable. Smith saw single males

hovering about females in the shrubbery, but no swarming was observed.

Aedes (Ochlerotatus) diantaeus
Howard, Dyar, and Knab

Aedes diantaeus Howard, Dyar, and Knab, 1917, Carnegie Inst. Wash. Pub., *159*, 4:758.
Aedes serus Martini, 1920, Ueber Stechmücken, p. 96.

ADULT FEMALE (pl. 56). Medium-sized species. *Head:* Proboscis black-scaled; palpi short, black-scaled. Occiput clothed dorsally with narrow curved pale-yellow scales and light-golden erect forked scales, with broad appressed cream-colored scales laterally. Tori light brown on outer surface, inner surface infuscated and clothed with small dark setae. *Thorax:* Integument of scutum black; scutum clothed with pale-golden scales, except for a pair of narrow median brown stripes (in some specimens these median stripes appear to be fused, thus forming one broad median stripe); posterior half stripes sometimes present, rarely distinct. Posterior pronotum with brown scales on dorsal half, becoming yellowish white ventrally. Scutellum with narrow pale-golden scales and light-brown setae on the lobes. Pleura with large rather well-defined patches of broad flat grayish-white scales. Scales on sternopleuron reaching to or near the anterior angle, narrowly separate from patch on prealar area. Mesepimeron densely clothed with scales to near lower margin. Hypostigial spot of scales absent. Lower mesepimeral bristles usually absent, one or two when present. *Abdomen:* First tergite with a few white scales on median area; remaining tergites dark purplish brown, with basolateral triangular spots of white scales sometimes joined by a narrow white band. Sternites white-scaled, with dark apical bands. *Legs:* Legs black with bronze and purplish reflections; femora largely pale-scaled on posterior surface; small grayish-white knee spots present. *Wing:* Length about 4.0 to 4.5 mm. Scales narrow, brown.

ADULT MALE. Coloration similar to that of the female. *Terminalia* (fig. 137). Lobes of ninth tergite (IX-L) narrowly separated, each bearing several spines. Tenth sternite (X-S) stout, heavily sclerotized and hooked at tip.

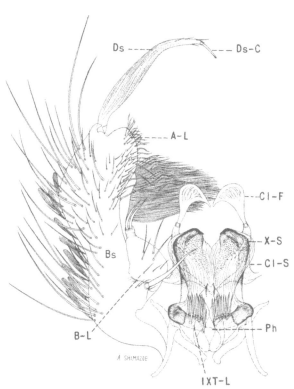

Fig. 137. Male terminalia of *Aedes diantaeus* Howard,
Dyar, and Knab.

LARVA (fig. 138). Antenna as long as the
head, slender, pale, nearly straight, spiculate;
antennal tuft 3- to 7-branched, barbed, inserted
near middle of shaft, reaching tip. Head hairs:
Postclypeal 4 small, branched; upper frontal 5
long, 3- or 4-branched (rarely 2- or 5-
branched), barbed; lower frontal 6 long,
double or triple, barbed; preantennal 7 long,
3- to 6-branched, barbed. Prothoracic hairs:
1 long, single; 2 short, single; 3 short, mul-
triple; 4 short, single or double; 5–6 long,
single; 7 long, single or double. Lateral abdominal hair 6 variable, double or triple on segment I, single to triple on II, and single on
III–VII. Comb of eighth segment with about
6 to 13 scales in a curved irregular single or
double row; individual scale thorn-shaped,

Phallosome (Ph) cylindrical, narrowed and
rounded at apex, open ventrally, closed dorsally. Claspette stem (Cl-S) pilose to near apex,
basal part stout, abruptly constricted and narrowed at apical fourth, shoulder at point of
constriction angulate and armed with several
small setae; claspette filament (Cl-F) with base
slender, broadly expanded and bladelike beyond base, tapered and recurved at tip. Basistyle (Bs) stout, outer surface concave, clothed
with scales and long and short setae, ventral
surface just beyond middle with a brushlike
close-set tuft of golden-brown to black setae
directed mesad; basal lobe (B-L) long and
conical, clothed with small setae, bearing at its
apex two stout curved dark spines, a long stout
dark dorsal spine arising from a sclerotized
protuberance near base of the lobe; apical lobe
(A-L) broadly rounded, bearing small retrorse
setae. Dististyle (Ds) about two-thirds as long
as the basistyle, expanded medially, pilose,
bearing three or four small subapical setae;
claw (Ds-C) slender, about one-sixth as long
as the dististyle.

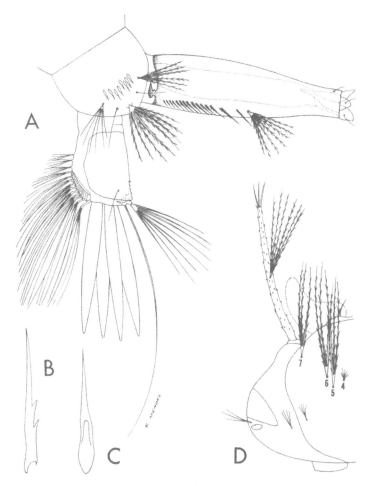

Fig. 138. Larva of *Aedes diantaeus* Howard, Dyar, and Knab.
A, Terminal segments. *B*, Pecten tooth. *C*, Comb scale. *D*,
Head.

without lateral spinules on basal part. Siphonal index about 3.5 to 4.0; pecten of about 13 to 22 teeth on basal half of siphon, with 1 to 3 distal teeth detached; siphonal tuft 4- to 8-branched, barbed, inserted beyond the pecten, about as long as basal diameter of siphon; dorsal preapical spine much shorter than the apical pecten tooth. Anal segment with the saddle extending three-fourths to seven-eighths down the sides, saddle spiculate on posterior part; lateral hair single, shorter than the saddle; dorsal brush bilaterally consisting of a long lower caudal hair and a much shorter multiple upper caudal tuft; ventral brush well developed, with three or four precratal tufts; gills about one and a half to two times as long as the saddle, pointed.

DISTRIBUTION. Forests of the northern United States from Maine to Montana and Wyoming, Canada, Alaska, northern Europe, and western Siberia. Smith (663) redescribed the species and summarized its distribution. *United States:* Maine (45, 663); Massachusetts (663, 746); Michigan (391, 663); Minnesota (533); Montana and New Hampshire (663); New York (31, 663); Vermont (663); Wyoming (221, 663). *Canada:* British Columbia (663, 708); Labrador (348, 663); Manitoba and Northwest Territories (289, 663); Nova Scotia (309, 708); Ontario (663, 708); Quebec (306, 708); Yukon (289, 663). *Alaska:* (396, 663).

BIONOMICS. The larvae are found mostly in cold shaded pools left from the melting snow in dense forests. Owen (533) states that the larvae appear in late May in Minnesota and are to be found as late as early July in dense swamps in coniferous forests. Jenkins (396) found the larvae usually associated with vegetation in pools in central Alaska. Natvig (523) observed that most of the larval habitats of this species in Norway were situated in woodlands, more or less shaded but on one occasion larvae were found in an open pool in a field.

According to Dyar (199) the males do not swarm but seize the females singly as they approach to bite. He observed the mating process in a woodland in British Columbia on a cloudy morning when the male grasped the female in the air and alighted on a near-by twig where copulation was completed.

Aedes (Ochlerotatus) dorsalis (Meigen)

Culex dorsalis Meigen, 1830, Syst. Beschr. Zweifl. Ins., 6:242.

Culex curriei Coquillett, 1901, Can. Ent., 33:259.

Culex onondagensis Felt, 1904, N.Y. State Mus. Bull., 79:278.

Culex lativittatus Coquillett, 1906, Ent. News, 17:109.

Aedes quaylei Dyar and Knab, 1906, Jour. N.Y. Ent. Soc., 14:191.

Grabhamia mediolineata Ludlow, 1907, Can. Ent., 39:129.

Grabhamia broquettii Theobald, 1913, Entomologist, 46:154.

Aedes grahami Ludlow, 1920, Ins. Ins. Mens., 7:154.

ADULT FEMALE (pl. 57). Medium-sized species. *Head:* Proboscis dark, speckled with pale scales on basal part; palpi short, dark, speckled with pale scales. Occiput with yellowish-white lanceolate scales and pale erect forked scales on median area; a small submedian patch of brown scales and a few black erect forked scales on either side; lateral region with broad appressed white scales and a few dark ones. Tori brown, darker on inner surface, with white scales on inner and dorsal surfaces. *Thorax:* Integument of scutum dark; scutum with a median longitudinal stripe of brown lanceolate scales (variable in width) or with a few brown scales on the median area; posterior brown half stripes may or may not be present; remainder of scutum, including the prescutellar space, with yellowish-white lanceolate scales. Posterior pronotum with narrow golden-brown scales on dorsal half, with grayish-white scales ventrally. Scutellum with narrow yellowish-white scales and light-brown setae on the lobes. Pleura with dense, poorly defined patches of broad appressed grayish-white scales. Patch on sternopleuron extending to anterior angle, separate from patch on prealar area. Mesepimeron entirely covered with scales. Hypostigial spot with many scales. Lower mesepimeral bristles about two to six. *Abdomen:* First tergite with a median patch of white scales; remaining tergites dark, with transverse segmental bands of white scales and a median dorsal stripe of white scales; last two tergites often almost entirely white. Venter white-scaled. *Legs:* Femora, tibiae, and first tarsal segment yellowish-white-scaled, speckled with dark scales; knee spots white. Hind tarsus with segments 1 to 3 ringed basally and apically with white, segment 4 with a basal white ring and a few white scales at apex, segment 5 almost entirely white. Middle tarsus with segments 1 to 3 narrowly ringed with white basally and apically, segment 4 dark except for a few white scales at base, segment 5 entirely dark. Front tarsus with segments 1 and 2 narrowly ringed with white basally and apically, segment 3 usually with a few white scales at base, segments 4 and 5 usually entirely dark. *Wing:* Length 4.0 to 4.5 mm. Scales narrow, dark brown and white intermixed, the white ones usually predominating; the costa and veins 1, 3, and 5 with more dark scales than on other veins.

ADULT MALE. Coloration similar to that of the female. *Terminalia* (fig. 139). Lobes of ninth tergite (IXT-L) about as long as broad, each bearing four to seven short spines. Tenth sternite (X-S) heavily sclerotized apically. Phallosome (Ph) conical, longer than broad, open ventrally, closed dorsally. Claspette stem (Cl-S) slender, pilose to near apex, extending to or a little beyond middle of basal lobe; claspette filament (Cl-F) nearly as long as the stem, broadly expanded at the middle on convex side, tapered to a recurved tip. Basistyle (Bs) cylindrical, about three and a half times as long as mid-width, clothed with scales and long and short setae, long setae numerous on inner ventral surface; basal lobe (B-L) prominent, constricted basally, somewhat expanded apically, bearing a short stout spine, a long strong recurved spine, and two or three long setae and shorter and weaker setae; apical lobe (A-L) short, rounded, bearing short curved setae. Dististyle (Ds) pilose, about three-fifths as long as the basistyle, slightly broader medially, bearing short setae before apex; claw (Ds-C) nearly one-fourth as long as the dististyle.

LARVA (fig. 140). Antenna about half as long as the head, spiculate; antennal tuft multiple, finely barbed, inserted near middle of shaft, not reaching tip. Head hairs: Postclypeal 4 small, multiple; upper frontal 5 single (occasionally double, rarely triple), weakly barbed; lower frontal 6 single (rarely double), weakly barbed; preantennal 7 multiple, barbed, reaching near insertion of antennal tuft. Prothoracic hairs: 1 long, usually single; 2 short, single; 3 short, 1- to 5-branched; 4 short, single or double; 5 long, double or triple; 6 long, single; 7 long, 2- to 4-branched. Lateral abdominal hair 6 usually double or triple on segments I–IV, double on V and single on VI. Comb of eighth segment with about 19 to 33 scales in a patch; individual scale fringed with subequal spinules. Siphonal index about 2.5 to 3.0; pecten of about 16 to 23 evenly spaced teeth reaching near middle of siphon; siphonal tuft 4- to 8-branched, inserted beyond the pecten; dorsal preapical spine shorter than the apical pecten tooth. Anal segment with the saddle extending one-half to two-thirds down the sides; lateral hair single (rarely double), much shorter than the saddle; dorsal brush bilaterally consisting of a long lower caudal hair and a shorter multiple upper caudal tuft; ventral brush large, with two to five precratal tufts; gills varying from short and budlike to nearly twice as long as the saddle.

DISTRIBUTION. The United States, Canada, Europe, and Asia. *United States:* California (287); Colorado (343); Connecticut (494); Delaware (167); Idaho (344, 672); Illinois

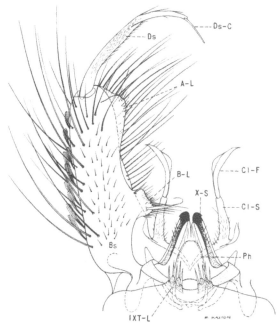

Fig. 139. Male terminalia of *Aedes dorsalis* (Meigen).

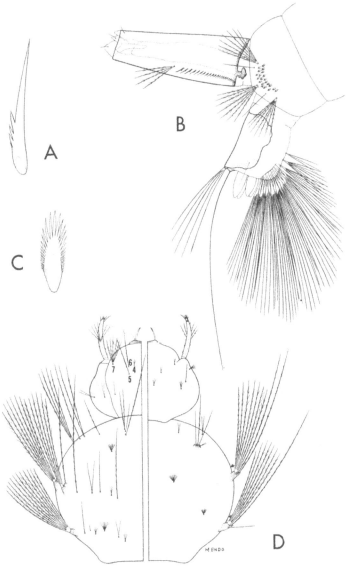

Fig. 140. Larva of *Aedes dorsalis* (Meigen). *A*, Pecten tooth. *B*, Terminal segments. *C*, Comb scale. *D*, Head and thorax (dorsal and ventral).

(268, 615); Iowa (625); Kansas (14, 530); Kentucky (570); Louisiana (213); Massachusetts (213, 267); Minnesota (533, 604); Mississippi (516); Missouri (6, 530); Montana (213, 475); Nebraska (213, 680); Nevada (213); New Mexico (20, 213); New York (31); North Dakota (213, 553); Ohio (731a); Oklahoma (628); Oregon (213, 672); Pennsylvania (213); South Dakota (14); Texas (249, 634); Utah (213, 575); Washington

(67, 672); Wisconsin (171, 213); Wyoming (578). *Canada:* Alberta (212, 598); British Columbia (212, 352); Manitoba (212, 598); Ontario (212, 289); Quebec (289, 708); Saskatchewan (212, 598).

BIONOMICS. The winter is passed in the egg stage, and the eggs hatch after flooding during the first warm weather in the spring. Breeding is continuous throughout the warm season, depending on the reflooding of the marshes. The larvae occur in a variety of habitats including both brackish and fresh water. They are often found in large numbers in tidal marshes along the Pacific coast and in saline pools along the margins of Great Salt Lake in Utah. The species also occurs in fresh-water marshes and in overflow from artesian wells and irrigation ditches. It seems to prefer alkaline water in grassy situations exposed to direct sunlight but is also occasionally found in more densely shaded pools. According to Rees and Nielsen (579) larvae have been known to complete development along the margins of the Great Salt Lake in water with as high as 12 per cent salt content.

The species is an important pest of man and animals in Utah, California, and often in other localities in western North America. According to Rempel (598) it is a most annoying prairie species in western Canada. The females are vicious biters, attacking any time during the day or night, but are particularly active toward evening or on calm cloudy days. Migratory and distributional flights of *Aedes dorsalis* have been traced for 22 miles in Utah and more than 30 miles in California (579). The males are known to accompany the females during these flights. Dyar (190) observed the males swarming over bushes or prominent objects. He describes a swarm in a low bush, about 3 feet from the ground, at Laurel, Montana, at sunset; also another swarm in a taller bush later when it was almost dark.

MEDICAL IMPORTANCE. Western equine encephalitis virus has been isolated from wild-caught *A. dorsalis* in the San Joaquin Valley, California (337), and in Colorado (687). California encephalitis virus has been isolated from wild-caught *A. dorsalis* from Kern County, California (593).

Aedes (Ochlerotatus) dupreei
(Coquillett)

Culex dupreei Coquillett, 1904, Can. Ent., *36*:10.

ADULT FEMALE (pl. 58). Small species. *Head:* Proboscis dark-scaled; palpi short, dark. Occiput with a median stripe of silver-white lanceolate scales bounded submedially by a patch of broad white scales which blend on either side with a large lateral patch of broad appressed brown scales; erect forked scales on central part pale. Tori light brown, darker on inner surface. *Thorax:* Integument of scutum dark brown; scutum with narrow bronze-brown scales except for a broad longitudinal stripe of silver-white scales extending the full length of the scutum (this longitudinal pale stripe is usually slightly more than one-third as wide as the scutum). Posterior pronotum with narrow bronze-brown scales on dorsal half. Scutellum with a median patch of narrow silver-white scales, setae brown. Pleura with small well-defined patches of appressed grayish-white scales. Scales on sternopleuron not reaching more than halfway to anterior angle, well separated from patch on prealar area. Mesepimeron with lower one-third bare. Hypostigial spot of scales absent. Lower mesepimeral bristles absent. *Abdomen:* First tergite with dark scales, usually mixed with pale scales; remaining tergites dark, with white basolateral patches. Venter white-scaled. *Legs:* Coxa of foreleg white-scaled. Legs dark-scaled, posterior surface of femora, tibiae, and first segment of tarsi with pale scales. *Wing:* Length about 2.5 mm. Scales narrow, brown.

ADULT MALE. Coloration similar to that of the female, but with pale scales more prevalent on the occiput. The longitudinal stripe on the scutum is broad and covers the entire dorsal surface. The lateral spots of the abdomen extend to near the middorsal line. *Terminalia* (fig. 141). Lobes of ninth tergite (IXT-L) as long as broad, each bearing three to five stout spines. Tenth sternite (X-S) heavily sclerotized apically. Phallosome (Ph) broadly conical, longer than broad, rounded apically, open ventrally, closed dorsally. Claspette stem (Cl-S) pilose, reaching a little beyond apex of basal lobe; claspette filament (Cl-F) ligulate, curved, nearly as long as the stem, gradually tapered to a point. Basistyle (Bs) about two and a half

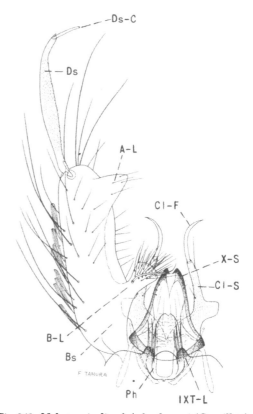

Fig. 141. Male terminalia of *Aedes dupreei* (Coquillett).

times as long as broad, clothed with large scales and long and short setae; basal lobe (B-L) longer than broad, expanded distally, bearing a large recurved dorsal spine near base, and many apical setae; apical lobe (A-L) distinct, slender, triangular, bearing a few small setae. Dististyle (Ds) about two-thirds as long as the basistyle, broader medially, pilose; claw (Ds-C) slender, about one-fourth as long as the dististyle.

LARVA (fig. 142). Antenna less than half as long as the head, sparsely spiculate; antennal tuft short, usually double, arising a little beyond middle of shaft. Head hairs: Postclypeal 4 small, branched; upper frontal 5 single, smooth or sparsely barbed apically; lower frontal 6 double or triple, smooth or barbed toward tip; preantennal 7 double or triple. Prothoracic hairs: 1–2 long, single; 3 short, usually double or triple; 4 short, single; 5–6 long, single; 7 long usually double. Lateral abdominal hair 6 usually single on segments I–VI. Comb of eighth segment with 7 to 10 scales in a single curved row; individual scale

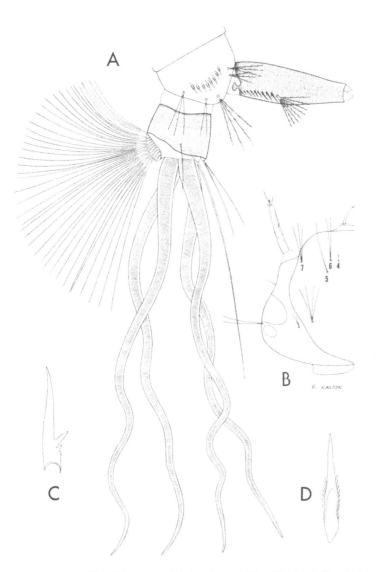

Fig. 142. Larva of *Aedes dupreei* (Coquillett). *A*, Terminal segments. *B*, Head. *C*, Pecten tooth. *D*, Comb scale.

tapered, with prominent darkly pigmented tracheae.

DISTRIBUTION. Southeastern United States north to New Jersey and west to Kansas and Texas. *United States:* Alabama (656); Arkansas (122, 213); Florida (135, 253); Georgia (135, 420); Illinois (615); Iowa (626); Kansas (14); Kentucky (570); Louisiana (213, 759); Mississippi (135, 546); Missouri (169, 530); New Jersey (213, 350); North Carolina (135, 646); Oklahoma (321); South Carolina (135, 742); Tennessee (135, 658); Texas (634); Virginia (176, 178).

BIONOMICS. The larvae are found in temporary rain-filled pools, particularly in woodlands following summer rains; they have a habit of hiding among the leaves and debris on the bottom, making it difficult to collect them. The eggs hatch whenever conditions are favorable. The larvae are easily recognized in the field by their extremely long anal gills. Michener (510) found them in a clear, partly shaded, grassy semipermanent pool at Camp Shelby, Mississippi. Pupae were also taken by Michener from a shaded woodland pool. Adults are often taken in light traps in the southern states but are rarely encountered in the field. Howard, Dyar, and Knab (382) state that the females do not appear to be attracted to man.

Aedes (Ochlerotatus) excrucians
(Walker)

Culex excrucians Walker, 1856, Ins. Saund., Dipt., p. 429.
Culex abfitchii Felt, 1904, N.Y. State Mus. Bull., 79:381.
Culex siphonalis Grossbeck, 1904, Can. Ent., *36*:332.
Aedes sansoni Dyar and Knab, 1909, Can. Ent., *41*:102.
Aedes euedes Howard, Dyar, and Knab, 1917, Carnegie Inst. Wash. Pub., *159*, 4:714.
Aedes aloponotum Dyar, 1917, Ins. Ins. Mens., *5*:98.
Aedes excrucians dytes Martini, 1922, Ent. Mitt., 2:164.

ADULT FEMALE (pl. 59). The females of some of the band-legged species, *A. excrucians, A. fitchii, A. increpitus, A. riparius,* and *A. stimulans,* are often difficult to identify with certainty; however, the characters given in the key will separate most specimens. Vockeroth (733) has shown that the tarsal claws of the

thorn-shaped, with a strong median spine and small lateral spinules on basal part. Siphonal index about 3.5 to 4.0; siphon much narrower than the anal segment; pecten of about 7 to 12 evenly spaced teeth on basal third of siphon; siphonal tuft large, multiple, strongly barbed, inserted beyond pecten. Anal segment ringed by the saddle; lateral hair single, a little longer than the saddle; dorsal brush bilaterally consisting of a long lower caudal hair and two or three shorter upper caudal hairs; ventral brush large, confined to the barred area; gills very long, eight times as long as the saddle or longer,

females of some of the species possess characters which can be used for separating them. Characters of the male terminalia offer more satisfactory means of distinguishing the species. Large species. *Head:* Proboscis dark, sprinkled with pale scales; palpi short, dark, lightly sprinkled with pale scales, a few pale scales at apices of segments. Occiput with a median patch of narrow whitish or pale-golden scales, bounded on either side by a patch of narrow brownish scales; lateral region with broad appressed white scales enclosing a dark-scaled patch; erect forked scales on median area pale, dark submedially. Tori yellow, darker on inner surface, with white scales on inner surface. *Thorax:* Integument of scutum dark brown to black; scutum with a varying pattern of brown and whitish scales, with a median brown stripe and yellowish white on the sides or completely reddish brown-scaled. Prescutellar space with yellowish-white scales. Posterior pronotum with narrow curved brown or golden scales on dorsal half, with white scales on ventral half. Scutellum with narrow yellowish-white scales and brown setae on the lobes. Pleura with distinct patches of grayish-white scales. Sternopleuron with scattered scales extending to near the anterior angle, separate from patch on the prealar area. Mesepimeron bare on lower one-third. Hypostigial spot of scales absent. Lower mesepimeral bristles absent or rarely one. *Abdomen:* First tergite with a large median patch of white scales; remaining tergites dark brown, with basal segmental white bands and with few to many scattered white scales posteriorly. Venter white-scaled, the median area with some black scales. *Legs:* Femora and tibiae with intermixed dark and pale scales, posterior surface pale; knee spots white. Tarsi dark, first segment with white scales intermixed, streaked with pale scales on posterior surface; hind tarsi with broad basal bands of white on all segments; middle tarsi with narrower basal bands on segments 1 to 4, obsolete or absent on segment 5; front tarsi with narrow basal white bands on segments 1 to 3, obsolete or absent on segments 4 and 5. Front tarsal claw as in figure 143, *B*. *Wing:* Length about 5.0 mm. Scales predominantly dark with pale scales intermixed.

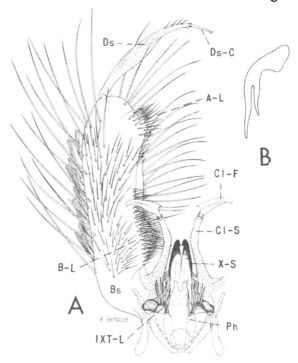

Fig. 143. *Aedes excrucians* (Walker). *A*, Male terminalia. *B*, Front tarsal claw of female.

ADULT MALE. Coloration similar to that of the female. *Terminalia* (fig. 143). Lobes of ninth tergite (IXT-L) rounded, each with a varying number of four or more rather slender long spines. Tenth sternite (X-S) heavily sclerotized beyond middle. Phallosome (Ph) longer than wide, open ventrally, closed dorsally, notched at apex. Claspette stem (Cl-S) long, slender, pilose on basal half; claspette filament (Cl-F) shorter than the stem, angularly expanded to a sharp point near base, tip recurved. Basistyle (Bs) about three times as long as mid-width, clothed with scales and long and short setae; basal lobe (B-L) represented by a large rugose area reaching to near base of apical lobe and clothed with numerous small setae arising from tubercles arranged in rows; apical lobe (A-L) large, broadly rounded, clothed with short setae. Dististyle (Ds) about three-fifths as long as the basistyle, broader medially, bearing a few subapical setae; claw (Ds-C) slender, about one-fifth as long as the dististyle.

LARVA (fig. 144). Antenna shorter than the head, slender, spiculate; antennal tuft usually 3- to 5-branched, barbed, inserted before middle of shaft, not reaching tip. Head hairs:

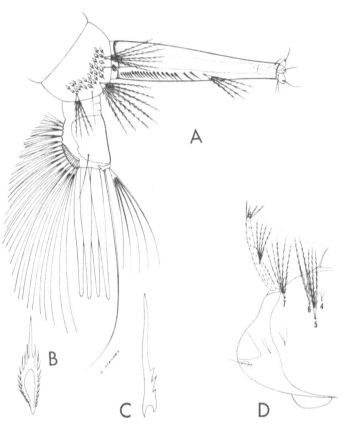

Fig. 144. Larva of *Aedes excrucians* (Walker). *A*, Terminal segments. *B*, Comb scale. *C*, Pecten tooth. *D*, Head.

Postclypeal 4 small, multiple; upper frontal 5 long, usually double (occasionally triple), barbed; lower frontal 6 long, usually double, barbed (either 5 or 6 occasionally single); preantennal 7 long, usually 4- to 8-branched. Prothoracic hairs: 1 long, single; 2 short, single; 3 short, single to triple; 4 short, single; 5–6 long, usually single; 7 long, single to triple. Lateral abdominal hair 6 usually double on segments I and II, single on III–VI. Comb of eighth segment with about 17 to 33 scales in a patch; individual scale thorn-shaped, with a long stout apical spine and smaller lateral spinules. Siphonal index about 5.0; siphon slender and tapered from near base; pecten of about 16 to 27 teeth not quite reaching middle of siphon, with 1 to 3 teeth detached; siphonal tuft 3- to 7-branched, inserted beyond pecten, individual hairs longer than the basal diameter of the siphon; dorsal preapical spine shorter than the apical pecten tooth. Anal segment with the saddle extending two-thirds to three-fourths

down the sides; lateral hair single, about as long as the saddle; dorsal brush bilaterally consisting of a long lower caudal hair and a shorter multiple upper caudal tuft; ventral brush large, with about three to five precratal tufts inserted to near base of segment; gills varying from about one to two and a half times as long as the saddle (shorter when found in brackish water), narrow, and tapered.

DISTRIBUTION. The species is widely distributed in the northern coniferous forest belt of North America. It occurs also in northern Europe and Asia. *United States:* Colorado (343); Connecticut (270, 494); Idaho (344, 672); Illinois (615); Maine (45, 270); Massachusetts (267, 270); Michigan (391, 529); Minnesota (213, 533); Montana (475); New Hampshire (66, 270); New Jersey (213, 350); New York (31); North Dakota (553); Ohio (482); Oregon (672); Pennsylvania (128, 298); Rhode Island (270, 436); Utah (575); Vermont (649); Washington (67, 672); Wisconsin (171, 213); Wyoming (578). *Canada:* Alberta (212, 598); British Columbia (352, 598); Manitoba (212, 466); Newfoundland (289); Northwest Territories (181, 289); Nova Scotia (309, 708); Ontario (212, 708); Prince Edward Island (311, 708); Quebec (289, 402); Saskatchewan (311, 598); Yukon (212, 289). *Alaska:* (213, 396).

BIONOMICS. The larvae are found in the spring in temporary pools in open swamps, along the edges of flooded grassy marshes, and in woodland pools. Frohne (293) collected the larvae in brackish water habitats in the coastal marshes of Alaska. As far as is known there is but a single generation a year, and hatching takes place during April, May, and early June throughout most of its range. In Norway, according to Natvig (523), the species breeds principally in pools in open areas exposed to the sun. Wesenberg-Lund (749) found that in Denmark the species breeds mostly in extremely shallow ponds on the plains, in the forests, or in the outskirts of the woods.

The females will bite in the woods any time during the day when their haunts are invaded, but they are most active in the evening. They often persist until late summer. Natvig (523) describes *Aedes excrucians* as being very annoying at times in Norway where they bite in the shade, rarely attacking in bright sunshine. Natvig further states that the adults have been found in houses in Norway, especially after rain. Dyar (197) describes the swarming of males of this species at Banff, Alberta, as follows: "They fly shortly after sunset in small swarms, very high in openings in the woods, roads or the tops of smaller trees. The swarms are loose and open, clustering for a few seconds in one spot, then dashing away to a distance. The swarming period does not seem to exceed half an hour."

Aedes (Ochlerotatus) fitchii
(Felt and Young)

Culex fitchii Felt and Young, 1904, Science, n.s., 20:312.
Aedes palustris Dyar, 1916, Ins. Ins. Mens., 4:89.
Aedes palustris var. *pricei* Dyar, 1917, Ins. Ins. Mens., 5:16.
Aedes mimesis Dyar, 1917, Ins. Ins. Mens., 5:116.

ADULT FEMALE (pl. 60). This species is often difficult to distinguish from *A. excrucians* and certain other band-legged *Aedes*; however, the characters given in the key will separate most specimens. Medium-sized species. *Head:* Proboscis dark, sparsely speckled with pale scales dorsally on basal half, speckled ventrally for entire length; palpi short, dark, sparsely speckled with pale scales, with irregular pale rings at bases of segments 3 and 4, terminal segment yellowish white at apex. Occiput with a median patch of pale-yellow lanceolate scales bounded by a submedian patch of narrow, golden-brown scales on either side; lateral region with broad appressed scales surrounding a dark patch; erect forked scales pale yellow on median area, those on submedian area dark. Tori yellow, darker on inner surface, with white scales on inner surface. *Thorax:* Integument of scutum dark brown; scutum with a broad median stripe of narrow brown scales and yellowish-white scales on anterior and lateral margins and on the prescutellar space (scutal pattern sometimes of brown and yellowish-white scales). Posterior pronotum with narrow curved golden-brown scales dorsally, broader, curved, and yellowish white ventrally. Scutellum with narrow yellowish-white scales and brown setae on the lobes. Pleura with extensive patches of rather broad grayish-white

scales. Scales on sternopleuron extending to anterior angle, distinctly separate from patch on prealar area. Mesepimeron bare on lower one-fourth to one-third. Hypostigial spot of scales absent. Lower mesepimeral bristles none to two, rarely three or four. *Abdomen:* First tergite with a median patch of white scales; remaining tergites dark bronze-brown with broad basal white bands. Some specimens have few to many pale scales medially and posteriorly, particularly on the posterior tergites. Venter pale with a few dark scales intermixed. *Legs:* Femora and tibiae with intermixed brown and white scales, posterior surface yellowish white; knee spots white. Tarsi dark, first segment with white scales intermixed on anterior surface and streaked with white on posterior surface; hind tarsi with broad basal white bands on all segments; middle tarsi with narrower basal white bands on segments 1 to 4, segment 5 usually all dark; front tarsi with narrow basal white bands on segments 1 to 3, segments 4 and 5 usually dark. Front tarsal claw as in figure 145, *B*. *Wing:* Length about 4.5 to 5.0 mm. Scales narrow to moderately broad, intermixed brown and yellowish white.

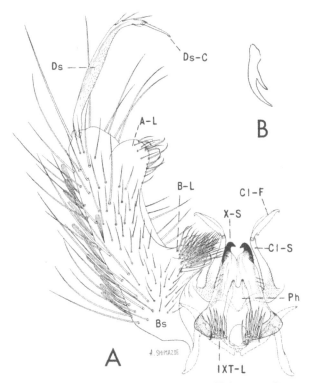

Fig. 145. *Aedes fitchii* (Felt and Young). *A*, Male terminalia. *B*, Front tarsal claw of female.

ADULT MALE. Coloration similar to that of the female. *Terminalia* (fig. 145). Lobes of ninth tergite (IXT-L) nearly as long as broad, each bearing several long strong spines. Tenth sternite (X-S) heavily sclerotized beyond middle. Phallosome (Ph) subcylindrical, open ventrally, closed dorsally, apex narrowed and deeply notched. Claspette stem (Cl-S) extending well beyond basal lobe, pilose on basal half; claspette filament (Cl-F) rather short, bladelike, notched on concave side near base, apex strongly narrowed and recurved. Basistyle (Bs) about three to three and a half times as long as mid-width, clothed with scales and short and long setae; basal lobe (B-L) triangular, densely clothed with moderate setae, a stout dorsal spine arising from near base; apical lobe (A-L) prominent, thumblike, clothed with moderate setae. Dististyle (Ds) a little broader medially, approximately three-fifths as long as the basistyle, finely pilose, bearing prominent subapical setae; claw (Ds-C) slender, about one-fifth as long as the dististyle.

LARVA (fig. 146). Antenna about half as long as the head, spiculate; antennal tuft multiple, inserted near middle of shaft, reaching near tip. Head hairs: Postclypeal 4 small, multiple; upper frontal 5, 3- or 4-branched (occasionally double), barbed, a little longer than hair 6; lower frontal 6 double or triple (occasionally single), sparsely barbed; preantennal 7 multiple, barbed, reaching beyond insertion of antennal tuft. Prothoracic hairs: 1 long, single to triple; 2 long, single; 3 medium, single or double; 4 medium, single; 5 long, double or triple; 6 long, single; 7 long, usually 3- or 4-branched. Lateral abdominal hair 6 double or triple on segments I and II, double on III–VI. Comb of eighth segment with 12 to 28 scales in a patch; individual scale with the apical spine three or four times as broad and twice as long as the preapical spines. Siphonal index 4.0 to 5.0; siphon gradually tapered from near base; pecten of 15 to 24 evenly spaced teeth not quite reaching middle of siphon; siphonal tuft 3- to 6-branched, sparsely barbed, inserted beyond pecten; dorsal preapical spine weak, much shorter than the apical pecten tooth. Anal segment with the saddle extending about two-thirds down the sides, saddle with spicules on posterior part; lateral hair single, longer than the saddle; dorsal brush bilaterally consisting of a long lower caudal hair and a shorter multiple upper caudal tuft; ventral brush large, with one or two precratal tufts; gills about one and a half to two times as long as the saddle, pointed.

DISTRIBUTION. Widely distributed in the forests of the northern United States, Canada, and Alaska. *United States:* California (287); Colorado (343); Connecticut (270, 494); Idaho (344, 672); Illinois (615); Iowa (625); Maine (45, 270); Massachusetts (267, 270); Michigan (391, 529); Minnesota (533); Montana (475); Nebraska (531); New Hampshire (66, 213); New Jersey (130, 350); New York (31); North Dakota (553); Ohio (482); Oregon (672); Rhode Island (35, 270); Utah (575); Vermont (649); Washington (67, 672); Wisconsin (171, 213); Wyoming (578). *Canada:* Alberta (212, 708); British Columbia (212, 352); Labrador (289); Manitoba (212, 466); Newfoundland (289); Northwest Territories (289, 708); Ontario (212, 464); Prince Edward Island and Quebec (708); Saskatchewan (212, 598); Yukon (212, 289). *Alaska:* (396, 699).

BIONOMICS. Freeborn (284) states that the eggs of *Aedes fitchii* are laid singly in batches of 4 to 10 on the mud at the edge of receding grassy pools, or in the crevices of sun-cracked earth in the beds of previous pools. The larvae develop in the spring in a wide variety of habitats, including pools of a temporary and semipermanent nature. They seem to prefer the deeper and less temporary pools. The larvae are frequently found in snow-water pools at the edge of snow banks. Mail (475) claims that there is a single generation a year and that late records are owing to late hatching.

Freeborn (284) found the species to 9,000 feet in the Sierra in Yosemite National Park where emergence was delayed until well into July. Freeborn states that the adults occur in the edges of the forested areas surrounding the mountain meadows and bite persistently both during the day and in the evening. According to Freeborn, the females ordinarily become quiet at dark but will continue their feeding throughout bright moonlight nights. The females sometimes cause great discomfort to sportsmen and cattle in forest areas.

Dyar (197) observed the males swarming at Banff, Alberta. He says, "They fly in small groups, after sunset, high in the openings in the forest, much as with *A. excrucians,* though they seem less wild and flighty." Dyar (219) also observed swarms of the males after sunset on a hillside between openings in the trees at Belton, Glacier National Park, Montana, and says "we caught them until we were tired of the sport."

Aedes (Ochlerotatus) flavescens (Müller)

Culex flavescens Müller, 1764, Fauna Ins. Fried., p. 87.

Culex lutescens Fabricius, 1775, Ent. Syst., p. 800.

Culex variegatus Schrank, 1781, Enum. Ins. Austr., p. 482.

Culex bipunctatus Robineau-Desvoidy, 1827, Mem. Soc. Nat. Hist. Paris, *3*:405.

Culex arcanus Blanchard, 1905, Les Moust., p. 303.

Culex fletcheri Coquillett, 1906, U.S. Bur. Ent. Tech. Ser., *11*:20.

ADULT FEMALE (pl. 61). Large yellowish species. *Head:* Proboscis dark brown, with scattered yellowish scales, yellowish scales less numerous on distal half; palpi short, dark brown, with yellowish scales intermixed. Occiput with narrow golden scales and golden and brown erect forked scales dorsally, with broad appressed cream-colored scales laterally. Tori brown, darker on inner surface, with dark setae and yellowish scales on inner surface. *Thorax:* Integument of scutum black; scutum clothed with narrow yellowish to light-brown scales and a broad median stripe of darker brown scales; the yellow and brown scales are blended on some specimens. Posterior pronotum with narrow light-brown scales on dorsal part, becoming a little broader and yellowish-white on ventral part. Scutellum with narrow yellow or light-brown scales and brown setae on the lobes. Pleura with extensive, poorly defined patches of rather narrow slightly curved, yellowish-white scales. Scales on sternopleuron extending to anterior angle, narrowly separated from patch on prealar area. Mesepimeron with scales extending to near lower margin. Hypostigial spot with many scales. Lower mesepimeral

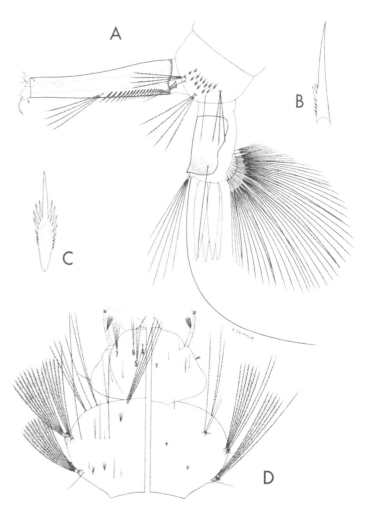

Fig. 146. Larva of *Aedes fitchii* (Felt and Young). *A,* Terminal segments. *B,* Pecten tooth. *C,* Comb scale. *D,* Head and thorax (dorsal and ventral).

bristles usually absent. *Abdomen:* First tergite with a median patch of dull yellow scales; remaining tergites entirely clothed with dull yellow scales. Venter with dull yellow scales. *Legs:* Femora and tibiae brown with yellowish scales intermixed, pale on posterior surface. First tarsal segment marked similar to tibiae but with apex dark; segments 2 to 5 of middle and hind tarsi with broad basal yellowish-white rings, dark apically (ring narrow or indistinct on segment 5 of middle tarsi); segments 2 to 4 of front tarsi with basal yellowish-white rings, dark apically, segment 5 entirely dark. *Wing:* Length about 5.5 to 6.0 mm. Scales narrow, predominantly yellow, with brown scales intermixed, those along the costa mostly pale.

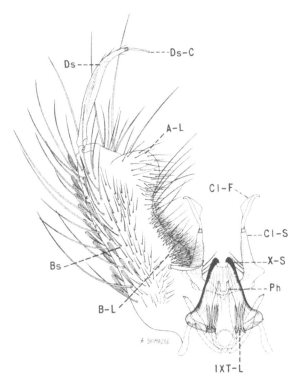

Fig. 147. Male terminalia of *Aedes flavescens* (Müller).

ADULT MALE. Coloration similar to that of the female. *Terminalia* (fig. 147). Lobes of ninth tergite (IXT-L) short, narrow, each bearing about six or seven long spines. Tenth sternite (X-S) heavily sclerotized and toothed apically. Phallosome (Ph) long, cylindrical, open ventrally, closed dorsally, notched at apex. Claspette stem (Cl-S) short, stout, tapered distally, pilose to near apex, with three spines on inner margin near base; claspette filament (Cl-F) about as long as the stem, basal third slender, followed by an angularly expanded broad blade tapered to a recurved point. Basistyle (Bs) about two and a half times as long as mid-width, clothed with scales and long and short setae; basal lobe (B-L) represented by a rugose area elevated basally and flattened distally, reaching beyond middle of basistyle, clothed with fine setae, a stout dorsal spine and several longer setae near base; apical lobe (A-L) prominent, rounded apically, clothed with many setae. Dististyle (Ds) about two-thirds as long as the basistyle, expanded medially, pilose, bearing three or four small setae before apex; claw (Ds-C) slender, about one-fifth as long as the dististyle.

LARVA (fig. 148). Antenna less than half as long as the head, spiculate; antennal tuft multiple, inserted near middle of shaft, reaching near tip. Head hairs: Postclypeal 4 small, multiple; upper front 5, 3- or 4-branched (rarely double), barbed; lower frontal 6 double or triple, barbed; preantennal 7 multiple, barbed, reaching insertion of antennal tuft. Prothoracic hairs: 1 long, single; 2 medium, single; 3 short, 2- to 4-branched; 4 short, single; 5 long, single or double; 6 long, single; 7 long, 2- to 4-branched. Lateral abdominal hair 6 double on segments I–VI. Comb of eighth segment with about 20 to 36 scales in a patch; individual scale with the subapical spines about half as long as the median spine. Siphonal index about 3.2 to 4.0; pecten of about 17 to 28 teeth on basal two-fifths of siphon, 1 or 2 distal teeth may or may not be detached; siphonal tuft 4- to 7-branched, inserted beyond pecten, about as long as the basal diameter of the siphon; dorsal preapical spine much shorter than the apical pecten tooth. Anal segment with the saddle extending about two-thirds down the sides; lateral hair single, about as long as the saddle; dorsal brush bilaterally consisting of a long lower caudal hair and a shorter multiple upper caudal tuft; ventral brush large, with several precratal tufts extending nearly to base of the segment; gills varying from shorter than the saddle to about twice as long.

DISTRIBUTION. Europe, Asia, the interior plains of the northern United States, Canada, and Alaska. *United States:* California (7); Colorado (343); Idaho (344, 672); Illinois (615); Iowa (625, 626); Kansas (531); Michigan (391, 529); Minnesota (224, 533); Missouri (14); Montana (213, 475); Nebraska (14, 680); New York (31); North Dakota (213, 553); Oregon (672); South Dakota (14); Utah (575); Washington (672); Wisconsin (171); Wyoming (14, 303). *Canada:* Alberta (212, 598); British Columbia (212, 352); Labrador (348); Manitoba (212, 466); Northwest Territories (289); Ontario (212, 289); Saskatchewan (212, 598); Yukon (162a). *Alaska:* (213, 396).

BIONOMICS. The larvae of *Aedes flavescens* appear very early in the spring in deep temporary pools in meadows and marshes on the open plains. Rempel (598) says the species

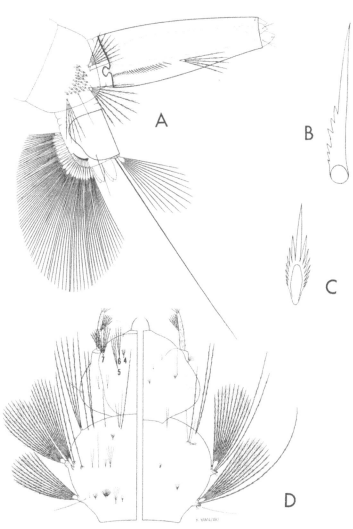

Fig. 148. Larva of *Aedes flavescens* (Müller). *A*, Terminal segments. *B*, Pecten tooth. *C*, Comb scale. *D*, Head and thorax (dorsal and ventral).

the forest. The females bite during the day and early evening, attacking cattle more often than man. According to Rempel (598), it is a serious pest in southern Saskatchewan.

Wesenberg-Lund (749) describes the mating habits of this species as observed in Denmark. He noticed males resting on the edges and tips of leaves of nettles. When he struck the nettles with his walking stick, both sexes arose, and mating took place during flight, requiring 50 to 70 seconds. Hearle (354) describes a mating swarm of this species on the prairie in Canada, 2½ to 3½ feet above the ground. Hearle gives a detailed account of the life history of the species.

Aedes (Ochlerotatus) fulvus pallens Ross

Aedes fulvus pallens Ross, 1943, Proc. Ent. Soc. Wash., 45:148.

Aedes fulvus, Dyar (in part), 1928, Carnegie Inst. Wash. Pub., 387:154.

ADULT FEMALE (pl. 62). A medium-sized to rather large bright orange-yellow species. *Head:* Proboscis yellow-scaled, dark at tip; palpi about one-fifth as long as the proboscis, yellow-scaled, with dark-scaled tips. Median area of occiput with narrow bright-yellow scales, yellow setae, and yellow erect forked scales; lateral region of occiput with broad flat yellow scales. Tori orange-colored, naked except for fine brown hairs on inner surface. *Thorax:* Scutum with integument yellow except for a pair of large dark-brown to black postero-lateral spots on either side of the prescutellar space; scutum sparsely clothed with narrow yellow scales on the anterior half and narrow dark-brown scales on the posterior half, particularly on the dark patches. Anterior pronotum, posterior pronotum, and pleura yellow; a dark spot present beneath the anterior spiracle. Pleuron with small patches of closely appressed silvery scales on prealar area and on upper half of the mesepimeron. Hypostigial spot of scales absent. Lower mesepimeral bristles absent. Scutellum with narrow dark-brown scales and brown setae on the lobes. *Abdomen:* First tergite with a median patch of dark scales; remaining tergites yellow scaled basally and laterally, dark apically; the sixth segment with the dark-scaled area greatly re-

breeds in heavily overgrown and partly shaded semipermanent pools in western Canada. Frohne (293) collected the larvae in brackish water habitats in the coastal marshes of Alaska. There is only one generation a year, but the females may survive until late in the season. Rees (575) says the larvae mature rather slowly, and in Utah the adults do not emerge until June.

The adults are seldom numerous but are bad biters when encountered on the plains. According to Wesenberg-Lund (749), the species inhabits the open meadows bordering lakes and seashores in Denmark and does not dwell in

duced, seventh segment almost entirely yellow-scaled. Venter yellow-scaled. *Legs:* Femur, tibia, and first tarsal segment of each leg yellow to orange-yellow, tipped with dark scales; remaining tarsal segments primarily dark-scaled, blended with yellow, the amount of yellow being variable. *Wing:* Length 4.5 to 5.0 mm. Scales yellow, rather broad on costa, subscosta, and vein 1, narrow and darker on remaining veins.

ADULT MALE. Coloration similar to that of the female. *Terminalia* (fig. 149). Lobes of ninth tergite (IXT-L) about as long as broad, each bearing several short spines. Tenth sternite (X-S) heavily sclerotized beyond middle. Phallosome (Ph) stout, conical, rounded apically, open ventrally, closed dorsally. Claspette stem (Cl-S) slender, moderately pilose, extending to or slightly beyond the distal margin of the basal lobe; claspette filament (Cl-F) nearly as long as stem, broadly expanded near base, bladelike, gradually tapered to a recurved point. Basistyle (Bs) about three times as long as mid-width, clothed with scales and long and short setae; basal lobe (B-L) prominent, constricted basally, expanded apically, bearing

numerous small apical setae and a large basal rod with tip flattened, broadened, and usually slightly retrorse; apical lobe (A-L) short, thumblike, bearing a few rather stout curved setae on apex. Dististyle (Ds) a little more than half as long as basistyle, broader medially, pilose, bearing about three small subapical setae; claw (Ds-C) slender, about one-fifth as long as the dististyle.

LARVA (fig. 150). Antenna less than half as long as the head, sparsely spiculate; antennal tuft multiple, inserted near middle of shaft, not reaching tip. Head hairs: Postclypeal 4 small, multiple; upper frontal 5 long, single; lower frontal 6 long, usually double (occasionally triple); preantennal 7 multiple, minutely barbed, reaching beyond insertion of antennal tuft. Prothoracic hairs: 1 long, single; 2–4 short, double or triple; 5–6 long, single; 7 long, usually triple. Lateral abdominal hair 6 usually double on segments I and II, single on III–V. Comb of eighth segment with 25 or more scales in a patch; individual scale rounded apically and fringed with subequal spinules. Siphonal index about 2.0; pecten of about 12 to 17 teeth reaching beyond middle of siphon, last tooth detached; siphonal tuft large, multiple, fan-shaped, heavily barbed, inserted within the pecten. Anal segment ringed by the saddle; lateral hair single or double, nearly as long as the saddle; dorsal brush bilaterally consisting of a long lower caudal hair and a shorter multiple upper caudal tuft; ventral brush large, heavy, restricted to the barred area; gills about twice as long as the saddle, pointed.

DISTRIBUTION. Southeastern United States, west to Oklahoma and Texas. *United States:* Alabama (84, 514); Arkansas (419, 683); Florida (135, 419); Georgia (135, 516); Illinois and Indiana (615); Kentucky (570); Louisiana (613, 759); Maryland (613); Mississippi (135, 516); North Carolina (135, 646); Oklahoma (628); South Carolina (135, 742); Tennessee (658); Texas (571); Virginia (176).

BIONOMICS. The larvae of this species have been collected only on a few occasions; however, the females are encountered somewhat more often, particularly in night-biting collec-

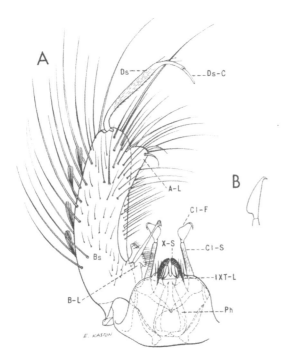

Fig. 149. *Aedes fulvus pallens* Ross. *A*, Male terminalia. *B*, Filament of claspette.

tions. Barret (34) describes a breeding place of this species observed in 1916 at the edge of the city of Charlotte, North Carolina. Larvae were found in a thicket in a sinkhole containing muddy water following heavy rain in July. The larvae were lying almost parallel to the surface of the water, and their dark bodies were easily seen against the slightly yellow muddy water. Root (608) found larvae of this species in floodpools of a creek in Lee County, Georgia, during June, 1923. Michener (510) captured adults of *A. fulvus pallens* in flight and resting on foliage at Camp Shelby, Mississippi, during 1943 and 1944.

Aedes (Ochlerotatus) grossbecki
Dyar and Knab

Aedes grossbecki Dyar and Knab, 1906, Jour. N.Y. Ent. Soc., *14*:201.

Culex sylvicola Grossbeck, 1906, Can. Ent., *38*:129.

ADULT FEMALE (pl. 63). Medium-sized to rather large species. *Head:* Proboscis dark, with a few white scales on basal half; palpi short, dark, speckled with white. Occiput with narrow white scales in a broad median patch, bounded by a small submedian patch of dark scales on either side; broad appressed white scales enclosing a dark-scaled patch laterally; erect forked scales on central part of occiput pale. Tori brown, with white scales on inner surface. *Thorax:* Integument of scutum dark brown or black; scutum with a more or less distinct broad median dark-brown stripe becoming golden brown anteriorly, posterior half-stripes present; the sides, anterior and lateral margins, and prescutellar space white. Posterior pronotum with narrow pale golden-brown scales on dorsal half, becoming broader and white on ventral half. Scutellum with narrow white scales and brown setae on the lobes. Pleura with small patches of broad grayish-white scales. Sternopleuron with a large patch of scales extending about halfway to anterior angle (a few detached scales at anterior angle), widely separated from patch of scales on pre-alar area. Mesepimeron bare on lower one-third. Hypostigial spot of scales absent. Lower mesepimeral bristles none to three. *Abdomen:* First tergite with a large median patch of white scales; remaining tergites each with a broad

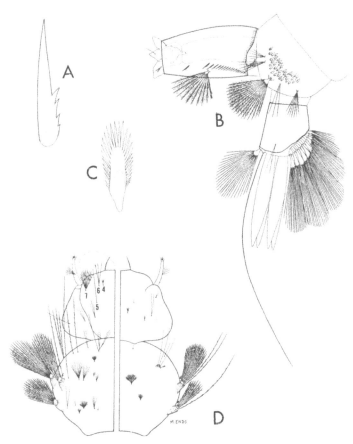

Fig. 150. Larva of *Aedes fulvus pallens* Ross. *A*, Pecten tooth. *B*, Terminal segments. *C*, Comb scale. *D*, Head and thorax (dorsal and ventral).

basal band of yellowish-white scales; the apical half of each tergite black-scaled, speckled with pale scales. Venter primarily white-scaled, spotted with dark. *Legs:* Femora, tibiae, and first segment of tarsi clothed with intermixed black and white scales; knee spots white. Segments of hind tarsi each with a broad basal white ring; segments 1 to 4 of front and middle tarsi with narrower basal white rings, segment 5 with the white ring greatly reduced or absent. *Wing:* Length about 4.5 to 5.0 mm. Scales very broad, triangular-shaped, intermixed dark and white.

ADULT MALE. Coloration similar to that of the female. *Terminalia* (fig. 151). Lobes of ninth tergite (IXT-L) about as long as broad, each bearing several long spines. Tenth sternite (X-S) heavily sclerotized apically. Phallosome

(Ph) conical, about twice as long as basal width, open ventrally, closed dorsally, and notched at apex. Claspette stem (Cl-S) pilose, extending beyond distal margin of basal lobe; claspette filament (Cl-F) about two-thirds as long as the stem, broad, expanded at middle, bladelike, tapered to a pointed curved tip. Basistyle (Bs) cylindrical, about three and a half times as long as mid-width, clothed with scales and long and short setae; basal lobe (B-L) prominent, conical, bearing numerous small apical setae and a stout slightly curved dorsal spine; apical lobe (A-L) prominent, thumblike, bearing fine setae. Dististyle (Ds) broader medially, nearly two-thirds as long as the basistyle, pilose, bearing two or three short setae before apex; claw (Ds-C) slender, nearly one-fourth as long as the dististyle.

LARVA (fig. 152). Antenna nearly half as long as the head, spiculate; antennal tuft multiple, inserted a little before middle of shaft, reaching near tip. Head hairs: Postclypeal 4 small, multiple; upper frontal 5 usually triple (sometimes double or 4-branched), barbed; lower frontal 6 double (sometimes triple), barbed; preantennal 7 multiple, barbed, reaching beyond insertion of antennal tuft. Prothoracic hairs: 1 long, single; 2 medium, single; 3 short, single or double; 4 short, single; 5 long, usually single or double; 6 long, single; 7 long, usually triple. Lateral abdominal hair 6 usually double on segments III–VI. Comb of eighth segment with about 30 to 35 scales in a patch; individual scale fringed with subequal spines, the median spine longer. Siphonal index about 3.5; pecten of numerous evenly spaced teeth on basal two-fifths of siphon; siphonal tuft rather large, multiple, barbed, inserted beyond pecten; dorsal preapical spine shorter than the apical pecten tooth. Anal segment with the saddle extending one-half to two-thirds down the sides, saddle spiculate; lateral hair single, a little shorter than the saddle; dorsal brush bilaterally consisting of a long lower caudal hair and a shorter multiple upper caudal tuft; ventral brush large, with about three to five precratal tufts; gills about as long as the saddle or a little longer, tapered.

DISTRIBUTION. This species is generally rare but occurs throughout most of the eastern United States. *United States:* Arkansas (88, 122); Delaware (167); Illinois (615); Kentucky (570); Louisiana (419, 759); Maryland (213); Mississippi (213, 546); Missouri (14); New Jersey (213, 350); New York (31, 492); Ohio (139); South Carolina (742); Tennessee (658); Vermont (270); Virginia (213).

BIONOMICS. The larvae of *A. grossbecki* are found in early spring pools. The senior author collected the larvae from woodland pools in several localities in northern New Jersey during the spring of 1947. Ross (615) says "although not abundant, this species is common throughout the post oak flats along the Mississippi River in extreme southern Illinois." Very little is known of the habits of adults of this species. Dr. Carl Venard, who collected a good number of male and female specimens during May, 1943, in Ohio, states that the females are persistent biters.

Aedes (Ochlerotatus) hexodontus Dyar

Aedes hexodontus Dyar, 1916, Ins. Ins. Mens., 4:83.
Aedes cyclocerculus Dyar, 1920, Ins. Ins. Mens., 8:23.
Aedes leuconotips Dyar, 1920, Ins. Ins. Mens., 8:24.
Aedes labradoriensis Dyar and Shannon, 1925, Jour. Wash. Acad. Sci., 15:78.
Aedes hexodontus, Knight, 1951 Ann. Ent. Soc. Amer., 44:93.

Knight (430) recognized "type *hexodontus*" variety and "tundra" variety in North America. These are based on larval and adult female characters. "Type *hexodontus*" variety has the scutum marked like that of *A. aboriginis*, "type *A. punctor*" variety, and *A. abserratus*.

ADULT FEMALE (pl. 64). Medium-sized species. *Head:* Proboscis dark-scaled, palpi short, dark-scaled. Occiput with narrow pale-yellow scales and yellow erect forked scales dorsally, with broad appressed white scales laterally. Tori usually dark, with grayish-white scales on dorsal and inner surfaces on some specimens. *Thorax:* In "type *hexodontus*" variety the scutum is clothed with yellow scales (occasionally yellowish white or yellowish brown) and a broad median or narrowly divided dark stripe as in "type *A. punctor*" variety; scales of the lateral margins and prescutellar

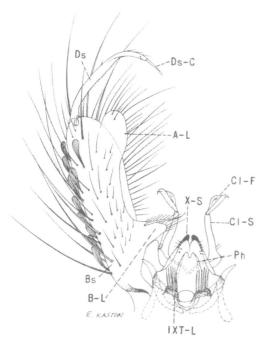

Fig. 151. Male terminalia of *Aedes grossbecki* Dyar and Knab.

punctor. Distinguishable from *A. abserratus* and *A. punctodes* on the shape of the basal lobe of the basistyle. Lobes of ninth tergite (IXT-L) sclerotized, a little longer than wide, each bearing several strong spines. Tenth sternite (X-S) heavily sclerotized and hooked apically. Phallosome (Ph) cylindrical, open ventrally, closed dorsally. Claspette stem (Cl-S) short, stout, curved near middle, pilose on basal half; claspette filament (Cl-F) dark, shorter than the stem, roundly expanded just before middle, tapered to a blunt curved point. Basistyle (Bs) about three to three and a half times as long as midwidth, clothed with scales and long and short setae; basal lobe (B-L) large and triangular-shaped, slightly detached basally, surface clothed with short setae except for a basal row of long setae preceded by a rather strong dorsal spine, dorsobasal angle of lobe with a slight protuberance; apical lobe (A-L) broadly rounded, clothed with many flattened curved setae. Dististyle (Ds) about two-thirds as long as the

space paler. The median dark band is absent or not well defined in "tundra" variety. Posterior pronotum with narrow curved yellow to yellowish-brown scales on dorsal part, scales on ventral part becoming a little broader and grayish white. Scutellum with narrow yellowish-brown scales and light-brown setae on the lobes. Pleura with extensive patches of grayish-white scales. Scales on sternopleuron extending to anterior angle, narrowly separate from patch on prealar area. Mesepimeron with scales extending to near lower margin. Hypostigial spot of scales absent. Lower mesepimeral bristles one to three. *Abdomen:* First tergite with yellowish-white scales on median area; remaining tergites dark-scaled, each with a basal band of white scales. Venter clothed with grayish-white scales. *Legs:* Femora dark-scaled, sprinkled with white, pale on posterior surface. Pale knee spots present. Tibiae dark, sprinkled with pale scales. Tarsi dark-scaled, the proximal segments sprinkled with pale scales. *Wing:* Length about 3.7 to 4.3 mm. Scales narrow, dark brown; a patch of white scales on base of costa.

ADULT MALE. Coloration similar to that of the female. *Terminalia* (fig. 153). Appears to be indistinguishable from *A. aboriginis* and *A.*

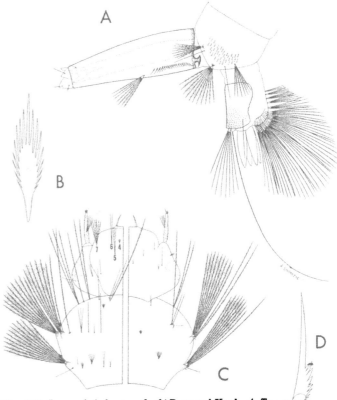

Fig. 152. Larva of *Aedes grossbecki* Dyar and Knab. *A,* Terminal segments. *B,* Comb scale. *C,* Head and thorax (dorsal and ventral). *D,* Pecten tooth.

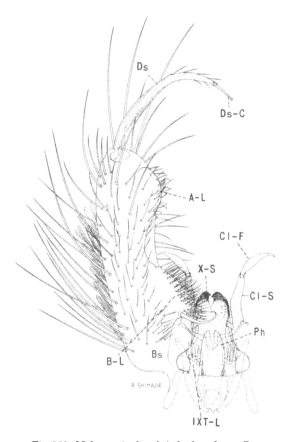

Fig. 153. Male terminalia of *Aedes hexodontus* Dyar.

basistyle, curved, swollen medially, pilose, bearing several subapical setae; claw (Ds-C) slender, about one-sixth as long as the dististyle.

LARVA (fig. 154). Antenna much shorter than the head, spiculate; antennal tuft small, multiple, inserted a little before middle of the shaft, not reaching tip. Head hairs: Postclypeal 4 small, branched; upper frontal 5 and lower frontal 6 double (sometimes single, rarely triple). "Type *hexodontus*" variety has head hairs 5 and 6 usually double; "tundra" variety has hairs 5 and 6 usually single, sometimes double. Preantennal 7, 3- to 6-branched, weakly barbed. Prothoracic hairs: 1 long, double or triple; 2 long, single; 3 medium, usually double or triple; 4 short, single; 5 long, 2- to 4-branched, rarely single; 6 long, single; 7 long, 3- or 4-branched, rarely double or 5-branched. Lateral abdominal hair 6 single to triple on segments I–VI. Comb of eighth seg-

ment with 5 to 9 scales in a single row; individual scale thorn-shaped, with small lateral spinules on basal part. Siphonal index about 3.0; pecten of about 10 to 20 evenly spaced teeth on basal third to two-fifths of siphon; siphonal tuft 3- to 8-branched, barbed, inserted beyond pecten; dorsal preapical spine shorter than the apical pecten tooth. Anal segment ringed by the saddle; lateral hair single, as long or a little longer than the saddle; dorsal brush bilaterally consisting of a long lower caudal hair and a shorter multiple upper caudal tuft; ventral brush large, confined to the barred area; gills about one and a half to three and a half times as long as the saddle, tapered.

DISTRIBUTION. This is primarily a species of the arctic tundra, but its range extends southward in the high mountains where its immature stages are found in alpine meadows. The extent of the range of this species is not known at this time, since it has been confused with other members of the *punctor* subgroup. The following distribution records are based largely on larvae and larval-associated adults. *United States:* California (287); Colorado (430); Idaho (672); Montana (430); Oregon and Washington (672). *Canada:* Baffin Island and British Columbia (289, 430); District of Keewatin, District of Mackenzie, and Manitoba (289); Quebec (289, 430); Yukon (162a, 289). *Alaska:* (430).

BIONOMICS. The larvae of *A. hexodontus* are found in various types of small pools in the mountains where some of the eggs hatch as soon as the snow begins to melt. Some hatching is delayed until long after the snow disappears. Jenkins and Knight (402) collected the larvae from rock pools and snow-water pools at Great Whale River, Quebec, in alpine habitats at about 1,000 feet elevation. These writers observed that the females emerge early in the season and are bad pests, biting during the day in sunlight or slight shade in the mountainous areas, in tundra meadows, and near snowbanks.

Aedes (Ochlerotatus) idahoensis
(Theobald)

Grabhamia spencerii var. *idahoensis* Theobald, 1903, Mon. Culic., 3:250.
Aedes idahoensis, Howard, Dyar, and Knab, 1917, Carnegie Inst. Wash. Pub., *159*, 4:727.

ADULT FEMALE (pl. 65). Similar to *A. spencerii* but has color differences in the scales on the abdominal tergites and the posterior pronotum. Medium-sized species. *Head:* Proboscis dark-scaled; palpi short, black, lightly speckled with white scales. Occiput clothed dorsally with curved grayish-white scales and pale erect forked scales, with broad appressed dark-gray scales laterally. Tori with silvery scales on inner and dorsal surfaces. *Thorax:* Integument of scutum black; scutum clothed with narrow curved gray scales, except for two narrowly separated submedian stripes and posterior half-stripes of narrow brown scales. (Posterior half-stripes are sometimes absent; the narrow submedian stripes are fused in many specimens forming a single broad brown stripe.) Posterior pronotum with scales on dorsal half mostly white or whitish brown, those on ventral half whitish. Scutellum with gray scales and light-brown setae on the lobes. Pleura with extensive poorly defined patches of broad grayish-white scales. Scales on sternopleuron extending to anterior angle, almost continuous with patch on prealar area. Mesepimeron with scales extending to near the lower margin. Hypostigial spot of scales present or absent. Lower mesepimeral bristles absent. *Abdomen:* First tergite with a large median patch of white scales; remaining tergites each brown, with broad basal white bands widening at the sides, with patches of pale scales apically on some segments. Venter densely clothed with gray scales. *Legs:* Femora with yellowish and brown scales intermixed, apical part mostly dark, posterior surface pale; white knee spots present. Tibiae pale scaled on posterior surface, with a mixture of dark and pale scales on outer surface, darker apically. Tarsal segments partly dark-scaled, proximal segments pale on posterior surface, distal segments mostly dark. *Wing:* Length about 3.9 to 4.2 mm. Scales narrow; costa, veins 1, 3, and 5 dark-scaled, other veins with pale scales.

ADULT MALE. Coloration similar to that of the female. *Terminalia* (fig. 155). Similar to *A. spencerii*. Lobes of ninth tergite (IXT-L) broader than long, each bearing several spines. Tenth sternite (X-S) sclerotized and hooked apically. Phallosome (Ph) cylindrical, longer than broad, open ventrally, closed dorsally, rounded and notched at apex. Claspette stem (Cl-S) short, pilose on basal half, bearing two to four small setae before apex; claspette filament (Cl-F) with a sharp angle on concave side near base, about two-thirds as long as slender part of stem. Basistyle (Bs) about three to three and a half times as long as mid-width, clothed with scales and long and short setae; basal lobe (B-L) constricted at base and expanded apically, bearing numerous setae and a large strong dorsal spine with recurved tip; apical lobe (A-L) prominent, broadly rounded, clothed with short retrorse setae. Dististyle (Ds) about two-thirds as long as the basistyle, curved, swollen medially, pilose, with two or three small setae before apex; claw (Ds-C) slender, about one-sixth as long as the dististyle.

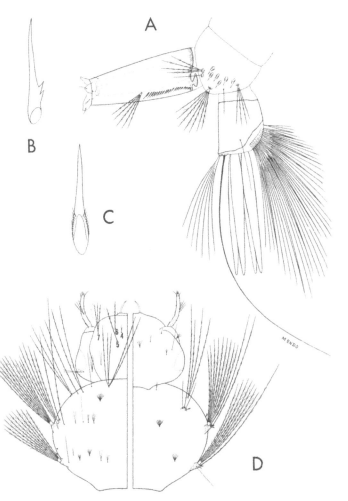

Fig. 154. Larva of *Aedes hexodontus* Dyar. *A*, Terminal segments. *B*, Pecten tooth. *C*, Comb scale. *D*, Head and thorax (dorsal and ventral).

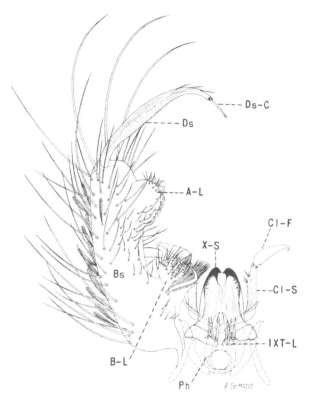

Fig. 155. Male terminalia of *Aedes idahoensis* (Theobald).

LARVA (fig. 156). Similar to *A. spencerii* but differs in the size and shape of the comb scales. The number of comb scales found in the two species overlap, although they average much higher in *A. idahoensis*. Antenna nearly half as long as the head, sparsely spiculate; antennal tuft multiple, inserted before middle of shaft, not reaching tip. Head hairs: Postclypeal 4 small, branched; upper frontal 5 and lower frontal 6 single (occasionally either 5 or 6 double); preantennal 7 about 2- to 7-branched, barbed, reaching beyond insertion of the antennal tuft. Prothoracic hairs: 1 medium, single; 2 short, single; 3 short, single or double; 4 short, single; 5 long, usually single; 6 long, single; 7 long, usually 3- or 4-branched. Body spiculate. Lateral abdominal hair 6 usually 2- to 4-branched on segments I–V, single or double on VI. Comb of eighth segment with about 13 to 29 scales in a patch; individual scale with spine of medium length and arising abruptly from the basal part, prominent lateral spinules on basal part. Siphonal index about 2.5; pecten of about 12 to 24 teeth reaching a little beyond middle of si-

phon, the distal 2 or 3 teeth detached; siphonal tuft usually 3- to 5-branched, inserted beyond the pecten, much shorter than the basal diameter of the siphon; dorsal preapical spine not more than half as long as the apical pecten tooth. Anal segment with the saddle extending to near midventral line; lateral hair single, much shorter than the saddle; dorsal brush bilaterally consisting of a long lower caudal hair and a shorter multiple upper caudal tuft; ventral brush well developed, with about two or three precratal tufts; gills about two to two and a half times as long as the saddle, pointed.

DISTRIBUTION. Northwestern United States and southwestern Canada. *United States:* Colorado (213, 343); Idaho (344, 672); Montana (213); Nebraska (680); Nevada (213); North Dakota (553); Oregon (492, 672); Utah (213, 575); Washington (67, 672); Wyoming (578). *Canada:* British Columbia (311, 708).

BIONOMICS. This species passes the winter in the egg stage, and there is a single generation a year. The larvae occur in pools filled by melting snow or by early spring rains. The species seems to favor water of high alkalinity. The females bite persistently during the day but are more active during the evening. Mail (475) states that *A. idahoensis* is a widely distributed and important pest of man and livestock in Montana, but it is only of local importance in many areas within its range. Rees (575) is of the opinion that the females may travel several miles from their place of emergence. The senior author observed the females in large numbers at Jackson Lake, Wyoming, in early July, 1951, where they were extremely annoying.

Aedes (Ochlerotatus) impiger (Walker)

Culex impiger Walker, 1848, List Dipt. Brit. Mus., 1:6.

Aedes nearcticus Dyar, 1919 (in part), Rept. Can. Arct. Exp., 3(C):32.

Aedes parvulus Edwards, 1921, Bull. Ent. Res., 12:314.

Aedes nearcticus, Dyar, 1922, Proc. U.S. Nat. Mus., 62:84.

Aedes alpinus (L.), Twinn, 1927 (in part), Can. Ent., 59:47.

Aedes (Ochlerotatus) nearcticus, Dyar, 1928, Carnegie Inst. Wash. Pub., 387:196.

Aedes (Ochlerotatus) nearcticus, Matheson, 1944, Handbook Mosq. No. Amer., p. 174.

Aedes (Ochlerotatus) nearcticus, Gjullin, 1946,
  Proc. Ent. Soc. Wash., *48*:215.
Aedes (Ochlerotatus) nearcticus, Natvig, 1948,
  Norsk. Ent. Tidsskr. Sup., *1*:311.
Aedes nearcticus, Rempel, 1950, Can. Jour. Res., D,
  28:225.
Aedes impiger (Walker), Vockeroth, 1954, Can.
  Ent., *86*:109.

Vockeroth (734a) found that the type female of *Culex impiger* Walker is identical with
the specimen of the species later described by
Dyar as *Aedes nearcticus*; therefore *impiger* is
the correct name.

ADULT FEMALE (pl. 66). Medium-sized
species. *Head:* Proboscis dark-scaled, palpi
short, dark. Occiput with a median area of
narrow golden scales and dark erect forked
scales, bounded by a submedian patch of dark-
brown and grayish-white scales on either side;
lateral region of occiput with broad flat appressed grayish-white and brown scales. Tori
brown or black, with white or brownish-white
scales on dorsal and inner surfaces. *Thorax:*
Integument of scutum black; scutum clothed
with bronze-brown scales; yellowish-white
scales usually present along the anterior and
lateral margins; prescutellar space with yellowish-white scales; the entire surface of the
scutum has many black setae, thus giving the
insect a hairy appearance. Posterior pronotum
with narrow bronze-brown or yellowish-brown
scales dorsally, becoming broader and paler
ventrally. Scutellum with pale yellowish-white
scales and black setae. Pleura with medium-
sized patches of broad grayish-white scales.
Scales on sternopleuron usually not reaching
the anterior angle, narrowly separate from
patch on prealar area. Mesepimeron with
scales usually reaching near lower margin.
Hypostigial spot of scales absent. Lower mesepimeral bristles about three to eight. *Abdomen:*
First tergite with a median patch of white
scales; remaining tergites brownish black,
with white basal bands. Venter white with
apices of segments more or less black-scaled.
Legs: Femora dark, with pale scales intermixed, posterior surface pale; knee spots
white. Tibiae dark, sprinkled with white, posterior surface streaked with white. Tarsi dark.
Wing: Length about 3.3 to 4.2 mm. Scales narrow, dark, usually with a patch of white scales
on base of costa.

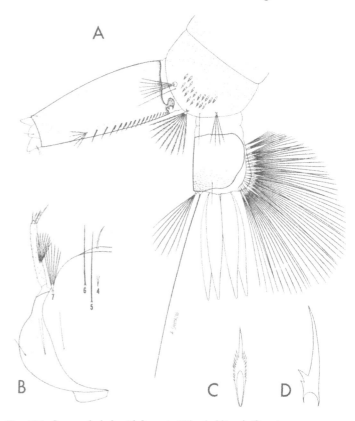

Fig. 156. Larva of *Aedes idahoensis* (Theobald). *A*, Terminal segments. *B*, Head. *C*, Comb scale. *D*, Pecten tooth.

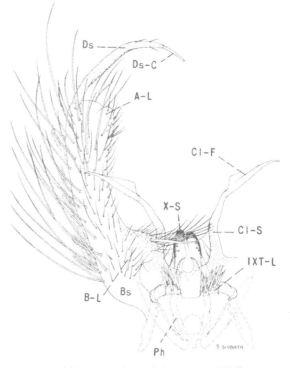

Fig. 157. Male terminalia of *Aedes impiger* (Walker).

ADULT MALE. Coloration similar to that of the female. *Terminalia* (fig. 157). Lobes of ninth tergite (IXT-L) short, rounded, sclerotized, each bearing several strong spines. Tenth sternite (X-S) heavily sclerotized beyond middle. Phallosome (Ph) long, narrow, cylindrical, open ventrally, closed dorsally, notched at apex. Claspette stem (Cl-S) slender, curved, pilose on basal half, reaching well beyond distal margin of basal lobe; claspette filament (Cl-F) long, angularly expanded near base, tapered and curved apically. Basistyle (Bs) about three and a half to four and a half times as long as mid-width, clothed with scales and long and short setae; basal lobe (B-L) narrow and conical, bearing a rather strong dorsal spine with recurved tip, followed by a row of weaker setae, with small setae on remainder of surface; apical lobe (A-L) rather prominent, thumblike, bearing a few small setae. Dististyle (Ds) about half as long as the basistyle, pilose, slightly expanded medially, bearing two to four small setae before apex; claw (Ds-C) about one-fifth as long as the dististyle.

LARVA (fig. 158). Antenna about half as long as the head, sparsely spiculate; antennal tuft small, usually triple, inserted a little before middle of shaft, not reaching tip. Head hairs: Postclypeal 4 small, usually branched; upper frontal 5 and lower frontal 6 long, single; preantennal 7, 2- to 4-branched, reaching insertion of antennal tuft. Prothoracic hairs: 1 long, single; 2 medium, single; 3 short, usually single or double; 4 short, single; 5 long, usually single or double; 6 long, single; 7 long, 3- or 4-branched. Lateral abdominal hair 6 variable, usually triple on segments I and II, double on III and IV, single on V and VI. Comb of eighth segment with about 8 to 16 scales in two irregular rows; individual scale with lateral spinules small, less than half as long as the median spine. Siphonal index about 3.0; siphon gradually tapered from near base; pecten of about 11 to 18 evenly spaced teeth on basal third of siphon; siphonal tuft multiple, barbed, inserted beyond pecten, before or near middle of siphon, longer than the basal diameter of the siphon; dorsal preapical spine shorter than the apical pecten tooth. Anal segment with the saddle extending halfway or a little more down the sides; lateral hair single, shorter than the saddle; dorsal brush bilaterally consisting of a long single upper caudal hair and a shorter 3- or 4-branched lower caudal tuft; ventral brush large, with one or two precratal tufts; gills variable in length, usually two to four times as long as the saddle, pointed.

DISTRIBUTION. This species is found in the treeless arctic regions of Alaska, Canada, Scandanavia, and Siberia. Its range is known to extend southward to Utah and Colorado at the higher elevations. *United States:* Colorado (492); Idaho (672); Montana (475); Oregon (672); Utah (580); Washington (67, 672); Wyoming (578). *Canada:* Alberta (353, 598); Manitoba (289, 711); Northwest Territories (212, 289); Ontario (492); Quebec (289, 402); Saskatchewan (310, 598); Yukon (212, 289). *Alaska:* (213, 396, 673).

BIONOMICS. The larvae of *Aedes impiger* are found in clear pools of water formed by melting snow at high elevations in mountains where alpine arctic conditions prevail. The rate of development of the larvae is very slow. Hearle

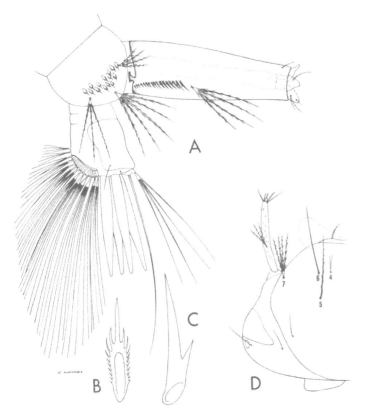

Fig. 158. Larva of *Aedes impiger* (Walker). *A*, Terminal segments. *B*, Comb scale. *C*, Pecten tooth. *D*, Head.

(353) describes a typical larval habitat in the latter part of June consisting of a pool in a mountain meadow in the Rockies near Banff at an elevation of 7,300 feet. Mail (475) collected the larvae of this species at Swift Current Pass in Montana at an elevation of 7,500 feet at the end of July. Rees and Nielson (580) collected larvae and adults in an alpine meadow at an elevation of 10,100 feet in the Uintah Mountains in Utah. Jenkins and Knight (402) found the larvae and pupae in pools at 500 to 1,000 feet elevation at Great Whale River, Quebec; however, they claim that the most typical habitats were small bog pools and snow pools above the timber line in alpine meadows. The larvae were found on a few occasions in rock pools containing vegetation. Jenkins (396) found larvae of this species only once in central Alaska, in a pool at the base of a snowbank at 3,000 feet elevation.

Mail (475) says that the females attack man, but he observed that the species is too rare and occurs at too high an altitude to be of much economic importance. Adults were also found by Jenkins and Knight (402) to be abundant in the open tundra area and in alpine meadows at Great Whale River, Quebec. They found the adults also in the mountains where they bit readily at all hours of the day, even in full sunlight. The females are described by these observers as fast flying and difficult to capture.

Aedes (Ochlerotatus) implicatus
Vockeroth

Aedes (Ochlerotatus) implicatus Vockeroth, 1954, Can. Ent., *86*:110.

Aedes impiger, Dyar, 1920 (not Walker, 1848), Ins. Ins. Mens., *8*:8.

Aedes (Heteronycha) impiger, Dyar, 1921 (not Walker), Trans. Roy. Can. Inst., *13*:100.

Aedes (Heteronycha) impiger, Dyar, 1922 (not Walker), Proc. U.S. Nat. Mus., *62*:63.

Aedes (Ochlerotatus) impiger, Matheson, 1944 (not Walker), Handbook Mosq. No. Amer., p. 169.

Aedes (Ochlerotatus) impiger, Gjullin, 1946 (not Walker), Proc. Ent. Soc. Wash., *48*:232.

Aedes impiger, Rempel, 1950 (not Walker), Can. Jour. Res., D., *28*:218.

Vockeroth (734a) has shown that the type of *Culex impiger* Walker is a specimen of the species generally known as *Aedes nearcticus*

Dyar; therefore, the name *impiger* has been misapplied. *Aedes impiger* (of authors) is a synonym of a species Vockeroth describes as *A. implicatus*.

ADULT FEMALE (pl. 67). Medium-sized species. *Head:* Proboscis dark-scaled, palpi short, dark. Occiput clothed dorsally with narrow white-curved scales, with broad appressed white scales enclosing a small black patch laterally; pale erect forked scales numerous on median area, those on sides dark. Tori yellow to black, usually with a few white scales on inner surface. *Thorax:* Integument of scutum black; scutum with brown scales in a more or less distinct broad stripe or paired stripes; anterior and lateral margins and prescutellar space with grayish-white scales. Posterior pronotum with curved grayish-white scales. Scutellum with narrow white scales and brown setae on the lobes. Pleura with dense poorly defined patches of broad grayish-white scales. Scales on sternopleuron reaching about halfway to anterior margin, narrowly separate from patch on prealar area. Mesepimeron with scales reaching near lower margin. Hypostigial spot of scales present or absent. Lower mesepimeral bristles one to three, rarely none. *Abdomen:* First tergite with a median patch of white scales; remaining tergites dark, each with a basal white band. Venter white-scaled. *Legs:* Femora dark, with pale scales intermixed, posterior surface pale, knee spots white. Tibiae dark, streaked with pale scales. Tarsi dark, first segment streaked with pale scales. *Wing:* Length about 3.5 to 4.3 mm. Scales narrow, dark, with a patch of a few to many white scales at base of costa.

ADULT MALE. Coloration similar to that of the female. *Terminalia* (fig. 159). Lobes of ninth tergite (IXT-L) broader than long, each bearing about four to eight rather long stout spines. Tenth sternite (X-S) heavily sclerotized apically. Phallosome (Ph) cylindrical, open ventrally except at base, closed dorsally, notched at apex. Claspette stem (Cl-S) slender, curved, pilose to near apex, extending beyond basal lobe; claspette filament (Cl-F) more than half as long as the stem, angularly expanded to a sharp point before middle, bladelike, tapered and curved apically. Basistyle (Bs) about three and a half to four times as long

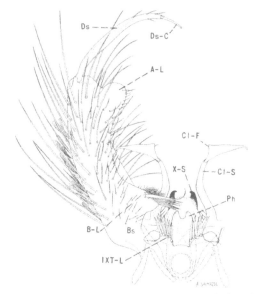

Fig. 159. Male terminalia of *Aedes implicatus* Vockeroth.

of siphon; dorsal preapical spine shorter than the apical pecten tooth. Anal segment with saddle extending about two-thirds down the sides, saddle spiculate posteriorly; lateral hair single, shorter than the saddle; dorsal brush bilaterally consisting of a long lower caudal hair and a shorter multiple upper caudal tuft; ventral brush large, with about three or four precratal tufts; gills about one and a half times as long as the saddle, pointed.

DISTRIBUTION. Northern United States, Canada, and Alaska. *United States:* Colorado (226); Idaho (672); Iowa (625); Massachusetts (701); Michigan (391); Minnesota (533); Montana (475); Nebraska (680); New Hampshire (404, 458); New York (31, 130); Utah (580); Washington (672); Wyoming (221, 303). *Canada:* Alberta (212, 598) British Columbia (212, 289); Manitoba (212, 708); Northwest Territories (289); Ontario (212, 289); Quebec (708); Saskatchewan (212, 598); Yukon (212, 289). *Alaska:* (213, 293).

as mid-width, clothed with scales and long and short setae; basal lobe (B-L) short, conical, clothed with slender setae and with an apical row of rather stout setae preceded by a large strong dorsal spine; apical lobe (A-L) slender, thumblike, clothed with small setae. Dististyle (Ds) pilose, about three-fifths as long as the basistyle, broader medially, bearing several small setae before apex; claw (Ds-C) slender, nearly one-fifth as long as the dististyle.

LARVA (fig. 160). Antenna about half as long as the head, spiculate; antennal tuft multiple, inserted before middle of shaft, not reaching tip. Head hairs: Postclypeal 4 small, usually single; upper frontal 5 single or double; lower frontal 6 long, single; preantennal 7 multiple, reaching insertion of antennal tuft. Prothoracic hairs: 1 long, single; 2 medium, single; 3 short, usually single or double; 4 short, single; 5 long, usually single; 6 long, single; 7 long, usually double. Lateral abdominal hair 6 usually double on segments I–IV, single on V and VI. Comb of eighth segment with about 25 to 35 scales in a patch; individual scale fringed with fairly stout spines, the median spine a little longer and stouter. Siphonal index 3.0 to 3.5; pecten of about 18 to 24 evenly spaced teeth on basal two-fifths of siphon; siphonal tuft usually triple (occasionally 2- or 4-branched), inserted beyond pecten at middle

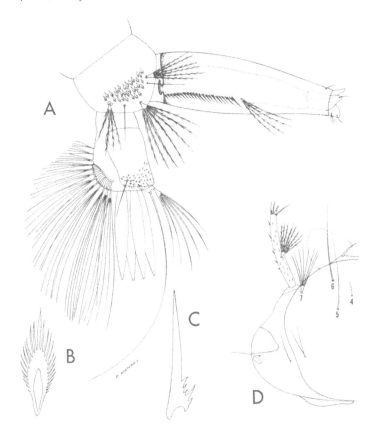

BIONOMICS. The larvae of *Aedes implicatus* are found in temporary pools, usually in forests, and are among the first to appear in the spring. According to Mail (475) the larvae are found in snow-water pools in the woods and meadows in wooded sections in Montana. Frohne (293) collected the larvae in brackish-water habitats in the coastal marshes of Alaska. Matheson (492) reports having taken the larvae at an elevation of 8,000 feet in Colorado.

Rees and Nielson (580) found the species to be rather common and an annoying pest in northern Utah at elevations ranging from 6,000 to 9,000 feet where it is found only along mountain streams and chiefly in or near willow growths. The females bite vigorously either in the shade during the day or in the evening. The adults may persist in woodlands until June or July. It is regarded as a rare species throughout most of its range.

Aedes (Ochlerotatus) increpitus Dyar

Aedes increpitus Dyar, 1916, Ins. Ins. Mens., 4:87.
Grabhamia vittata Theobald (not Bigot), 1903, Can. Ent., 35:313.
Aedes mutatus Dyar, 1919, Ins. Ins. Mens., 7:24.
Aedes hewitti Hearle, 1923, Can. Ent., 55:5.

ADULT FEMALE (pl. 68). Medium-sized species. *Head:* Proboscis dark-scaled; palpi short, dark-scaled, with apices of the segments white. Occiput with narrow curved white scales on median area; a submedian patch of narrow brown scales on either side, with broad appressed grayish-white scales surrounding a small area of broad appressed bluish-black scales laterally; erect forked scales yellowish white and brown. Tori brown, darker on inner surface, with an inconspicuous patch of white scales on inner surface. *Thorax:* Integument of scutum dark brown; scutum usually with a broad median stripe of narrow brown scales and a varied pattern of brown and yellowish-white scales on the submedian areas; anterior margin and prescutellar space with narrow yellowish-white scales. Posterior pronotum with narrow brown or a mixture of brown and white scales on dorsal half, with white scales ventrally. Scutellum with grayish-white scales and light-brown setae on the lobes. Pleura with rather distinct patches of broad white scales.

Patch of scales on sternopleuron extending about halfway to anterior angle (often with scattered scales extending to anterior angle), separate from patch on prelar area. Mesepimeron with lower one-fourth to one-third bare. Hypostigial spot of scales absent. Lower mesepimeral bristles one to five. *Abdomen:* First tergite with a median patch of white scales; remaining tergites black, each with a basal band of white scales widening laterally. Venter white-scaled, with black patches on the sides of each segment. *Legs:* Femora and tibiae black-scaled with many pale scales intermixed, posterior surface pale; knee spots white. Tarsi dark, first segment streaked with white on posterior surface; hind tarsi broadly ringed with white basally; middle tarsi with narrower white basal bands on segments 1 to 4, white band reduced or absent on 5; front tarsi with white basal bands on segments 1 to 3, white reduced or absent on segment 4, segment 5 entirely dark. *Wing:* Length about 4.0 to 4.5 mm. Scales narrow, brown, with some white ones intermixed; pale scales confined almost entirely to front part of wing.

ADULT MALE. Coloration similar to that of the female. *Terminalia* (fig. 161). Lobes of ninth tergite (IXT-L) about as broad as long, each bearing several strong spines. Tenth sternite (X-S) prominent, heavily sclerotized toward apex. Phallosome (Ph) conical, longer

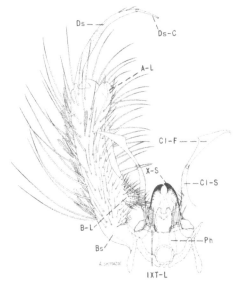

Fig. 161. Male terminalia of *Aedes increpitus* Dyar.

than wide, open ventrally, closed dorsally, notched at apex. Claspette stem (Cl-S) long, slender, curved, pilose to near apex; claspette filament (Cl-F) as long as the stem, angularly expanded to a sharp, slightly retrorse point near middle, tapered to a recurved point. Basistyle (Bs) about three and a half to four times as long as wide, clothed with scales and long and short setae; basal lobe (B-L) represented by a small rugose elevated area reaching nearly halfway to the apical lobe, bearing short setae becoming somewhat longer basally, but there is no single prominent large seta; apical lobe (A-L) prominent, thumblike, clothed with small scattered setae. Dististyle (Ds) a little more than half as long as the basistyle, curved, expanded medially, pilose, bearing two or three small setae before apex; claw (Ds-C) slender, about one-fourth as long as the dististyle.

LARVA (fig. 162). Antenna about half as long as the head, spiculate; antennal tuft multiple, barbed, inserted near middle of shaft, not reaching tip. Head hairs: Postclypeal 4 small, branched; upper frontal 5 double or triple (sometimes single, rarely 4-branched); lower frontal 6 single or double (sometimes triple); preantennal 7 about 5- to 11-branched, barbed, reaching insertion of antennal tuft. Prothoracic hairs: 1 long, usually double (rarely single); 2 medium, single; 3 short, double; 4 medium, single; 5 long, usually double or triple; 6 long, single; 7 long, usually 3- or 4-branched. Lateral abdominal hair 6 usually double on segments I–VI. Comb of eighth segment usually with 20 or more scales in a patch; individual scale fringed laterally with small spinules and apically with subequal spines, the median spine is a little stronger. Siphonal index 3.0 to 3.5; pecten of about 16 to 24 evenly spaced teeth on basal two-fifths of siphon; siphonal tuft 3- to 7-branched, barbed, inserted beyond pecten. Anal segment with the saddle extending about two-thirds to three-fourths down the sides; spicules toward apex of saddle well developed, more than twice as long as those toward base; lateral hair single, shorter than the saddle; dorsal brush bilaterally consisting of a long lower caudal hair and a shorter multiple upper caudal tuft; ventral brush well developed, with two to four precratal tufts; gills varying from about as long as the saddle to about twice as long, pointed.

DISTRIBUTION. Known from the western United States and southwestern Canada. *United States:* California (287); Colorado (226, 343); Idaho (344, 672); Montana (475); Nevada (213); New Mexico (273, 492); Oregon (672); Utah (575); Washington (67, 672); Wyoming (578). *Canada:* British Columbia (352, 708); Saskatchewan (598, 708).

BIONOMICS. The larvae of *A. increpitus* are found in pools along streams left when floodwaters subside, in depressions filled by irrigation water, heavy rains, and melting snow. The species is usually confined to river valleys and to moderate elevations in mountainous areas. The eggs usually hatch in March, April, or May, depending somewhat on the altitude.

The adults are often encountered in large numbers in wooded areas near their larval

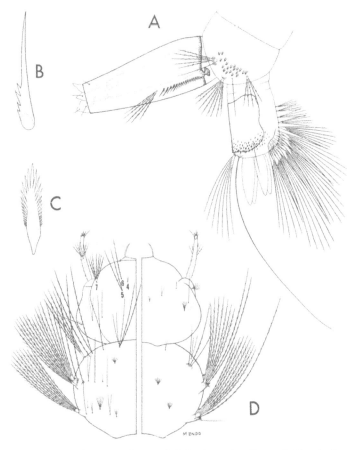

Fig. 162. Larva of *Aedes increpitus* Dyar. *A*, Terminal segments. *B*, Pecten tooth. *C*, Comb scale. *D*, Head and thorax (dorsal and ventral).

habitats in the late spring and early summer months. The females persist well into July and early August in the mountain areas. This is the predominant species of the Yosemite Valley in California. It is an important pest mosquito throughout most of its range. Mail (475) claims that it is a persistent biter and is often annoying to cattle on summer range and to tourists and sportsmen in the woods in certain areas in Montana.

Aedes (Ochlerotatus) infirmatus
Dyar and Knab

Aedes infirmatus Dyar and Knab, 1906, Jour. N.Y. Ent. Soc., *14*:197.

ADULT FEMALE (pl. 69). Medium-sized species. *Head:* Proboscis dark-scaled; palpi short, dark. Median area of occiput with white lanceolate scales, bounded on either side by a large patch of broad brownish-yellow scales surrounding a dark-scaled patch, with broad appressed dingy-yellow to brownish-yellow scales laterally; erect forked scales on central part of occiput pale. Tori yellow, darker on inner surface. *Thorax:* Integument of scutum dark brown; scutum with a broad silver-white to pale-yellow median stripe extending posteriorly to a little beyond the middle; a small area of pale scales on the prescutellar space; remainder of scutum with fine dark bronze-brown scales. Posterior pronotum with narrow dark-brown scales on dorsal half. Scutellum with dark bronze-brown scales and dark-brown setae on the lobes (a few white scales on middle lobe). Pleura with small dense patches of broad flat grayish-white scales. Scales on sternopleuron reaching about halfway to anterior angle, separate from patch on prealar area. Lower half of mesepimeron bare. Hypostigial spot of scales absent. Lower mesepimeral bristles absent. *Abdomen:* Tergites dark-scaled, with conspicuous basal triangular patches of white scales laterally. Venter white-scaled, occasionally speckled with a few dark scales. *Legs:* Femora dark-scaled, posterior surface of front and middle femora pale, all aspects of hind femora pale on basal two-thirds. Tibiae and tarsi dark. *Wing:* Length about 3.3 to 3.8 mm. Scales narrow, dark.

ADULT MALE. Coloration similar to that of

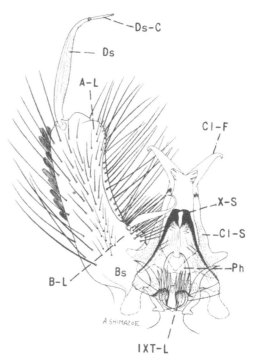

Fig. 163. Male terminalia of *Aedes infirmatus* Dyar and Knab.

the female, but with narrow basal white bands usually present on some of the abdominal segments. *Terminalia* (fig. 163). Lobes of ninth tergite (IXT-L) about as long as broad, each bearing four to seven spines. Tenth sternite (X-S) heavily sclerotized apically. Phallosome (Ph) conical, about two-thirds as broad as long, rounded apically, open ventrally, closed dorsally. Claspette stem (Cl-S) pilose, slender, extending well beyond basal lobe; claspette filament (Cl-F) about three-fourths as long as the stem, convex side with a large sharp retrorse projection which bears one or more smaller accessory retrorse spines, tapered and curved apically. Basistyle (Bs) about three and a half times as long as mid-width, clothed with scales and long and short setae, with numerous long setae on inner ventral margin; basal lobe (B-L) prominent, bluntly conical, bearing numerous short setae and a large dorsal curved spine with recurved tip; apical lobe (A-L) small, broadly rounded, bearing short setae. Dististyle (Ds) nearly two-thirds as long as the basistyle, broader medially, pilose; claw (Ds-C) slender, about one-fifth as long as the dististyle.

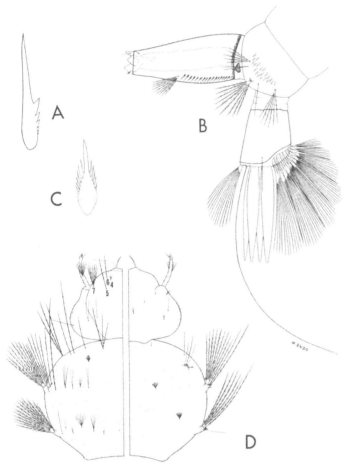

Fig. 164. Larva of *Aedes infirmatus* Dyar and Knab. *A*, Pecten tooth. *B*, Terminal segments. *C*, Comb scale. *D*, Head and thorax (dorsal and ventral).

LARVA (fig. 164). Antenna a little less than half as long as the head, sparsely spiculate; antennal tuft small, multiple, inserted near middle of shaft, not reaching tip. Head hairs: Postclypeal 4 small, multiple; upper frontal 5 and lower frontal 6 single, weakly barbed, reaching beyond preclypeus; preantennal 7 about 5- to 10-branched, barbed, reaching beyond insertion of antennal tuft. Prothoracic hairs: 1 medium, single; 2–4 short, usually single; 5–6 long, single; 7 long, usually 3- or 4-branched. Body moderately spiculate (spicules generally shorter and more scattered than in *A. scapularis*). Lateral abdominal hair 6 single on segments III–V. Comb of eighth segment with about 15 to 24 scales in a patch; individual scale with the median spine three to four times as broad and about twice as long as

the subapical spinules. Siphonal index about 2.5; pecten of about 11 to 17 dark evenly spaced teeth reaching slightly beyond middle of siphon; siphonal tuft multiple, inserted beyond pecten; dorsal preapical spine much shorter than the apical pecten tooth. Anal segment ringed by the saddle; lateral hair single, a little shorter than the saddle; dorsal brush bilaterally consisting of a long lower caudal hair and a much shorter multiple upper caudal tuft; ventral brush large, confined to the barred area; gills much longer than the saddle, usually about twice as long, pointed.

DISTRIBUTION. Known from the southeastern United States, west to Texas. *United States:* Alabama (135, 516); Arkansas (88, 213); Florida (135, 213); Georgia (135, 516); Kentucky (570); Louisiana (213, 759); Mississippi (135, 546); Missouri (14, 531); North Carolina (135, 646); South Carolina (135, 742); Tennessee (516, 658); Texas (213, 463).

BIONOMICS. The larvae of *Aedes infirmatus* develop from early spring to late fall in temporary pools following rains. The females are persistent biters attacking during the day in or near wooded areas. They are occasionally encountered at night near dwellings, but seldom enter the buildings. Beyer (60) describes this species as driving cattle from Louisiana woodlands by its attacks.

Aedes (Ochlerotatus) intrudens Dyar

Aedes intrudens Dyar, 1919, Ins. Ins. Mens., 7:23.

ADULT FEMALE (pl. 70). Medium-sized species. *Head:* Proboscis dark-scaled; palpi short, dark, sprinkled with grayish-white scales. Occiput with narrow curved pale-golden or yellowish-white scales and erect forked scales on median area (the erect forked scales are darker submedially), with broad appressed yellowish-white scales laterally. Tori yellow or light brown, inner surface dark brown and often bearing a few pale scales. *Thorax:* Integument of scutum black; scutum uniformly clothed with bronze-brown narrow curved scales, shading to light brown or yellowish white on anterior and lateral margins and on the prescutellar space, occasionally with indications of paired median brown stripes. Posterior pronotum with narrow curved brownish or yellowish-white scales dorsally, becoming broader and paler ventrally. Scutellum with yellowish-white scales and light-brown setae on the lobes. Pleura with medium-sized, well-defined patches of broad appressed grayish-white scales. Scales on sternopleuron extending one-half to two-thirds way to anterior angle, distinctly separate from patch on prealar area. Mesepimeron with lower one-fourth to one-third bare. Hypostigial spot of scales present or absent. Lower mesepimeral bristles one to five, rarely none. *Abdomen:* First tergite with a median patch of white scales; remaining tergites black, each with a broad basal white band somewhat widened laterally. Venter grayish-white-scaled. *Legs:* Femora dark brown, with pale scales intermixed, posterior surface pale. Tibiae and tarsi dark; posterior surface of tibiae and first tarsal segment with pale scales. *Wing:* Length about 4.0 to 4.2 mm. Scales narrow, dark brown, with or without a small patch of pale scales at base of costa.

ADULT MALE. Coloration similar to that of the female. *Terminalia* (fig. 165). Lobes of ninth tergite (IXT-L) short, rounded, each bearing five to eight short stout spines. Tenth

sternite (X-S) sclerotized and hooked apically. Phallosome (Ph) stout, a little longer than broad, constricted apically, open ventrally and along the median dorsal line. Claspette stem (Cl-S) with basal half pilose, a sharp projection on inner surface at the middle, this projection bears a small seta, part beyond projection slender; claspette filament (Cl-F) angularly expanded to a sharp point near middle, pointed and curved apically. Basistyle (Bs) about three and one-half to four times as long as mid-width, clothed with scales and long and short setae, a dense tuft of setae on ventral surface near apex; basal lobe (B-L) elongate, bearing two rather stout spines at apex and a large dorsal spine at base; apical lobe (A-L) rounded apically and clothed with rather long setae. Dististyle (Ds) short, expanded medially, pilose, bearing several small setae before apex; claw (Ds-C) about one-fifth as long as the dististyle.

LARVA (fig. 166). Antenna a little shorter than the head, spiculate, much darker on distal half; antennal tuft multiple, inserted near middle of shaft, not reaching tip. Head hairs: Postclypeal 4 small, multiple; upper frontal 5, 3- or 4-branched, barbed; lower frontal 6 double or triple, barbed; preantennal 7 multiple, barbed, extending beyond insertion of antennal tuft. Prothoracic hairs: 1 long, single; 2 short, single; 3 short, 2- to 4-branched; 4 short, single; 5–6 long, single; 7 long, usually double. Lateral abdominal hair 6 single or double on segments I and II, usually single on III–VI. Comb of eighth segment with about 12 to 16 scales in an irregular double row; individual scale thorn-shaped, with minute lateral spinules on basal part. Siphonal index about 3.0; pecten of about 13 to 18 teeth on basal half to three-fifths of siphon, with 1 to 3 distal teeth detached; siphonal tuft about 4- to 10-branched, inserted beyond pecten (occasionally near base of distal tooth); dorsal preapical spine much shorter than the apical pecten tooth. Anal segment about three-fourths to seven-eighths ringed by the saddle; ventral margin of saddle deeply incised; lateral hair single, much shorter than the saddle; dorsal brush bilaterally consisting of a long lower caudal hair and a shorter multiple upper caudal tuft; ventral brush well developed, with

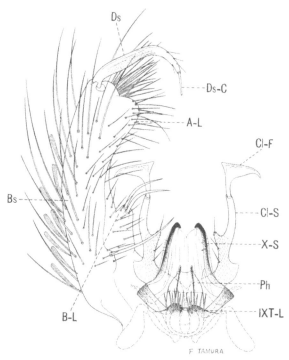

Fig. 165. Male terminalia of *Aedes intrudens* Dyar.

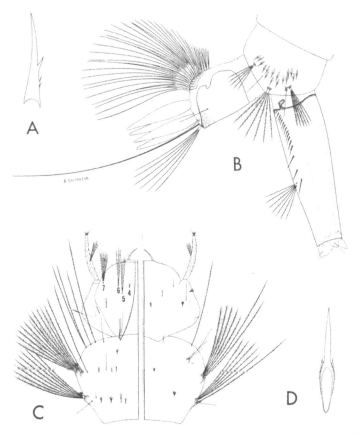

Fig. 166. Larva of *Aedes intrudens* Dyar. *A*, Pecten tooth. *B*, Terminal segments. *C*, Head and thorax (dorsal and ventral). *D*, Comb scale.

habitats, including woodland pools, open bogs, and marshes, particularly in water from melting snow. Natvig (523) found that the larval habitats of this species in Norway are principally shallow pools and water-filled ditches, usually at the border of pine woodlands. Jenkins (396) observed that *Carex* swamps containing deep water, semipermanent pools, and margins of small lakes are typical larval habitats of this species in Alaska.

This is a forest mosquito and is a persistent biter both by day and by night. The females often persist in woodlands throughout most of the summer. According to Rempel (598) the species is widely distributed over Western Canada where it is common in the northern wooded areas but rare on the open prairie. Natvig (523) says he has been attacked by this mosquito in woodlands on a sunny day, and that it has been found in dwellings, cow stables, and hoghouses in Norway. Dyar (197) observed that the adults are fond of entering houses at Banff, Alberta, and White River, Ontario, differing in this habit from most other forest mosquitoes. Owen (533) also took adults of this species in cottages in Minnesota.

two or three precratal tufts; gills longer than the saddle, pointed.

DISTRIBUTION. Northern United States, southern Canada, and northern Europe. *United States:* Colorado (226); Connecticut (494); Idaho (672); Maine (45, 564); Massachusetts (267, 270); Michigan (391, 529); Minnesota (533); Montana (475); New Hampshire (66, 213); New York (31); North Dakota (14, 553); Oregon (672); Pennsylvania (298); Rhode Island (436); South Dakota (14); Utah (575); Washington (672); Wisconsin (171); Wyoming (578). *Canada:* Alberta (212, 598); British Columbia (212, 352); Labrador (289, 348); Manitoba (311, 466); New Brunswick and Nova Scotia (708); Ontario (212, 289); Prince Edward Island (710); Quebec (212, 289); Saskatchewan (311, 598). *Alaska:* (396).

BIONOMICS. The larvae occur in a variety of

Aedes (Ochlerotatus) melanimon Dyar

Aedes melanimon Dyar, 1924, Ins. Ins. Mens., *12*: 126.

Aedes klotsi Matheson, 1933, Proc. Ent. Soc. Wash., *35*:69.

Dr. Alan Stone of the United States National Museum has compared specimens of *A. klotsi* with *A. melanimon* and says that he believes they are the same. The senior author has also compared these two species and is of the same opinion; therefore, *klotsi* is listed here as a synonym of *melanimon*.

ADULT FEMALE. Similar to *A. dorsalis* but with wing scales predominantly dark. Medium-sized species. *Head:* Proboscis long, dark brown, with a few pale scales on basal part; palpi short, dark, lightly speckled with pale scales. Occiput with yellowish-white lanceolate scales and pale erect forked scales dorsally, with broad brown and white scales laterally. Tori brown, darker on inner surface, with white scales on dorsal and inner surfaces. *Thorax:* Integument of scutum grayish-black; scutum with a median stripe of brown lanceolate scales, often with scattered yellowish-white scales; remainder of scutum clothed with yellowish-white lanceolate scales or with patches of brown scales and yellowish-white scales; prescutellar space with yellowish-white scales. Posterior pronotum with narrow brown scales on dorsal half, with broader grayish-white scales on ventral half. Scutellum with narrow yellowish-white curved scales and light-brown setae on the lobes. Pleura with dense poorly defined patches of broad grayish-white scales. Patch of scales on sternopleuron extending to the anterior angle, nearly continuous with patch on prealar area. Mesepimeron entirely covered with scales. Hypostigial spot with many scales. Lower mesepimeral bristles about one to five. *Abdomen:* First tergite with a median patch of white scales; remaining tergites black, with transverse segmental bands of white scales and a median longitudinal stripe of white scales (this median stripe usually narrower than in *A. dorsalis*). Venter with grayish-white scales. *Legs:* Femora with brown and white scales intermixed, posterior surface pale yellow; knee spots white. Tibia dark, posterior surface pale. Hind tarsus with segments 1 to 3 ringed basally and apically with white, segment 4 with a white basal ring and a few white scales at apex, segment 5 almost entirely white. Middle tarsus with segments 1 to 3 marked with white basally and apically, segment 4 dark except for a few white scales at base, segment 5 entirely dark. Front tarsus with segments 1 and 2 narrowly ringed with white basally and apically, segment 3 with a few white scales at base, segments 4 and 5 usually entirely dark. *Wing:* Length about 3.8 to 4.3 mm. Scales narrow, predominantly brown to brownish black, with pale scales intermixed, lacking the contrasting dark- and light-scaled veins as in *dorsalis*.

ADULT MALE. Coloration similar to that of female. *Terminalia* (fig. 167). Lobes of ninth tergite (IXT-L) about as broad as long, separated by a little more than the width of one lobe, each bearing several short spines. Tenth sternite (X-S) sclerotized apically. Phallosome (Ph) conical, rounded apically, open ventrally, closed dorsally. Claspette stem (Cl-S) slender, pilose, bearing a few short setae sub-

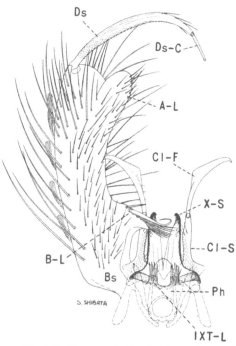

Fig. 167. Male terminalia of *Aedes melanimon* Dyar.

apically; claspette filament (Cl-F) sharply expanded near middle, sickle-shaped and curved apically. Basistyle (Bs) three to three and one-half times as long as mid-width, with scales and long setae on outer aspect; basal lobe (B-L) conical, bearing a short straight spine, a long strong spine with recurved tip, followed by a row of long setae (the strong spine of *A. dorsalis* is more widely separated from the distal spine); apical lobe (A-L) prominent, thumblike, with numerous short setae on dorsal surface. Dististyle (Ds) slender, about two-thirds as long as the basistyle, bearing a few setae near apex; claw (Ds-C) about one-fifth as long as the dististyle.

LARVA. Dr. Stone says, "We have no larval material that I can be sure is *melanimon*." After a review of the description of the larva as given by Dyar (237), it appears that it is probably very similar to *A. dorsalis*.

DISTRIBUTION. Dyar (228) described *Aedes melanimon* from specimens collected at Bakersfield, California. Matheson (490) described a collection of males and females of *A. klotsi* (= *melanimon* Dyar) obtained by Dr. A. B. Klots and Dr. Elsie B. Klots at Fort Garland,

Colorado, July 20–25, 1932, at an elevation of 8,300 feet. The United States National Museum has adult specimens of *A. melanimon* collected by C. S. Richards at Chinook, Montana, and Smith Valley, Nevada.

BIONOMICS. Very little is known of the biology of this species except that the specimens described by Matheson were found along the edge of a small cold clear mountain stream. The adults are described as being abundant in the tall grass in the meadow just before the stream emptied into Mountain Home Lake. According to Dyar (237) the larvae occur in temporary pools filled by occasional rains.

Aedes (Ochlerotatus) mitchellae (Dyar)

Culex mitchellae Dyar, 1905, Jour. N.Y. Ent. Soc., 13:74.

ADULT FEMALE (pl. 71). Medium-sized species. *Head:* Proboscis dark-scaled, with a broad white-scaled ring at middle; palpi short, dark, often lightly speckled with white scales, tips white. Median area of occiput with pale-yellow to golden-yellow lanceolate scales and pale erect forked scales; a submedian triangular patch of narrow dark bronze-brown scales and dark erect forked scales on either side; occiput with broad whitish scales surrounding a patch of broad dark scales laterally. Tori brown or black, with pale scales on inner surface. *Thorax:* Integument of scutum black; scutum clothed with narrow pale-golden scales dorsally, becoming dark bronze-brown laterally; anterior margin and prescutellar space pale golden; a pair of narrow, rather indefinite, submedian lines of pale-golden scales often present and extending the entire length of the scutum. Posterior pronotum clothed with narrow bronze-brown scales, a few white scales on ventral margin. Scutellum with pale-golden scales and brown setae on the lobes. Pleura with rather small sparse patches of moderately broad grayish-white scales. Scales on sterno-pleuron extending to anterior angle, separate from patch on prealar area. Mesepimeron bare on lower one-third to one-half. Hypostigial spot of scales absent. Lower mesepimeral bristles zero or one, rarely two. *Abdomen:* First tergite with a median patch of white scales; remaining tergites dark-scaled, with narrow basal bands

of white to yellowish white and median patches of yellowish-white scales (the median patches frequently extending to the apical margin of the segments), basolateral patches of white scales present. Venter primarily white-scaled, with dark scales intermixed. *Legs:* Femora black-scaled, speckled with white, posterior surface pale; knee spots white. Tibiae and first tarsal segment speckled with white, streaked with white on posterior surface. Hind tarsi with broad basal white rings on segments 2 to 4, segment 5 entirely white; front and middle tarsi with narrower basal white rings on segments 1 to 3, segments 4 and 5 entirely dark. *Wing:* Length about 3.0 to 4.0 mm. Scales narrow, dark.

ADULT MALE. Coloration similar to that of female. *Terminalia* (fig. 168). Similar to *A. sollicitans* and *A. nigromaculis*; however, the basal lobe is a little more prominent in *A. mitchellae.* Lobes of ninth tergite (IXT-L) shorter than wide, each bearing four to seven short spines. Tenth sternite (X-S) heavily sclerotized beyond the middle. Phallosome (Ph) stoutly conical, about two-thirds as broad as long, open ventrally, closed dorsally, and rounded apically. Claspette stem (Cl-S) rather

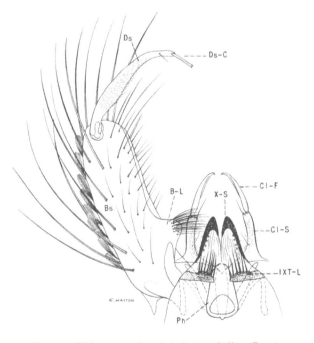

Fig. 168. Male terminalia of *Aedes mitchellae* (Dyar).

stout, pilose, bearing a short seta near apex arising from a prominent tubercle; claspette filament (Cl-F) as long as the stem, slender, curved. Basistyle (Bs) about three times as long as the mid-width, clothed with scales and long and short setae; basal lobe (B-L) prominent and rounded at apex, bearing many slender setae; apical lobe absent. Dististyle (Ds) about two-thirds as long as the basistyle, broader on basal half, pilose; claw (Ds-C) slender, about one-fifth as long as the dististyle.

LARVA (fig. 169). Antenna about half as long as the head, spiculate; antennal tuft multiple, barbed, inserted near middle of shaft, not reaching tip. Head hairs: Postclypeal 4 small, branched; upper frontal 5 and lower frontal 6 long, barbed, usually single; preantennal 7 usually 5- to 10-branched, barbed, reaching insertion of antennal tuft. Prothoracic hairs: 1 long, single; 2 short, single; 3 short, single or double; 4 short, usually single; 5–6 long, single; 7 long, usually 3- or 4-branched. Lateral abdominal hair 6 usually double on segments III–V. Comb of eighth segment with about 14 to 20 scales in a patch; individual scale thorn-shaped, with smaller lateral spinules on basal part. Siphonal index 3.0 to 3.5; pecten of about 12 to 22 evenly spaced teeth on basal two-fifths of siphon; siphonal tuft multiple, barbed, inserted beyond pecten; dorsal preapical spine as long as the apical pecten tooth. Anal segment ringed by the saddle; lateral hair single, a little shorter than the saddle; dorsal brush bilaterally consisting of a long lower caudal hair and a shorter multiple upper caudal tuft; ventral brush large, confined to the barred area; gills variable in length, the dorsal pair about one and a half times as long as the saddle, pointed.

DISTRIBUTION. Known from the southeastern United States, north to New York and west to New Mexico. *Aedes mitchellae* reaches its greatest abundance in the Atlantic and Gulf coastal plains. *United States:* Alabama (135, 213); Arkansas (88); Delaware (120, 167); District of Columbia (318); Florida (135, 213); Georgia (135, 213); Illinois (615); Louisiana (420, 759); Maryland (64, 161); Mississippi (135, 546); New Jersey (350, 643); New Mexico (273); New York (31); North Carolina (135, 646); Oklahoma (321);

South Carolina (135, 742); Tennessee (658); Texas (463, 634); Virginia (176).

BIONOMICS. The larvae are found in temporary rain-filled pools. Both adults and larvae occur in the extreme south throughout the year, following rains. Although the immature stages are rarely if ever found in salt marshes, its distribution seems to be limited largely to the coastal plains. Michener (510) says that in Mississippi the larvae were usually in shallow, clear, partly shaded rain-water pools which held water for moderately long periods. The pools usually contained grass and other emergent vegetation. King *et al.* (419) state that the females are rather severe biters. The adults are frequently captured in light traps operated in the coastal plains.

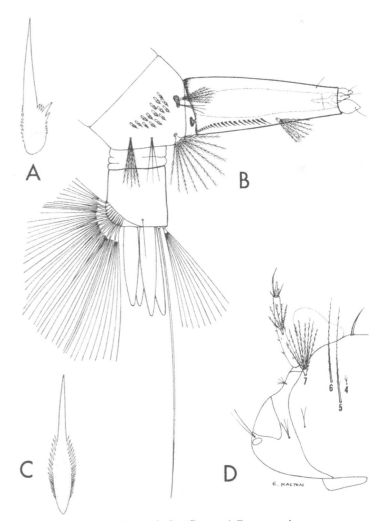

Fig. 169. Larva of *Aedes mitchellae* (Dyar). *A*, Pecten tooth. *B*, Terminal segments. *C*, Comb scale. *D*, Head.

206

Tribe Culicini

Aedes (Ochlerotatus) muelleri Dyar

Aedes (Heteronycha) muelleri Dyar, 1920, Ins. Ins. Mens., *8*:81.
Aedes iridipennis Dyar, 1922, Ins. Ins. Mens., *10*:92.

ADULT FEMALE. Medium-sized species. *Head:* Proboscis long, dark; palpi short, dark. Occiput with narrow white scales dorsally, white to light-brown erect forked scales posteriorly, a patch of rather broad white scales laterally. Tori dark brown, with white scales on inner surface. *Thorax:* Integument of scutum brown, with frosted lines or stripes; scutum with a rather narrow median stripe of white lanceolate scales, followed on either side with brown lanceolate scales; posterior half stripes present; the anterior and lateral margins with white scales. Scutellum with white lanceolate scales and light-brown setae on the lobes. Pleura with dense patches of white scales. Lower mesepimeral bristles absent. *Abdomen:* First tergite with a median patch of white scales; remaining tergites bronzy brown, with narrow basal segmental white bands (sometimes absent on posterior segments); basolateral triangular patches of white scales on segments V–VII, as seen from dorsal view. Venter with dingy-white and brown scales (brown scales more numerous medially and posteriorly on the sternites). *Legs:* Legs dark brown or black, with metallic luster in some lights, posterior surface pale; knee spots white. *Wing:* Scales narrow, dark brown or black.

ADULT MALE. Description adapted from Dyar (237). Coloration similar to that of the female. *Terminalia.* Lobes of ninth tergite (IXT-L) small, rounded, each bearing a few spines. Tenth sternite (X-S) prominent, slender. Claspette stem (Cl-S) curved, uniform, moderate length; claspette filament (Cl-F) as long or longer than the stem, sickle-shaped. Basistyle (Bs) almost three times as long as wide, rounded at apex; basal lobe (B-L) with a stout spine arising from near basal margin; apical lobe (A-L) small. Dististyle (Ds) distinctly swollen near middle; claw (Ds-C) long.

LARVA. Unknown.

DISTRIBUTION. The male and female of this species were described by Dyar from specimens obtained near Mexico City, Mexico. It has also been collected by C. H. T. Townsend (August 17, 1917), at the head of Indian Creek in the Chiricahua Mountains in Arizona at 6,100 feet elevation, where it was biting by day in a cave. The female description given here is based on a single, partly brushed, paratype specimen from Mexico City, Mexico.

BIONOMICS. It appears to be a rare species, and nothing is known of its habits except that it occurs in the mountains.

Aedes (Ochlerotatus) nigripes (Zetterstedt)

Culex nigripes Zetterstedt, 1838, Ins. Lapp., p. 807.
Aedes innuitus Dyar and Knab, 1917, Ins. Ins. Mens., *5*:166.
Aedes alpinus Dyar (not Linnaeus), 1920, Ins. Ins. Mens., *8*:53.

ADULT FEMALE (pl. 72). Medium-sized to rather large species. *Head:* Proboscis dark-scaled; palpi short, dark, scattered pale scales sometimes present. Occiput with long golden-brown lanceolate scales dorsally, broad brownish-white apressed scales mixed with a few dark ones laterally. Dark-brown setae numerous on anterior part of occiput and brown erect forked scales present on posteromedian region. Tori black, with grayish-white scales on inner and dorsal surfaces. *Thorax:* Integument of scutum black; scutum clothed with narrow golden-brown scales and dense, long brownish-black setae; scales somewhat paler on lateral margins and on the prescutellar space. Posterior pronotum clothed with narrow golden-brown scales and brown setae. Scutellum densely clothed with narrow golden-brown scales and brown setae. Pleura with dense patches of narrow, usually curved, light-yellow scales. Scales on sternopleuron reaching near anterior angle, separate from patch on prealar area. Mesepimeron usually bare on lower one-fourth. Hypostigial spot of scales absent. Lower mesepimeral bristles one to five. *Abdomen:* First tergite with a median patch of white scales; remaining tergites black, with broad basal bands of white to cream-white scales. Venter white-scaled. *Legs:* Femora and tibia dark brown, sprinkled with white scales, posterior surface with cream-white scales. Tarsi dark brown, segment 1 streaked with pale scales. *Wing:* Length 4.0 to 4.3 mm. Scales dark brown; patches of white scales on bases of costa and veins 1 and 6.

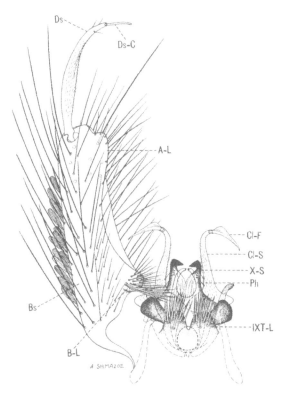

Fig. 170. Male terminalia of *Aedes nigripes* (Zetterstedt).

LARVA (fig. 171). Antenna less than half as long as the head, spiculate; antennal tuft short, single to triple, inserted near middle of shaft. Head hairs: Postclypeal 4 small, multiple; upper frontal 5 and lower frontal 6 single, weakly barbed; preantennal 7 double or triple (rarely 4-branched), weakly barbed, reaching insertion of antennal tuft. Prothoracic hairs: 1 long, 2- to 4-branched; 2 medium, single; 3 short, single to 4-branched; 4 short, single; 5 long, 2- to 4-branched; 6 long, usually double or triple; 7 long, about 5- to 10-branched. Lateral abdominal hair 6 single or double on segments I–VI. Comb of eighth segment of about 10 to 17 scales arranged in a partly double row; individual scale thorn-shaped, with minute lateral spinules on basal part.

ADULT MALE. Coloration similar to that of the female. *Terminalia* (fig. 170). Lobes of ninth tergite (IXT-L) short, rounded, heavily sclerotized, each bearing many dark stout spines. Tenth sternite (X-S) heavily sclerotized and hooked apically. Phallosome (Ph) cylindrical, open ventrally, closed dorsally, broad and deeply incised apically, with the points curved inward. Claspette stem (Cl-S) slender, pilose, reaching well beyond basal lobe; claspette filament (Cl-F) long, roundly expanded at basal third, bladelike, tapered and curved apically, sclerotized. Basistyle (Bs) about three and a half to four times as long as mid-width, clothed with scales and long and short setae, long setae rather dense on ventral aspect; basal lobe (B-L) small, rounded, clothed with long setae, the dorsal one is usually somewhat stronger (the stoutness of this seta varies in different specimens); apical lobe (A-L) rather small, rounded, bearing a few short setae. Dististyle (Ds) a little more than half as long as the basistyle, pilose, bearing two or three small setae subapically; claw (Ds-C) about one-sixth as long as the dististyle.

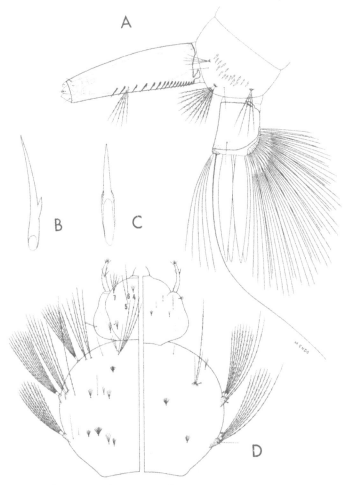

Fig. 171. Larva of *Aedes nigripes* (Zetterstedt). *A*, Terminal segments. *B*, Pecten tooth. *C*, Comb scale. *D*, Head and thorax (dorsal and ventral).

Siphonal index 2.5 to 3.0; pecten of about 13 to 17 teeth on basal half to three-fifths of siphon, the distal 1 to 3 teeth detached; siphonal tuft 2- to 6-branched, weakly barbed, inserted either before or beyond the apical pecten tooth, about as long or a little longer than the basal diameter of the siphon; dorsal preapical spine much shorter than the apical pecten tooth. Anal segment ringed by the saddle or with the saddle narrowly incomplete at midventral line; lateral hair single or double, shorter than the saddle; dorsal brush bilaterally consisting of a long lower caudal hair and a shorter multiple upper caudal tuft; ventral brush well developed, confined to the barred area; gills about one to two and a half times as long as the saddle, tapered.

DISTRIBUTION. This species is largely restricted to the treeless arctic regions of North America, Northern Europe, and Siberia. *Canada:* Manitoba (289, 711); Northwest Territories (181, 289); Quebec (289, 402); Yukon (162a). *Alaska:* (213, 396, 673).

BIONOMICS. The larvae of *Aedes nigripes* develop in early spring pools formed from melting snow and ice. Twinn *et al.* (711) found that the typical larval habitats of this species at Churchill, Manitoba, were shallow snow-water pools among marsh grasses and dwarf willows and birch. First and second instar larvae were first noted by these observers during 1947 on June 4, and fourth instar larvae began to appear in sheltered pools on June 13. The first pupae were found on June 16 and the first adults emerged on June 20 and 21. Jenkins (396) found the larvae in snow-water pools near a glacier and in alpine pools in the mountains above 2,400 feet in central Alaska. Jenkins and Knight (402) observed that the larvae occurred in rock pools, alpine pools, and in tundra bog pools from sea level to 1,000 feet elevation at Great Whale River, Quebec.

The adults have been reported as occurring in immense numbers in the arctic regions. Twinn *et al.* (711) observed that the adult emergence was increasingly heavy the last 10 days of June and was largely completed during the first week of July. These writers noted that there was a marked decline in the abundance of adults after mid-July, although they persisted in diminishing numbers into August.

The females are strong fliers and aggressive biters. Twinn *et al.* (711) observed several mating swarms over the tundra during an early afternoon in July. The swarms are described as darting, dancing clouds of mosquitoes around the margin of a shallow lake. The swarms extended 3 to 9 feet above the ground. Jenkins and Knight (402) observed that the females feed during all hours of the day at temperatures from 48° to 65° F, even in bright sunlight and in winds to 20 m.p.h.

Aedes (Ochlerotatus) nigromaculis
(Ludlow)

Grabhamia nigromaculis Ludlow, 1907, Geo. Wash. Univ. Bull., 5:85.
Grabhamia grisea Ludlow, 1907, Can. Ent., 39:130.

ADULT FEMALE (pl. 73). Medium-sized species. *Head:* Proboscis dark-scaled, with a white ring near middle (greatly reduced or absent on some specimens); palpi short, dark. Occiput with pale-yellow lanceolate scales and erect pale forked scales on median area; a submedian triangular patch of narrow brown scales and a few brown erect forked scales on either side, with dingy-yellow broad appressed scales often surrounding a small patch of dark scales laterally. Tori brown to black, with white scales on inner and dorsal surfaces. *Thorax:* Integument of scutum dark; scutum with a broad median stripe of golden-brown lanceolate scales; most of the remainder of the dorsal surface of the scutum, including the prescutellar space, with varying shades of yellow lanceolate scales; lateral margins dark bronze-brown-scaled, particularly on anterior half. Scales of posterior pronotum narrow bronze-brown. Scutellum with yellow lanceolate scales and darker setae on the lobes. Pleura with large poorly defined patches of broad appressed cream-white scales. Scales on sternopleuron extending to anterior angle, continuous with patch on prealar area. Mesepimeron with lower one-third bare. Hypostigial spot of scales present. Lower mesepimeral bristles zero or rarely one. *Abdomen:* First tergite with a median patch of yellowish scales; remaining tergites each basally, laterally, and medially yellowish-scaled, nearly surrounding large patches of dark scales. The pale median scales of the tergites forming a longitudinal stripe. Venter yellowish-scaled. *Legs:* Femora and tibiae dark, liberally speckled with pale scales, posterior surface pale. Hind tarsal segments each with a broad white basal band (varying on segment 1 from a rather narrow basal band to one covering most of the segment); segment 5 of hind tarsus all white on some specimens; front and middle tarsi with segments 4 and 5 entirely dark; basal white bands of other segments narrower than the corresponding ones of the hind tarsus. *Wing:* Length about 3.2 to 4.2 mm. Scales narrow, dark, sprinkled with pale scales.

ADULT MALE. Coloration similar to that of the female. *Terminalia* (fig. 172). The terminalia appears to be almost identical with that of *A. sollicitans* and similar to that of *A. mitchellae*. The most reliable character for separating them is found in the basal lobe which is somewhat more prominent in *A. mitchellae*. Lobes of ninth tergite (IXT-L) about two-thirds as long as wide, each bearing three to seven short spines. Tenth sternite (X-S) heavily sclerotized. Phallosome (Ph) conical, longer than broad, open ventrally, closed dorsally. Claspette stem (Cl-S) stout, pilose, extending to near middle of basal lobe; claspette filament

(Cl-F) as long as the stem, slender, and curved apically. Basistyle (Bs) about three times as long as mid-width, clothed with scales and long and short setae; basal lobe (B-L) slightly raised, bearing numerous fine setae; apical lobe absent. Dististyle (Ds) nearly two-thirds as long as the basistyle, broader at basal third, densely pilose; claw (Ds-C) slender, about one-fifth as long as the dististyle.

LARVA (fig. 173). Antenna short, less than one-third as long as the head, sparsely spiculate; antennal tuft single to triple, smooth, inserted near middle of shaft, reaching near tip. Head hairs: Postclypeal 4 small, single or branched; upper frontal 5 and lower frontal 6 usually single (occasionally one or more double), smooth or weakly barbed, a little longer than the antennae; preantennal 7 multiple, finely barbed, reaching beyond insertion of antennal tuft. Prothoracic hairs: 1 medium, single; 2 short, single; 3 short, single or double; 4 short, single; 5 long, usually single; 6 long, single; 7 long, 2- to 4-branched. Lateral abdominal hair 6 usually double or triple on segments I–V. Comb of eighth segment with about 6 to 12 scales in an uneven single row or an irregular patch; individual scale thorn-shaped, with small lateral spinules on basal part. Siphonal index about 2.0 to 2.5; pecten of about 16 to 23 teeth reaching apical third of siphon, with 2 to 4 distal teeth detached; siphonal tuft shorter than apical diameter of the siphon, smooth, 2- to 5-branched, inserted beyond pecten at apical fourth of siphon (rarely inserted before the apical pecten tooth); dorsal preapical spine about as long as the apical pecten tooth, inserted well before apex of siphon. Anal segment ringed by the saddle; lateral hair usually single, shorter than the saddle; dorsal brush bilaterally consisting of a single long lower caudal hair and a shorter multiple upper caudal tuft; ventral brush confined to the barred area; gills usually about one and a half to two times as long as the saddle, bluntly pointed.

DISTRIBUTION. Southern Canada, central and western United States, and Mexico. *United States:* California (287); Colorado (343); Idaho (344, 672); Illinois (615); Iowa (625, 626); Kansas (14, 360); Kentucky (570); Minnesota (530, 533); Missouri (6, 324);

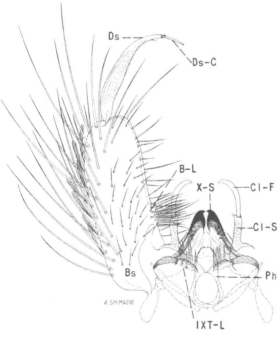

Fig. 172. Male terminalia of *Aedes nigromaculis* (Ludlow).

season during which they are laid. The larvae occur mostly in alkaline waters in rain-filled depressions and in irrigation ditches. According to Rees (575), hatching begins in May in Utah, and successive broods occur throughout the season until late September. According to Mail (475), *Aedes nigromaculis* is an important mosquito of the arid plains of Montana where it seems to favor alkaline waters in irrigated sections.

The females bite readily during the day, but are more active in the evening. Tate and Gates (680) regard the species as a major pest in Nebraska where it seems to reach its greatest abundance during the spring and early summer. They are strong fliers, having been taken several miles from their breeding places. The species reaches its greatest abundance in irrigated pastures throughout the more arid plains of the Middle West and in California. Dyar (190) observed swarming of the males of this species at Laurel, Montana, where they were flying above prominent objects on the prairie. On one occasion they gathered above his head, and on another occasion males of *A. nigromaculis* were observed high above a bush.

Aedes (Ochlerotatus) niphadopsis
Dyar and Knab

Aedes niphadopsis Dyar and Knab, 1918, Ins. Ins. Mens., 5:166.

ADULT FEMALE (pl. 74). Medium-sized grayish-colored species. *Head:* Proboscis dark, with white scales intermixed; palpi short, dark, with intermixed white scales. Median area of occiput with white lanceolate scales bounded by a submedian area of narrow brown scales on either side, with broad appressed white scales laterally; erect forked scales on central area yellow, dark on submedian areas. Tori brown, with white scales on inner and dorsal surfaces. *Thorax:* Integument of scutum black; scutum clothed with narrow curved dark-brown scales on median area; sides and prescutellar space with broader curved white scales; paired narrow dark lines and white spots present posteriorly on some specimens. Posterior pronotum with curved whitish-brown scales dorsally, a few broader white scales ventrally. Scutellum clothed with narrow white scales and pale

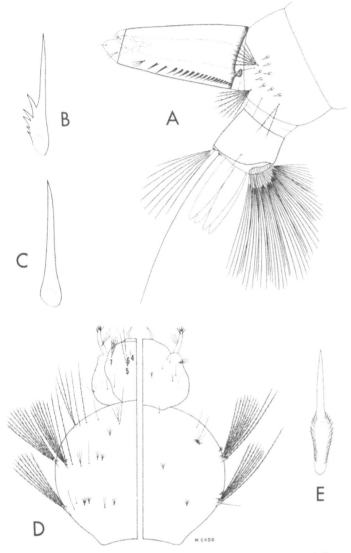

Fig. 173. Larva of *Aedes nigromaculis* (Ludlow). *A*, Terminal segments. *B*, and *C*, Pecten teeth. *D*, Head and thorax (dorsal and ventral). *E*, Comb scale.

Montana (213, 475); Nebraska (14, 680); New Mexico (213, 273); North Dakota (553); Oklahoma (628); Oregon (672); South Dakota (14, 213); Texas (463, 634); Utah (575); Washington (67, 672); Wyoming (213, 578). *Canada:* Alberta (212, 598); Manitoba (311, 466); Saskatchewan (464, 598).

BIONOMICS. The winter is passed in the egg stage, and several broods may be produced during the year. Eggs laid well before the end of the breeding season may hatch the same

setae. Pleura densely clothed with poorly defined patches of broad white scales. Scales on sternopleuron extending to anterior angle, continuous with patch on prealar area. Mesepimeron bare on lower one-fourth. Hypostigial spot of few scales. Lower mesepimeral bristles two or three, rarely none. *Abdomen:* First tergite white-scaled; remaining tergites dark-scaled, with broad basal white bands, a median dorsal white stripe sometimes visible, thus reducing the black to submedian spots. Venter almost entirely white-scaled. *Legs:* Legs with a mixture of pale and dark scales, thus giving them a grayish appearance. *Wing:* Length about 4.0 mm. Wings with white and dark scales intermixed; the white scales are numerous on costa, subcosta, and vein 1.

ADULT MALE. Coloration similar to that of the female. *Terminalia* (fig. 174). Lobes of ninth tergite (IXT-L) about as broad as long, each bearing eight to twelve strong dark spines. Tenth sternite (X-S) heavily sclerotized apically. Phallosome (Ph) longer than wide, cylindrical, apex rounded, open ventrally, closed dorsally. Claspette stem (Cl-S) long, reaching beyond basal lobe, slender, pilose to near apex; claspette filament (Cl-F) long, broadly expanded near basal third, tapered to a long curved point. Basistyle (Bs) about three times as long as mid-width, clothed with scales and numerous long and short setae; basal lobe

(B-L) low, triangular, bearing short setae and a strong dorsal spine followed by several subequal setae; apical lobe (A-L) thumb-shaped, bearing several setae. Dististyle (Ds) about two-thirds as long as the basistyle, curved, somewhat swollen medially, pilose, bearing two or three small subapical setae; claw (Ds-C) slender, about one-fifth as long as the dististyle.

LARVA (fig. 175). Antenna short, about half as long as the head, spiculate; antennal tuft multiple, inserted at middle of shaft, reaching near tip. Head hairs: Postclypeal 4 small, single; upper frontal 5 and lower frontal 6 long, single (occasionally double); preantennal 7 multiple, barbed, reaching near insertion of antennal tuft. Prothoracic hairs: 1 long, usually double; 2 medium, single; 3 short, usually double; 4 medium, single; 5 long, 2- to 4-branched; 6 long, single; 7 long, usually triple. Lateral abdominal hair 6 usually double on segments II–VI. Comb of eighth segment with about 8 to 14 scales in an irregular single or partly double row; individual scale with the subapical spinules nearly half as long as the apical spine. Siphonal index 3.5 to 4.0; pecten of about 8 to 14 teeth on basal third of siphon, the distal 1 or 2 teeth detached; siphonal tuft about 3- to 7-branched, sparsely barbed, inserted beyond pecten before middle of siphon; dorsal preapical spine about as long as the apical pecten tooth. Anal segment with the saddle extending half to two-thirds down the sides; lateral hair single or double, shorter than the saddle; dorsal brush bilaterally consisting of a long lower caudal hair and a shorter multiple upper caudal tuft; ventral brush large, with two or three precratal tufts; gills budlike, much shorter than the saddle.

DISTRIBUTION. Known to occur on the plains and low foothills of the northwestern states. *United States:* Idaho (344, 672); Nevada (492, 575); Oregon (672); Utah (213, 575).

BIONOMICS. The larvae of *Aedes niphadopsis* are found in small, scattered surface pools derived from melting snow. The females are vicious biters attacking man or animals any time during the day but are more active toward evening. According to Rees (575) this species is important in certain localities in the foothills and plains of Utah for a short period in the spring, generally during May. Rees be-

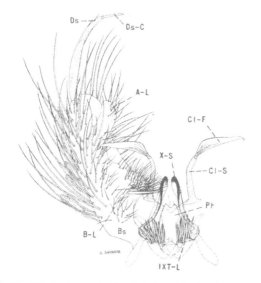

Fig. 174. Male terminalia of *Aedes niphadopsis* Dyar and Knab.

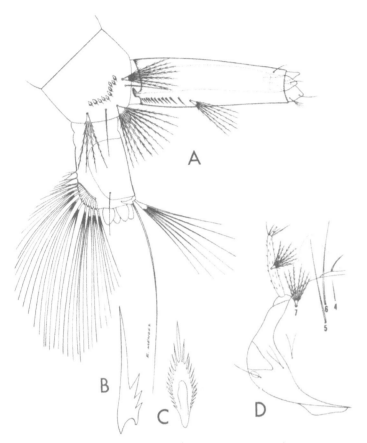

Fig. 175. Larva of *Aedes niphadopsis* Dyar and Knab. *A,* Terminal segments. *B,* Pecten tooth. *C,* Comb scale. *D,* Head.

lieves that it probably has a flight range of several miles since the adults are at times widely dispersed over the plains.

Rees (575) observed the males swarming at Wales, Utah, at sundown when they occurred in large swarms hovering over bushes. The females were among the vegetation near the males. The females entered the swarm for copulation with the males, and the pairs left the swarm and alighted on the vegetation where copulation was completed.

Aedes (Ochlerotatus) pionips Dyar

Aedes pionips Dyar, 1919, Ins. Ins. Mens., 7:19.

ADULT FEMALE (pl. 75). Similar to *A. communis* but has differences in the scutal pattern and in the shape of the tarsal claws as described by Vockeroth (734). Medium-sized species (usually a little larger than *A. communis*). *Head:* Proboscis dark-scaled; palpi short,

dark. Occiput with narrow yellow or white scales surrounding submedian patches of narrow brown scales; yellow erect forked scales numerous; occiput clothed laterally with broad flat appressed yellow or white scales. Tori light brown on outer surface, dark brown and with yellow or white scales on inner surface. *Thorax:* Integument of scutum dark brown or black; scutum clothed with yellow or yellowish-white scales, except for two well-defined rather broad submedian dark-brown stripes and posterior half stripes; the two dark-brown submedian stripes are separated by a very narrow median line of yellow scales. (The scutal pattern is less variable than in *A. communis*, and the yellow-scaled line separating the two dark submedian stripes is narrower.) Prescutellar space with grayish-yellow scales. Posterior pronotum with curved brown or brownish-white scales dorsally, becoming cream-white ventrally. Scutellum with narrow curved yellow or white scales and light-brown setae on the lobes. Pleura with extensive poorly defined patches of broad grayish-white or cream-white scales. Scales on sternopleuron extending to anterior angle, not distinctly separated from patch on prealar area. Mesepimeron with scales extending to near lower margin. Hypostigial spot of scales absent. Lower mesepimeral bristles one to four, rarely none. *Abdomen:* First tergite with a median patch of white scales; remaining tergites each dark-scaled, with a basal white band widening at the sides. Venter clothed with grayish-white scales, each segment with a more or less distinct apical band of dark scales. *Legs:* Femora dark brown, sprinkled with pale scales, posterior surface pale; knee spots white. Tibiae and tarsi dark brown. The front tarsal claw is a little more elongate than that of *A. communis*, and the main claw continues for some distance beyond the lateral tooth before it bends (fig. 176, *B*). *Wing:* Length about 4.5 to 5.0 mm. Scales narrow, brown, a small patch of white scales at base of costa and occasionally at base of vein 1.

ADULT MALE. Similar to *A. communis*. The palpi of *A. pionips* are slightly shorter and subequal to the proboscis, but are distinctly longer than the proboscis in *A. communis*. Coloration similar to that of the female. *Terminalia* (fig.

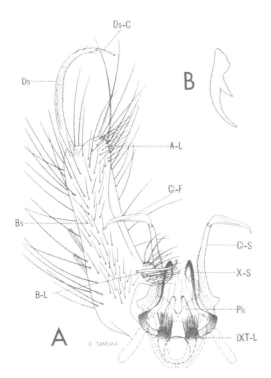

Fig. 176. *Aedes pionips* Dyar. *A*, Male terminalia. *B*, Front tarsal claw of female.

LARVA (fig. 177). Antenna shorter than the head, slender, curved, spiculate; antennal tuft multiple, barbed, inserted near basal two-fifths of the shaft, not reaching tip. Head hairs: Postclypeal 4 small, multiple; upper frontal 5 usually 4- to 6-branched (rarely 3-branched), barbed; lower frontal 6, 3- to 6-branched, barbed; preantennal 7 multiple, barbed, reaching insertion of antennal tuft. Prothoracic hairs: 1 long, single; 2–3 medium, single; 4 short, single; 5 long, double or triple (rarely 4-branched); 6 long, single; 7 long, triple (rarely 4-branched). Lateral abdominal hair 6 usually double on segments I and II, single or

176). Vockeroth (734) found that slight differences exist in the shape of the claspette filament and the basal lobe of the basistyle of *A. communis* and *A. pionips*. Lobes of ninth tergite (IXT-L) a little longer than wide, each bearing several stout spines. Tenth sternite (X-S) heavily sclerotized apically. Phallosome (Ph) stout, conical, open ventrally, closed dorsally, rounded and notched at apex. Claspette stem (Cl-S) long, reaching well beyond distal margin of basal lobe, pilose on basal half; claspette filament (Cl-F) shorter than the stem, expanded near base, flat and bladelike, tapered to a long recurved tip. Basistyle (Bs) about three and a half to four times as long as mid-width, clothed with scales and long and short setae; basal lobe (B-L) small, slightly detached basally, bearing prominent setae, and with a strong dorsal spine arising from near base; apical lobe (A-L) prominent, rounded apically, bearing many moderate setae. Dististyle (Ds) about three-fifths as long as the basistyle, pilose, with two or three small setae before apex; claw (Ds-C) slender, about one-sixth as long as the dististyle.

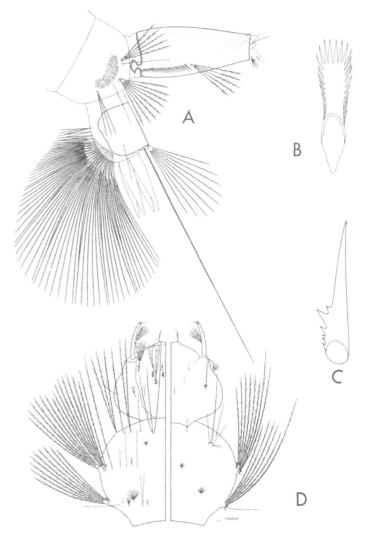

Fig. 177. Larva of *Aedes pionips* Dyar. *A*, Terminal segments. *B*, Comb scale. *C*, Pecten tooth. *D*, Head and thorax (dorsal and ventral).

double on III–VI. Comb of eighth segment of many scales (more than 60) in a patch; individual scale fringed apically with subequal spinules. Siphonal index about 2.5; siphon stout and tapered; pecten of about 20 to 33 evenly spaced teeth on basal two-fifths of siphon; siphonal tuft 4- to 8-branched, barbed, inserted beyond pecten near middle of siphon; dorsal preapical spine shorter than the apical pecten tooth. Anal segment with the saddle extending about two-thirds to three-fourths down the sides; lateral hair single, a little shorter than the saddle; dorsal brush bilaterally consisting of a lower caudal hair and a short, multiple upper caudal tuft; ventral brush well developed with one to three precratal tufts; gills usually one and a half to two times as long as the saddle, pointed.

DISTRIBUTION. Alaska, Canada, and the Rocky Mountain region of the United States. *United States:* Colorado (226); Idaho (672); Montana (213, 475); North Dakota (14); Wyoming (221). *Canada:* Alberta (212, 598); British Columbia (212, 289); Labrador (289, 348); Manitoba (289); Northwest Territories (289, 708); Ontario (212, 289); Quebec (289, 402); Saskatchewan (212, 598); Yukon (212, 289). *Alaska:* (213, 396).

BIONOMICS. The winter is passed in the egg stage and the larvae occur in early spring pools. Jenkins (396) found *Aedes pionips* to be one of the most abundant and widespread species in central Alaska where it occurs from sea level to about 3,000 feet. The larvae were found in a variety of larval habitats including excavations, vehicle tracks, moose tracks, depressions in bogs, oxbow lakes, and in extensive deep ponds. The species is said to be a pioneer in utilizing new habitats. The larvae are said to be slow in developing, and the adults usually do not emerge until midsummer. Twinn (708) states that this mosquito is found in the forests and scattered woodlands over a large part of Canada south of the region where the ground is permanently frozen. Dyar (197) observed the males swarming after sunset in a spruce forest at Prince Albert, Saskatchewan.

Aedes (Ochlerotatus) pullatus
(Coquillett)

Culex pullatus Coquillett, 1904, Proc. Ent. Soc. Wash., 6:168.

Aedes acrophilus Dyar, 1917, Ins. Ins. Mens., 5:127.

Culex jugorum Villeneuve, 1919, Bull. Soc. Ent. France, p. 59.

Aedes gallii Martini, 1920, Arch. Schiffs-Trop. Hyg., 24, Beih., 1:110.

Aedes metalepticus Dyar, 1920, Ins. Ins. Mens., 8:51.

Aedes pearyi Dyar and Shannon, 1925, Jour. Wash. Acad. Sci., 15:78.

Aedes seguyi Apfelbeck, 1929, Konowia, 8:288.

ADULT FEMALE (pl. 76). Medium-sized species. *Head:* Proboscis black-scaled; palpi short, black-scaled, with a few pale scales intermixed on basal half. Occiput clothed with narrow pale-yellow scales and yellow erect forked scales dorsally, with broad appressed yellowish-white scales laterally. Tori dark brown to black, clothed with grayish-white scales. *Thorax:* Integument of scutum black; scutum clothed with yellowish-brown scales, with a narrow bare faint median line bordered on each side by narrow parallel stripes of light-brown scales, each of these stripes bordered laterally by a broader bare stripe; bare curved posterior half stripes may or may not be present; scales on prescutellar space yellow. Posterior pronotum with narrow curved yellowish-white to yellowish-brown scales, becoming a little broader and grayish white ventrally. Scutellum with yellow scales and light-brown setae on the lobes. Pleura with extensive, poorly defined patches of grayish-white scales. Sternopleuron heavily scaled to near middle, usually with scattered scales extending to anterior angle; scales on sternopleuron separate from patch on prealar area. Mesepimeron with scales reaching near lower margin. Hypostigial spot of many scales. Lower mesepimeral bristles one to five. *Abdomen:* First tergite white-scaled; remaining tergites bluish-black-scaled, each with a basal band of white scales. Venter mostly white-scaled, with apical patches of dark scales. *Legs:* Femora with dark-brown and pale scales intermixed, darker apically, posterior surface pale; knee spots white. Tibiae dark brown, sprinkled with white. First tarsal segment dark brown, sprinkled with white, remaining tarsal segments dark brown. *Wing:* Length about 3.5 to 4.0 mm. Scales narrow, blackish brown, with patches of pale scales at bases of costa, veins 1 and 6.

ADULT MALE. Coloration similar to that of the female. *Terminalia* (fig. 178). Lobes of ninth tergite (IXT-L) short, each bearing about five to seven short stout spines. Tenth sternite (X-S) heavily sclerotized apically. Phallosome (Ph) longer than wide, open ventrally, closed dorsally. Claspette stem (Cl-S) long, abruptly bent forming an angle near middle, the basal half stout, pilose, the distal half slender; claspette filament (Cl-F) shorter than stem, bladelike and roundly expanded before middle, tapered to a slightly recurved tip. Basistyle (Bs) about three and a half times as long as mid-width, clothed with scales and long and short setae, the ventral surface densely clothed with long setae; basal lobe (B-L) small at the dorsal side, bearing a stout dark apically curved dorsal spine and several smaller setae; on the ventral side arises a wartlike projection bearing two unequal spines; apical lobe (A-L) large, thumblike, bearing setae in the form of a patch near tip. Dististyle (Ds) about half as long as the basistyle, curved, pilose, expanded before middle, bearing a few small setae before apex; claw (Ds-C) slender, about one-sixth as long as the dististyle.

LARVA (fig. 179). Antenna a little more than half as long as the head, spiculate; antennal tuft multiple, barbed, inserted before middle of shaft, not reaching tip. Head hairs: Postclypeal 4 minute, branched; upper frontal 5 usually 5- to 8-branched, barbed; lower frontal 6 usually 3- to 5-branched, barbed; preantennal 7 usually 8- to 13-branched, barbed, reaching near insertion of antennal tuft. Prothoracic hairs: 1 long, usually double; 2 medium, single; 3 medium, double; 4 medium, single; 5 long, usually triple; 6 long, single; 7 long, 3- or 4-branched. Lateral abdominal hair 6 usually double on segments II–V. Comb of eighth segment with many scales in a large triangular patch; individual scale fringed with subequal spines, the apical ones of about equal size. Siphonal index 3.0 to 3.5; siphon uniformly tapered from base; pecten of about 13 to 25 evenly spaced teeth on basal two-fifths of siphon; siphonal tuft large, usually 5- to 8-branched, barbed, inserted beyond pecten at middle of siphon, longer than the basal diameter of the siphon; dorsal preapical spine much shorter than the apical pecten tooth.

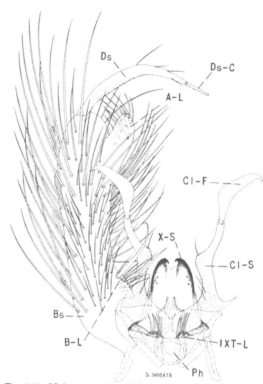

Fig. 178. Male terminalia of *Aedes pullatus* (Coquillett).

Anal segment with saddle extending about two-thirds down the sides; lateral hair single, shorter than saddle; dorsal brush bilaterally consisting of a long lower caudal hair and a shorter multiple upper caudal tuft; ventral brush large, with two or three precratal tufts; gills about one and a half times as long as saddle, pointed.

DISTRIBUTION. The mountainous regions of the western United States, Canada, Alaska, and Europe. Natvig (523) states that this species has a coastal distribution in northern Norway, where it is found principally on the more or less mountainous islands. It occurs in the mountainous regions in central Europe and at very high altitudes in southern Europe. Jenkins (396) found the species principally at the higher elevations in central Alaska, but he states that it does occur down to sea level. *United States:* California (71, 406); Colorado (226, 343); Idaho (344, 672); Michigan (391); Montana (213, 475); Oregon (672); Utah (575); Washington (67, 672); Wyoming (578). *Canada:* Alberta (212, 598); British

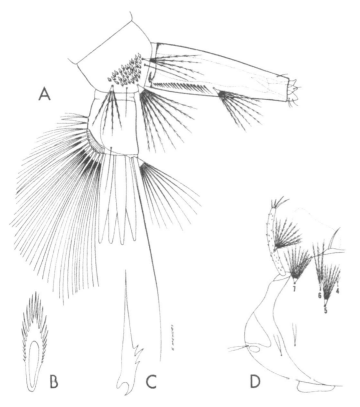

Fig. 179. Larva of *Aedes pullatus* (Coquillett). *A*, Terminal segments. *B*, Comb scale. *C*, Pecten tooth. *D*, Head.

Columbia (212, 289); Northwest Territories (289, 708); Quebec (289, 402); Yukon (212, 289). *Alaska:* (213, 396, 670).

BIONOMICS. The larvae occur mostly in pools formed by melting snow or by the overflow from mountain streams, usually at high eleva-tions and often to 10,000 feet approximately. Jenkins and Knight (402) collected the larvae at Great Whale River, Quebec, principally in rock pools from sea level to 1,000 feet eleva-tion. The larvae were found in small sunlit pools formed by melting snow at Yosemite National Park, California, by Johnson and Thurman (406), at an elevation of 8,600 feet. Bohart (71) obtained the larvae in meadow pools in the Sierra in California at 9,500 feet, and Rees (575) found the larvae in Utah at elevations ranging from 8,000 to 12,000 feet. Larvae may be found from spring until late summer, depending on the season and eleva-tion.

The females are persistent biters, attacking any time during the day in wooded areas. The adults are often abundant in timbered regions of the higher mountains, generally far removed from human habitations. Rees (575) states that the adults emerge in Utah from the latter part of May until the end of July and that females have been taken on wing from May until near the end of August. Dyar (213) states that the males swarm after sunset in openings of the forest or over willows. Jenkins and Knight (402) describe in considerable detail similar swarming and mating of the males with the females at Great Whale River, Quebec.

Aedes (Ochlerotatus) punctodes Dyar

Aedes (Ochlerotatus) sp., Dyar, 1919, Rept. Can. Arctic. Exp. 3, Part C, p. 33.
Aedes punctodes Dyar, 1922, Ins. Ins. Mens., *10*:1.
Aedes (Ochlerotatus) punctodes, Knight, 1951, Ann. Ent. Soc. Amer., *44*:97.

ADULT FEMALE. Similar to the "tundra" varieties of *A. hexodontus* and *A. punctor*, both of which are known to overlap the range of *A. punctodes*.

ADULT MALE. Coloration similar to that of the female. *Terminalia.* Appears to be indis-tinguishable from *A. abserratus* (fig. 115), but separable from *A. punctor*, *A. hexodontus*, and *A. aboriginis* by the shape of the basal lobe of the basistyle. Lobes of ninth tergite (IXT-L) as long as broad, sclerotized, each bearing sev-eral strong spines. Tenth sternite (X-S) heavily sclerotized and hooked apically. Phallosome (Ph) long, cylindrical, open ventrally, closed dorsally, rounded and notched at apex. Clasp-ette stem (Cl-S) short, densely pilose on basal half; claspette filament (Cl-F) nearly as long as the stem, weakly sclerotized, roundly ex-panded before middle, tapered and recurved at tip. Basistyle (Bs) about three to three and a half times as long as mid-width, clothed with scales and long and short setae; basal lobe (B-L) medium-sized, irregularly triangle-shaped, slightly detached basally, surface clothed with short setae except for a basal row of large setae preceded by a weak dorsal spine, the dorsobasal angle of the lobe has a prom-inent cone-shaped protuberance; apical lobe (A-L) broadly rounded, bearing short flattened curved setae. Dististyle (Ds) nearly two-thirds as long as the basistyle, pilose, bearing several subapical setae; claw (Ds-C) about one-sixth as long as the dististyle.

LARVA (fig. 180). Most larvae of *A. punctodes* are distinct from *A. punctor* but intergrades do occur. Antenna about half as long as the head, spiculate; antennal tuft multiple, inserted a little before middle of shaft, not reaching tip. Head hairs: postclypeal 4 small, branched; upper frontal 5 single or double, very lightly barbed; lower frontal 6 single, sparsely barbed; preantennal 7, 3- to 5-branched, sparsely barbed. Prothoracic hairs: 1 long, single or double; 2 long, single; 3 long, usually single; 4 short, single; 5 long, usually single; 6 long, single; 7 long, double or triple. Lateral abdominal hair 6 single or double on segments I–VI. Comb of eighth segment with about 10 to 16 scales in an irregular single or partly double row; individual scale thorn-shaped, with small lateral spinules on basal part. Siphonal index about 3.0; pecten of numerous evenly spaced teeth on basal two-fifths of siphon; siphonal tuft mutliple, barbed, inserted beyond the pecten; dorsal preapical spine shorter than the apical pecten tooth. Anal segment with the saddle incomplete, extending to near mid-ventral line (rarely incomplete in *A. punctor*); lateral hair single, as long or longer than the saddle; dorsal brush bilaterally consisting of a long lower caudal hair and a shorter multiple upper caudal tuft; ventral brush well developed, with about three precratal tufts; gills varying from much shorter to a little longer than the saddle, tapered.

DISTRIBUTION. At present *Aedes punctodes* is known only from the coastal region of Alaska, with its distribution extending up the larger river valleys (293, 429).

BIONOMICS. Frohne (293) claims that *A. punctodes* is the principal Alaskan salt-marsh mosquito and that it predominates in the more saline waters, whereas the closely related *punctor* prefers less brackish water. Frohne found newly hatched larvae in the upper cool inlet marshes in Alaska about the end of April, and a few larvae were still present in August.

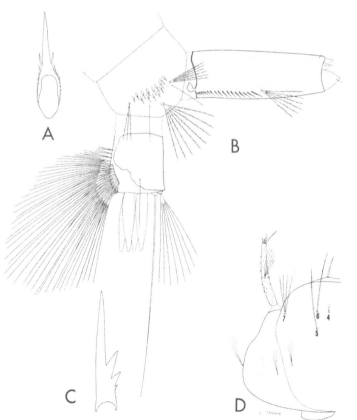

Fig. 180. Larva of *Aedes punctodes* Dyar. *A*, Comb scale. *B*, Terminal segments. *C*, Pecten tooth. *D*, Head.

Aedes (Ochlerotatus) punctor (Kirby)

Culex punctor Kirby, 1837, in Richardson's Fauna Bor.-Amer., 4:309.

Culex implacabilis Walker, 1848, List Dipt. Brit. Mus., 1:7.

Culex provocans Walker, 1848, List Dipt. Brit. Mus., 1:7.

Culicelsa auroides Felt, 1905, N.Y. State Mus. Bull., 97:449.

Aedes punctor, Knight, 1951, Ann. Ent. Soc. Amer., 44:90.

Knight (430) recognized two general varieties in North America, "type *punctor*" and "tundra." These two varieties are based on markings of the scutum of the female and on lesser significant larval characters. The scutum of "type *punctor*" variety is marked similar to *A. aboriginis*, "type *A. hexodontus*" variety, and *A. abserratus*.

ADULT FEMALE (pl. 77). Medium-sized species. *Head:* Proboscis long, dark-scaled; palpi short, dark-scaled. Occiput with narrow curved pale-yellow scales and yellow erect forked scales dorsally, with broad appressed cream-white scales laterally. Tori light brown, dark brown on inner surface. *Thorax:* "Type *punctor*" variety has the scutum dark brown, clothed with yellow scales (occasionally yellowish white or yellowish brown), with a dark-brown or black broad median longitudinal stripe or narrowly divided lines; usually with paired submedian dark areas on posterior half of scutum; scales on anterior and lateral margins and prescutellar space paler. "Tundra" variety has the scutum brown, the darker markings are absent or only faintly defined. Posterior pronotum with narrow curved yellow or yellowish-brown scales dorsally, broader and grayish-white ventrally. Scutellum with yellowish-brown scales and light-brown setae on the lobes. Pleura with extensive patches of grayish-white scales. Scales on sternopleuron extending to anterior angle, narrowly separate from patch on prealar area. Mesepimeron with scales extending to lower margin. Hypostigial spot of scales absent. Lower mesepimeral bristles one to five. *Abdomen:* First tergite with a large median patch of yellowish-white scales; remaining tergites dark-scaled, each with a basal band of white scales. Venter clothed with grayish-white scales, apices of some of the segments black-scaled medially. *Legs:* Front and hind femora dark-scaled, sprinkled with white; middle femur entirely dark or lightly sprinkled with white; all femora mostly pale on posterior surface. Pale knee spots present. Tibiae dark-scaled, sprinkled with white. Tarsi dark. *Wing:* Length about 4.0 to 4.7 mm. Scales narrow, dark brown; a small to large patch of pale scales at base of costa of "tundra" variety. "Type *punctor*" variety has all wing scales dark or rarely a few pale scales at base of costa.

ADULT MALE. Coloration similar to that of the female. *Terminalia* (fig. 181). Appears to be indistinguishable from *A. aboriginis* and *A. hexodontus*, but is separable from *A. abserratus* and *A. punctodes* by the shape of the basal lobe of the basistyle. Lobes of ninth tergite (IXT-L) sclerotized, a little longer than wide, each bearing several strong spines. Tenth sternite (X-S) heavily sclerotized and hooked apically. Phallosome (Ph) cylindrical, open ventrally, closed dorsally. Claspette stem (Cl-S) short, stout, curved near middle, pilose on basal half; claspette filament (Cl-F) shorter

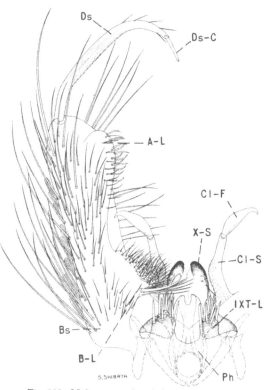

Fig. 181. Male terminalia of *Aedes punctor* (Kirby).

than the stem, heavily sclerotized, broadly rounded before middle, and tapered to a recurved point. Basistyle (Bs) about three and a half times as long as mid-width, clothed with scales and long and short setae; basal lobe (B-L) large and triangular-shaped, slightly detached basally, surface densely clothed with short setae except for a basal row of longer setae preceded by a rather strong dorsal spine, dorsobasal protuberance small; apical lobe (A-L) broadly rounded, clothed with flattened curved setae. Dististyle (Ds) about two-thirds as long as the basistyle, slightly swollen medially, pilose, bearing several small setae before apex; claw (Ds-C) slender, about one-sixth as long as the dististyle.

LARVA (fig. 182). Antenna much shorter than the head, spiculate; antennal tuft multiple, inserted a little before middle of shaft, not reaching tip. Head hairs: Postclypeal 4 small, multiple; upper frontal 5 and lower frontal 6 single or double (rarely triple); weakly barbed; preantennal 7 usually 3- to 6-branched, weakly barbed. Prothoracic hairs: 1 long, single or double; 2 long, single; 3 long, usually single; 4 short, single; 5 long, single to triple; 6 long, single; 7 long, usually 3- or 4-branched. Lateral abdominal hair 6 usually single or double on segments I–VI. ("Type *punctor*" variety has lateral abdominal hair 6 nearly always single on segment II and always single on IV–VI; in "type tundra" variety this hair is double or triple on segment II and is single or double on IV–VI.) Comb of eighth segment with about 10 to 19 scales in an irregular single or double row; individual scale thorn-shaped, with small lateral spinules on basal part. Siphonal index about 3.0; pecten of numerous evenly spaced teeth on basal third or two-fifths of siphon; siphonal tuft 3- to 6-branched, weakly barbed, inserted beyond pecten; dorsal preapical spine shorter than apical pecten tooth. Anal segment ringed by the saddle (narrowly incomplete in some specimens found along the coastal belt of Alaska where *A. punctodes* occurs); lateral hair single, as long or a little longer than the saddle; dorsal brush bilaterally consisting of a long lower caudal hair and a shorter multiple upper caudal tuft; ventral brush large, confined to the barred area or with one or two precratal tufts; gills about one and one-half to three times as long as the saddle, tapered.

DISTRIBUTION. The northern United States, Canada, Alaska, northern Asia, and northern Europe. Its extension southward to Colorado is only at higher elevations. *United States:* Colorado (226, 343); Illinois (615); Maine (45, 564); Maryland (64, 161); Massachusetts (267, 270); Michigan (391); Minnesota (213, 533); Montana (475); New Hampshire (66, 430); New Jersey (448a); New York (31); North Dakota (553); Utah (575); Vermont (237); Wisconsin (171, 213); Wyoming (578). *Canada:* Alberta (212, 598); British Columbia (289, 430); Labrador (289, 348); Manitoba (289, 708); New Brunswick (212, 708); Northwest Territories (289, 430); Nova Scotia (289); Ontario (289, 430); Prince Edward Island (311, 708); Quebec (289, 402, 430); Saskatchewan (598, 708); Yukon (289, 708). *Alaska:* (293, 396, 430).

BIONOMICS. The larvae develop mostly in pools filled by melting snow. Rees (575) says

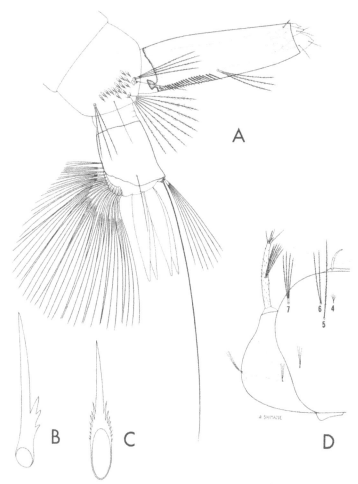

Fig. 182. Larva of *Aedes punctor* (Kirby). *A*, Terminal segments. *B*, Pecten tooth. *C*, Comb scale. *D*, Head.

the larvae appear late in the spring or in early summer in Utah in pools in the forests and in mountain meadows, especially pools with mossy bottoms. Owen (533) found the larvae from early May until early July in Minnesota in cold shaded pools in coniferous and deciduous forests and occasionally in open bogs. Twinn *et al.* (711) began finding the larvae of *A. punctor* at Churchill, Manitoba, in grass-bottomed snow-water pools in the open and among sparse tree growth and in open places in the forest during early June.

Larvae were collected by Jenkins (396) in central Alaska from early May through July from bogs, marshes, snow-melt and cold seepage pools, permanent pools, excavations, and wheel tracks. The habitats were usually open and unshaded. The larvae were active in central Alaska in water at 33° F with thin surface ice above and permafrost below. Jachowski and Schultz (394) claim to have allowed water containing larvae of this species to freeze solid at Umiat, Alaska, and that after thawing, most of the larvae continued their development. Frohne (293) found the larvae to be about as common as *A. punctodes* in the Alaskan coastal marshes but generally less common in the saltier water.

Rees (575) found *A. punctor* in greatest abundance in timbered regions in Utah at altitudes from 9,000 to 13,000 feet where it was a persistent biter, attacking man and animals. Owen (533) found this mosquito to be one of the most annoying species occurring in the dense forested areas in the northern half of Minnesota.

Aedes (Ochlerotatus) purpureipes Aitken

Aedes (Kompia) purpureipes Aitken, 1941, Pan-Pac. Ent., *17*:82.

Aedes (Ochlerotatus) purpureipes, Vargas, 1949, Rev. Inst. de Salub. y Enf. Trop. Mexico, *10*: 261.

Vargas (723) studied 1 male and 5 females of *A. purpureipes* from reared material collected in Mexico and concluded that *Kompia* is a synonym of *Ochlerotatus*.

ADULT FEMALE. Medium-sized species. *Head:* Proboscis long, black with metallic blue iridescence in some lights; palpi short, black with metallic blue iridescence. Occiput with golden narrow curved scales and golden forked scales on median area followed by submedian patches of broad dark appressed scales with metallic iridescence, and golden and dark-forked scales, with silvery appressed scales surrounding a dark patch laterally. Tori black, with a patch of silvery scales on inner surface. *Thorax:* Integument of scutum orange; scutum with a narrow median stripe of narrow golden scales, followed on either side by a broader stripe of narrow black scales; remainder of scutum clothed with golden scales except for posterolateral half-stripes of narrow black scales. Prescutellar space bare, with few silvery scales on either side. Posterior pronotum with

a patch of broad appressed silvery scales on posterior half. Scutellum with broad silvery scales on median lobe, setae dark. Pleura with small patches of broad silvery scales. Postspiracular and lower mesepimeral bristles absent. *Abdomen:* First tergite with a median patch of dark scales; remaining tergites blackish-brown scaled, with metallic purple reflections in certain lights. There are indications of narrow basal segmental pale bands which widen laterally into spots of silvery scales on the posterior segments. Venter with blackish-brown scales. *Legs:* Femora dark, with metallic iridescence except for basal part which is orange. Silvery white knee spots present. Tibiae and tarsi black with purple and blue iridescence. *Wing:* Scales narrow, dark brown.

ADULT MALE. Vargas (723) briefly describes the male terminalia as possessing a simple phallosome, without lateral teeth; claspette well developed and with the filament almost as long as the stem; basal lobe of the basistyle slightly marked and bearing a strong dorsal spine. No specimens were seen by us.

LARVA. Vargas (723) describes and figures the larva of a tree-hole breeding mosquito which he states is most probably that of *A. purpureipes*, since it was collected from the same tree hole in which the pupae of *A. purpureipes* were found. The following description is based on larvae collected by W. A. McDonald in the Santa Rita Mountains, Arizona, and agrees in most part with the description and figures given by Vargas. Antenna less than half as long as the head, smooth; antennal tuft represented by a single hair inserted near middle of shaft and reaching tip. Head hairs: Postclypeal 4 small, multiple; upper frontal 5 and lower frontal 6 smooth, single; preantennal 7 medium-sized, multiple. Thorax and abdomen densely spiculate. Lateral abdominal hair 6 double on segments I–VI. Comb of eighth segment with 5 or 6 scales in a single row; individual scale long and thornlike, with minute spinules on basal part. Siphonal index about 2.6 to 3.0; pecten of about 15 to 21 evenly spaced teeth reaching near middle of siphon; siphonal tuft 2- to 4-branched, inserted beyond the pecten. Anal segment with the saddle reaching half to two-thirds down the sides; lateral hair 3- or 4-branched, a little longer than the saddle, inserted at ventrolateral margin of the saddle; dorsal brush bilaterally consisting of a long lower caudal hair and a shorter multiple upper caudal tuft; ventral brush consisting of sparse tufts restricted to the barred area; gills with the dorsal pair about two to two and a half times as long as the saddle, the dorsal pair longer than the ventral pair, each bluntly rounded.

DISTRIBUTION. Female specimens listed by Aitken (10) are from El Triunfo, San Bartolo and Miraflores, Distrito Sur de Baja California, Mexico, and Patagonia, Santa Cruz County, Arizona. Vargas (723) reared males and females from pupae collected in San Blas, Sinaloa, Mexico. W. A. McDonald collected larvae of *A. purpureipes* from tree holes in the Santa Rita Mountains, Santa Cruz County, Arizona, on August 22 and 24, 1954. Female and larval specimens were generously provided by Mr. McDonald and Dr. J. N. Belkin.

BIONOMICS. Very little is known of the habits of this species. The specimens reported by Aitken (10) from Lower California, Mexico, were taken at night, and the Arizona specimen was taken during the daytime in the act of biting. The specimens reported by Vargas (723) were reared from pupae collected from a tree hole. Larvae were collected from tree holes in sycamore and oak in the Santa Rita Mountains in Arizona by W. A. McDonald.

Aedes (Ochlerotatus) rempeli Vockeroth

Aedes (Ochlerotatus) rempeli Vockeroth, 1954, Can. Ent., *86*:112.

Vockeroth (734a) described *A. rempeli* from specimens collected at Great Whale River, Quebec, and Padley, Northwest Territories. The following descriptions of the male terminalia and larva are based on a larval exuvia and a reared male, from Padley, Northwest Territories, loaned to us by Dr. J. R. Vockeroth. We have not seen a female of this species; therefore the description given here follows that of Vockeroth (1954). Specimens of *A. rempeli* were not received in time for illustrating; however, Vockeroth (734a) has illustrated the male terminalia and the terminal segments of the larva.

ADULT FEMALE (adapted from Vockeroth, 1954). Medium-sized species. *Head:* Proboscis black; palpi short, black, with a few white scales below at base of penultimate segment. Occiput with whitish scales, those on the median part narrow, those on the lateral parts broad and appressed; black erect forked scales present. *Thorax:* Integument of scutum blackish brown; scutum clothed with grayish-yellow scales, except for a submedian longitudinal line of dark-brown scales extending from near the anterior margin of the scutum to the prescutellar space; poorly defined posterior half lines of dark-brown scales present. Scutellum with long, slender, pale scales and black setae. Pleura with patches of broad whitish scales. Mesepimeron with scales reaching lower third. Hypostigial spot of scales absent. Lower mesepimeral bristles two or three. *Abdomen:* Tergites black with basal white bands. Venter black-scaled, with white scales on the posterior corners of the sternites. *Legs:* Femora dark-scaled, posterior surface pale; knee spots pale. Tibiae dark, with a few white scales anteriorly and many posteriorly. Tarsi dark, with a few white scales posteriorly on basal segment. *Wing:* Scales narrow, dark, a few white scales at base of costa.

ADULT MALE. Coloration similar to that of the female (Vockeroth, 1954). *Terminalia.* Lobes of ninth tergite (IXT-L) about as long as broad, each bearing several spines. Tenth sternite (X-S) heavily sclerotized apically. Phallosome (Ph) stout, nearly as broad as long, open ventrally, closed dorsally, notched at apex. Claspette stem (Cl-S) slightly enlarged and bent just before the middle, lacking distinct setae; claspette filament (Cl-F) slightly shorter than the stem, roundly expanded on convex margin near basal third, tapered and curved apically. Basistyle (Bs) about three to three and a half times as long as mid-width, with long setae on inner ventral margin and on outer aspect; basal lobe (B-L) large, conical, bearing long setae that decrease in length posteriorly, lacking a strong dorsal spine; apical lobe (A-L) small, nearly bare, bearing only a few small setae. Dististyle (Ds) curved, slightly expanded medially, finely pilose; claw (Ds-C) slender, about one-fifth as long as the dististyle.

LARVA. Antenna about half as long as the head, spiculate; antennal tuft multiple, inserted slightly before middle of shaft, not reaching tip. Inner preclypeal spines long, slender, darkly pigmented, separated by more than the length of one spine. Head hairs: Postclypeal 4 small, triple; upper frontal 5 and lower frontal 6 single, smooth; preantennal 7 multiple, weakly barbed. Prothoracic hairs: 1 long, single; 2–4 medium, single; 5 long, triple; 6 long, single; 7 long triple. Thorax and abdomen smooth. Lateral abdominal hair 6 usually double on segments I–III, double or single on IV and V, single on VI. Comb of eighth segment with about 33 scales in a triangular patch; individual scale with small basal spinules and fringed apically with long subequal spinules. Siphonal index about 4.0; pecten of numerous evenly spaced teeth on basal two-fifths of siphon; siphonal tuft multiple, barbed, inserted beyond the pecten at middle of the siphon; dorsal preapical spine longer than the apical pecten tooth. Anal segment completely ringed by the saddle; lateral hair single, a little longer than the saddle; dorsal brush bilaterally consisting of a long upper caudal hair and a shorter triple lower caudal tuft of medium length; ventral brush large, with one precratal tuft; gills about three times as long as the anal segment, slender, tapered to a point.

DISTRIBUTION. Known from a single male reared from a larva collected at Great Whale River, Quebec, and two adult specimens (male and female) reared from larvae collected at Padley, Northwest Territories (Vockeroth, 734a).

BIONOMICS. According to Vockeroth (734a) the larvae of *Aedes rempeli* collected at Padley, Northwest Territories, came from a large, apparently permanent, pool with the bottom mostly bare rock with a few clumps of grass, where they were associated with the larvae of *A. communis*, *A. nigripes*, and *A. impiger*. The single larva collected at Great Whale River, Quebec, was from a rather small unshaded rock-crevice pool. The larva was collected on June 17, and the adult emerged on June 23.

Aedes (Ochlerotatus) riparius
Dyar and Knab

Aedes riparius Dyar and Knab, 1907, Jour. N.Y. Ent. Soc., *15*:213.

ADULT FEMALE (pl. 78). A medium-sized to rather large species. *Head:* Proboscis dark, with light-brownish scales intermixed, pale scales more numerous toward the middle; palpi short, dark, with white scales at apices of segments. Occiput with narrow yellowish scales on the median area bounded by submedian patches of brown, lateral region with cream-white scales partly surrounding a black patch; erect forked scales on median area yellowish, those on sides dark. Tori light brown, inner surface darker and with a few pale scales. *Thorax:* Integument of scutum dark brown; scutum with fine golden-brown scales forming a broad median stripe, posterior half stripes present; the sides of the scutum and the prescutellar space with yellowish-white scales. Posterior pronotum with narrow curved golden scales, becoming broader and white in the ventral corner. Scutellum with narrow curved yellowish-white or pale-golden scales and brown setae on the lobes. Pleura with poorly defined patches of white scales. Scales on sternopleuron extending to the anterior angle, narrowly separate from patch on the prealar area. Mesepimeron with lower one-fourth to one-third bare. Hypostigial spot with few to many scales. Lower mesepimeral bristles absent. *Abdomen:* First tergite with a large median patch of white scales; remaining tergites dark, with white scales intermixed and with white basal bands. Venter with white and black scales rather evenly distributed. *Legs:* Femora, tibiae, and first segment of tarsi with dark and pale scales intermixed, posterior surface pale; knee spots white; remaining tarsal segments dark, with basal white bands, except on segment 5 on front and middle legs; the white tarsal bands are much broader on the hind legs. Front tarsal claw as in figure 183,*B. Wing:* Length about 4.5 to 5.0 mm. Scales narrow, predominantly dark, with white scales intermixed.

ADULT MALE. Coloration similar to that of the female. *Terminalia* (fig. 183). Lobes of ninth tergite (IXT-L) short, rounded, each bearing several rather long setae. Tenth sternite (X-S) sclerotized and hooked apically. Phal-

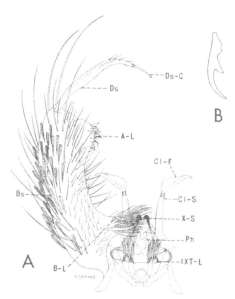

Fig. 183. *Aedes riparius* Dyar and Knab. *A*, Male terminalia. *B*, Front tarsal claw of female.

usually single; 7 long, double or triple. Lateral abdominal hair 6 usually double on segments I–VI. Comb of eighth segment with about 6 to 9 scales in an irregular single row; individual scale with a long median spine, and small lateral spinules on basal part. Siphonal index about 3.0 to 3.5; pecten of about 14 to 17 teeth on basal half to three-fifths of siphon, the distal 2 or 3 teeth detached; siphonal tuft about 3- to 7-branched, weakly barbed, inserted beyond pecten, nearly as long as basal diameter of siphon; dorsal preapical spine shorter than the apical pecten tooth. Anal segment with the saddle extending about three-fourths down the sides; lateral hair single, shorter than the saddle; dorsal brush bilaterally consisting of a long lower caudal hair and a shorter multiple

losome (Ph) cylindrical, longer than wide, open ventrally, closed dorsally, notched at apex. Claspette stem (Cl-S) slender, extending beyond distal margin of basal lobe, pilose to near apex; claspette filament (Cl-F) roundly expanded from near base, sickle-shaped, nearly as long as the stem. Basistyle (Bs) about two and a half to three times as long as mid-width, clothed with rather short setae and many scales on outer aspect; basal lobe (B-L) long, conical, densely clothed with short setae, a long stout seta on dorsobasal margin followed by progressively shorter and weaker setae; apical lobe (A-L) prominent, broad, bearing many short flat recurved setae. Dististyle (Ds) a little more than half as long as the basistyle, broader medially, pilose, with several small setae before apex; claw (Ds-C) slender, about one-fifth as long as the dististyle.

LARVA (fig. 184). Antenna much shorter than the head, spiculate; antennal tuft multiple, barbed, inserted a little before middle of shaft, not reaching tip. Head hairs: Postclypeal 4 small, multiple; upper frontal 5 and lower frontal 6 usually double (one of 5 or 6 occasionally triple), weakly barbed; preantennal 7, 4- to 9-branched, barbed, reaching near insertion of antennal tuft. Prothoracic hairs: 1 long, usually single; 2 short, single; 3 short, usually double or triple; 4 short, single; 5–6 long,

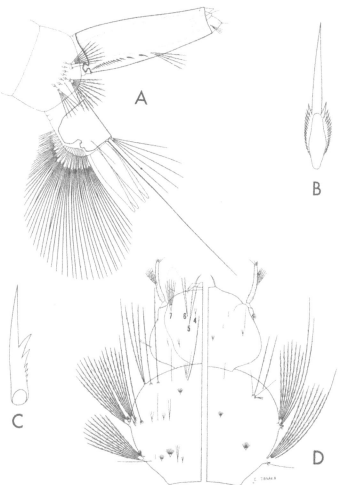

Fig. 184. Larva of *Aedes riparius* Dyar and Knab. *A*, Terminal segments. *B*, Comb scale. *C*, Pecten tooth. *D*, Head and thorax (dorsal and ventral).

upper caudal tuft; ventral brush well developed, with about four to six precratal tufts; gills about one and a half times as long as the saddle, pointed.

DISTRIBUTION. Known from the prairie regions of the northern United States from western New York to Colorado and Montana, and from Ontario to Alberta in Canada. It also occurs in northern Europe (523). *United States:* Colorado (226); Iowa (625); Michigan (391, 529); Minnesota (223, 533); Montana (223); New York (492); North Dakota (223, 553); Wisconsin (171); Wyoming (303). *Canada:* Alberta (223, 708); British Columbia (289); Manitoba (223, 289, 466); Northwest Territories (289); Ontario (223, 708); Saskatchewan (223, 598); Yukon (162a). *Alaska:* (673).

BIONOMICS. The larvae of *Aedes riparius* are found in early spring pools on the prairies. Dyar (223) states that this species seems to prefer spring pools under scrub oaks and other shrubbery where the forest and prairie meet. The females bite either by day or night, usually in the open country and in the less shaded forests. Dyar describes his observations at Warroad, Minnesota, as follows: "The pools had just dried out, and the adults were starting on their migration flight. Under the trees, in leaves, in the grass along the moister parts of fields and ditches, they were present in numbers, both males and females in nearly equal proportion not flying much in the day, but on the alert and easily flushed up, only to settle on the grass or leaves again in a few seconds." He also found a number of males perched after sunset in the top of a dead willow bush in an opening of the woods.

Aedes (Ochlerotatus) scapularis
(Rondani)

Culex scapularis Rondani, 1848, Studi. Ent., Baudie Truqui, p. 109.
Ochlerotatus confirmatus Lynch Arribálzaga, 1891, Rev. Mus. de la Plata, *2*:146.
Aedes hemisurus Dyar and Knab, 1906, Jour. N.Y. Ent. Soc., *14*:199.
Aedes indolescens Dyar and Knab, 1907, Jour. N.Y. Ent. Soc., *15*:11.

ADULT FEMALE (pl. 79). The female appears to be almost identical in appearance with *A.*

infirmatus, except that the hind tibia is pale on the posterior surface and broadly dark basally and apically. The hind tibia of *A. infirmatus* is usually entirely dark. Small to medium-sized species. *Head:* Proboscis dark-scaled; palpi short, dark. Median area of occiput with yellowish-white curved scales and straw-colored erect forked scales, with broad appressed brown and dingy-white scales laterally. Tori brown, darker on inner surface. *Thorax:* Integument of scutum dark brown; scutum with a broad silver-white median stripe extending posteriorly to a little beyond middle; prescutellar space margined with yellowish-white scales, remainder of scutum with fine bronze-brown scales. Scutellum with pale scales on the median lobe. Pleura with small patches of broad appressed dingy-white scales. Scales on sternopleuron not reaching anterior angle, separate from patch on the prealar area. Mesepimeron with lower one-third to one-half bare. Hypostigial spot of scales absent. Lower mesepimeral bristles absent. *Abdomen:* Tergites bronze-brown-scaled, with basolateral patches of white scales. Venter white-scaled. *Legs:* Femora dark-scaled; posterior surface of front and middle femora pale, basal two-thirds of hind femur pale on all aspects. Tibiae dark, posterior surface pale, hind tibia with base and apex dark. Tarsi dark, the proximal segments streaked with pale scales. *Wing:* Length about 3.0 to 3.5 mm. Scales narrow, dark.

ADULT MALE. Coloration similar to that of the female. *Terminalia* (fig. 185). Lobes of ninth tergite (IXT-L) about as broad as long, each bearing four to eight stout spines. Tenth sternite (X-S) heavily sclerotized beyond middle. Phallosome (Ph) longer than broad. stoutly conical, rounded apically, open ventrally, closed dorsally. Claspette stem (Cl-S) slender, pilose, extending well beyond basal lobe; claspette filament (Cl-F) about three-fourths as long as the stem, expanded blade-like, tapered and curved apically, a sharp retrorse projection arising from the convex side near middle. Basistyle (Bs) about three times as long as mid-width, clothed with scales and long and short setae; basal lobe (B-L) rounded or bluntly conical, bearing short setae apically and a large dorsal spine with recurved tip; apical lobe (A-L) small, rounded, bearing a

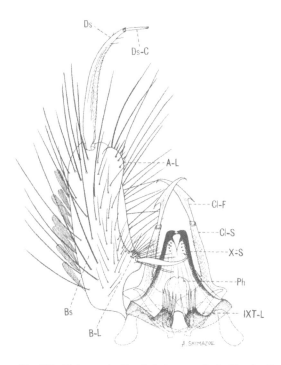

Fig. 185. Male terminalia of *Aedes scapularis* (Rondani).

pecten tooth. Anal segment ringed by the saddle; lateral hair single, shorter than the saddle; dorsal brush bilaterally consisting of a long lower caudal hair and a shorter multiple upper caudal tuft; ventral brush large, confined to the barred area; gills about two to two and a half times as long as the saddle, pointed.

DISTRIBUTION. Florida and Texas, south to Argentina and the Antilles. *United States:* This species is known to occur in the Rio Grande Valley of Texas (249, 571). Pritchard *et al.* (562) report the finding of a single larva in a temporary pool on Vaca Key, Florida, November 15, 1945. This specimen was reported as *Aedes euplocamus* Dyar and Knab; however, Dr. Alan Stone has examined the specimen and is of the opinion that it is *A. scapularis.*

BIONOMICS. The larvae of *A. scapularis* are found mostly in temporary pools, marshes, and

few small straight setae, nearly bare. Dististyle (Ds) about two-thirds as long as the basistyle, slightly broader medially, pilose, bearing two to four small setae before apex; claw (Ds-C) slender, about one-fifth as long as the dististyle.

LARVA (fig. 186). Antenna about half as long as the head, sparsely spiculate; antennal tuft small, multiple, inserted near middle of shaft, not reaching tip. Head hairs: Postclypeal 4 small, branched; upper frontal 5 and lower frontal 6 single, weakly barbed; preantennal 7 multiple, sparsely barbed, reaching beyond insertion of antennal tuft. Prothoracic hairs: 1 long, single; 2 medium, single; 3 short, usually single; 4 medium, single; 5 long, usually single; 6 long, single; 7 long, 2- to 4-branched. Body densely spiculate. Lateral abdominal hair 6 usually double on segments I and II, single on III–VI. Comb of eighth segment with about 12 to 27 scales in a patch; individual scale fringed apically with spines of nearly equal size. Siphonal index 2.0 to 2.5; pecten of about 11 to 18 evenly spaced dark teeth reaching slightly beyond middle of siphon; siphonal tuft multiple, much shorter than the basal diameter of the siphon, inserted beyond pecten; dorsal preapical spine much shorter than the apical

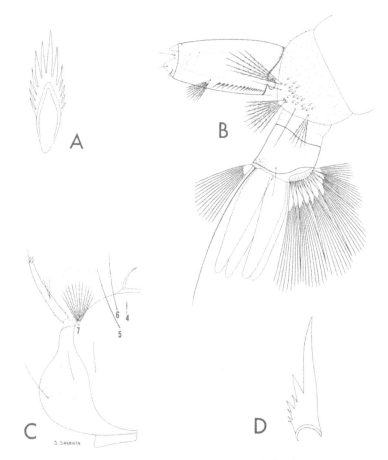

Fig. 186. Larva of *Aedes scapularis* (Rondani). *A,* Comb scale. *B,* Terminal segments. *C,* Head. *D,* Pecten tooth.

pools of muddy water in stream beds. The adults are especially active during the afternoon and early evening. Causey and Kumm (145) marked specimens of this species with bronzing powder in a forest in Brazil, and by means of recaptures, recorded a flight distance of at least 4 kilometers within 11 days.

Aedes (Ochlerotatus) schizopinax Dyar

Aedes schizopinax Dyar, 1929, Proc. U.S. Nat. Mus., 75(23) :1.

ADULT FEMALE (pl. 80). Small- to medium-sized species. *Head:* Proboscis dark, yellowish gray on ventral side becoming darker apically; palpi short, dark, speckled with white. Occiput with narrow yellowish scales and yellow erect forked scales on median area, with broad yellow appressed scales laterally. Tori light brown, darker and with yellowish scales on inner surface. *Thorax:* Integument of scutum brown; scutum with a median golden-brown stripe, a bare median line, and two submedian bronze-brown stripes (stripes sometimes obsolete); sides with narrow curved yellowish-white or brownish-white scales; prescutellar space with yellowish-white scales. Posterior pronotum with narrow curved yellowish-white scales. Scutellum with narrow yellowish-white scales and brown setae on the lobes. Pleura with dense poorly defined patches of medium broad grayish-white scales. Scales on sternopleuron reaching anterior angle, almost continuous with patch on prealar area. Scales on mesepimeron extending to lower margin. Hypostigial spot of scales absent. Lower mesepimeral bristles about three. *Abdomen:* First tergite with a median patch of white scales; remaining tergites bronze-brown, with broad basal white bands widening at the sides and forming lateral patches; seventh segment mostly pale. Venter with white scales. *Legs:* Femora brown, speckled with pale scales, posterior surface pale; knee spots pale. Tibiae dark, with pale scales intermixed. Tarsi dark, basal segments streaked with pale scales. *Wing:* Length about 3.0 to 4.0 mm. Scales narrow, dark, with a patch of a few to many pale scales on base of costa.

ADULT MALE. Coloration similar to that of the female. *Terminalia* (fig. 187). Lobes of

ninth tergite (IXT-L) about as long as broad, each bearing about four to seven stout spines. Tenth sternite (X-S) sclerotized and hooked apically. Phallosome (Ph) much longer than broad, cylindrical, open ventrally, closed dorsally, notched at apex. Claspette stem (Cl-S) rather stout, pilose to near apex; claspette filament (Cl-F) nearly as long as the stem, sickle-shaped, expanded near middle, tapered and curved at tip. Basistyle (Bs) about three and a half times as long as mid-width, clothed with scales and long and short setae; basal lobe (B-L) prominent, bearing numerous short setae arising from tubercles, with an enlargement at base bearing a curved dorsal spine followed by a row of long slender setae; apical lobe (A-L) prominent, bearing short, broadened, curved setae. Dististyle (Ds) about three-fifths as long as the basistyle, broader medially, pilose, bearing three or four short setae before apex; claw (Ds-C) slender, about one-sixth as long as the dististyle.

LARVA (fig. 188). Antenna about half as long as the head, spiculate; antennal tuft multiple, barbed, inserted near middle of shaft, not reaching tip. Head hairs: Postclypeal 4 small, branched; upper frontal 5 triple (occasionally 4-branched), barbed; lower frontal 6 double or triple, barbed; preantennal 7 multiple, barbed, reaching insertion of antennal tuft. Prothoracic hairs: 1 long, 3- or 4-branched; 2 long, single; 3 long, double or triple; 4 short, single; 5 long, usually double or triple; 6 long, single; 7 long, usually 3- or 4-branched. Lateral abdominal hair 6 usually double on segments I–VI. Comb of eighth segment with many scales in a patch; individual scale thorn-shaped, with a long median spine, and short lateral spinules on basal part. Siphonal index about 3.0; pecten of about 14 to 19 evenly spaced teeth on basal two-fifths of siphon; siphonal tuft usually 3- to 5-branched, barbed, inserted beyond pecten near middle of siphon, about as long as the basal diameter of the siphon; dorsal preapical spine much shorter than the apical pecten tooth. Anal segment with the saddle extending about seven-eighths down the sides, saddle spiculate on apical part; lateral hair single, longer than the saddle; dorsal brush bilaterally consisting of a long lower caudal hair and a shorter multiple upper

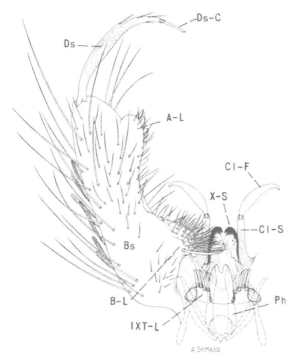

Fig. 187. Male terminalia of *Aedes schizopinax* Dyar.

caudal tuft; ventral brush large, with about two precratal tufts; gills usually about as long as the saddle, pointed.

DISTRIBUTION. Known from Gallatin and Jefferson counties, Montana, and Yellowstone National Park, Wyoming (475).

BIONOMICS. Mail (475) states that the larvae were found in small depressions and in cattle tracks near the margins of permanent pools. The pools in which the larvae occurred were rather foul, covered with green scum, and frequently highly alkaline. Adults were described from reared specimens; therefore nothing is known of their habits.

Aedes (Ochlerotatus) sollicitans
(Walker)

Culex sollicitans Walker, 1856, Ins. Saund., Dipt., p. 427.

ADULT FEMALE (pl. 81). Medium-sized species. *Head:* Proboscis dark-scaled, ringed with white near middle; palpi short, dark, with a few white scales at tips. Occiput with a broad median patch of narrow golden-yellow scales, bounded on either side with a submedian patch

of narrow dark bronze-brown scales, with broad appressed yellowish scales surrounding a patch of dark scales laterally; erect forked scales on central part of occiput pale, those on lateral region dark. Tori brown, with pale scales on inner and dorsal surfaces. *Thorax:* Integument of scutum black; scutum with narrow golden to golden-brown scales dorsally, becoming dark bronze-brown laterally; anterior margin and prescutellar space with somewhat paler scales; often a pair of narrow, rather indefinite, submedian lines of pale-yellow to golden-yellow scales extending nearly the full length of the scutum. Posterior pronotum with narrow curved bronze-brown scales, grayish white on ventral part. Scutellum with light-golden scales

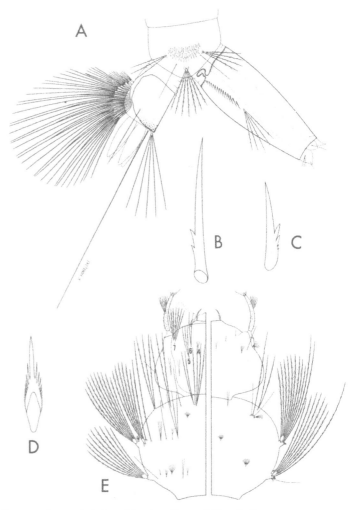

Fig. 188. Larva of *Aedes schizopinax* Dyar. *A*, Terminal segments. *B*, and *C*, Pecten teeth. *D*, Comb scale. *E*, Head and thorax (dorsal and ventral).

and darker setae on the lobes. Pleura with dense, poorly defined patches of appressed grayish-white scales. Scales on sterno-pleuron extending to anterior angle, not distinctly separate from patch on prealar area. Mesepimeron bare on lower one-third to one-half. Hypostigial spot of few to many scales. Lower mesepimeral bristles zero to one. *Abdomen:* First tergite with a median patch of yellowish-white scales; remaining tergites dark, each white laterally, pale yellow basally and medially. Venter whitish to pale-yellow-scaled, speckled with dark scales. *Legs:* Femora and tibiae dark, liberally speckled with pale scales, posterior surface pale, knee spots white. Hind tarsus with segment 1 ringed with white at base and with a yellow ring at middle, segments 2 to 4 with broad white basal rings, segment 5 entirely white; front and middle tarsi similarly marked, but with bands narrower on segments 1 to 3, absent on 4; segment 5 of front tarsus varies from entirely dark to nearly all white; segment 5 of middle tarsus mostly white, blended with dark scales. *Wing:* Length about 3.5 to 4.5 mm. Scales broad, mixed brown and white.

ADULT MALE. Coloration similar to that of

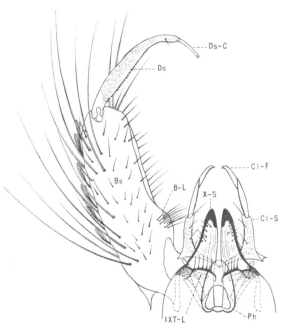

Fig. 189. Male terminalia of *Aedes sollicitans* (Walker).

the female. *Terminalia* (fig. 189). Appears to be almost identical with *A. nigromaculis* and similar to *A. mitchellae*. However, the basal lobe is less prominent in *A. sollicitans* and *A. nigromaculis* than in *A. mitchellae*. Lobes of ninth tergite (IXT-L) short, wider than long, each bearing four to seven short spines. Tenth sternite (X-S) heavily sclerotized. Phallosome (Ph) stoutly conical, about two-thirds as broad as long, open ventrally, closed dorsally, rounded at apex. Claspette stem (Cl-S) rather stout (usually a little stouter than in *A. mitchellae*), pilose, extending to or slightly beyond basal lobe, bearing a short seta near apex arising from a prominent tubercle; claspette filament (Cl-F) as long as the stem, slender, curved. Basistyle (Bs) about three times as long as mid-width, clothed with scales and numerous long and short setae; basal lobe (B-L) only slightly raised, bearing numerous short setae; apical lobe absent. Dististyle (Ds) about two-thirds as long as the basistyle, broader at basal third, pilose; claw (Ds-C) slender, one-fifth as long as the dististyle.

LARVA (fig. 190). Antenna about half as long as the head, spiculate; antennal tuft multiple, barbed, inserted near middle of shaft. Head hairs: Postclypeal 4 small, branched; upper frontal 5 and lower frontal 6 long, usually single, smooth or finely barbed; preantennal 7 about 6- to 10-branched, barbed, reaching insertion of antennal tuft. Prothoracic hairs: 1 long, single; 2 medium, single; 3 short, usually single or double; 4 short, single; 5–6 long, single; 7 long, usually double or triple. Lateral abdominal hair 6 usually triple on segments I and II, double on III–V, single on VI. Comb of eighth segment with about 11 to 21 scales in a patch; individual scale thorn-shaped, with a long apical spine and smaller lateral spinules. Siphonal index 2.0 to 2.5; pecten of about 15 to 27 evenly spaced teeth reaching middle of siphon or slightly beyond (the distal tooth occasionally detached on one side); siphonal tuft about 4- to 8-branched, inserted beyond pecten, much shorter than basal diameter of siphon; dorsal preapical spine as long as the apical pecten tooth. Anal segment ringed by the saddle; lateral hair single, shorter than the saddle; dorsal brush bilaterally consisting of a long lower caudal hair and a shorter multiple

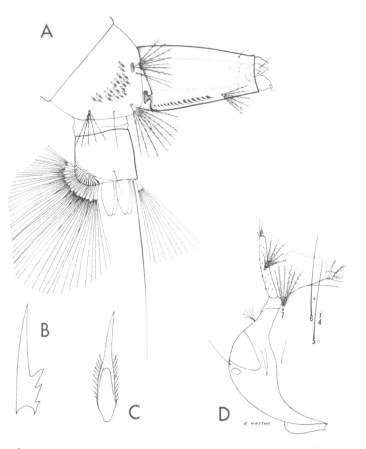

Fig. 190. Larva of *Aedes sollicitans* (Walker). *A*, Terminal segments. *B*, Pecten tooth. *C*, Comb scale. *D*, Head.

upper caudal tuft; ventral brush large, confined to the barred area; gills variable in length, usually short, but may be as long or longer than the saddle, bluntly rounded.

DISTRIBUTION. The Gulf and Atlantic coasts of North America from Texas to New Brunswick, and the West Indies. It also occurs in several inland areas where brackish water is found. *United States:* Alabama (135, 419); Arizona (268); Arkansas (88, 121); Connecticut (213, 494); Delaware (120, 167); District of Columbia (318); Florida and Georgia (135, 213); Illinois (268, 615); Indiana (152, 268); Kansas (14); Kentucky (65); Louisiana (213, 759); Maine (45, 270); Maryland (64, 161); Massachusetts (267, 270); Mississippi (135, 546); Missouri (6, 14); Nebraska (14, 530); New Hampshire (66, 270); New Jersey (213, 350); New Mexico (20, 273); New York (31); North Carolina (135, 213); North Dakota (14);

Ohio (731a); Oklahoma (628); Pennsylvania (298, 756); Rhode Island (35, 270); South Carolina (135, 742); Texas (463, 634); Virginia (176, 178). *Canada:* New Brunswick (707, 708); Nova Scotia (304, 708); Prince Edward Island (710).

BIONOMICS. The larvae of *Aedes sollicitans* occur mostly in salt marshes in coastal areas. The species has been found also in brackish-water swamps in many of the inland states, particularly in the oil fields. The larvae and adults may be found any time during the year in the extreme south. It occurs from April or May to October in the marshes along the north Atlantic seaboard.

The adults are strong fliers and often migrate in large numbers to communities many miles from the salt-water marshes in which they breed. Carpenter and Middlekauff (138) list adult records of this species from inland communities in the southeastern states far from the coastal marshes. An occasional female specimen has been captured in light traps at distances as great as 100 miles from the salt-water pools in which the larvae are known to develop. The females are persistent biters and will attack any time during the day or night. The adults rest in the vegetation during the daytime and will attack anyone invading their haunts, even in full sunlight.

Aedes (Ochlerotatus) spencerii
(Theobald)

Culex spencerii Theobald, 1901, Mon. Culic., 2:99.

ADULT FEMALE (pl. 82). Similar to *A. idahoensis* but has color differences in the scales on the abdominal tergites and the posterior pronotum. Medium-sized species. *Head:* Proboscis dark-scaled; palpi short, dark-scaled, lightly speckled with white scales. Occiput clothed dorsally with narrow curved yellowish scales and yellow erect forked scales, with broad appressed gray scales enclosing an area of dark scales laterally. Tori largely black, with grayish-white scales on dorsal and inner surfaces. *Thorax:* Integument of scutum black; scutum clothed with narrow curved yellowish-white scales, except for a broad median longitudinal stripe of narrow brown scales extending from near anterior margin posteriorly to the prescutellar space (this stripe divided by a narrow line of pale scales on some specimens); posterior half stripes of brown scales present or absent. Scutal pattern somewhat variable. Posterior pronotum largely clothed with narrow brown scales, with white scales usually present in the lower corner. Scutellum with yellowish-white scales and light-golden setae. Pleura with extensive poorly defined patches of grayish-white scales. Scales on sternopleuron extending to anterior angle, almost continuous with patch on prealar area. Mesepimeron with scales extending to near the lower margin. Hypostigial spot of scales present or absent. Lower mesepimeral bristles absent. *Abdomen:* First tergite white-scaled; remaining tergites with a median white stripe and apical and basal white bands, or the dorsum may be almost entirely white-scaled. When many dark scales are present the white scales usually do not form uniform transverse bands. Venter pale-scaled. *Legs:* Femora with yellowish and brown scales intermixed, apical part mostly dark, posterior surface pale; white knee spots present. Tibiae pale-scaled on posterior surface, darker apically. Tarsal segments partly dark-scaled, largely pale on posterior surface, darker on distal segments. *Wing:* Length of wing about 3.8 to 4.1 mm. Scales narrow; costa, veins 1, 3, and 5 dark-scaled, other veins with pale scales.

ADULT MALE. Coloration similar to that of the female but with the abdomen darker. *Terminalia* (fig. 191). Similar to *A. idahoensis*. Lobes of ninth tergite (IXT-L) broader than long, heavily sclerotized, each bearing several spines. Tenth sternite (X-S) heavily sclerotized apically. Phallosome (Ph) cylindrical, longer than broad, open ventrally, closed dorsally, rounded apically. Claspette stem (Cl-S) pilose, reaching to distal margin of basal lobe, bearing two or three small setae before apex; claspette filament (Cl-F) shorter than stem, expanded bladelike from near base, tapered to a slightly recurved tip, a small but pronounced notch near base on convex side. Basistyle (Bs) about three to three and a half times as long as mid-width, with scales and long setae on outer margin; basal lobe (B-L) large, constricted at base and expanded apically, bearing numerous setae and a large strong dorsal spine with recurved tip; apical lobe (A-L) broadly rounded, bearing short

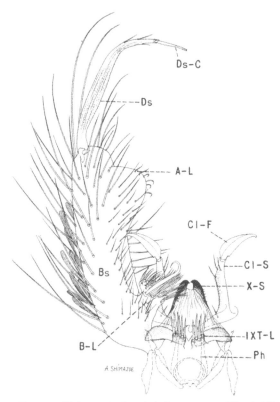

Fig. 191. Male terminalia of *Aedes spencerii* (Theobald).

retrorse setae. Dististyle (Ds) about two-thirds as long as the basistyle, expanded medially, pilose, with two or three small setae before apex; claw (Ds-C) slender, about one-sixth as long as the dististyle.

LARVA (fig. 192). Similar to *A. idahoensis* but differs in the size and shape of the comb scales. Although the number of comb scales found in the two species overlap, the average number is distinctly lower in *A. spencerii*. Antenna much less than half as long as the head, spiculate; antennal tuft multiple, inserted near middle of shaft, reaching near tip. Head hairs: Postclypeal 4 small, multiple; upper frontal 5 and lower frontal 6 single (occasionally one of them double); preantennal 7 multiple, barbed, reaching insertion of antennal tuft. Prothoracic hairs: 1 medium, usually single; 2 short, single; 3 short, single or double; 4 short, single; 5–6 long, usually single; 7 long, 2- to 4-branched. Body spiculate, especially on thorax. Lateral abdominal hair 6 usually triple on segment I, double on II and III, and single on IV–VI. Comb of eighth segment with about 7 to

13 scales in an irregular, partly double row; individual scale with spine long and gradually tapered from basal part, with minute lateral spinules on basal part (total length of scale about one-third greater than in *A. idahoensis*). Siphonal index 2.0 to 2.5; pecten of about 13 to 17 teeth on basal half to three-fifths of siphon, with 1 to 3 distal teeth detached; siphonal tuft usually 2- to 4-branched, inserted beyond pecten, shorter than the basal diameter of the siphon; dorsal preapical spine shorter than the apical pecten tooth. Anal segment with saddle extending about seven-eighths down the sides; lateral hair single, much shorter than the saddle; dorsal brush bilaterally consisting of a long lower caudal hair and a shorter multiple upper caudal tuft; ventral brush well developed, with about two or three precratal tufts; gills about two to three times as long as the saddle, pointed.

DISTRIBUTION. The prairie regions of the northern United States and southern Canada. *United States:* Illinois (615); Iowa (625); Kansas (14); Michigan (391); Minnesota (213, 533); Montana (475); Nebraska (680); New York (31); North Dakota (213, 553); South Dakota (14); Utah (575); Wisconsin (171); Wyoming (578). *Canada:* Alberta (212, 708); British Columbia (352, 708); Manitoba (311, 466); Saskatchewan (310, 708).

BIONOMICS. The larvae of *Aedes spencerii* are found in temporary rain pools, pools formed by melting snow, marshes, roadside ditches, irrigated meadows, and other similar habitats. According to Rempel (598), the larvae generally appear in western Canada during the first week in May. Rees (575) observed that the larvae appear in April or early May in northern Utah and that the adults emerge in May and are found on wing until late July. Rees further states that the pools in which the larvae develop are frequently alkaline. Owen (533) has taken the adults in Minnesota from May to September 15.

The females are diurnal and from about mid-May to mid-June are often the most abundant and annoying mosquitoes of the prairie regions, where they inhabit the open plains, avoiding the timber lands. Although this species is generally a spring mosquito, Philip (547) ob-

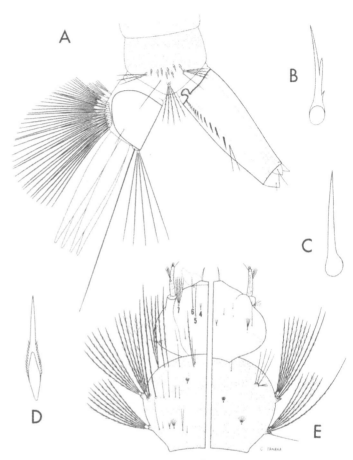

Fig. 192. Larva of *Aedes spencerii* (Theobald). *A*, Terminal segments. *B*, and *C*, Pecten teeth. *D*, Comb scale. *E*, Head and thorax (dorsal and ventral).

served the adults emerging in large numbers in the Dakotas following a heavy rain about the middle of September. Rempel (598) observed that there may be a second or even a third brood when the weather is favorable. He also says that this is the most common and widespread species on the prairies of western Canada. Mail (475) describes this mosquito as being very bloodthirsty and, when numerous, making life a torture in rural communities.

Aedes (Ochlerotatus) squamiger
(Coquillett)

Culex squamiger Coquillett, 1902, Proc. U.S. Nat. Mus., 25:85.

Grabhamia deniedmannii Ludlow, 1904, Can. Ent., 36:234.

ADULT FEMALE (pl. 83). Medium-sized to rather large species. *Head:* Proboscis dark, with grayish-white scales intermixed to near apex; palpi short, dark, with a few white scales intermixed, the terminal segment white-scaled at apex. Occiput with a broad median area of grayish-white scales and dark erect forked scales, bounded by a submedian area of dark scales intermixed with a few white ones and dark erect forked scales, with broad white appressed scales laterally. Tori dark brown to black, with grayish-white scales on dorsal and inner surfaces. *Thorax:* Integument of scutum black; scutum with a more or less distinct median brown stripe abruptly widening behind the middle; this median stripe is followed submedially by mottled areas of brown and white scales, the white scales predominating. Prescutellar space with grayish-white scales. Scutellum with narrow curved grayish-white scales and pale-brown setae on the lobes. Posterior pronotum with broad whitish scales on ventral part, gradually fading to narrow whitish-brown scales dorsally. Pleura with extensive, rather well-defined patches of very broad dingy-white scales. Scales on sternopleuron extending about halfway to anterior angle, distinctly separate from patch on prealar area. Mesepimeron with lower one-third to one-half bare. Hypostigial spot of scales absent. Lower mesepimeral bristles zero to three. *Abdomen:* First tergite with a median patch of white scales; remaining tergites black except for medially expanded basal white bands and a few scattered white scales on the dark areas. Venter with grayish-white scales. *Legs:* Femora, tibiae, and first tarsal segment with black and white scales intermixed, posterior surface of femora pale; knee spots white. Hind tarsi with broad basal white bands on all segments; middle tarsi with basal white bands on segments 1 to 4, narrow or indistinct on segment 5; front tarsi with basal white bands on segments 1 to 3, white band narrow or indistinct on segment 4 and indistinct or absent on segment 5. *Wing:* Length about 4.5 to 5.0 mm. Scales broad, triangular-shaped, brown and white scales rather evenly intermixed.

ADULT MALE. Coloration similar to that of the female but with dorsal abdominal white bands broader. *Terminalia* (fig. 193). Lobes of ninth tergite (IXT-L) short, sclerotized, each bearing four to eight rather strong spines. Tenth sternite (X-S) heavily sclerotized apically. Phallosome (Ph) cylindrical, rounded and notched at apex, open ventrally, closed dorsally. Claspette stem (Cl-S) rather short, pilose to near tip; claspette filament (Cl-F) as long as the stem, narrowly bladelike, with tip recurved, the bladelike part bearing a sharp retrorse projection medially. Basistyle (Bs) about three times as long as mid-width, clothed with numerous scales and long and short setae; basal lobe (B-L) prominent, rounded, clothed with small setae and a strong pale spine followed by long setae on dorsal margin; apical lobe (A-L) thumblike, bearing a few rather short setae. Dististyle (Ds) about two-thirds as long as the basistyle, pilose, bearing two or three small setae before apex; claw (Ds-C) slender, about one-fifth as long as the dististyle.

LARVA (fig. 194). Antenna nearly half as long as the head, spiculate; antennal tuft multiple, inserted a little before middle of the shaft, not reaching tip. Head hairs: Postclypeal 4 small, single or branched; upper frontal 5 double or triple (occasionally single); lower frontal 6 usually double (occasionally single or triple); preantennal 7 about 6- to 11-branched, barbed, reaching beyond insertion of the antennal tuft. Prothoracic hairs: 1 long, single or double; 2 short, single; 3 short, usually double; 4 short, single; 5 long, single or double; 6 long, single; 7 long, usually double or triple. Lateral abdominal hair 6 usually double on segments I–VI. Comb of eighth segment with about 20 to 32 scales in an irregular double row or patch; individual scale with subapical spines nearly as long as the median spine. Siphonal index about 3.0 to 3.5; pecten of about 14 to 22 evenly spaced teeth not reaching middle of siphon; siphonal tuft large, about 3- to 8-branched, barbed, inserted beyond pecten; dorsal preapical spine shorter than the apical pecten tooth. Anal segment with the saddle extending two-thirds or a little more down the sides; lateral hair single or double, as long as the saddle or longer; dorsal brush bilaterally consisting of a long lower caudal hair and a shorter multiple upper caudal tuft; ventral brush large, with two to four precratal tufts; gills budlike, much shorter than the saddle.

DISTRIBUTION. Known to occur along the

Pacific coast from Sonoma County, California, south to Baja California, Mexico (287).

BIONOMICS. The larvae of *Aedes squamiger* are found primarily in salt-marsh pools that are diluted by fresh water from the winter and early spring rains. Ditches, puddles, and extensive pickleweed marshes along the California coast provide habitats for the immature stages. The larvae of *A. dorsalis* are frequently found in association with those of *A. squamiger*. The larval period of *A. squamiger* ranges from about November to March, the cooler season of the year, and the adults emerge from about March to May. The eggs overwinter in the marshes, and the large spring brood hatches the following February and March.

Freeborn (284) states that the females migrate inland from the salt marshes by following the wooded fresh-water creeks and spreading laterally along their courses. The females are vicious biters, attacking man and animals any time during the day, but are more annoying just before dusk. Following its annual spring migration this species is sometimes quite annoying to the residents of the San Francisco Bay area.

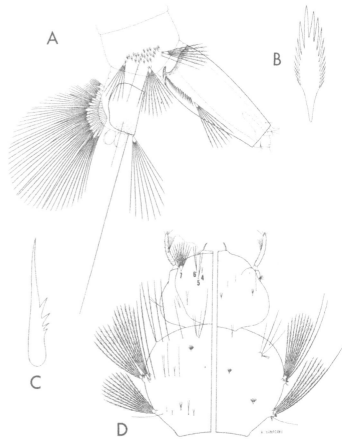

Fig. 194. Larva of *Aedes squamiger* (Coquillett). *A*, Terminal segments. *B*, Comb scale. *C*, Pecten tooth. *D*, Head and thorax (dorsal and ventral).

Aedes (Ochlerotatus) sticticus (Meigen)[16]

Culex sticticus Meigen, 1838, Syst. Beschr. Zweifl. Ins., 7:1.

Culex hirsuteron Theobald, 1901, Mon. Culic., 2:98.

Culex aestivalis Dyar, 1904, Jour. N.Y. Ent. Soc., 12:245.

Culex pretans Grossbeck, 1904, Ent. News, 15:332.

Aedes aldrichii Dyar and Knab, 1908, Proc. U.S. Nat. Mus., 25:57.

Aedes gonimus Dyar and Knab, 1918, Ins. Ins. Mens., 5:165.

Aedes vinnipegensis Dyar, 1919, Ins. Ins. Mens., 7:34.

Aedes lateralis (Meigen), 1932, Edwards *in* Wytsman, Gen. Ins., *194*:144.

ADULT FEMALE (pl. 84). Medium-sized species. *Head:* Proboscis dark-scaled; palpi short,

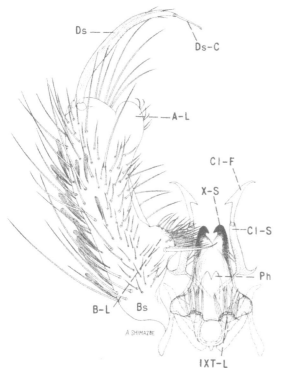

Fig. 193. Male terminalia of *Aedes squamiger* (Coquillett).

[16]Consult Edwards (1932) for additional synonyms. *Aedes sticticus* has been retained as the name for this species because of the uncertainty regarding the true identity of *A. lateralis*.

dark. Occiput with narrow pale-yellow scales and yellowish erect forked scales on median area, with broad appressed yellowish scales surrounding a dark-scaled patch laterally. Tori yellow, darker on inner surface and with dingy-white scales. *Thorax:* Integument of scutum dark brown to black; scutum with narrow yellowish-white scales; two median stripes of fine narrow golden-brown scales extending to the pale-scaled prescutellar space (these two stripes separated by a narrow median line of pale scales which is sometimes indistinct or absent); posterior half stripes of golden-brown scales present. Posterior pronotum with narrow golden-brown scales dorsally, with a few paler scales ventrally. Scutellum with narrow yellowish-white scales and brown setae on the lobes. Pleura with medium-sized well-defined patches of broad appressed grayish-white scales. Sternopleuron with a narrow line of scales reaching near anterior angle; scales on sternopleuron narrowly separate from patch on prealar area. Mesepimeron bare on lower one-fourth to one-third. Hypostigial spot of scales absent. Lower mesepimeral bristles absent. *Abdomen:* First tergite with a median patch of white scales; remaining tergites dark-scaled, each with a narrow basal white band which broadens on either side into a basal triangular white patch. Venter primarily white-scaled, the apices of the terminal segments usually speckled with dark scales. *Legs:* Front and middle femora dark, speckled with pale scales, the posterior surface pale; hind femur mostly pale-scaled, dark apically; knee spots white. Tibiae and tarsi dark, but with tibiae and some of the tarsal segments streaked with pale scales. *Wing:* Length about 3.2 to 4.0 mm. Scales narrow, dark, with or without pale scales on base of costa.

ADULT MALE. Coloration similar to that of the female, but with dorsal abdominal white bands broader, and the dark scales usually more prevalent on the venter. *Terminalia* (fig. 195). Lobes of ninth tergite (IXT-L) about as long as broad, each bearing three to seven short spines. Tenth sternite (X-S) heavily sclerotized beyond middle. Phallosome (Ph) broadly conical, nearly twice as long as broad, rounded apically, open ventrally, closed dorsally, lightly sclerotized. Claspette stem (Cl-S)

rather stout, pilose, extending to posterior margin of basal lobe, bearing one to three small setae before apex; claspette filament (Cl-F) much shorter than the stem, roundly expanded at basal third, bladelike, tapered to a recurved tip, with a small notch on concave side near base. Basistyle (Bs) about three to three and a half times as long as the mid-width, clothed with scales and long and short setae; basal lobe (B-L) large, quadrate, semidetached, bearing many short setae on apical part and a dense tuft of longer setae subapically (this subapical tuft is adjacent to a large recurved dorsal spine); apical lobe (A-L) broadly rounded, bearing many short, flat, retrorse setae. Dististyle (Ds) about three-fifths as long as the basistyle, broader medially, pilose on inner surface, bearing two or three short setae before apex; claw (Ds-C) slender, nearly one-fourth as long as the dististyle.

LARVA (fig. 196). Antenna nearly half as long as the head, spiculate; antennal tuft multiple, inserted before middle of shaft,

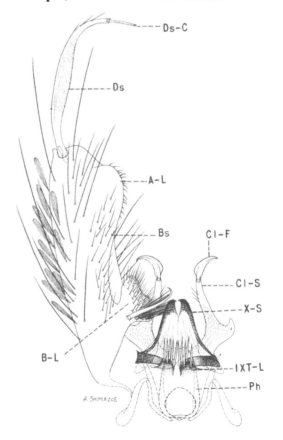

Fig. 195. Male terminalia of *Aedes sticticus* (Meigen).

reaching near tip. Head hairs: Postclypeal 4 small, branched; upper frontal 5, 2- to 4-branched (rarely one single), weakly barbed; lower frontal 6 double (sometimes single or triple), weakly barbed; preantennal 7 multiple, sparsely barbed, reaching insertion of antennal tuft. Prothoracic hairs: 1 long, single; 2 medium, single; 3 short, single or double; 4 short, single; 5 long, usually single; 6 long, single; 7 long, usually 3- or 4-branched. Lateral abdominal hair 6 usually double on segments I–VI. Comb of eighth segment with about 18 to 25 scales in a patch; individual scale thorn-shaped, with a long median spine and lateral spinules which are usually very short but are nearly half as long as the median spine in some specimens. Siphonal index 2.5 to 3.0; pecten of about 15 to 20 evenly spaced teeth reaching near middle of siphon; siphonal tuft small, 4- to 6-branched, inserted beyond pecten; dorsal preapical spine shorter than the apical pecten tooth. Anal segment with the saddle extending about seven-eighths down the sides; lateral hair single, shorter than the saddle; dorsal brush bilaterally consisting of a long lower caudal hair and a shorter multiple upper caudal tuft; ventral brush large, with two to four precratal tufts; gills usually two to two and a half times as long as the saddle, pointed.

DISTRIBUTION. United States, Canada, Alaska, northern Europe, and Siberia. *United States:* Alabama (135, 420); Arkansas (88, 121); California (287); Colorado (343); Connecticut (270, 494); Delaware (167); District of Columbia (318); Florida (135, 516); Georgia (135); Idaho (344, 672); Illinois (615); Indiana (152, 345); Iowa (625); Kansas (14, 530); Kentucky (134, 570); Louisiana (420, 759); Maine (45, 270); Maryland (213); Massachusetts (213, 267); Michigan (391, 529); Minnesota (533); Mississippi (135, 546); Missouri (6, 213); Montana (475); Nebraska (14, 680); New Hampshire (66, 213); New Jersey (213, 350); New York (31); North Carolina (135, 646); North Dakota (553); Ohio (482, 731a); Oklahoma (213, 628); Oregon (672); Pennsylvania (298, 756); South Carolina (135, 742); South Dakota (14, 213); Tennessee (135, 658); Texas (213, 463); Utah

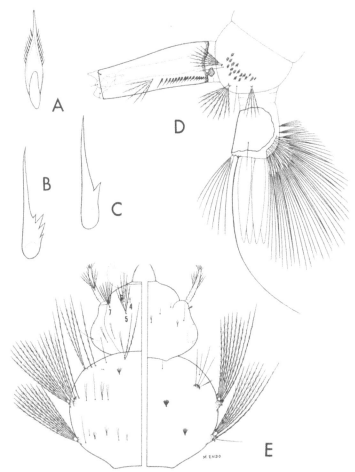

Fig. 196. Larva of *Aedes sticticus* (Meigen). *A*, Comb scale. *B*, and *C*, Pecten teeth. *D*, Terminal segments. *E*, Head and thorax (dorsal and ventral).

(575); Vermont (270); Virginia (176, 213); Washington (67, 672); Wyoming (578). *Canada:* Alberta (708); British Columbia (212, 708); Manitoba (212, 466); New Brunswick (708); Ontario (289, 311); Quebec (304, 708); Saskatchewan (310, 598).

BIONOMICS. The larvae of *Aedes sticticus* are found mostly in floodwater pools in river valleys both in woodlands and open country, but they also occur in rain-filled pools containing dead leaves or other vegetable matter. Rees (575) says it is apparent that the eggs will remain viable, in the absence of flooding, for at least three seasons and perhaps longer. Similar observations are reported by Dyar (237) and Twinn (706). The maximum period of emergence is during late March, April, and early May in the southern part of its range, and May and early June farther north. Smaller broods may develop later in the season following the flooding of breeding pools.

The adults are serious pests in many areas during May, June, and early July. They may be found from May to September in Canada (706), although they occur as early as February in the southern United States. The females are persistent biters, attacking during daylight hours and in early evening in woodlands and thickets near their breeding places. Twinn (706) states that they are most active in eastern Canada after sundown and the first thing in the morning, but they will attack in the daytime in shade or in the open, even in bright sunshine and with a breeze blowing. Matheson (492) says the females are known to migrate several miles, and Twinn (706) reports migrations as far as four miles at Ottawa, Canada.

Aedes (Ochlerotatus) stimulans (Walker)

Culex stimulans Walker, 1848, List Dipt. Brit. Mus., *1*:4.

Culicada subcantans Felt, 1905, N.Y. State Mus. Bull., *97*:448.

Aedes mercurator Dyar, 1920, Ins. Ins. Mens., *8*:13.

Aedes stimulans classicus Dyar, 1920, Ins. Ins. Mens., *8*:113.

Aedes stimulans mississippii Dyar, 1920, Ins. Ins. Mens., *8*:113.

Aedes stimulans albertae Dyar, 1920, Ins. Ins. Mens., *8*:115.

ADULT FEMALE (pl. 85). Medium-sized species. *Head:* Proboscis dark, sprinkled with white scales; palpi short, dark, speckled with white scales. Occiput with a median patch of pale-yellow lanceolate scales; the submedian area behind the eye with narrow golden-brown scales; lateral region with broad appressed yellowish-white scales surrounding a small patch of dark scales; erect forked scales on central part of occiput pale. Tori yellow, darker on inner surface, with or without a few white scales on inner surface. *Thorax:* Integument of scutum black; scutum with narrow yellowish-white to light-brown scales and a broad median longitudinal stripe of brown scales, or with a varying pattern of

brown and paler scales. Posterior pronotum with narrow curved golden or brown scales on dorsal half, with white scales on ventral half. Scutellum with narrow yellowish-white scales and brown setae on the lobes. Pleura with distinct patches of broad white scales. Patch of scales on sternopleuron reaching about halfway to anterior angle (a few scattered scales reaching anterior angle on some specimens), separate from patch on prealar area. Mesepimeron with lower one-third bare. Hypostigial spot of scales absent. Lower mesepimeral bristles usually one to four. *Abdomen:* First tergite with a median patch of white scales; remaining tergites each with a broad basal band of white to pale-yellow scales; apical half of each tergite dark-scaled, frequently speckled with a few pale scales; apices of terminal segments pale-scaled. Venter primarily white-scaled, frequently spotted with dark. *Legs:* Femora with intermixed dark and white scales, posterior surface pale. Tibiae with intermixed dark and white scales, posterior surface streaked with white. Segments of hind tarsi each dark, with a broad basal white ring; segments 1 to 4 of middle tarsi with basal white rings, segment 5 usually entirely dark; segments 1 to 3 of front tarsi with basal white rings, segments 4 and 5 usually entirely dark. Front tarsal claw appears to be indistinguishable from that of *A. fitchii* (fig. 145, *B*). *Wing:* Length about 4.3 to 4.8 mm. Scales narrow, intermixed brown and white along the anterior veins or entirely dark.

ADULT MALE. Coloration similar to that of the female, except for wider basal segmental white bands on abdomen. *Terminalia* (fig. 197). Lobes of ninth tergite (IXT-L) about as long as broad, each bearing several spines. Tenth sternite (X-S) heavily sclerotized beyond middle. Phallosome (Ph) cylindrical, narrowed and notched apically, about twice as long as basal width, open ventrally, closed dorsally. Claspette stem (Cl-S) slender, curved, pilose on basal part; claspette filament (Cl-F) about as long as stem or slightly longer, angularly expanded near middle, gradually tapered to a pointed curved tip. Basistyle (Bs) cylindrical, about three to three and a half times as long as mid-width, clothed with scales and long and short setae, with many long setae on

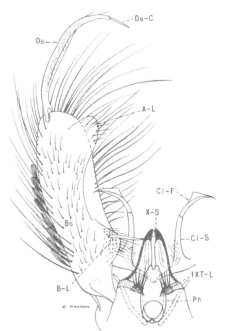

Fig. 197. Male terminalia of *Aedes stimulans* (Walker).

spaced teeth on basal third of siphon; siphonal tuft usually 3- or 4-branched, inserted beyond pecten. Anal segment with the saddle extending about two-thirds down the sides, saddle spiculate posteriorly; lateral hair single (rarely double on one side), a little shorter than the saddle; dorsal brush bilaterally consisting of a long lower caudal hair and a shorter multiple upper caudal tuft; ventral brush large, with two to four precratal tufts; gills usually a little longer than the saddle, each tapering to a blunt point.

DISTRIBUTION. Canada, Alaska, and the northern United States. Its range is known to extend as far south as Mississippi. *United States:* Colorado (343); Connecticut (270, 494); Delaware (120, 167); Idaho (344); Illinois (615); Iowa (625, 626); Kansas (14); Maine (45, 564); Massachusetts (267, 270); Michigan (391, 529); Minnesota (533, 604); Mississippi (419); Missouri (14);

inner ventral margin; basal lobe (B-L) low, bluntly rounded, bearing small apical setae and a stout curved, dorsal spine; apical lobe (A-L) prominent, thumb-shaped, bearing a few short setae. Dististyle (Ds) approximately two-thirds as long as the basistyle, broader medially, pilose, bearing short setae before apex; claw (Ds-C) slender, about one-fifth as long as the dististyle.

LARVA (fig. 198). Antenna less than half as long as the head, spiculate; antennal tuft small, multiple, inserted near middle of shaft. Head hairs: Postclypeal 4 small, branched; upper frontal 5 usually double (occasionally single or triple, rarely 4-branched); lower frontal 6 single (rarely double); preantennal 7 multiple, weakly barbed, reaching near insertion of antennal tuft. Prothoracic hairs: 1 medium, single; 2 short, single; 3 short, usually single or double; 4 short, single; 5 long, usually single; 6 long, single; 7 long, double or triple. Lateral abdominal hair 6 double or triple on segment I, usually double on II–V, single on VI. Comb of eighth segment with about 25 to 35 scales in a patch; individual scale with the apical spine about one and one-half times as long as the subapical spines. Siphonal index 3.0 to 3.5; pecten of about 20 to 25 evenly

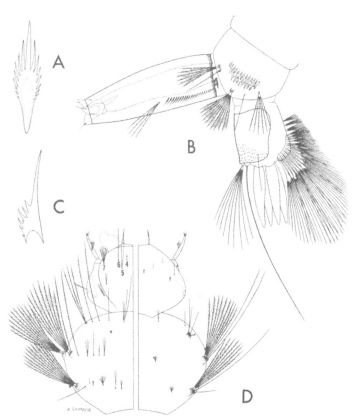

Fig. 198. Larva of *Aedes stimulans* (Walker). *A,* Comb scale. *B,* Terminal segments. *C,* Pecten tooth. *D,* Head and thorax (dorsal and ventral).

Montana (475); Nebraska (14, 680); New Hampshire (66); New Jersey (350, 643); New York (31); Ohio (482, 731a); Pennsylvania (756); Rhode Island (270, 436); South Dakota (14); Utah (575); Vermont (404, 648); Wisconsin (171); Wyoming (578). *Canada:* Alberta (212, 708); British Columbia (289, 352); Manitoba (309, 708); New Brunswick (536, 708); Northwest Territories (289); Nova Scotia (308, 708); Ontario (289, 708); Prince Edward Island (311, 710); Quebec (306, 708); Saskatchewan (212, 464); Yukon (212, 289). *Alaska:* (699).

BIONOMICS. The larvae of *Aedes stimulans* develop in temporary pools formed by overflow of streams and in surface pools filled by snow water and early spring rains. It is particularly addicted to woodland pools. The larvae are among the first to appear in the spring, and the adults may be found on wing in April, May, or June, depending on the locality and the temperature.

The adults are long-lived and may persist in the woods in numbers until June and July, and may occasionally be encountered as late as September. Matheson (492) claims to have taken adults at least two miles from any known breeding place. The females are persistent biters and feed readily in the woods at all hours, the bite being very painful. Owen (533) claims that *A. stimulans* is an important pest species in the eastern half of Minnesota. It is believed to be one of the most abundant and annoying woodland mosquitoes of the northeastern United States and eastern Canada, frequently invading villages and parks located near woodlands.

Aedes (Ochlerotatus) taeniorhynchus
(Wiedemann)

Culex taeniorhynchus Wiedemann, 1821, Dipt. Exot., p. 43.
Culex damnosus Say, 1823, Jour. Acad. Nat. Sci. Phila., 3:11.
Taeniorhynchus niger Giles, 1904, Jour. Trop. Med., 7:382.
Culex portoricensis Ludlow, 1905, Can. Ent., 37:386.
Aedes epinolus Dyar and Knab, 1914, Ins. Ins. Mens., 2:61.

ADULT FEMALE (pl. 86). Medium-sized to rather small species. *Head:* Proboscis dark-scaled, with a white ring near middle; palpi short, dark, with white scales at tips. Occiput with a median patch of golden-yellow to pale golden-brown lanceolate scales, bounded on either side by a few dark scales, followed laterally by a large patch of broad appressed white scales enclosing a small dark-scaled area; erect forked scales on central part pale. Tori brown, with white scales on inner surface. *Thorax:* Integument of scutum dark brown; scutum clothed with narrow golden-brown scales becoming pale yellow to nearly silver-white on the anterior margin, the prescutellar space, and immediately above the wing bases. Posterior pronotum with narrow dark-brown scales. Scutellum with yellowish to silver-white scales and brown setae on the lobes. Pleura with small patches of broad flat grayish-white scales. Scales on sternopleuron extending about halfway to anterior angle (often with scattered scales to near angle), separate from patch on prealar area. Mesepimeron with a patch of scales on upper half, lower half bare. Hypostigial spot of scales absent. Lower mesepimeral bristles zero or one. *Abdomen:* First tergite with a median patch of dark scales, a few white scales often intermixed; remaining tergites dark-scaled, with narrow basal white bands dorsally and conspicuous white patches laterally; apices of the distal segments with a few pale scales. Sternites white-scaled basally, dark-scaled or speckled with white apically. *Legs:* Femora and tibiae dark-scaled, pale on posterior surface; knee spots white. Hind tarsi dark, segments 1 to 4 each with a broad basal white ring, segment 5 usually entirely white; front and middle tarsi dark, with narrower basal white rings on segments 1 to 3, segments 4 and 5 with rings reduced or absent. *Wing:* Length about 2.8 to 3.2 mm. Scales narrow, dark.

ADULT MALE. Coloration similar to that of the female. *Terminalia* (fig. 199). Lobes of ninth tergite (IXT-L) about as broad as long, each bearing three to five stout setae. Tenth sternite (X-S) sclerotized apically. Phallosome (Ph) stout, cylindrical, rounded apically, open ventrally, closed dorsally. Claspette stem (Cl-S) slender, pilose, reaching a little beyond

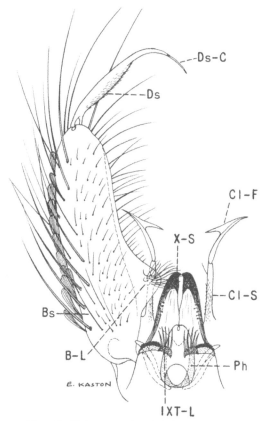

Fig. 199. Male terminalia of *Aedes taeniorhynchus*
(Wiedemann).

hairs; 1 medium, single; 2 short, single; 3 short, usually double; 4 short, single; 5 long, usually single; 6 long, single; 7 long, usually double. Body spiculate. Lateral abdominal hair 6 usually double on segments I and II, with three or more branches on III–V, single on VI. Comb of eighth segment with about 9 to 20 scales in a patch; individual scale small, rounded apically, and fringed with subequal spinules. Siphonal index usually a little less than 2.0; pecten of about 11 to 17 short evenly spaced teeth reaching middle of siphon or slightly beyond; siphonal tuft multiple, barbed, inserted beyond pecten, not more than half as long as basal diameter of the siphon; dorsal preapical spine as long as the apical pecten tooth. Anal segment ringed by the saddle, saddle spiculate posteriorly; lateral hair single, shorter than the saddle; dorsal brush bilaterally consisting of a long lower caudal hair and a shorter multiple upper caudal tuft; ventral brush large, confined to the barred area; gills usually much shorter than the saddle, bluntly rounded.

DISTRIBUTION. This species is found along the Atlantic coast from Brazil to New England, and is known to occur along the Pacific coast from Peru to California. It is also found in several inland localities where brackish water exists. *United States:* Alabama (135, 419); Arkansas (88, 121); California (287); Connecticut (494); Delaware (120, 167); District of Columbia (318); Florida and Georgia (135, 419); Louisiana (213, 759); Maryland (64, 161); Massachusetts (267, 270); Mississippi (135, 546); New Jersey (350); New York (31); North Carolina (135, 419); Pennsylvania (756); Rhode Island (404); South Carolina (135, 742); Texas (463, 634); Virginia (176, 178).

BIONOMICS. The larvae of *Aedes taeniorhynchus* develop mostly in salt marshes in coastal areas and occasionally in near-by freshwater pools. They have been found also in inland brackish-water swamps, particularly in oil fields, in areas far removed from the coast. The larvae and adults may be found any time during the year in the extreme south, but populations are usually heavier following high tides or a combination of high tides and heavy rains during the summer and early fall. The species reaches its greatest abundance along the coastal

apex of basal lobe, bearing a short seta before apex; claspette filament (Cl-F) nearly as long as the stem, curved, tapered to a point, and bearing a prominent simple sharp retrorse projection medially on the convex side. Basistyle (Bs) about three and a half times as long as mid-width, clothed with large scales and numerous long and short setae; basal lobe (B-L) broadly rounded, bearing many slender setae on apex; apical lobe absent. Dististyle (Ds) about half as long as the basistyle, broadened medially, pilose; claw (Ds-C) slender, a little more than one-fourth as long as the dististyle.

LARVA (fig. 200). Antenna less than half as long as the head, sparsely spiculate; antennal tuft small, double or triple, inserted slightly before middle of shaft, not reaching tip. Head hairs: Postclypeal 4 small, branched; upper frontal 5 and lower frontal 6 long, single; preantennal 7 short, multiple, barbed. Prothoracic

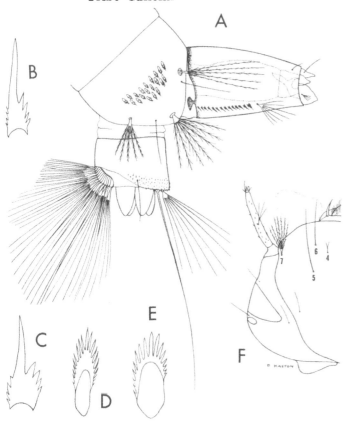

Fig. 200. Larva of *Aedes taeniorhynchus* (Wiedemann). *A*, terminal segments. *B*, and *C*, Pecten teeth. *D*, and *E*, Comb scales. *F*, Head.

regions of the southern United States and the Caribbean region.

The females are persistent biters and will attack any time during the day or night. The adults rest in the vegetation during the daytime and will attack anyone invading their haunts, even in bright sunlight. They are strong fliers and often migrate in large numbers to communities where they become serious pests, even many miles from the salt-water marshes. Carpenter and Middlekauff (138) list adult collections of this species taken from many inland communities in the southeastern United States.

Aedes (Ochlerotatus) thelcter Dyar

Aedes (Taeniorhynchus) thelcter Dyar, 1918, Ins. Ins. Mens., *6*:129.

Aedes keyensis Buren, 1947, Proc. Ent. Soc. Wash., *49*:228.

ADULT FEMALE (pl. 87). Medium-sized species. *Head:* Proboscis dark-scaled; palpi short, dark. Occiput clothed dorsally with curved grayish-white scales and pale erect forked scales, with broad appressed yellowish-brown scales surrounding a small area of dark-brown scales laterally. Tori brown, darker on inner surface and bearing a few small dark setae. *Thorax:* Integument of scutum dark brown; scutum clothed with yellowish scales, with or without a broad median stripe of light-brown scales. Posterior pronotum with narrow curved brown scales. Scutellum with yellowish scales and dark setae on the lobes. Pleura with small patches of broad flat grayish-white scales. Scales on sternopleuron not reaching anterior angle, separate from patch on the prealar area. Mesepimeron with lower one-third bare. Hypostigial spot of scales absent. Lower mesepimeral bristles absent or rarely one. *Abdomen:* First tergite with a median patch of white scales; remaining tergites dark, with median basal triangular white spots prolonged medially; prominent large basolateral white spots present. Venter grayish-white-scaled except for a few dark scales at apices of some of the segments. *Legs:* Legs dark-brown-scaled, with femora, tibiae, and first tarsal segment pale scaled on posterior surface. *Wing:* Length about 3.0 to 3.7 mm. Scales narrow, brown.

ADULT MALE. Coloration similar to that of the female. *Terminalia* (fig. 201). Lobes of ninth tergite (IXT-L) short, about as wide as long, each bearing several short spines. Tenth sternite (X-S) heavily sclerotized and hooked apically. Phallosome (Ph) conical, longer than wide, open ventrally, closed dorsally. Claspette stem (Cl-S) short, slender, pilose; claspette filament (Cl-F) about as long as the stem, expanded and curved bladelike, with a short retrorse tooth on outer curved surface. Basistyle (Bs) stout, a little more than twice as long as mid-width, clothed with scales and long and short setae; basal lobe (B-L) prominent, rounded apically, constricted basally, bearing a strong curved dorsobasal spine and many shorter setae; apical lobe (A-L) short, rounded, bearing a few short setae. Dististyle (Ds) about two-thirds as long as the basistyle, expanded medially, pilose on inner surface; claw (Ds-C) about one-sixth as long as the dististyle.

LARVA (fig. 202). Antenna less than half as

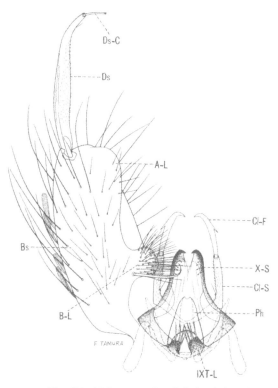

Fig. 201. Male terminalia of *Aedes thelcter* Dyar.

long as the head, spiculate; antennal tuft small, multiple, inserted before middle of shaft, not reaching tip. Head hairs: Postclypeal 4 small, branched; upper frontal 5 and lower frontal 6 single, sparsely barbed; preantennal 7 large, multiple, sparsely barbed. Prothoracic hairs: 1 long, single; 2 medium, single; 3 short, usually double or triple; 4 short, single or double; 5–6 long, single; 7 long, usually double or triple. Lateral abdominal hair 6 double on segments I and II, single on III–VI. Comb of eighth segment with about 15 to 24 scales in two irregular rows; individual scale with the median spine three to four times as broad and nearly twice as long as the subapical spinules. Siphonal index 2.0 to 2.5; pecten of about 14 to 20 teeth extending nearly to tip of siphon, with 1 to 3 teeth detached; siphonal tuft multiple, inserted within the pecten near middle of siphon, shorter than basal diameter of the siphon. Anal segment ringed by the saddle; lateral hair usually double, shorter than the saddle; dorsal brush bilaterally consisting of a long lower caudal hair and a much shorter heavy brushlike upper caudal tuft; ventral

brush large, usually with two precratal tufts; gills usually longer than the saddle, pointed.

DISTRIBUTION. Known from Mexico (481), the southwestern United States, and the Florida Keys. *United States:* Florida (688); Oklahoma (321); Texas (249, 463, 634).

BIONOMICS. The larvae of *Aedes thelcter* may be found throughout the spring, summer, and fall in overflow pools from irrigation ditches and in temporary rain-filled pools, particularly in the Rio Grande Valley. Thurman *et al.* (688) found the larvae in temporary pools in limestone depressions in the Florida Keys. The largest number of adults collected around Corpus Christi, Texas, was captured in light traps.

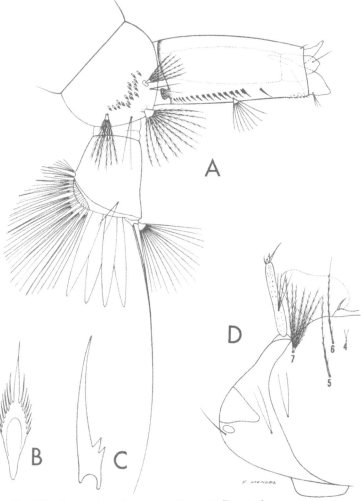

Fig. 202. Larva of *Aedes thelcter* Dyar. *A,* Terminal segments. *B,* Comb scale. *C,* Pecten tooth. *D,* Head.

Aedes (Ochlerotatus) thibaulti
Dyar and Knab

Aedes thibaulti Dyar and Knab, 1910, Proc. Ent. Soc. Wash., *11*:174.

ADULT FEMALE (pl. 88). Medium-sized species. *Head:* Proboscis black-scaled; palpi short, black. Occiput with narrow yellow scales on median area; a small submedian patch of broad dark scales on either side, with broad appressed pale scales laterally; pale erect forked scales numerous on dorsal surface. Tori yellow, darker and with a few small setae on inner surface. *Thorax:* Integument of scutum black; scutum with narrow bronze-brown scales in a wide median stripe widened beyond the middle to cover almost the entire posterodorsal surface of the scutum; golden lanceolate scales on the sides; a few golden scales present on the margins of the prescutellar space. Posterior pronotum with narrow golden scales. Scutellum with pale-golden scales and brown setae on the lobes. Pleura with small well-defined patches of broad grayish-white scales. Scales on sternopleuron not reaching anterior angle or with scattered scales anteriorly, well separated from patch on prealar area. Mesepimeron with a patch of scales on anterior two-thirds, lower one-third bare. Hypostigial spot of scales absent. Lower mesepimeral bristles usually absent, rarely one. *Abdomen:* First tergite with scattered pale scales; remaining tergites blue-black-scaled, with conspicuous basolateral patches of white to yellowish-white scales. Venter primarily white-scaled, the apices of the terminal segments dark. *Legs:* Legs blue-black-scaled; posterior surface of the femora pale; knee spots white. *Wing:* Length about 3.7 to 4.2 mm. Scales narrow, dark.

ADULT MALE. Coloration similar to that of the female. *Terminalia* (fig. 203). Lobes of ninth tergite (IXT-L) about as long as broad, each bearing several stout spines. Tenth sternite (X-S) heavily sclerotized apically. Phallosome (Ph) cylindrical, about two-thirds as broad as long, bluntly rounded and pointed at apex, open ventrally, closed dorsally. Claspette stem (Cl-S) stout, pilose, bearing a short stout branch just beyond middle; claspette filament (Cl-F), large, contorted, leaflike, about as long as the claspette stem, arising from the tip of the short median branch of

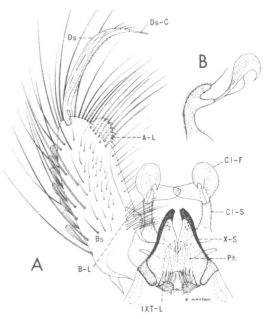

Fig. 203. *Aedes thibaulti* Dyar and Knab. *A*, Male terminalia. *B*, Claspette (side view).

the stem. Basistyle (Bs) nearly three times as long as mid-width, clothed with scales and long and short setae; basal lobe (B-L) rounded, bearing many rather long setae, each arising from a distinct tubercle; apical lobe (A-L) thumb-shaped, bearing short curved setae. Dististyle (Ds) about two-thirds as long as the basistyle, pilose, bearing several small setae before apex; claw (Ds-C) slender, about one-sixth as long as the dististyle.

LARVA (fig. 204). Antenna nearly as long as the head, slender, spiculate; antennal tuft large, multiple, sparsely barbed, inserted near middle of shaft, reaching near tip. Inner preclypeal spines long, slender, lightly pigmented, separated by more than the length of one spine. Head hairs: Postclypeal 4 small, branched; upper frontal 5 large, usually 5- to 8-branched, barbed; lower frontal 6 large, 3- to 5-branched (rarely double), barbed; preantennal 7 large, multiple, barbed. Hairs 5, 6, and 7 inserted nearly in a straight line. Prothoracic hairs: 1 long, single; 2 medium, single; 3 short, usually double; 4 short, single; 5–6 long, single; 7 long, double or triple. Lateral abdominal hair 6, 3- or 4-branched on segments I and II, double on segments III–VI. Comb of eighth segment with about 24 to 35 scales in a patch; individual scale fringed with fairly stout subequal spines, the median spine somewhat longer and stronger. Siphonal index about 4.5 to 5.0; pecten of about 20 to 32 evenly spaced teeth on basal two-fifths of siphon; siphonal tuft 4- to 8-branched, barbed, inserted beyond pecten; dorsal preapical spine shorter than the apical pecten tooth. Anal segment with the saddle extending two-thirds to three-fourths down the sides; saddle spiculate toward apex; lateral hair single, slightly longer than the saddle; dorsal brush bilaterally consisting of a long lower caudal hair and a shorter multiple upper caudal tuft; ventral brush large, with two or three precratal tufts; gills about 0.7 to 1.2 times as long as the saddle, pointed.

DISTRIBUTION. Known from the southeastern United States, north to Ohio and west to Texas. *United States:* Alabama (135, 516); Arkansas (121, 376); Florida (135, 514); Georgia (516); Illinois (615); Kentucky (570); Louisiana (81, 759); Mississippi (135, 546); Missouri (14, 213); North Carolina (135, 646); Ohio (482, 731a); South Carolina (135, 742); Tennessee (136); Texas (571).

BIONOMICS. The larvae of *Aedes thibaulti* are found in the flooded bases of sweet- and tupelo-gum trees growing in low areas subject to flooding. They are rarely found in the hollow bases of other kinds of trees in similar locations. The larvae are most numerous during late February and March; however, they have been collected in Arkansas from December to May by the senior author. The larvae and pupae remain for the most part in the hollow bases of roots and other parts of the tree cavity away from the light. Horsfall (376) believes that the eggs may remain viable for more than one year when their breeding places are not flooded.

The adults do not seem to fly far but remain near their breeding places, resting among vegetation or in hollow bases of trees and stumps. The females are persistent biters, attacking even at midday when their haunts are invaded. The bites are painful and the wheals may remain a day or more.

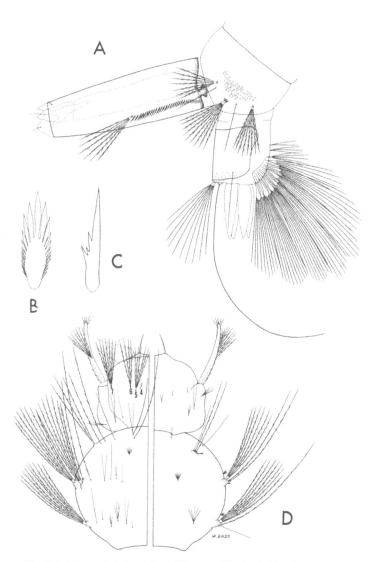

Fig. 204. Larva of *Aedes thibaulti* Dyar and Knab. *A*, Terminal segments. *B*, Comb scale. *C*, Pecten tooth. *D*, Head and thorax (dorsal and ventral).

Aedes (Ochlerotatus) tormentor
Dyar and Knab

Aedes tormentor Dyar and Knab, 1906, Jour. N.Y. Ent. Soc., *14*:191.

ADULT FEMALE (pl. 89). The female appears to be indistinguishable from *Aedes atlanticus*. Medium-sized species. *Head:* Proboscis dark brown; palpi short, dark brown. Occiput with white lanceolate scales on median area, bounded on either side by a submedian patch of dark scales, with broad appressed white scales laterally; erect forked scales pale on median area, brown submedially. Tori yellow, darker on inner surface. *Thorax:* Integument of scutum dark brown; scutum with dark bronze-brown scales, except for a narrow, well-defined, median stripe of silver-white to pale-yellow scales, extending posteriorly to cover the median lobe of the scutellum. Posterior pronotum with narrow curved dark-brown scales on dorsal half. Pleura with small patches of broad flat grayish-white scales. Scales on sternopleuron reaching a little more than half-way to anterior angle, well separated from patch on the prealar area. Mesepimeron with lower one-half bare. Hypostigial spot of scales absent. Lower mesepimeral bristles absent. *Abdomen:* Tergites dark-scaled, with basolateral triangular patches laterally. Venter with white scales, often with dark scales at tips of segments. *Legs:* Legs dark, posterior surface of femora pale scaled. *Wing:* Length about 3.5 mm. Scales narrow, dark.

ADULT MALE. Coloration similar to that of the female. *Terminalia* (fig. 205). Lobes of ninth tergite (IXT-L) about three-fourths as broad as long, each bearing several short spines. Tenth sternite (X-S) heavily sclerotized beyond middle. Phallosome (Ph) stout, about three-fifths as broad as long, open ventrally, closed dorsally, rounded apically. Claspette stem (Cl-S) long, slender, pilose on basal half, extending a little beyond middle of basistyle; claspette filament (Cl-F) about two-fifths as long as stem, expanded at middle, striated, curved and tapered apically. Basistyle (Bs) about three and a half to four times as long as mid-width, clothed with scales and long and short setae; basal lobe (B-L) cylindrical, nearly four times as long as wide, bearing a long curved spine near base and many short stout setae on apex; apical lobe (A-L) prominent, triangular, bearing many short setae. Dististyle (Ds) approximately three-fifths as long as the basistyle, broader medially, pilose; claw (Ds-C) slender, about one-fourth as long as the dististyle.

LARVA (fig. 206). Antenna a little less than half as long as the head, sparsely spiculate; antennal tuft multiple, inserted near middle of shaft, not reaching tip. Head hairs: Postclypeal 4 small, branched; upper frontal 5 and lower frontal 6 long, single; preantennal 7 multiple, sparsely barbed, reaching insertion of antennal tuft. Prothoracic hairs: 1 long, single; 2 medium, single; 3 short, single or double; 4 short, single; 5–6 long, single; 7 long, triple. Lateral abdominal hair 6 single or double on segments I and II, usually single on III–V. Comb of eighth segment with about 9 to 12 scales in a single curved or irregular row; individual scale thorn-shaped, with a long median spine, and short lateral spinules on basal part. Siphonal index about 2.0; pecten of numerous evenly spaced dark teeth, extending to near distal fourth of siphon; siphonal tuft multiple, barbed, inserted within the pecten; dorsal preapical spine much shorter than the apical pecten tooth. Anal segment ringed by the saddle; lateral hair single or double, shorter than the saddle; dorsal brush bilaterally consisting of a very long lower caudal hair and a much shorter multiple upper caudal tuft; ventral brush large, with one or

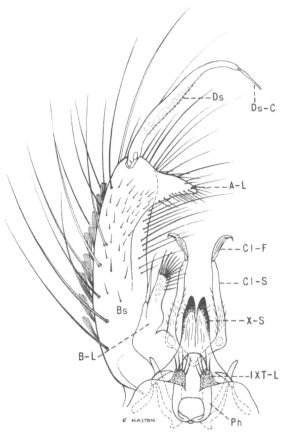

Fig. 205. Male terminalia of *Aedes tormentor* Dyar and Knab.

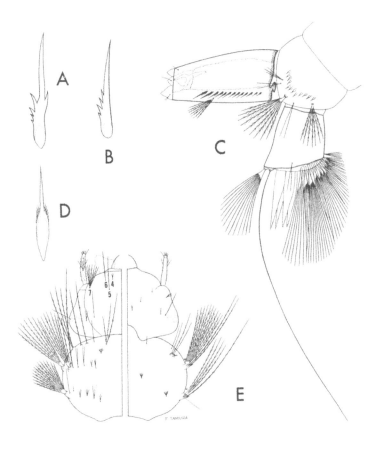

Fig. 206. Larva of *Aedes tormentor* Dyar and Knab. *A*, and *B*, Pecten teeth. *C*, Terminal segments. *D*, Comb scale. *E*, Head and thorax (dorsal and ventral).

two precratal tufts; gills about one and a half times as long as the saddle, pointed.

DISTRIBUTION. Mexico (481) and the southeastern United States. *United States:* Alabama (419, 516); Arkansas (213, 683); Florida and Georgia (419, 516); Louisiana (419, 759); Mississippi (213, 516); Missouri (14); North Carolina (135, 646); Ohio (731a); Oklahoma (321); South Carolina (135, 742); Texas (463, 634).

BIONOMICS. The larvae of *Aedes tormentor* are found in temporary pools following summer rains but seem to be rare throughout most of its range. The larvae are somewhat darker than those of *A. infirmatus* with which they are often associated. Very little is known of the habits of the adults of *A. tormentor*, since the females cannot be separated at this time from those of *A. atlanticus*. Adult records are based wholly on males determined by an examination of the terminalia.

Aedes (Ochlerotatus) tortilis (Theobald)

Culex tortilis Theobald, 1903, Entomologist, *26*:281.
Aedes auratus Grabham, 1906, Can. Ent., *38*:313.
Culex bracteatus Coquillett, 1906, Proc. Ent. Soc. Wash., 7:184.
Aedes habanicus Dyar and Knab, 1906, Jour. N.Y. Ent. Soc., *14*:198.
Aedes balteatus Dyar and Knab, 1907, Jour. N.Y. Ent. Soc., *15*:9.
Aedes plutocraticus Dyar and Knab, 1907, Jour. N.Y. Ent. Soc., *15*:11.
Aedes tortilis virginensis Dyar, 1922, Ins. Ins. Mens., *10*:56.
Aedes (Ochlerotatus) tortilis, Dyar, 1928, Carnegie Inst. Wash. Bull., *387*:169.

ADULT FEMALE (pl. 90). Small to medium-sized species. *Head:* Proboscis dark-scaled; palpi short, dark. Occiput clothed dorsally with narrow curved yellowish-white scales and straw-colored erect forked scales, with broad dull gray appressed scales laterally. Tori light brown, darker on inner surface. *Thorax:* Integument of scutum dark brown; scutum with dark-brown narrow curved scales; a large broad median patch of dull golden scales, broken into three stripes or diffused; prescutellar space bordered with dull golden scales. Posterior pronotum with dark-brown scales on dorsal half. Scutellum with narrow curved golden scales and brown setae on the lobes. Pleura with small patches of broad appressed grayish-white scales. Scales on sternopleuron reaching about halfway to anterior angle, well separated from patch on the prealar area. Mesepimeron with lower one-third bare. Hypostigial spot of scales absent. Lower mesepimeral bristles absent. *Abdomen:* First tergite with a few pale scales on the median area; remaining tergites black to dull violet, with narrow basal yellowish-white bands, widening at the sides to form lateral triangular patches; tergite VII usually entirely dark. Venter clothed with grayish-white scales. *Legs:* Femora dark brown, posterior surface pale. Tibiae dark, streaked with pale scales. Tarsi dark, first segment streaked with pale scales. *Wing:* Length about 3.0 to 3.5 mm. Scales narrow, brown.

ADULT MALE. Coloration similar to that of the female. *Terminalia* (fig. 207). Lobes of ninth tergite (IXT-L) small, a little longer

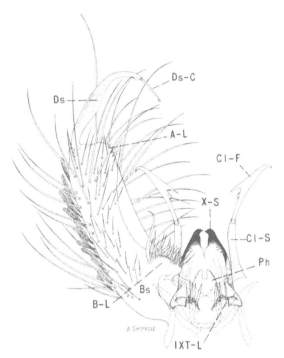

Fig. 207. Male terminalia of *Aedes tortilis* (Theobald).

preantennal 7 multiple, weakly barbed. Thorax sparsely spiculate. Prothoracic hairs: 1 medium, single; 2–4 short, single; 5 long, single; 6 medium, single; 7 long, usually double or triple. Lateral abdominal hair 6 usually double on segments I and II, single on III–VI. Comb of eighth segment with many scales in a small patch; individual scale with the apical spine a little longer and stronger than the subapical spines. Siphonal index 2.5 to 3.0; pecten of about 12 to 15 dark teeth reaching near middle of siphon; siphonal tuft multiple, inserted beyond pecten, individual hairs shorter than the basal diameter of the siphon; dorsal preapical spine much shorter than the apical pecten tooth. Anal segment ringed by the saddle; lateral hair single, shorter than the saddle; dorsal brush bilaterally consisting of a long lower caudal

than wide, each bearing several stout spines. Tenth sternite (X-S) sclerotized and hooked at apex. Phallosome (Ph) short, conical, rounded and notched at apex, open ventrally, closed dorsally in median area. Claspette stem (Cl-S) slender, pilose to near tip, reaching to distal margin of basal lobe; claspette filament (Cl-F) nearly as long as the stem, expanded near middle on convex side and bears a large retrorse projection which has one or more small accessory retrorse spines, tapered and recurved apically. Basistyle (Bs) about three times as long as mid-width, clothed with long and short setae and many scales; basal lobes (B-L) small, triangular-shaped, bearing a strong dorsal spine followed by a row of fine setae on basal part; apical lobe (A-L) rather small, rounded, bearing a few minute setae. Dististyle (Ds) about two-thirds as long as the basistyle, slightly expanded medially, pilose; claw (Ds-C) about one-sixth as long as the dististyle.

LARVA (fig. 208). Antenna much shorter than the head, spiculate; antennal tuft multiple, inserted near middle of shaft. Head hairs: Postclypeal 4 small, multiple; upper frontal 5 and lower frontal 6 single, weakly barbed;

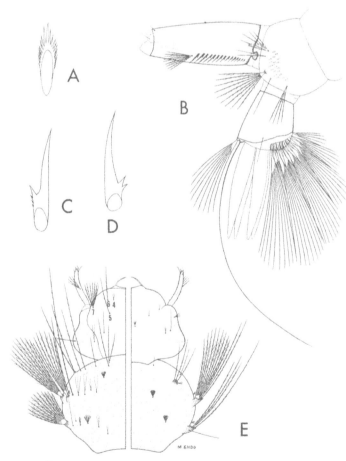

Fig. 208. Larva of *Aedes tortilis* (Theobald). *A*, Comb scale. *B*, Terminal segments. *C*, and *D*, Pecten teeth. *E*, Head and thorax (dorsal and ventral).

hair and a much shorter multiple upper caudal tuft; ventral brush well developed, confined to the barred area or with one precratal tuft; gills usually about one and a half times as long as the saddle.

DISTRIBUTION. This species is known to occur in the Bahamas, the Virgin Islands, and the Greater Antilles. A female was captured in a light trap in 1945 at Key West, Florida (669). Additional female specimens have been reported by Pritchard *et al.* (562) from Vaca Key, Cudjoe Key, and Stock Island, Florida, and one male from Clewiston, Florida, in the everglades.

BIONOMICS. According to Dyar (237), the larvae occur in ground pools. Tulloch (700) took females at sea level and at higher altitudes to 3,000 feet in Puerto Rico. The adult throws the hind legs over the head when resting, much as the Sabethines do. According to Hill and Hill (361), the bite is very irritating. The widespread occurrence of this species in the Keys and southern Florida leads us to believe that it is endemic.

Aedes (Ochlerotatus) trichurus (Dyar)

Culex trichurus Dyar, 1904, Jour. N.Y. Ent. Soc., *12*:170.

Culex cinereoborealis Felt and Young, 1904, Science, n.s., *20*:312.

Aedes pagetonotum Dyar and Knab, 1909, Smithson. Misc. Coll., *52*:253.

Aedes poliochros Dyar, 1919, Ins. Ins. Mens., *7*:35.

ADULT FEMALE (pl. 91). Medium-sized species. *Head:* Proboscis dark-scaled; palpi, short, dark. Occiput with narrow curved white scales and both pale and dark erect forked scales on median area; lateral area with broad appressed grayish scales enclosing a dark patch. Tori brown on outer surface, inner surface darker and with a patch of white scales. *Thorax:* Integument of scutum dark brown; scutum with a median broad brown stripe widening behind the middle; anterior margin, sides, and prescutellar space with grayish-white scales. In some specimens the entire dorsum is gray. Posterior pronotum with curved grayish-white scales. Scutellum with grayish-white scales and dark setae on the lobes. Pleura clothed

with moderately broad grayish-white scales in poorly defined patches. Scales on sternopleuron reaching the anterior angle, not separate from patch on the prealar area. Mesepimeron with scales extending to near lower margin. Hypostigial spot of scales present. Lower mesepimeral bristles three to six. *Abdomen:* First tergite with a median patch of white scales; remaining tergites dark brown with basal white bands widening at the sides. Venter mostly pale-scaled, apical part of segments dark. *Legs:* Femora dark-scaled, lightly sprinkled with white, posterior surface pale. Tibiae dark, lightly sprinkled with pale scales. Tarsi dark. *Wing:* Length about 4.4 to 4.7 mm. Scales narrow, brown, a patch of few to many white scales on base of costa.

ADULT MALE. Coloration similar to that of the female. *Terminalia* (fig. 209). Lobes of ninth tergite (IXT-L) bluntly rounded, each bearing numerous setae; a small rounded projection between the lobes. Tenth sternite (X-S) heavily sclerotized apically. Phallosome (Ph) long, cylindrical, open ventrally, closed dorsally, fringed at apex. Claspette stem (Cl-S) long, curved, reaching well beyond basal lobe, pilose on basal half, somewhat expanded near

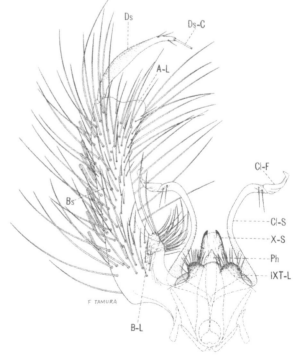

Fig. 209. Male terminalia of *Aedes trichurus* (Dyar).

apex and bearing two setae; claspette filament (Cl-F) very short, not more than one-fourth to one-fifth as long as the stem, striated, expanded from near base, pointed at tip. Basistyle (Bs) about three and a half to four times as long as wide, clothed with scales and long and short setae; basal lobe (B-L) with two or three rather strong long setae on dorsal surface and with a ventral point bearing many fine shorter setae; apical lobe (A-L) prominent, rounded, bearing a few setae. Dististyle (Ds) about half as long as the basistyle, expanded medially, pilose, bearing three or four small setae before apex; claw (Ds-C) slender, about one-sixth as long as the dististyle.

LARVA (fig. 210). Antenna less than half as long as the head, sparsely spiculate; antennal tuft multiple, inserted before middle of shaft, reaching near tip. Head hairs: Postclypeal 4 small, branched; upper frontal 5 double or triple; lower frontal 6 single, reaching beyond preclypeus; preantennal 7 usually 5- or 6-branched, sparsely barbed, reaching beyond insertion of antennal tuft. Prothoracic hairs: 1 long, single; 2 short, single; 3 short, double or multiple; 4 short, single; 5 long, usually triple; 6 long, single; 7 long, about 5- or 6-branched. Lateral abdominal hair 6 usually triple on segment I, double on II and III, and single on IV–VI. Comb of eighth segment with about 14 to 16 scales in a double row; individual scale thorn-shaped, with smaller lateral spinules on basal part. Siphonal index about 3.0; pecten of about 16 to 22 teeth reaching distal fourth of siphon, the distal 4 or 5 teeth detached; siphonal tuft large, multiple, barbed, inserted within the pecten near basal third of siphon; three or four small lateral tufts inserted above the pecten; four or five 2- to 4-branched subdorsal tufts present; dorsal preapical spine much shorter than the apical pecten tooth. Anal segment with the saddle extending about seven-eighths down the sides; lateral hair single, shorter than the saddle; dorsal brush bilaterally consisting of a long lower caudal hair and a shorter multiple upper caudal tuft; ventral brush well developed, with about three precratal tufts; gills longer than the saddle, pointed.

DISTRIBUTION. Known from the forested regions of the northern United States and south-ern Canada. *United States:* Connecticut (270, 494); Idaho (672); Maine (45); Massachusetts (267, 270); Michigan (391); Minnesota (533); Montana (213, 475); New Hampshire (66); New York (31); Rhode Island (436); Vermont (270); Washington (672); Wisconsin (171). *Canada:* Alberta (598, 708); British Columbia (352, 708); Manitoba (212, 598); New Brunswick (708); Nova Scotia (309, 708); Ontario (213, 311); Prince Edward Island (311, 710); Quebec (212, 304); Saskatchewan (310, 598).

BIONOMICS. The larvae of *Aedes trichurus* are found in early spring pools formed by snow and rain in the northern swamps. Owen

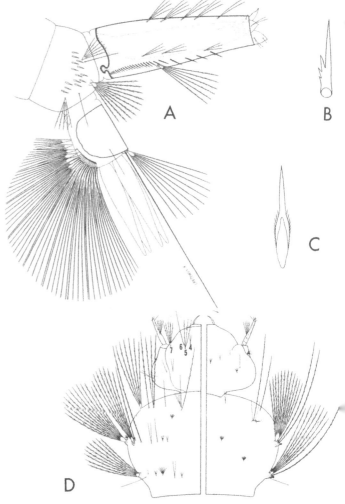

Fig. 210. Larva of *Aedes trichurus* (Dyar). *A,* Terminal segments. *B,* Pecten tooth. *C,* Comb scale. *D,* Head and thorax (dorsal and ventral).

(533) took the larvae from partly shaded pools in a swamp and from an open bog with moss in Minnesota. Mail (475) says the larvae are found in melting snow pools in, and at the edges of, the forests in Montana.

According to Owen (533) the females feed readily when encountered in the northern forests of Minnesota but are less voracious than *A. communis* with which they are frequently associated. Dyar (222) gives an account of the swarming of this species at Warroad, Minnesota, where he observed several small swarms of males, numbering about 50 each, drifting from the woods at dusk, out over a meadow, high enough from the ground to be out of reach, except in one case where he was able to capture a few in a net.

Aedes (Ochlerotatus) trivittatus
(Coquillett)

Culex trivittatus Coquillett, 1902, Jour. N.Y. Ent. Soc., *10*:193.

Culex inconspicuus Grossbeck, 1904, Ent. News, *15*: 333.

ADULT FEMALE (pl. 92). Medium-sized species. *Head:* Proboscis dark-scaled; palpi short, dark. Median area of occiput with white lanceolate scales and pale erect forked scales; occiput with a few dark erect forked scales and numerous broad appressed dingy-white scales surrounding a small patch of broad brown scales laterally. Tori brown, with small setae on inner surface. *Thorax:* Integument of scutum dark brown; scutum with a pair of conspicuous submedian stripes of narrow white to yellowish-white scales separated by a median brown stripe of about equal width; sides with bronze-brown scales; anterior margin and prescutellar space with yellowish-white scales. Posterior pronotum with narrow bronze-brown scales, becoming broader and grayish-white ventrally. Scutellum with yellowish-white lanceolate scales and brown setae on the lobes. Pleura with medium-sized rather distinct patches of appressed grayish-white scales. Scales on sternopleuron usually not extending to the anterior angle, distinctly separate from patch on prealar area. Mesepimeron bare on lower one-third. Hypostigial spot of scales absent. Lower mesepimeral bristles absent. *Abdomen:* First tergite dark-scaled; remaining tergites dark, with basolateral patches of white scales, often with small median basal patches of white on some of the tergites. Venter white-scaled. *Legs:* Legs dark-scaled; posterior surface of femora, tibiae, and first tarsal segment pale. *Wing:* Length 3.5 to 4.0 mm. Scales narrow, dark brown.

ADULT MALE. Coloration similar to that of the female, except that the abdomen usually has narrow white basal bands dorsally, at least on some segments. *Terminalia* (fig. 211). Lobes of ninth tergite (IXT-L) about as long as broad, each bearing four to six short stout spines. Tenth sternite (X-S) sclerotized beyond middle. Phallosome (Ph) stoutly conical, longer than broad, rounded and notched apically, open ventrally, closed dorsally. Claspette stem (Cl-S) pilose, extending to distal part of basal lobe or slightly beyond; claspette filament (Cl-F) nearly as long as the stem, expanded bladelike at basal third and bearing a large sharp retrorse projection, occasionally one or more minute accessory retrorse spines present, tapered and curved apically. Basistyle (Bs) about three and a half to four times as long as mid-width, clothed with scales and

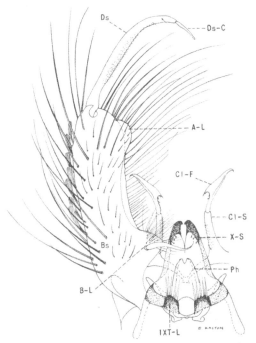

Fig. 211. Male terminalia of *Aedes trivittatus* (Coquillett).

long and short setae; basal lobe (B-L) prominent, bluntly conical, bearing numerous short setae apically and a large curved dorsal spine associated with a group of long fine setae near base; apical lobe (A-L) short, rounded, bearing a few short setae. Dististyle (Ds) about two-thirds as long as the basistyle, broader medially, pilose; claw (Ds-C) slender, about one-fifth as long as the dististyle.

LARVA (fig. 212). Antenna less than half as long as the head, sparsely spiculate; antennal tuft multiple, inserted near middle of shaft, not reaching tip. Head hairs: Postclypeal 4 small, branched; upper frontal 5 and lower frontal 6 single, sparsely barbed; preantennal 7 about 4- to 9-branched, sparsely barbed. Prothoracic hairs: 1 medium, single; 2 short,

single; 3 short, usually single; 4 short, single; 5–6 long, single; 7 long, usually triple. Body spiculate. Lateral abdominal hair 6 usually double on segments I and II, single on III–V. Comb of eighth segment with about 17 to 26 scales in a patch; individual scale thorn-shaped, with the median spine about twice as broad and generally about one and one-third times as long as the subapical spines. Siphonal index about 2.0; pecten of about 14 to 19 dark evenly spaced teeth, reaching beyond middle of siphon; siphonal tuft usually 4- to 9-branched, inserted beyond pecten, shorter than basal diameter of the siphon; dorsal preapical spine shorter than the apical pecten tooth. Anal segment ringed by the saddle; lateral hair small, single, nearly as long as the saddle; dorsal brush bilaterally consisting of a long lower caudal hair and a shorter multiple upper caudal tuft; ventral brush large, confined to the barred area; gills about two and a half to three times as long as the saddle, pointed.

DISTRIBUTION. Southern Canada and the eastern United States. Its range is known to extend west to Idaho and New Mexico. *United States:* Arkansas (88, 122); Colorado (343); Connecticut (270, 494); Delaware (167); District of Columbia (318); Georgia (419, 608); Idaho (344); Illinois (615); Indiana (152, 345); Iowa (625, 626); Kansas (14, 530); Kentucky (570); Louisiana (237); Maine (45, 270); Maryland (161, 213); Massachusetts (213, 267); Minnesota (213, 533); Missouri (6, 14); Montana (475); Nebraska (14, 680); New Jersey (213, 350); New Mexico (213, 273); New York (31); North Carolina (420); North Dakota (553); Ohio (2, 731a); Oklahoma (628); Pennsylvania (298, 756); Rhode Island (436); South Carolina (84, 742); South Dakota (14); Tennessee (420); Texas (463, 634); Virginia (176, 213); West Virginia (237); Wisconsin (171); Wyoming (578). *Canada:* Nova Scotia (309, 708); Ontario (708).

BIONOMICS. The larvae of *Aedes trivittatus* may be found any time during the summer following rains, in floodwater pools in meadows, swamps, and woodlands. The senior author found the larvae of this species to be about as numerous as those of *A. vexans* in the Passaic meadows in New Jersey following a summer

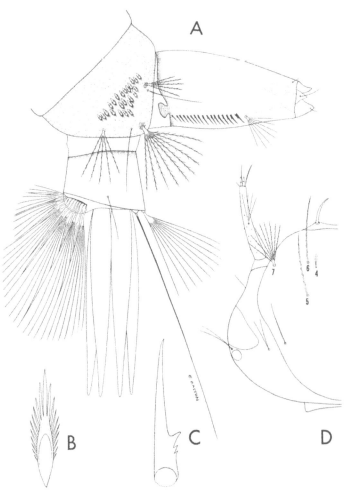

Fig. 212. Larva of *Aedes trivittatus* (Coquillett). *A*, Terminal segments. *B*, Comb scale. *C*, Pecten tooth. *D*, Head.

flood in 1947. The females are persistent biters, attacking at dusk and during the daytime, even in bright sunshine, when the vegetation in which they hide is disturbed. The bite of this mosquito is very painful. The females normally leave their hiding places in the vegetation at dusk. It is particularly a mosquito of open or lightly wooded areas and is seldom found in large numbers in dense woodlands. The adults do not seem to move far from their breeding places. Abdel-Malek (2, 3) gives excellent acounts of the biology and morphology of the immature stages of this species.

MEDICAL IMPORTANCE. According to Hammon *et al.* (340), Dr. Carl Eklund of the Rocky Mountain Spotted Fever Laboratory has recently isolated a virus, closely related but probably not identical to California encephalitis virus, from *A. trivittatus* of North Dakota.

Aedes (Ochlerotatus) ventrovittis Dyar

Aedes ventrovittis Dyar, 1916, Ins. Ins. Mens., 4:84.
Aedes fisheri Dyar, 1917, Ins. Ins. Mens., 5:19.

ADULT FEMALE (pl. 93). A small to medium-sized species. *Head:* Proboscis black-scaled; palpi short, dark. Occiput with narrow curved yellowish-white scales dorsally, with appressed white scales surrounding a dark patch laterally; erect forked scales pale on median area, those on submedian area dark. Tori dark brown or black, with narrow white scales on inner surface. *Thorax:* Integument of scutum black; scutum with narrow brown scales suggesting a broad median stripe; the anterior and lateral margins and prescutellar space with yellowish scales. Posterior pronotum with narrow curved brownish scales becoming yellowish white ventrally. Scutellum with narrow yellowish scales and pale setae on the lobes. Pleura with medium-sized patches of broad grayish-white scales. Scales on sternopleuron usually extending to near the anterior angle, narrowly separate from patch on the prealar area. Mesepimeron with lower one-third bare. Hypostigial spot of scales absent. Lower mesepimeral bristles absent. *Abdomen:* First tergite with white scales; remaining tergites metallic brownish black, with narrow basal white bands which widen laterally to form triangular patches (bands may be absent medially on

some segments). Venter white-scaled or with black scales apically on the segments. *Legs:* Femora dark, sprinkled with white scales, posterior surface pale; knee spots white. Tibiae dark, sprinkled with white. Tarsi dark, first segment streaked with pale scales. *Wing:* Length about 3.0 to 3.5 mm. Scales narrow, dark, with or without intermixed white scales extending outward from bases of veins.

ADULT MALE. Coloration similar to that of the female. *Terminalia* (fig. 213). Similar to *A. sticticus.* Lobes of ninth tergite (IXT-L) sclerotized, about as long as broad, each bearing about four to six short stout spines. Tenth sternite (X-S) heavily sclerotized apically. Phallosome (Ph) lightly sclerotized, conical, much longer than broad, open ventrally, closed dorsally, rounded and notched at apex. Claspette stem (Cl-S) pilose on basal half, extending to near distal margin of basal lobe, bearing one or two short setae before apex; claspette filament (Cl-F) shorter than stem, sclerotized, broadly expanded bladelike at basal third, tapered and recurved at tip, with a small notch near base on concave side. Basistyle (Bs) about three times as long as mid-width, clothed with

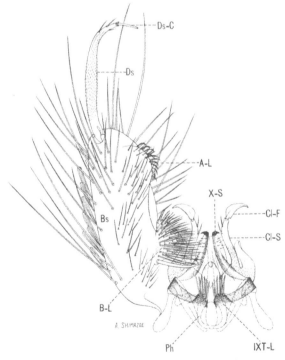

Fig. 213. Male terminalia of *Aedes ventrovittis* Dyar.

scales and long and short setae; basal lobe (B-L) large, semidetached at base, rounded apically, bearing many setae on apical part and a tuft of longer setae on basal part, a large stout recurved dorsal spine arising from near base; apical lobe (A-L) broadly rounded, bearing many short flat retrorse setae. Dististyle (Ds) about three-fifths as long as the basistyle, pilose, swollen medially, bearing two or three short setae before apex; claw (Ds-C) slender, about one-fifth as long as the dististyle.

LARVA (fig. 214). Antenna about half as long as the head, spiculate; antennal tuft multiple, inserted near middle of shaft. Head hairs: Post-clypeal 4 minute, branched distally; upper frontal 5 and lower frontal 6 single (hair 5 rarely double), weakly and sparsely barbed; preantennal 7 multiple, reaching insertion of antennal tuft. Prothoracic hairs: 1 long, single; 2 short, single; 3 short, usually double or triple; 4 short, single; 5 long, double or triple; 6 long, single or double; 7 long, usually triple. Lateral abdominal hair 6 usually multiple on segments I and II, double or triple on III, double on IV, single or double on V, single on VI. Comb of eighth segment with about 8 to 12 scales in an irregular single or partly double row; individual scale thorn-shaped with a long slender median spine, and minute lateral spinules on basal part. Siphonal index 2.5 to 3.0; pecten of about 11 to 18 teeth on basal half or three-fifths of siphon, with 1 to 4 distal teeth detached; siphonal tuft about 3- to 6-branched, barbed, inserted beyond the pecten, shorter than the basal diameter of the siphon; dorsal preapical spine much shorter than the apical pecten tooth. Anal segment with the saddle extending about seven-eighths down the sides; lateral hair single, shorter than the saddle; dorsal brush bilaterally consisting of a long lower caudal hair and a shorter multiple upper caudal tuft; ventral brush well developed, with one to three precratal tufts; gills about two and a half to three times as long as the saddle, pointed.

DISTRIBUTION. Known from the mountains of the western United States. *United States:* California (284, 287); Idaho (672); Washington (213, 672).

BIONOMICS. The larvae of *Aedes ventrovittis* are found in shallow, clear pools formed by melting snow in the High Sierra of California at elevations of about 6,000 to 11,000 feet. The larvae have been found in pools with frozen edges and surrounded by snowbanks, where their development is rapid. Bohart (71) found the larvae in open shallow pools, associated with larvae of *A. hexodontus*, in water which was warm to the touch. The adults are among the first mosquitoes to follow the melting of the snow in the meadows. The females are persistent biters and annoying to man in mountain meadows and in woods.

Freeborn (284) describes swarms of *A. ventrovittis* encountered in the High Sierra meadows at Yosemite National Park, California, in 1920, when the females were so abundant that white horses literally became black with them, and as many as 30 or 40 could be seen sucking

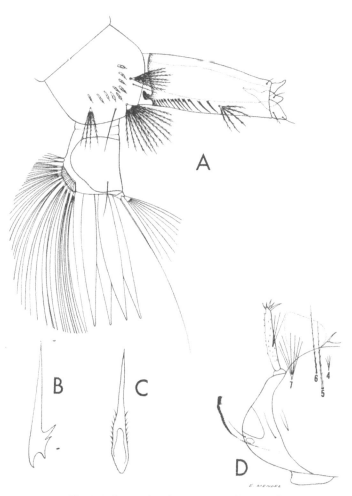

Fig. 214. Larva of *Aedes ventrovittis* Dyar. *A*, Terminal segments. *B*, Pecten tooth. *C*, Comb scale. *D*, Head.

blood on the back of one hand at the same time. During the following year these mosquitoes were encountered in numbers about equal to the preceding year. At that time the meadows were entirely under snow, which was from 10 to 15 feet deep in some places. Bohart (71) observed the males in singing swarms of several hundred individuals overhead at noon in July in Alpine County, California.

Aedes (Finlaya) atropalpus (Coquillett)

Culex atropalpus Coquillett, 1902, Can. Ent., *34*:292.
Aedes epactius Dyar and Knab, 1908, Proc. U.S. Nat. Mus., *35*:53.
Aedes perichares Dyar, 1921, Ins. Ins. Mens., *9*:36.

Dyar (237) recognized two races, *A. atropalpus atropalpus* from the eastern states and *A. atropalpus epactius* from Arizona and Mexico. The pale scales on the scutum are yellowish golden in specimens from the eastern states. The broad dark median stripe is irregular in shape in specimens from Arkansas and Texas, and the pale scales of the scutum are lighter in color.

ADULT FEMALE (pl. 94). Small- to medium-sized species. *Head:* Proboscis dark-scaled; palpi short, dark. Occiput with a median patch of yellowish-white lanceolate scales, bounded submedially by a large area of broad appressed white scales extending laterally down the sides and enclosing a lateral patch of broad dark scales; erect forked scales on central part of occiput pale, those on submedian area dark. Tori light brown to black, usually with a sprinkling of white scales on inner surface. *Thorax:* Integument of scutum black; scutum with a broad median stripe of fine dark bronze-brown scales; this dark stripe becomes much broader posteriorly behind the scutal angle; scutum with yellowish-white or light-golden scales anteriorly, submedially, and laterally (color pattern somewhat variable). Prescutellar space margined with pale-golden scales. Posterior pronotum with narrow dark scales dorsally, with broader pale scales ventrally. Scutellum with narrow yellow scales and dark setae on the lobes. Pleura with small well-defined patches of broad flat silver-white scales. Scales on sternopleuron usually reaching a little more than halfway to anterior

angle, well separated from patch on prealar area. Mesepimeron bare on basal one-third. Hypostigial spot of scales absent. Lower mesepimeral bristles absent. *Abdomen:* First tergite dark-scaled, with a few white scales intermixed; remaining tergites dark, with basal white bands. Venter white-scaled basally, speckled or entirely covered with dark scales apically. *Legs:* Front femur dark, posterior surface pale; middle and hind femora pale on basal half, dark on distal half; knee spots white. Tibiae dark, tipped with white basally and apically. Hind tarsi with basal and apical rings on segments 1 to 4, segment 5 almost entirely white; front and middle tarsi similarly marked, but with white rings narrower on segments 1 and 2; greatly reduced or lacking on 3 and 4, segment 5 entirely dark. *Wing:* Length about 3.0 to 3.5 mm. Scales narrow, all dark except for a patch of white scales on base of costa.

ADULT MALE. Coloration similar to that of the female. *Terminalia* (fig. 215). Lobes of

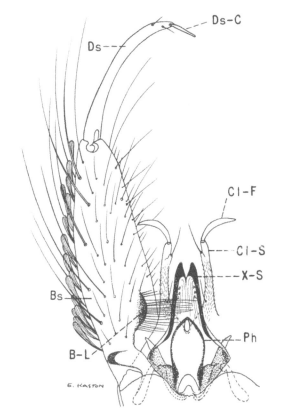

Fig. 215. Male terminalia of *Aedes atropalpus* (Coquillett).

ninth tergite inconspicuous, without spines or setae. Tenth sternite (X-S) heavily sclerotized beyond middle. Phallosome (Ph) longer than broad, slightly expanded on distal third and rounded apically, open ventrally, closed dorsally. Claspette stem (Cl-S) slender, extending a little beyond basal third of basistyle, pilose to near apex, and with a seta arising from a prominent tubercle before apex; claspette filament (Cl-F) shorter than the stem, slender, sickle-shaped. Basistyle (Bs) about three to three and a half times as long as mid-width, clothed with long and short setae and many large scales; basal lobe (B-L) represented by a small slightly raised darkly pigmented area covered by a dense patch of setae; apical lobe absent. Dististyle (Ds) about two-thirds as long as the basistyle, not expanded medially, glabrous except for two or three small setae near apex; claw (Ds-C) slender, about one-fifth as long as the basistyle.

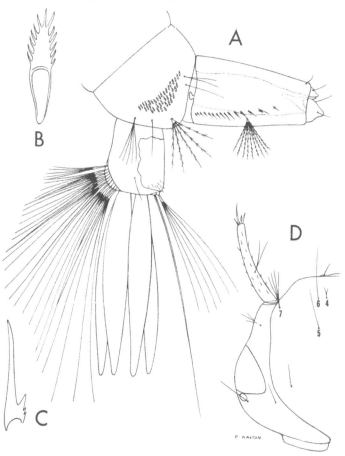

Fig. 216. Larva of *Aedes atropalpus* (Coquillett). *A*, Terminal segments. *B*, Comb scale. *C*, Pecten tooth. *D*, Head.

LARVA (fig. 216). Antenna nearly half as long as the head, sparsely spiculate; antennal tuft small, double or triple, inserted near middle of shaft. Head hairs: Postclypeal 4 small, branched; upper frontal 5 and lower frontal 6 single; preantennal 7, 2- to 5-branched, barbed. Prothoracic hairs: 1 long, usually double, rarely single; 2 short, single; 3 short, single to triple; 4 short, single; 5 long, single to triple; 6 long, single, rarely double; 7 long, 2- to 4-branched. Lateral abdominal hair 6 usually multiple on segments I–V, double on VI. Comb of eighth segment with about 21 to 58 scales in a patch; individual scale fringed apically with subequal spinules. Siphonal index about 1.6 to 2.0; pecten of about 12 to 24 teeth extending nearly to tip of siphon, the last 2 to 4 teeth detached; siphonal tuft about 4- to 9-branched, barbed, inserted within the pecten. Anal segment with the saddle extending nearly halfway down the sides, saddle spiculate posteriorly; lateral hair single or double, much shorter than the saddle; dorsal brush bilaterally consisting of a long lower caudal hair and a shorter multiple upper caudal tuft; ventral brush well developed, confined to the barred area; gills usually two and a half to four times as long as the saddle, pointed.

DISTRIBUTION. Known from the eastern United States west to New Mexico, from southern Canada, Mexico (106, 382, 481), and El Salvador (447). *United States:* Arizona (213); Arkansas (121); Connecticut (494); District of Columbia (213, 318); Georgia (130, 677); Kansas (14); Maine (45, 270); Maryland (130, 213); Massachusetts (213, 267); Minnesota (533); Missouri (324); New Hampshire (66, 213); New Jersey (350, 643); New Mexico (20, 213); New York (31, 130); North Carolina (130, 646); Oklahoma (628); Pennsylvania (213); Rhode Island (270, 436); South Carolina (130); Tennessee (656, 658); Texas (463, 634); Vermont (270, 404); Virginia (130, 213); West Virginia (130); Wisconsin (171). *Canada:* Labrador (289); Ontario (307, 708); Quebec (128).

BIONOMICS. The larvae may be found throughout the summer in overflow pools in rockholes along mountain streams and occasionally in rain-filled rockholes well removed from the stream. Several broods are produced each year in warmer climates, although the larvae are slow in development. The eggs are laid singly above the water level and are apparently attached firmly to the rock, to withstand winter floods. At least some of the eggs hatch within twenty-four hours after flooding in the summer. Larvae were collected by the senior author in Arkansas during February, March, and September, through December. The species reaches its greatest abundance along rocky streams in the Appalachian Mountain region.

The females of *Aedes atropalpus* are rather persistent biters and are frequently annoying near rocky streams. Carpenter (121) found adults, both males and females, on Petit Jean Mountain in Arkansas in 1938, resting during the daytime under rock ledges near their larval habitats, and several females were taken biting during daylight hours in the same area. Owen (533) states that the females feed freely on man in Minnesota when encountered but were never observed far from their larval habitats. Because of the limited number of suitable aquatic habitats for the immature stages, the species is seldom abundant and is troublesome only locally.

Aedes (Finlaya) triseriatus (Say)

Culex triseriatus Say, 1823, Jour. Acad. Nat. Sci. Phila., *3*:12.

Finlaya nigra Ludlow, 1905, Can. Ent., *37*:387.

Aedes triseriatus var. *hendersoni* Cockerell, 1918, Jour. Econ. Ent., *11*:199.

ADULT FEMALE (pl. 95). Medium-sized species. *Head:* Proboscis black; palpi short, black. Occiput with curved silver-white lanceolate scales and white erect forked scales on median area, with broad appressed white scales laterally. Tori brown with fine dark setae on inner surface. *Thorax:* Integument of scutum brown to black; scutum with a wide median stripe of narrow dark-brown scales; the stripe becomes much broader distally and covers most of the posterior half of the scutum; sides with white lanceolate scales; prescutellar space bordered with white scales. Scutellum with dark-brown setae on the lobes; a patch of white scales on the middle lobe. Posterior pronotum clothed with broad appressed silver-white

scales. Pleura with dense, well-defined patches of broad closely appressed silver-white scales. Scales on sternopleuron not extending to anterior angle, distinctly separate from patch on prealar area. Mesepimeron with lower one-fifth bare. Hypostigial spot of scales absent. Lower mesepimeral bristles absent. *Abdomen:* Tergites blue-black-scaled, with basal patches of white scales laterally. Venter white-scaled, with the apices of the terminal segments black. *Legs:* Hind femur yellowish white on basal half to two-thirds, dark apically; front and middle femora black, posterior surface pale; knee spots white. Tibiae and tarsi dark. *Wing:* Length about 3.5 to 4.0 mm. Scales narrow, dark.

ADULT MALE. Coloration similar to that of the female. *Terminalia* (fig. 217). Lobes of ninth tergite (IXT-L) distinct, well separated, each bearing about two to five rather long setae. Tenth sternite (X-S) prominent, sclerotized apically. Phallosome (Ph) stout, cylindrical, truncate apically, open ventrally, closed dorsally. Claspette stem (Cl-S) long, pilose to near tip, with small setae on inner surface; claspette filament (Cl-F) a little longer and narrower than the stem, gradually tapered to a pointed

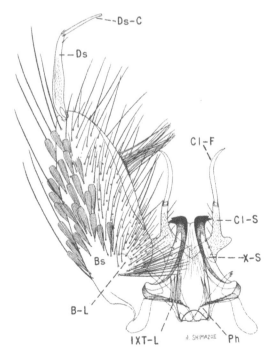

Fig. 217. Male terminalia of *Aedes triseriatus* (Say).

somewhat recurved tip, extending beyond middle of basistyle. Basistyle (Bs) nearly three times as long as mid-width, clothed with large scales and long and short setae; basal lobe (B-L) not well defined, represented by a dense patch of setae near base of basistyle; apical lobe not well defined, but probably represented by a small dense patch of long setae on inner face of basistyle at apical third. Dististyle (Ds) pilose, less than half as long as the basistyle, broader medially; claw (Ds-C) slender, curved, about half as long as the dististyle.

LARVA (fig. 218). Antenna about half as long as the head, smooth; antennal tuft represented by a single hair inserted near middle of shaft. Head hairs: Postclypeal 4 well developed, multiple; upper frontal 5 single (rarely double), longer than hair 6; lower frontal 6, 2- to 4-branched; preantennal 7 short, multiple. Prothoracic hairs: 1 long, usually 2- to 4-branched; 2 medium, single; 3–4 short, 2- to 4-branched; 5 long, usually 2- to 4-branched; 6 long, single; 7 long, usually 2- to 4-branched. Lateral abdominal hair 6 triple, sometimes double on segments I and II, double on III–VI. Comb of eighth segment with about 8 to 15 scales in a single or partly double row; individual scale long, gradually tapered, and evenly fringed with short spinules (scale a little more slender than in *A. zoosophus*). Siphonal index about 2.5 to 3.0; pecten of about 17 to 22 evenly spaced teeth reaching near middle of siphon; siphonal tuft represented by a single or double (occasionally triple) barbed hair, inserted beyond pecten. Anal segment with the saddle extending about two-thirds down the sides; lateral hair represented by a multiple tuft, a little shorter than the saddle, barbed, inserted near ventrolateral margin of saddle; dorsal brush bilaterally consisting of a long lower caudal hair and a shorter multiple upper caudal tuft; ventral brush large, composed of rather sparse tufts and with two or three precratal tufts; gills with the dorsal pair varying from about as long as the saddle to a little longer, the dorsal pair somewhat longer than the ventral pair, stout, bluntly rounded.

DISTRIBUTION. Southern Canada and the eastern United States. Its range is known to extend south to the Florida Keys and west to

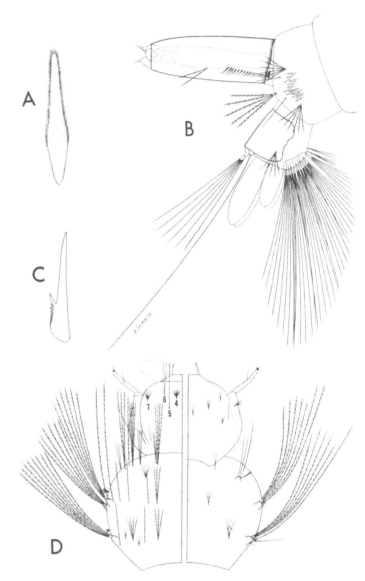

Fig. 218. Larva of *Aedes triseriatus* (Say). *A*, Comb scale. *B*, Terminal segments. *C*, Pecten tooth. *D*, Head and thorax (dorsal and ventral).

Utah and Idaho. *United States:* Alabama (135, 419); Arkansas (88, 121); Colorado (343); Connecticut (494); Delaware (120, 167); District of Columbia (213, 318); Florida and Georgia (135, 213); Idaho (344); Illinois (615); Indiana (152, 345); Iowa (213, 625); Kansas (213, 360); Kentucky (134, 570); Louisiana (213, 759); Maine (270); Maryland (64, 161); Massachusetts (267, 270); Michigan (391, 529); Minnesota (533); Mississippi (135, 546); Missouri (6, 324); Montana (213); Nebraska (14, 680); New Hampshire (66, 213); New Jersey (213, 350); New York (31); North Carolina (135, 213); North Dakota (553); Ohio (482, 731a); Oklahoma (213, 628); Pennsylvania (298,756); Rhode Island (404, 436); South Carolina (135, 742); South Dakota (14); Tennessee (135, 658); Texas (463, 634); Vermont (270); Virginia (176, 178); Wisconsin (171); Wyoming (578). *Canada:* British Columbia (352, 708); Ontario (212, 464); Quebec (704, 708); Saskatchewan (708).

BIONOMICS. This species is the most widely distributed treehole-breeding mosquito in North America. The larvae develop in holes in many kinds of deciduous trees and occasionally in artificial containers such as wooden tubs, barrels, and watering troughs. The winter is passed in the egg stage in the north, and there are several broods each year. The eggs are laid upon the sides of the cavity just above the water line, singly or in groups of two to five, and hatch only when covered with water at favorable temperature. Larvae may be found any time during the year in the extreme southern part of the range, and from about May to September in the northern United States.

The adults are crepuscular and fly mostly during the early morning and evening hours, but probably not much at night since they are rarely if ever taken in night-biting catches. They are occasionally serious pests in residential areas in or near woodlands since they are readily disturbed in the daytime and are rather persistent biters. The bite is quite painful, and the irritation may last for a long time. Repass (599) reports the successful laboratory colonization of this species.

Aedes (Finlaya) varipalpus (Coquillett)

Culex varipalpus Coquillett, 1902, Can. Ent., *34*:292.
Taeniorhynchus sierrensis Ludlow, 1905, Can. Ent., *37*:231.

ADULT FEMALE (pl. 96). Small- to medium-sized species. *Head:* Proboscis dark-scaled; palpi short, dark, with apices of segments white, the terminal segment tipped with white. Occiput with a median stripe of white lanceolate scales and pale erect forked scales, submedian patches of dark-brown to black lanceolate scales behind the eyes; lateral region with dark-brown and white appressed scales. Tori dark brown or black, densely clothed with white scales on inner and dorsal surfaces. *Thorax:* Integument of scutum dull black; scutum with a median stripe of pale-yellow or golden-curved scales extending to middle; narrow posterior curved lines of golden scales present; pale-yellow or golden scales above wing base, on anterior and lateral margins and around the prescutellar space; remainder of scutum with curved dark-brown scales. Posterior pronotum with narrow curved dark-brown scales on dorsal half, with broad white recumbent scales on ventral half. Scutellum with broad white scales and long dark setae on the lobes. Pleura with small distinct dense patches of broad silver-white scales. Scales on sternopleuron extending about one-half to two-thirds way to the anterior angle, widely separated from patch on prealar area. Mesepimeron bare on lower one-third. Hypostigial spot of scales absent. Lower mesepimeral bristles absent. *Abdomen:* First tergite with a median patch of white scales; remaining tergites dark bronze-brown to black-scaled, with basal white bands curved posteriorly, narrowed laterally, and attached or detached from basolateral white patches. Venter white; segments II and III with dark lateral patches, IV and V white basally, dark medially, and with narrow bands of white scales apically. *Legs:* Femora dark, sparsely speckled with white, posterior surface pale; knee spots white. Tibiae dark, narrowly marked with white basally. Tarsi with white bands extending more or less over the joints; front tarsi with segment 1 narrowly ringed with white apically and basally, 2 and sometimes 3 very narrowly marked with white, 4 and 5 usually entirely dark; middle tarsi with segment 1 ringed with white basally and apically, 2 and sometimes 3 narrowly marked with white, 4 and 5 usually entirely dark; hind tarsi with segments 1 to 4 broadly banded with white, segment 5 entirely white. *Wing:* Length about 3.0 to 3.8 mm. Scales narrow, dark, a small patch of white scales at base of costa.

ADULT MALE. Coloration similar to that of the female. *Terminalia* (fig. 219). Lobes of ninth tergite (IXT-L) about as broad as long, each bearing about four to seven spines. Tenth sternite (X-S) heavily sclerotized apically.

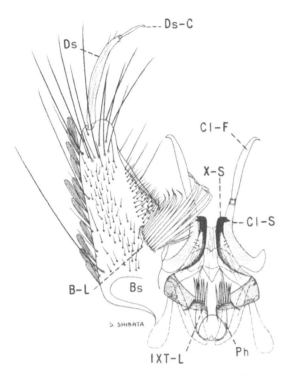

Fig. 219. Male terminalia of *Aedes varipalpus* (Coquillett).

Phallosome (Ph) cylindrical, much longer than wide, open ventrally, closed dorsally, notched at apex. Claspette stem (Cl-S) curved, pilose to near apex; claspette filament (Cl-F) narrow, ligulate, tip curved, about as long as the stem. Basistyle (Bs) about three times as long as mid-width, clothed with scales and long and short setae, the long setae more numerous distally; basal lobe (B-L) represented by a slight elevation bearing many stout apically curved spines basally, followed by short setae which extend to distal fourth of basistyle; apical lobe absent. Dististyle (Ds) a little more than half as long as the basistyle, pilose on inner surface, expanded medially, bearing two or three small setae before apex; claw (Ds-C) slender, about one-fourth as long as the dististyle.

LARVA (fig. 220). Antenna about half as long as the head, smooth; antennal tuft represented by a single hair inserted beyond middle of shaft, reaching near tip. Head hairs: Postclypeal 4 small, multiple; upper frontal 5 single or double; lower frontal 6 usually double, sometimes single; preantennal 7 short,

multiple, weakly barbed. Prothoracic hairs: 1 long, double or triple; 2 medium, single; 3 short, usually double or triple; 4 small, single to triple; 5 long, single or double; 6 long, single; 7 long, usually double. Lateral abdominal hair 6 usually double or triple on segments I–III, usually double on IV and V and single on VI. Comb of eighth segment with about 12 to 28 scales in an irregular double row or patch; individual scale long, gradually tapered, and evenly fringed with short spinules. Siphonal index 3.0 to 3.5; pecten of about 11 to 24 evenly spaced teeth on basal fourth or third of siphon; siphonal tuft triple (sometimes 4- or 5-branched), barbed, inserted beyond pecten before middle of siphon; dorsal preapical spine as long or longer than the apical pecten tooth. Anal segment with the saddle extending halfway down the sides; lateral hair single or double, much longer than the saddle; dorsal brush bilaterally consisting of a long lower caudal hair and a shorter multiple upper caudal tuft; ventral brush sparse, with one to three precratal hairs; gills about three to four and a half times as long as the saddle, each stout, sides parallel, bluntly rounded at apex, the dorsal and ventral pairs of about equal length.

DISTRIBUTION. Western United States and southwestern Canada. *United States:* Arizona (213); California (287); Nevada (213); Oregon (213, 672); Washington (67, 672). *Canada:* British Columbia (352, 708).

BIONOMICS. The female of *Aedes varipalpus*, the western treehole mosquito, deposits her eggs just above the water level where they will be covered by water rising in the tree cavity. Freeborn (284) states that in the fall the females deposit eggs which produce larvae that overwinter; adults emerge from these larvae in May or June. It is believed that the eggs from this early summer generation are responsible for the fall generation; however, some straggling emergence occurs throughout the summer. The larval period is quite long, lasting from one to several months.

The larvae have been collected most commonly in oak trees, particularly live oak, also in rot cavities in California laurel, western cottonwood, sycamore, willow, maple, and eucalyptus. They are occasionally found in

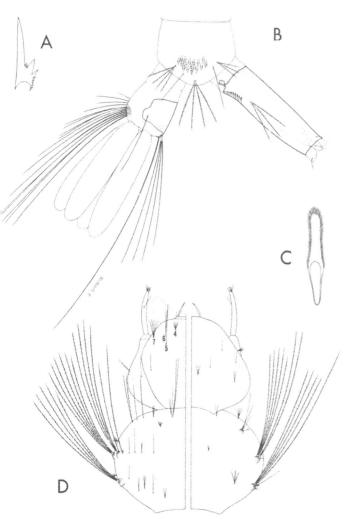

Fig. 220. Larva of *Aedes varipalpus* (Coquillett). *A*, Pecten tooth. *B*, Terminal segments. *C*, Comb scale. *D*, Head and thorax (dorsal and ventral).

artificial receptacles such as cemetery urns, water barrels, and tubs containing leaves or wood debris. The larvae are slow in their motions and can remain at the bottom of the container with their heads buried in the sediment for a long time.

The adults fly in the daytime, usually in wooded areas, and the females feed readily on man and warm-blooded animals. This species is often an important pest in recreation areas and near homes in or near woodlands. The males have been observed to congregate around warm-blooded animals where they remain in flight while awaiting the females that come to

bite. When a female arrives, the male was observed to seize her, and copulation generally took place in flight.

Aedes (Finlaya) zoosophus
Dyar and Knab

Aedes zoosophus Dyar and Knab, 1918, Ins. Ins. Mens., 5:165.
Aedes alleni Turner, 1924, Ins. Ins. Mens., *12*:84.

ADULT FEMALE (pl. 97). Medium-sized species. *Head:* Proboscis black; palpi short, black. Occiput with narrow silver-white scales and yellowish erect forked scales on median area, with broad appressed white scales laterally. Tori brown, inner surface dark brown and with a few fine dark setae. *Thorax:* Integument of scutum dark brown; scutum silver-scaled on anterior half, with a median golden-brown stripe, narrowly divided on some specimens; posterior half dark brown with silvery scales bordering the prescutellar space and forming a narrow line on either side of the prescutellar space. Posterior pronotum with narrow curved yellowish-white to cream-white scales. Scutellum with broad white scales and brown setae on the lobes. Pleura with small dense patches of broad silver-white scales. Scales of sternopleuron not reaching anterior angle, separate from patch on prealar area. Mesepimeron with lower one-fifth to one-fourth bare. Hypostigial spot of scales absent. Lower mesepimeral bristles absent. *Abdomen:* First tergite with a median patch of dark scales; remaining tergites bronze-brown, with narrow basal cream-white bands and silver-white lateral patches. Venter dark, with broad basal white bands. *Legs:* Hind femora with pale scales on all aspects of basal half, distal half brownish black; middle and front femora dark, with pale scales on posterior surface of basal half; knee spots white. Tibiae dark, narrowly marked with white basally. Hind tarsi dark, with broad basal white bands on segments 1 to 4; middle and front tarsi dark, with basal white bands on segments 1 to 3. *Wing:* Length about 3.0 to 3.5 mm. Scales narrow, dark except for a patch of white scales on base of costa, subcosta, and vein 1.

ADULT MALE. Coloration similar to that of the female. *Terminalia* (fig. 221). Lobes of ninth tergite (IXT-L) widely separated, each

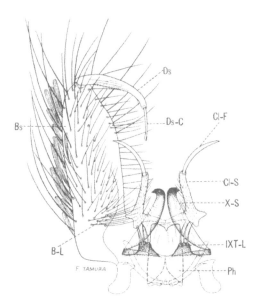

Fig. 221. Male terminalia of *Aedes zoosophus* Dyar and Knab.

clypeal 4 well developed, multiple; upper frontal 5 single (occasionally double), longer than hair 6; lower frontal 6, 2- to 4-branched; pre-antennal 7 multiple, barbed. Prothoracic hairs: 1 medium, 3- to 5-branched; 2 short, single; 3 short, 2- to 4-branched; 4 short, 3- or 4-branched; 5 long, 3- to 6-branched; 6 long, single; 7 long, 3- to 5-branched. Lateral abdominal hair 6 usually multiple on segments I and II and double on segments III–V. Comb of eighth segment with about 6 to 12 scales arranged in an uneven or partly double row; individual scale long, gradually tapered, and evenly fringed with short spinules (usually a little stouter than in *A. triseriatus*). Siphonal index about 2.0 to 2.5; pecten of about 17 to 25 evenly spaced teeth reaching near middle of siphon; siphonal tuft double (occasionally triple), barbed, inserted beyond pecten. Anal segment with the saddle extending two-thirds

bearing three or four spines. Tenth sternite (X-S) heavily sclerotized and toothed apically. Phallosome (Ph) cylindrical, a little longer than wide, truncate apically. Claspette stem (Cl-S) stout, reaching basal third of basistyle, pilose, bearing several small setae on inner surface; claspette filament (Cl-F) ligulate, narrower and a little longer than the stem, gradually tapered to a point. Basistyle (Bs) about three times as long as broad, clothed with scales and long setae on outer aspect, inner aspect rather densely clothed with smaller setae; basal lobe (B-L) poorly defined but represented by an area of rather dense setae near base of basistyle; apical lobe absent. Dististyle (Ds) about two-fifths as long as the basistyle, expanded on basal half, sparsely pilose on inner surface; claw (Ds-C) slender, about half as long as the dististyle.

LARVA (fig. 222). Similar to *A. triseriatus* except for a small depression on either side of the saddle and the position of the lateral hair on the anal segment. The larvae are almost white in color when alive whereas *A. triseriatus* larvae are usually darker. Antenna about half as long as the head, smooth; antennal tuft represented by a single hair inserted a little beyond middle of shaft. Head hairs: Post-

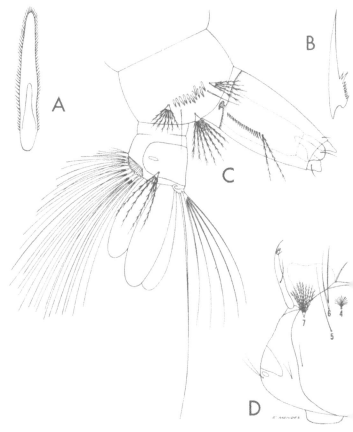

Fig. 222. Larva of *Aedes zoosophus* Dyar and Knab. *A*, Comb scale. *B*, Pecten tooth. *C*, Terminal segments. *D*, Head.

or a little more down the sides; saddle with a light-colored depression on either side near ventral margin; lateral hair longer than saddle, 3- to 7-branched, barbed, inserted near center on posterior border of saddle; dorsal brush bilaterally consisting of a long lower caudal hair and a shorter multiple upper caudal tuft; ventral brush large, with one to three precratal tufts; gills usually one and a half to two times as long as the saddle, stout, tapered, and bluntly rounded, the dorsal pair distinctly longer than the ventral pair.

DISTRIBUTION. Known only from the southwestern United States. *United States:* Kansas (14, 400); Oklahoma (628); Texas (97, 463, 634).

BIONOMICS. The larvae of *Aedes zoosophus* are found in rot cavities of trees, particularly willows, and occasionally in artificial containers. They are often associated with the larvae of *Orthopodomyia signifera* and *A. triseriatus* in their breeding places in Texas. The females readily attack man and are similar to *A. triseriatus* and *A. varipalpus* in this respect. Adults are occasionally captured in light traps.

Aedes (Stegomyia) aegypti (Linnaeus)[17]

Culex aegypti Linnaeus, 1762, Hass. Pal. Reise, p. 470.

ADULT FEMALE (pl. 98). Small- to medium-sized species. *Head:* Proboscis dark-scaled; palpi short, dark, tipped with silver-white scales. Clypeus with broad appressed silver-white scales. Occiput with a rather narrow median stripe of broad white scales, bounded on either side by a large submedian patch of broad dark scales, with broad appressed white scales and a few dark scales laterally; erect forked scales pale on posterior part of occiput; a line of silver-white scales along the margin of the eye. Tori black, bearing silver-white scales. *Thorax:* Integument of scutum dark brown to black; scutum with narrow dark bronze-brown scales except for a small patch of narrow silver-white scales on the anterior margin and a conspicuous lyre-shaped pattern of pale scales on the dorsum. (The outer part of the "lyre" is com-

posed of rather broad silver-white scales; the "strings" of the "lyre" are represented by a pair of narrow submedian lines composed of slender yellowish-white scales.) The prescutellar space is surrounded by white scales. Anterior pronotal lobe and posterior pronotum with broad appressed silver-white scales. Scutellum with broad appressed silver-white scales and brown setae on the lobes. Pleura with small distinct patches of broad flat silver-white scales. Scales on sternopleuron extending about halfway to anterior angle. Prealar area rarely with more than 5 or 6 scales. Mesepimeron with two small patches of scales, lower one-third bare. Hypostigial spot of scales absent. Lower mesepimeral bristles absent. *Abdomen:* First tergite largely pale-scaled; remaining tergites dark-scaled, with narrow basal white bands dorsally and silver-white basal patches laterally. Venter white-scaled, the last two segments predominantly dark. *Legs:* Femora dark, pale on posterior surface on basal half; front and middle femora with a white line extending from the base to near the tip; a similar white line on distal half of hind femora; knee spots white. Tibiae all dark-scaled. Hind tarsi with broad basal white rings on segments 1 to 4, segment 5 white; front and middle tarsi with narrower basal white rings on segments 1 and 2, segments 3 to 5 entirely dark. *Wing:* Length about 2.5 to 3.0 mm. Scales rather narrow, dark.

ADULT MALE. Coloration similar to that of female. *Terminalia* (fig. 223). Lobes of ninth tergite (IXT-L) large, triangular, each bearing a few very small setae at apex. Tenth sternite (X-S) large, moderately sclerotized, bluntly rounded apically, bearing a ventral branch at basal third. Phallosome (Ph) conical, a little more than twice as long as basal width, open ventrally, closed dorsally at apical third; each plate with a row of strong ventral teeth on apical part, those at tip more numerous and forming a crown. Claspette absent; interbasal fold (I-F) prominent, pilose. Basistyle (Bs) short, about one and a half times as long as wide, clothed with large scales and long setae on outer aspect; basal lobe (B-L) represented by a few stout apically bent spines. and a dense patch of setae covering much of the inner surface of the basistyle (this may not

---

[17]Consult Dyar (1928) or Edwards (1932) for synonyms.

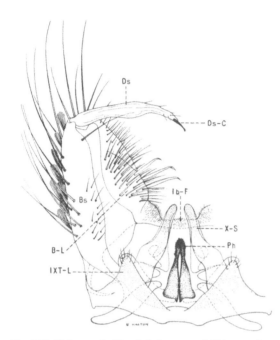

Fig. 223. Male terminalia of *Aedes aegypti* (Linnaeus).

be a true basal lobe); apical lobe absent. Dististyle (Ds) about two-thirds as long as the basistyle, slightly expanded near middle, bearing several small setae before apex; claw (Ds-C) pointed, about one-seventh as long as the dististyle.

LARVA (fig. 224). Antenna a little more than one-third as long as the head, smooth; antennal tuft represented by a single hair inserted near middle of shaft. Head hairs: Postclypeal 4 rather large, 3- to 5-branched; upper frontal 5, lower frontal 6, and preantennal 7 single. Hair 6 placed almost directly in front of hair 5. Prothoracic hairs: 1 medium, 2- to 5-branched; 2 short, single; 3 short, double (occasionally triple); 4 short, single or double; 5 long, usually double; 6 long, single; 7 long, double or triple. Lateral abdominal hair 6 variable, usually double or triple on segments I–V. Comb of eighth segment with 7 to 12 scales in a single curved row; individual scale thorn-shaped, with a strong median spine and several shorter stout lateral spines. Siphonal index about 2.0; pecten of about 10 to 19 evenly spaced teeth reaching middle of siphon; siphonal tuft 2- to 5-branched (usually triple),

inserted beyond pecten. Anal segment with the saddle reaching about seven-eighths down the sides; lateral hair single or double, about as long as the saddle; dorsal brush bilaterally consisting of a long lower caudal hair and a 2- to 4-branched upper caudal tuft; ventral brush composed of about 7 to 10 long double hairs confined to the barred area; gills about three times as long as the saddle, each broad and bluntly rounded at tip.

DISTRIBUTION. This species is found throughout most of the tropical and subtropical regions of the world. *United States:* Alabama (135, 419); Arizona (57); Arkansas (88, 121); District of Columbia (213, 318); Florida and Georgia (135, 419); Illinois (615); Indiana (152, 345); Kansas (14); Kentucky (213, 570); Louisiana (419); Mississippi (135, 546); Missouri (6, 14); New Mexico (20, 273); North Carolina (135, 419); Oklahoma (371, 628); South Carolina (135, 742); Tennessee (135, 658); Texas (463, 571); Virginia (176, 213).

BIONOMICS. The eggs are deposited in artificial containers, either just above the water level or on the surface of the water. Winter is passed in the egg stage; however, breeding may be continuous throughout the year in warmer climates. Johnson (405) has shown that the eggs require ripening for several days before hatching. LeVan (456) demonstrated that eggs will hatch after being held for one year. The immature stages occur mostly in artificial water containers in and near residences in the range; however, the senior author has taken the larvae on a few occasions in the southern United States from rot cavities of shade trees near residences. The larvae are often found in rot cavities of trees in the African jungle far from human habitations.

The females are wary in feeding, often attacking around the ankles. They may even crawl short distances under the clothing to find a favorable spot to feed. Feeding generally takes place in the shade during the daytime, but the females will feed in lighted rooms at night. Human blood seems to be preferred to that of domestic animals. The adults frequently rest inside houses, in closets, cupboards, cabinets, behind doors, and even behind picture frames, and probably never fly more than a few hun-

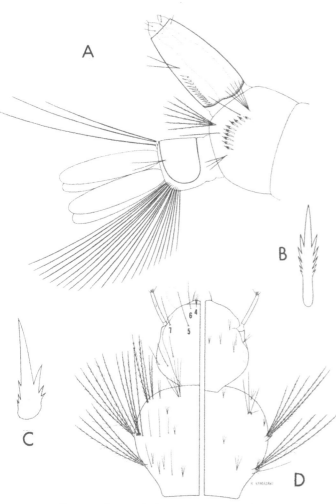

Fig. 224. Larva of *Aedes aegypti* (Linnaeus). *A*, Terminal segments. *B*, Comb scale. *C*, Pecten tooth. *D*, Head and thorax (dorsal and ventral).

dred feet from the water container in which the immature stages developed.

MEDICAL IMPORTANCE. This species is an important vector of yellow fever and dengue.

Aedes (Aedimorphus) vexans (Meigen)

Culex vexans Meigen, 1830, Syst. Beschr. Zweifl. Ins., *6*:241.
? *Culex parvus* Macquart, 1834, Suites a Buffon, *1*: 36.
Culex articulatus Rondani, 1872, Bull. Soc. Ent. Ital., *4*:30.
Culex malariae Grassi, 1898, Atti Acc. Lincei, *7*:168.
Culex sylvestris Theobald, 1901, Mon. Culic., *1*:406.
Culex nocturnus Theobald, 1903, Mon. Culic., *3*:159.
Culex montcalmi Blanchard, 1905, Les Moust., p. 307.

? *Culex arabiensis* Patton, 1905, Jour. Bombay Nat. Hist. Soc., *16*:633.
Culicada nipponii Theobald, 1907, Mon. Culic., *4*: 337.
Culicada minuta Theobald, 1907, Mon. Culic., *4*:338.
? *Culicada eruthrosops* Theobald, 1910, Mon. Culic., *5*:299.
Aedes euochrus Howard, Dyar, and Knab, 1917, Carnegie Inst. Wash. Pub., *159*, *4*:716.

ADULT FEMALE (pl. 99). Medium-sized species. *Head:* Proboscis dark-scaled, lightly sprinkled with light-brown scales; palpi short, dark, the fourth segment with a few white scales at base and tip. Occiput with narrow pale-yellow to golden-brown scales and dark and pale erect forked scales on median area, with broad whitish or pale-yellow scales laterally; a small submedian dark-scaled patch near margin of eye on either side. Tori light brown or dark brown, with a few grayish-white scales on dorsal and inner surfaces. *Thorax:* Integument of scutum brown; scutum clothed with narrow golden-brown scales, with those on the anterior and posterolateral margins and the prescutellar space paler. Posterior pronotum with curved golden-brown scales. Scutellum with narrow golden scales and light-brown setae on the lobes. Pleura with rather small distinct patches of broad flat grayish-white scales. Scales on sternopleuron reaching about halfway to anterior angle, separate from patch on prealar area. Mesepimeron with a patch of scales on upper part, lower one-half bare. Hypostigial spot of scales absent. Lower mesepimeral bristles absent. *Abdomen:* First tergite with dark and pale scales intermixed on median area; tergites II to VI dark-scaled, each with a conspicuous indented basal white band and basolateral white patches (basal bands usually not joining the lateral patches); white scales present on apical margins of the last three tergites. Venter white-scaled, each segment with dark scales, often in a V-shaped patch, with the base of the V directed anteriorly. *Legs:* Femora dark, with scattered pale scales, posterior surface pale; knee spots pale. Tibiae dark, pale on posterior surface. Hind tarsi dark, with narrow basal white rings on all segments; front and middle tarsi similarly marked, but with white rings on segments 4 and 5 greatly reduced or absent. *Wing:* Length 3.5 to 4.0 mm. Scales narrow, dark.

Fig. 225. Male terminalia of *Aedes vexans* (Meigen).

ADULT MALE. Coloration similar to that of the female. *Terminalia* (fig. 225). Lobes of ninth tergite (IXT-L) indistinct, only slightly elevated, widely separated, each bearing a few weak setae. Tenth sternite (X-S) lightly sclerotized. Phallosome (Ph) small, narrow near base, broader on apical half, consisting of two heavily sclerotized plates bearing strong apical teeth. Claspette stem (Cl-S) stout, pilose, fused to basistyle, rounded apically, and crowned with many slender curved setae; claspette filament absent. Basistyle (Bs) about two and a half times as long as mid-width, clothed with scales and long setae on outer aspect, with long setae on inner ventral margin; basal and apical lobes absent. Dististyle (Ds) broad, flattened, about two-thirds as long as the basistyle, bearing scattered small setae, tip densely pilose

and bluntly pointed; claw (Ds-C) stout, arising from a small projection at apical fifth of the dististyle.

LARVA (fig. 226). Antenna about half as long as the head, spiculate; antennal tuft multiple, barbed, inserted near middle of shaft, nearly reaching tip. Head hairs: Postclypeal 4 small, multiple; upper frontal 5, 3- to 5-branched, barbed; lower frontal 6 double or triple, barbed; preantennal 7 multiple, barbed, reaching beyond insertion of antennal tuft. Prothoracic hairs: 1 medium, single; 2 short, single; 3 short, single to triple; 4 short, single or branched; 5 long, usually single; 6 long, single; 7 long, double or triple. Lateral abdominal hair 6 double or triple on segments I–V. Comb of eighth segment with about 9 to 12 scales in an irregular single or double row; individual scale thorn-shaped, with a long apical spine, and short lateral spinules on basal part. Siphonal index 3.0 to 3.5; pecten of about 14 to 20 dark teeth reaching middle of siphon or beyond, with 1 to 3 distal teeth detached; siphonal tuft small, 3- to 6-branched, inserted beyond pecten. Anal segment with the saddle extending about seven-eighths down the sides; lateral hair single or double, shorter than the saddle; dorsal brush bilaterally consisting of a long lower caudal hair and a shorter multiple upper caudal tuft; ventral brush large, with four or five precratal tufts; gills about two to two and a half times as long as the saddle, pointed.

DISTRIBUTION. This species is found throughout most of the Palearctic, Nearctic, and Oriental regions. It is common in southern Canada and throughout most of the United States, but is less abundant in the extreme south. *United States:* Alabama (135, 419); Arizona (213); Arkansas (88, 121); California (287); Colorado (343); Connecticut (270, 494); Delaware (120, 167); District of Columbia (213, 318); Florida and Georgia (135, 213); Idaho (344, 672); Illinois (615); Indiana (152, 345); Iowa (625, 626); Kansas (14, 360); Kentucky (134, 570); Louisiana (213, 759); Maine (45, 270); Maryland (64, 161); Massachusetts (267, 270); Michigan (391, 529); Minnesota (533, 604); Mississippi (135, 546); Missouri (6, 324); Montana (475); Nebraska (14, 680); New Hampshire (66);

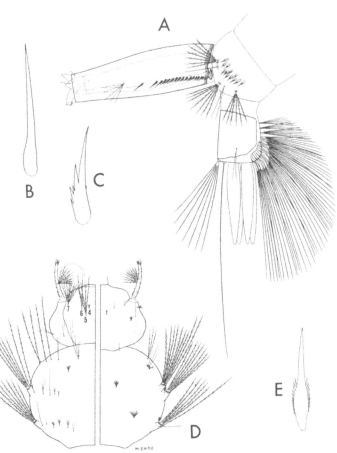

Fig. 226. Larva of *Aedes vexans* (Meigen). *A*, Terminal segments. *B*, and *C*, Pecten teeth. *D*, Head and thorax (dorsal and ventral). *E*, Comb scale.

vexans are found in temporary rain-filled pools and pools formed by floodwater. The larvae are frequently found in irrigation seepage water. Emergence of the adults takes place throughout the year in the extreme southern United States but is more common from May to October farther north. The winter is passed in the egg stage, and there is, according to some observers, only a single generation a year. Hearle (351) observed that all the eggs do not hatch with a single flooding, but that larvae appear periodically after alternate flooding and drying of the eggs during the course of the summer. Freeborn (284) says that "unlike eggs of many *Aedes*, those of *vexans* hatch if moisture is supplied during the same season." Rees (575) says that there is apparently more than one brood produced each year in Utah, and Ross (615) believes the species may have several generations each year.

Aedes vexans is generally abundant and a troublesome biter wherever it occurs. The females will feed in shady places during the day and can be particularly annoying at dusk and after dark. The adults are known to migrate long distances from their breeding places. Rees (575) found that the species has a flight range of 5 to 8 miles in Utah, and Hearle (351) claims that in British Columbia it migrates a distance of ten miles.

New Jersey (213, 350); New Mexico (213, 273); New York (31); North Carolina (135, 213); North Dakota (553); Ohio (213, 731a); Oklahoma (628); Oregon (672); Pennsylvania (298, 756); Rhode Island (35, 270); South Carolina (135, 742); South Dakota (14, 530); Tennessee (135, 658); Texas (463, 634); Utah (575); Vermont (270); Virginia (176, 178); Washington (67, 672); West Virgina (213); Wisconsin (171); Wyoming (578). *Canada:* Alberta (212, 598); British Columbia (352, 708); Manitoba (212, 466); New Brunswick (212, 708); Nova Scotia (309, 708); Ontario (212, 708); Prince Edward Island (710); Quebec (306, 708); Saskatchewan (212, 598); Yukon (708).

BIONOMICS. The immature stages of *Aedes*

Aedes (Aedes) cinereus Meigen

? *Culex ciliaris* Linnaeus, 1767, Syst. Nat., *1*:1002.

Aedes cinereus Meigen, 1818, Syst. Beschr. Zweifl. Ins., *1*:13.

Culex rufus Glimmerthal, 1845, Bull. Soc. Imp. Nat. Moscou, *18*:295.

Culex nigritulus Zetterstedt, 1850, Dipt. Scand., *9*:345.

Aedes fuscus Osten Sacken, 1877, Bull. U.S. Geol. and Geog. Surv. of the Terrt., *3*:191.

Aedes leucopygus Eysell, 1903, Abh. Ver. Nat. Kassel, *48*:285.

Culex pallidohirta Grossbeck, 1905, Can. Ent., *37*:359.

Culex pallidocephala Theobald (error), 1910, Mon. Culic., *5*:612.

Aedes cinereus fuscus Dyar, 1924, Ins. Ins. Mens., *12*:179.

Aedes cinereus hemiteleus Dyar, 1924, Ins. Ins. Mens., *12*:179.

ADULT FEMALE (pl. 100). Medium-sized to rather small species. *Head:* Proboscis dark-scaled; palpi short, dark. Occiput with narrow pale-yellow to light golden-brown scales on median area, bounded on either side by a large submedian patch of broad appressed brown scales; lateral region of occiput with broad yellowish-white scales; erect forked scales on dorsal surface brown. Tori brown, darker on inner surface and bearing small setae. *Thorax:* Integument of scutum brown; scutum evenly clothed with fine narrow reddish-brown scales, paler on anterior margin, on lateral margin above wing base, and on the prescutellar space. Posterior pronotum with narrow curved golden-brown scales dorsally, paler ventrally. Scutellum with narrow pale-golden scales and brown setae on the lobes. Pleura with small well-defined patches of broad grayish-white scales. Scales on sternopleuron extending about halfway to anterior angle, usually separate from patch on the prealar area. Mesepimeron bare on lower one-third. Hypostigial spot of scales absent. Lower mesepimeral bristles absent. *Abdomen:* First tergite with a median patch of brown scales, a few pale scales intermixed; remaining tergites brown, without pale bands or with narrow partial or complete bands; basolateral spots usually joined forming a continuous line. Venter pale-scaled. *Legs:* Femora dark, posterior surface pale; knee spots pale. Tibiae dark, with pale scales on posterior surface. Tarsi dark. *Wing:* Length about 3.2 to 3.8 mm. Scales narrow, brown.

ADULT MALE. Coloration similar to that of the female. Palpi minute, even shorter than those of the female. Antennae extremely bushy. *Terminalia* (fig. 227). Lobes of ninth tergite (IXT-L) about as long as broad, rounded apically, widely separated, each bearing several slender spines arising from distinct tubercles. Tenth sternite, (X-S) heavily sclerotized, slender, rodlike, tip narrow. Phallosome (Ph) heavily sclerotized, closed at base and apex, open dorsally and ventrally; swollen on apical half, rounded apically. Claspette stem (Cl-S) pilose, unequally bifurcate; the shorter inner branch with one or two apical and one or two subapical setae; the longer branch generally with two apical and two to four subapical spines; claspette filament absent. Basistyle (Bs) about twice

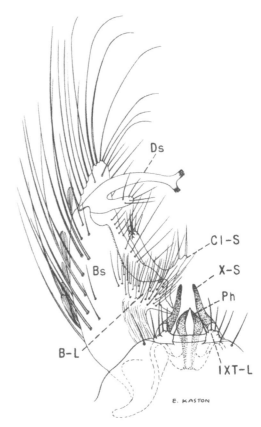

Fig. 227. Male terminalia of *Aedes cinereus* Meigen.

as long as mid-width, cone-shaped, clothed with scales and long setae on outer aspect; basal lobe (B-L) broad at base and with the narrow apical part directed mesad, densely covered with slender setae, those on apex stouter; apical lobe absent. Dististyle (Ds) inserted well before apex of basistyle, unequally bifurcate; the short inner branch bearing several small setae at apex; the main branch bifurcate at apex, with each small branch blunt and crowned with small spines.

LARVA (fig. 228). Antenna more than half as long as the head, slender, spiculate; antennal tuft multiple, inserted before middle of shaft. Head hairs: Postclypeal 4 small, multiple; upper frontal 5, 5- to 9-branched, barbed; lower frontal 6, 4- to 8-branched, barbed; preantennal 7 long, multiple, barbed. Hairs 5, 6, and 7 inserted nearly in a straight line. Prothoracic hairs: 1 medium, single; 2 short, single; 3 short, 2- to 4-branched; 4 short, double or triple; 5 long, usually single; 6 long, single; 7 long, usually double or triple. Lateral abdominal hair 6 usually double on segments I and II, single on segments III–V. Comb of eighth segment with about 9 to 16 scales in a partly double row; individual scale thorn-shaped, with a strong apical spine and small lateral spinules. Siphonal index 4.0 to 4.5; pecten of about 12 to 21 teeth reaching beyond middle of siphon, with 1 to 3 distal teeth detached; siphonal tuft small, with three to five branches, inserted beyond pecten. Anal segment with the saddle extending about three-fourths down the sides; lateral hair usually double (sometimes triple), shorter than the saddle; dorsal brush bilaterally consisting of a long lower caudal hair and a shorter multiple upper caudal tuft; ventral brush well developed, with about three or four precratal tufts; gills with the dorsal pair about two and a half times as long as the saddle, the dorsal pair slightly longer than the ventral pair, pointed.

DISTRIBUTION. United States, Canada, Europe, and Asia. *United States:* Alabama (516); Arkansas (88, 121); California (71, 287); Colorado (343); Connecticut (270, 494); Delaware (167); District of Columbia (318); Florida (136); Georgia (420, 516); Idaho (344, 672); Illinois (615); Indiana (133); Iowa (625, 626); Kansas (14); Maine (45, 270); Maryland (64, 161); Massachusetts (267, 270); Michigan (391, 529); Minnesota (533, 604); Mississippi (136, 546); Missouri (14); Montana (213, 475); Nebraska (14, 680); New Hampshire (66, 270); New Jersey (213, 350); New York (31); North Carolina (514); North Dakota (14); Ohio (482, 731a); Oklahoma (321); Oregon (213, 672); Pennsylvania (298, 756); Rhode Island (270, 436); South Carolina (420); South Dakota (14); Tennessee (514); Utah (575); Vermont (270); Washington (67, 672); Wisconsin (171, 213); Wyoming (578). *Canada:* Alberta (212, 598); British Columbia (352, 598); Labrador (289, 348); Manitoba (212, 711); Northwest Territories (289); Nova Scotia (707, 708); Ontario (212, 708); Prince Edward Island (311, 710); Quebec (304, 708); Saskatchewan (212, 598); Yukon (212, 289). *Alaska:* (293, 396, 673).

BIONOMICS. *Aedes cinereus* is apparently

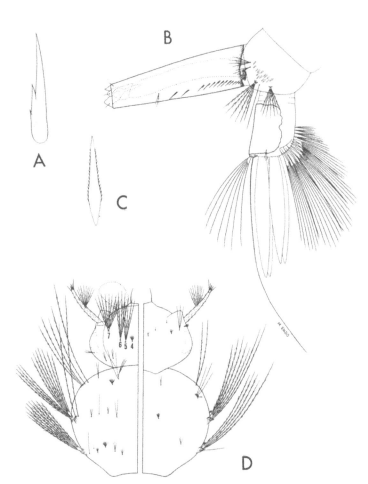

Fig. 228. Larva of *Aedes cinereus* Meigen. *A*, Pecten tooth. *B*, Terminal segments. *C*, Comb scale. *D*, Head and thorax (dorsal and ventral).

single-brooded, overwintering in the egg stage and hatching mostly in the late spring. The larvae are found in woodland pools, unshaded temporary rain-filled pools, and occasionally in marshes. Rees (575) observed that the larvae are found principally along the margins of mountain streams in wooded sections in Utah; however, Jenkins (396) found the larvae in bogs, grass marshes, in emergent vegetation at the margins of lakes, permanent pools, and abandoned gold prospect holes in Alaska at elevations of less than 1,800 feet. Mail (475) states that the first brood hatches from overwintered eggs in snow pools in the woods in Montana, and that later in the season the larvae develop in rain pools but these are also from overwintered eggs. The senior author observed that this was one of the more common species of mosquitoes found in woodland pools in northern New Jersey in late April and May. Hearle (351) found the larvae chiefly in cottonwood flood swamps in the Frazer Valley in British Columbia.

Hearle (351), Owen (533), and Marshall (480) claim that the females are troublesome biters on occasions. Mail (475) found that the females are bad biters in Montana but that they do not always attack when encountered. He says that the species is quite localized, does not fly far from its breeding places, and will not enter houses. Freeborn and Bohart (287) observed that the females seem to prefer walking to flying and have some difficulty in piercing the human skin. The adults are rarely encountered in the southern United States except in light-trap collections; therefore, very little is known of their habits in this region.

Genus CULEX Linnaeus[18]

Culex Linnaeus, 1758, Syst. Nat. Ed., *10*:602.

ADULT. Proboscis of more or less uniform thickness (slightly swollen at tip in subgenus *Melanoconion*). Palpi of male long or short; when longer than the proboscis the last two segments are slender, hairy, and upturned; palpi of female rarely more than one-fifth as long as the proboscis. Antennae of male usually distinctly plumose. Pharyngeal bar of female distinctly toothed. Posterior pronotal bristles present; spiracular and postspiraculars absent. Pleura usually with a few small patches of scales, sometimes bare. Tip of abdomen of female bluntly rounded, the eighth segment not retracted. Squama fringed. A pair of padlike pulvilli at the bases of the tarsal claws. Tarsal claws of female without teeth. *Male terminalia.* Tenth sternite crowned with numerous short spines or a comblike row of teeth. Claspette absent. Phallosome formed of a pair of heavily sclerotized variously divided plates. Basistyle somewhat conical; basal lobe absent; subapical lobe present and bearing modified spines and setae.

PUPA. Respiratory trumpet usually of moderate length, opening small. Dendritic tuft (hair 2) on abdominal segment I well developed. Paddle with smooth margin, without fringe; usually two very short apical hairs (7 and 8) placed side by side or a subapical hair in addition to the apical hair.

LARVA. Head wider than long. Mouth bristles normal and unmodified except in the predacious subgenus *Lutzia*. Antennal tuft inserted beyond middle of shaft of antenna in most species. Comb of eighth segment usually consisting of many scales in a triangular patch; in a single row in some species. Siphon usually long or very long, short in some species; pecten and several pairs of siphonal tufts present. Anal segment completely ringed by the saddle.

EGGS. Usually long, narrowed at one end, without prominent markings on integument. The eggs are fixed together in raftlike masses, except for a few species, and deposited on the surface of the water.

BIONOMICS. The larvae are found chiefly in ground pools of a permanent or semipermanent nature; however, a few species utilize artificial containers, treeholes, bamboo sections, and leaf axils of *Bromeliaceae*. The females feed chiefly at dusk or during the night. Some species seek the blood of mammals; others attack birds; some feed on cold-blooded vertebrates; and the feeding habits of many still remain unknown.

CLASSIFICATION. The genus includes a large number of species, most of which are restricted to the tropical and subtropical regions of the world. Subgenera are largely defined by characters of the male terminalia. Edwards (251) lists the recognized subgenera of *Culex*. North America north of Mexico has comparatively few species belonging to three subgenera, *Culex*, *Melanoconion*, and *Neoculex*.

DISTRIBUTION. Largely restricted to the tropical and subtropical regions of the world.

KEYS TO THE SPECIES
ADULT FEMALES[19]

1. Wing scales narrow on vein 2; occiput usually lacking broad appressed scales dorsally 2
   Wing scales slightly or distinctly broadened on vein 2; occiput usually with broad appressed scales dorsally, sometimes limited to a narrow border behind the eyes (subgenus *Melanoconion*) 19

2 (1). Tarsal segments with rather distinct white rings, particularly on hind legs......3
   Tarsal segments without distinct white rings (when pale rings are present, they are narrow and brownish[20]) 8

3 (2). Proboscis with a contrasting white ring near middle 4
   Proboscis without a complete contrasting white ring near middle......6

4 (3). White tarsal rings on hind legs very narrow *bahamensis* D. and K., p. 274
   White tarsal rings on hind legs rather broad 5

5 (4). Femora and tibiae each with a longitudinal line of white scales or a row of white spots on outer surface; a V-shaped dark marking on venter of each abdominal segment......*tarsalis* Coq., p. 294

---

[18]Consult Dyar (1928) or Edwards (1932) for synonyms.

[19]Adults of some of the dull-colored species cannot be identified with certainty except by characters of the male terminalia.

[20]California specimens of *C. restuans* Theobald sometimes have rather distinct white rings on the tarsi.

Femora and tibiae lacking a longitudinal line of white scales or a row of white spots on outer surface; an oval dark spot on venter of each abdominal segment
stigmatosoma Dyar, p. 292

6 (3). Venter of abdominal segments II to VI with triangular dark markings
thriambus Dyar, p. 296
Venter of abdominal segments II to VI lacking triangular dark markings................7

7 (6). Fifth segment of hind tarsi white basally and apically, narrowly dark at middle
coronator D. and K., p. 277
Fifth segment of hind tarsi dark except for a narrow white basal ring
virgultus Theob., p. 298

8 (2). Abdominal segments each with an apical triangular patch of white scales on either side (at least on the posterior segments), usually joined by a narrow dorsoapical band of similar scales (subgenus *Neoculex*) 9
Abdominal segments with pale scales basal when present................13

9 (8). Abdominal segments II to IV all dark as seen from above......*reevesi* Wirth, p. 320
Abdominal segments II to IV with apical bands or apicolateral spots as seen from above 10

10 (9). Hind femur with pale posterior stripe complete; palpi about twice as long as flagellar segment 4 of antenna................11
Hind femur with pale posterior stripe ending shortly before the apex; palpi about two and a half to three times as long as flagellar segment 4 of antenna............12

11 (10). Pale scales of occiput ashy white; abdominal segment V about one and three-tenths times as broad as long in unengorged dried specimens..*territans* Walker, p. 322
Pale scales of occiput with a yellowish tinge; abdominal segment V about 1.5 to 1.7 times as broad as long in unengorged dried specimens
boharti Brookman and Reeves, p. 319

12 (10). Palpi usually with pale scales at base of terminal segment and apex of subterminal segment; erect forked scales on occiput pale to light brownish
apicalis Adams, p. 315
Palpi all dark; erect forked scales on occiput dark brown
arizonensis Bohart, p. 317

13 (8). Scutum bright reddish-brown, clothed with narrow hairlike golden-brown scales: pleura and coxae reddish brown
erythrothorax Dyar, p. 279
Scutum light brown, brown or dark brown (never reddish brown), clothed with narrow

row curved scales; pleura and coxae never bright reddish brown................14

14 (13). Abdominal segments each with a rather broad basal band of whitish scales dorsally 15
Abdominal segments lacking broad basal whitish band dorsally (pale scaling when present in narrow bands or restricted to lateral patches) 17

15 (14). Abdominal bands broadly rounded on posterior margin and constricted laterally, rather narrowly joining or entirely disconnected from the lateral patches; scales of scutum somewhat coarse, golden
pipiens Linn., p. 284;
quniquefasciatus Say, p. 286
Abdominal bands with posterior margin nearly straight and broadly joining the lateral patches (particularly on segments 3 to 5); scales of scutum fine, golden brown 16

16 (15). Small species (wing length about 2.5 to 2.8 mm.); scutum lacking a pair of small pale-scaled submedian spots near middle
interrogator D. and K., p. 280
Medium-sized species (wing length about 4.0 to 4.4 mm.); scutum usually with a pair of pale-scaled submedian spots near middle................*restuans* Theob., p. 288

17 (14). Pleuron with few or no scales (when present, rarely more than 5 or 6 in a single group)..........*nigripalpus* Theob., p. 288
Pleuron with several groups of broad pale scales, each group usually comprised of more than 6 scales................18

18 (17). Abdominal segments usually with narrow dingy-yellow basal bands dorsally and with the apices of the segments more or less blended with yellowish scales, segment VII either primarily or entirely clothed with dingy-yellow scales
salinarius Coq., p. 290
Abdominal segments with basal bands of white scales, apices of segments dark, segment VII primarily clothed with dark scales................*chidesteri* Dyar, p. 275

19 (1). Hind tarsal segments ringed with white
opisthopus Komp, p. 310
Hind tarsal segments entirely dark
abominator D. and K., p. 300; *anips* Dyar, p. 301; *atratus* Theob., p. 303; *erraticus* (D. and K.), p. 305; *iolambdis* Dyar, p. 307; *mulrennani* Basham, p. 308; *peccator* D. and K., p. 312; *pilosus* (D. and K.), p. 313

MALE TERMINALIA

1. Tenth sternite crowned apically with a comblike row of blunt teeth................2

Tenth sternite crowned apically with a dense tuft of spines................................15

2 (1). Subapical lobe of basistyle divided into two main divisions; lateral plates of phallosome not columnar (subgenus *Melanoconion*) 3

Subapical lobe of basistyle not divided into two main divisions; lateral plates of phallosome columnar, blunt, and usually denticulate at apices (subgenus *Neoculex*) 11

3 (2). Dististyle simple, tapered from base, not enlarged, or otherwise modified distally
atratus (Theob.), p. 303

Dististyle with distal part divided, expanded, or snout-shaped................................4

4 (3). Basal division of subapical lobe of basistyle about equally divided and cleft to near basistyle so that the subapical lobe appears in three large divisions................5

Basal division of subapical lobe of basistyle not divided, or if divided, the cleft does not reach the basistyle and the two arms are unequal in size................................6

5 (4). Lobes of ninth tergite long and slender, widely separated basally; dististyle with tip pointed and slightly reflexed
pilosus (D. and K.), p. 314

Lobes of ninth tergite large and ovate, approximate at bases, apical part extending laterally; dististyle constricted subapically and expanded apically
mulrennani Basham, p. 308

6 (4). Apical division of subapical lobe of the basistyle with a broadened leaf..............7

Apical division of the subapical lobe of the basistyle without a leaf........................10

7 (6). Dististyle with apical half quadrately expanded, bearing conspicuous processes and with a prominent hirsute crest.......8

Dististyle with apical half roundly expanded (never quadrate), bearing minute setae on outer crest
erraticus (D. and K.), p. 305

8 (7). Crest of dististyle with a prominent fused horn; basistyle with an angular projection between the subapical lobe and the dististyle..*abominator* D. and K., p. 300

Crest of dististyle without a prominent fused horn; basistyle lacking an angular projection between the subapical lobe and the dististyle................................9

9 (8). Apex of dististyle, below the subterminal claw, forming an acute point, a thornlike point on the inner surface of dististyle at basal third......*peccator* D. and K., p. 312

Apex of dististyle broadly rounded below the subterminal claw, without a thornlike point on inner surface at basal third
anips Dyar, p. 301

10 (6). Lobes of ninth tergite ovoid; inner plate of phallosome with 2 or 3 blunt teeth in addition to a long lateral tooth
iolambdis Dyar, p. 307

Lobes of ninth tergite finger-shaped and wrinkled apically; inner plate of phallosome rounded apically and lacking small blunt teeth........*opisthopus* Komp, p. 310

11 (2). Plates of phallosome joined at base and also connected by a sclerotized bridge subapically 12

Plates of phallosome connected only at base, without a connecting bridge subapically................*apicalis* Adams, p. 316

12 (11). Basistyle with long setae on inner margin before the subapical lobe
arizonensis Bohart, p. 317

Basistyle without long setae on inner margin before the subapical lobe (setae short when present) 13

13 (12). Plates of phallosome crowned with denticles or small teeth apically; dististyle gradually narrowed toward apex........14

Plates of phallosome without denticles or teeth apically; dististyle strongly narrowed on distal one-third
reevesi Wirth, p. 321

14 (13). Plates of phallosome strongly narrowed and heavily sclerotized apically, subapical bridge stout
boharti Brookman and Reeves, p. 319

Plates of phallosome broadly rounded apically, usually not heavily sclerotized, subapical bridge narrow
territans Walker, p. 322

15 (1). Subapical lobe arising near middle of basistyle, thumblike, about twice as long as wide; subapical lobe bearing stout rods, shorter than the lobe
bahamensis D. and K., p. 274

Subapical lobe arising near apical third of basistyle, little if any longer than wide; subapical lobe bearing several appendages, all longer than the lobe............16

16 (15). Tenth sternites pointed and clothed on apical part with fine setae; subapical lobe of basistyle bearing rods and setae but without a leaflike appendage
coronator D. and K., p. 277

Tenth sternites crowned with pointed or blunt spines; subapical lobe of basistyle bearing rods, setae and a leaflike appendage (narrow in *C. tarsalis*)..................17

17 (16). Subapical lobe of basistyle with eight or more appendages; basal arm of tenth sternite represented by a short protuberance, never long and curved.................18

Subapical lobe of basistyle with five or six appendages; basal arm of tenth sternite long and curved............................19

18 (17). Dorsal arm of phallosome pointed or narrowly rounded apically, directed posteriorly, and crossing over the ventral arm nearly at right angle to its winglike outward extension, the distance between the tips of the dorsal and ventral arms and between the tips of the dorsal arms are about equal to one another
quinquefasciatus Say, p. 286
Dorsal arm of phallosome truncate or bluntly rounded apically, directed posterolaterally, and obliquely crossing over the winglike lateral extension of the ventral arm; the distance between the tips of the dorsal and ventral arms much less than the distance between the tips of the dorsal arms..............*pipiens* Linn., p. 284

19 (17). Leaflike filament of subapical lobe narrow, clublike; crown of tenth sternite with outer spines blunt....*tarsalis* Coq., p. 294
Leaflike filament of subapical lobe rather broad; crown of tenth sternite either with all spines pointed or with the outer spines blunt 20

20 (19). Subapical lobe of basistyle with only five appendages (three rods, a leaf, and a seta) 21
Subapical lobe of basistyle with six appendages 23

21 (20). Each plate of the phallosome with 4 strong teeth between the dorsal and ventral arms 22
Each plate of phallosome without strong teeth between the dorsal and ventral arms (a row of short teeth present)
chidesteri Dyar, p. 276

22 (21). Crown of tenth sternite with outer spines blunt; apical margin of eighth tergite with numerous short stout setae (shorter than those on the lobes of the ninth tergite)..............*nigripalpus* Theob., p. 282
Crown of tenth sternite with all spines pointed; apical margin of eighth tergite with numerous rather long slender setae (longer than those on lobes of ninth tergite) ....*thriambus* Dyar, p. 297

23 (20). Each plate of phallosome with a single short stout triangular tooth
restuans Theob., p. 288
Each plate of phallosome with 3 or more strong teeth or denticles.....................24

24 (23). Subapical lobe of basistyle with three rods, a flattened filament, a leaf, and a long slender seta........*virgultus* Theob., p. 298
Subapical lobe of basistyle with three rods, a hooked spine, a leaf, and a seta........25

25 (24). Each plate of phallosome with a series of denticles, without a group of strong teeth..........*interrogator* D. and K., p. 281
Each plate of phallosome with a group of strong teeth (small teeth may or may not be present) 26

26 (25). Each plate of phallosome with about 4 stout pointed teeth adjacent to a group of compact slender teeth which arise from the inner side of the ventral arm
stigmatosoma Dyar, p. 292
Each plate of phallosome with about 6 to 12 strong teeth, without teeth on inner side of ventral arm.................................27

27 (26). Dorsal arm of phallosome bent medially at a right angle; the most apical tooth on the phallosome very small in comparison with the adjacent teeth, straight, and conical.................*salinarius* Coq., p. 290
Dorsal arm of phallosome nearly straight; the most apical tooth on phallosome about the same size or larger than the adjacent teeth, bent, and pointed
erythrothorax Dyar, p. 279

LARVAE (FOURTH INSTAR)

1. Siphon spinose; anal gills two, thick, bulbous..............*bahamensis* D. and K., p. 275
Siphon not spinose; anal gills four, normal 2

2 (1). Antenna of nearly uniform shape, with antennal tuft inserted near middle of shaft
restuans Theob., p. 289
Antennal tuft placed in a constriction near outer third of shaft, the part beyond the tuft more slender.................................3

3 (2). Siphon with several single, irregularly placed hairs, and at most a single pair of 2- or 3-haired subapical tufts
thriambus Dyar, p. 297
Siphonal hair tufts multiple (occasionally double or single) not irregularly placed, although the subapical tuft may be laterally out of line.................................4

4 (3). Lower frontal head hair 6 long, single or double 5
Lower frontal head hair 6 and upper frontal 5 long, with three or more branches....18

5 (4). Siphon with two or three pairs of small subdorsal tufts in addition to siphonal tufts (subdorsal tufts absent in *C. opisthopus*); pecten tooth fringed on one side nearly to tip (subgenus *Melanoconion*) 6
Siphon lacking small subdorsal tufts; pecten tooth with 1 to 4 long coarse side teeth 14

6 (5). Comb scales long, thorn-shaped, arranged in a curved irregular single or double row 7

Comb scales rounded apically and fringed with subequal spinules, arranged in a triangular patch 8

7 (6). Siphonal index about 3.0 to 4.0; siphon with about eight pairs of very long siphonal tufts, the proximal pair about as long as the siphon
pilosus (D. and K.), p. 314
Siphonal index about 6.0 to 7.0; siphon with five or six pairs of siphonal tufts, none of which are more than one-half the length of the siphon
erraticus (D. and K.), p. 306

8 (6). Siphonal tufts all short (length about equal to or slightly longer than the diameter of the siphon at the point of insertion of each tuft); antennal shaft uniformly dark brown......*opisthopus* Komp, p. 311
Proximal siphonal tufts long (at least one and a half to two times as long as the diameter of the siphon at the point of insertion of each tuft); antennal shaft whitish on basal two-thirds.....................9

9 (8). Individual comb scale short, broad, free part about as short as the base
abominator D. and K., p. 300
Individual comb scale long and narrow, free part longer than the base...............10

10 (9). Upper frontal head hair 5 usually 5- to 7-branched; siphonal tufts not more than twice as long as the diameter of the siphon at point of insertion of each tuft
atratus Theob., p. 304
Upper frontal head hair 5 single to triple; siphonal tufts long, proximal tufts two to three times as long as the diameter of the siphon at point of insertion of each tuft 11

11 (10). Upper frontal head hair 5 less than one-half as long as lower frontal 6
peccator D. and K., p. 312
Upper frontal head hair 5 one-half to three-fourths as long as lower frontal 6........12

12 (11). Upper frontal head hair 5 usually double 13
Upper frontal head hair 5 usually triple
mulrennani Basham, p. 309

13 (12). Pecten tooth long, slender and with fine lateral teeth; inner preclypeal spines long, about one-fifth the length of the head............................*anips* Dyar, p. 302
Pecten tooth shorter, stouter and with coarser lateral teeth; inner preclypeal spines much less than one-fifth the length of the head..........*iolambdis* Dyar, p. 307

14 (5). Upper frontal head hair 5 and lower frontal 6 usually double and about equal in length; siphon with a broad basal darkly pigmented area or band
arizonensis Bohart, p. 317

Upper frontal head hair 5 and lower frontal 6 not of equal length and usually not both double; siphon without a broad basal darkly pigmented area or band...........15

15 (14). Basal diameter of siphon about twice its apical diameter; siphonal tufts relatively short, individual hairs of the proximal tuft rarely more than one-seventh as long as the siphon........*apicalis* Adams, p. 316
Basal diameter of siphon distinctly less than twice its apical diameter; siphonal tufts relatively long, individual hairs of the proximal tuft one-fifth as long as the siphon or longer.....................................16

16 (15). Upper frontal head hair 5 triple, lower frontal 6 double......*reevesi* Wirth, p. 321
Upper frontal head hair 5 single or double, lower frontal 6 usually single..............17

17 (16). Abdominal segment IV much paler than segments III or V in living specimens; upper frontal head hair 5 double, rarely triple
boharti Brookman and Reeves, p. 319
Abdominal segments rather evenly pigmented; upper frontal head hair 5 single (occasionally double, usually double in California specimens)
territans Walker, p. 323

18 (4). Siphonal tufts inserted in a straight line..19
One or more siphonal tufts inserted laterally out of line....................................20

19 (18). Siphonal index about 4.5 to 5.5; usually five pairs of large siphonal tufts, all inserted in a straight line
tarsalis Coq., p. 295
Siphonal index about 8.0; usually seven to nine pairs of small siphonal tufts, all inserted laterally in a straight line
chidesteri Dyar, p. 276

20 (18). Siphon with three pairs of tufts
virgultus Theob., p. 299
Siphon with four or five pairs of tufts......21

21 (20). Pecten reaching distal three-fourths of siphon; siphonal index 3.0 to 3.5
interrogator D. and K., p. 281
Pecten on basal third of siphon; siphonal index at least 4.0.................................22

22 (21). Siphonal index 4.0 to 5.0; lower frontal head hair 6 usually with five or more branches 23
Siphonal index 6.0 to 8.0; lower frontal head hair 6 usually with three or four branches 24

23 (22). Dorsal microsetae toward apex of saddle of anal segment conspicuously larger than those at dorsal middle; lateral hairs of abdominal segments III and IV usually triple..............*stigmatosoma* Dyar, p. 293
Dorsal microsetae toward apex of saddle not conspicuously larger than those at

dorsal middle; lateral hairs of abdominal
segments III and IV usually double
pipiens Linn., p. 285;
quinquefasciatus Say, p. 287

24 (22). Siphon with strong subapical spines
coronator D. and K., p. 278
Siphon lacking strong subapical spines..25

25 (24). Thorax densely spiculate (spicules dark);
lateral hair of anal segment usually single.................*nigripalpus* Theob., p. 283
Thorax glabrous or sparsely spiculate at
most; lateral hair of anal segment usually
double 26

26 (25). Siphon usually with five pairs of siphonal
tufts; anal gills usually about as long as
the saddle of the anal segment
erythrothorax Dyar, p. 279
Siphon usually with four pairs of siphonal
tufts; anal gills usually a little longer
than the saddle of the anal segment
salinarius Coq., p. 291

Culex (Culex) bahamensis Dyar and Knab

Culex bahamensis Dyar and Knab, 1906, Jour. N.Y.
Ent. Soc., *14*:210.

Culex eleuthera Dyar, 1917, Ins. Ins. Mens., *5*:184.

Culex petersoni Dyar, 1920, Ins. Ins. Mens., *8*:27.

ADULT FEMALE (pl. 101). Small- to medium-sized species. *Head:* Proboscis dark-scaled,
with a rather broad white median band; palpi
short, dark, pale-scaled at tips. Occiput with
narrow curved pale-golden scales and dark
erect forked scales dorsally, with broad white
scales laterally. *Thorax:* Integument of scutum
brown; scutum clothed with narrow curved
golden-brown scales, paler on anterior and
lateral margins and around the prescutellar
space, pale scales sometimes present at the
scutal angle. Scutellum with narrow pale-
golden scales and brown setae on the lobes.
Pleura with small patches of broad white
scales. *Abdomen:* First tergite with a median
patch of dark scales; remaining tergites dark-
brown-scaled, each with a basal band of white
scales. Venter predominantly pale-scaled.
Legs: Legs dark-scaled except for the following: femora, tibiae, the first tarsal segment, and
often the second, pale-scaled on posterior surface; femora and tibiae tipped apically with
white; tarsi with narrow basal and apical rings
at the joints, rings on front and middle tarsi
sometimes obsolete. *Wing:* Length about 3.0
mm. Scales narrow, dark.

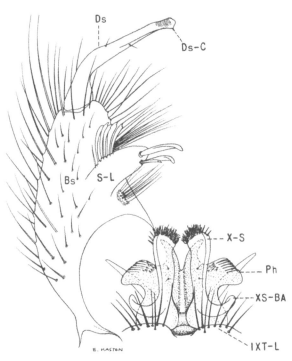

Fig. 229. Male terminalia of *Culex bahamensis* Dyar and
Knab.

ADULT MALE. Coloration similar to that of
the female. *Terminalia* (fig. 229). Lobes of
ninth tergite (IXT-L) slightly elevated, each
bearing several slender setae. Tenth sternite
(X-S) densely crowned with short pointed
spines; basal arm (XS-BA) rather stout,
slightly curved. Phallosome (Ph) consists of
two plates. Each plate with a slender outwardly curved apical process and a large
hatchet-shaped lateral lobe; the apical half of
the lateral lobe has flattened appressed teeth.
Claspette absent. Basistyle (Bs) about twice as
long as mid-width, clothed with numerous
setae, longer on outer aspect; two dense patches
of moderately long setae present on inner surface, one anterior and one posterior to the subapical lobe. Subapical lobe (S-L) stout, a little
more than twice as long as wide, bearing two
long and one short slightly flattened apical rods
and a row of several flattened setae on the posterior side. Dististyle (Ds) about two-thirds
as long as the basistyle, broad at base and
gradually tapered distally, finely crenulate
near apex, truncate at tip, bearing two small
setae on distal half; claw (Ds-C) minute, inserted a little before apex of dististyle.

LARVA (fig 230). Antenna shorter than the head, constricted beyond antennal tuft, with part before constriction spiculate, part beyond constriction sparsely spiculate; antennal tuft large, multiple, barbed, inserted at outer third of shaft, extending well beyond tip. Head hairs: Postclypeal 4 short, single; upper frontal 5 and lower frontal 6 long, multiple, barbed; preantennal 7 multiple, barbed, reaching near insertion of antennal tuft. Prothoracic hairs: 1–3 long, single; 4 long, usually double; 5–6 long, single; 7 long, triple. Body finely and sparsely spiculate. Lateral abdominal hair 6 double or triple on segments III–VI. Comb of eighth segment with many scales in a patch; individual scale rounded apically and fringed with subequal spinules. Siphonal index 3.5 to 4.5; siphon finely and densely spiculate basally, progressively more coarsely spiculate apically; pecten of about 10 to 14 teeth progressively more widely spaced, reaching a little beyond basal third of siphon; individual tooth with coarse teeth on one side; six or seven smooth multiple siphonal tufts present, two or three of the basal pairs inserted within the pecten and about as long as the diameter of the siphon at the point of their insertion. Anal segment completely ringed by the saddle; the saddle darkly pigmented dorsally; lateral hair double or triple, shorter than the saddle; dorsal brush bilaterally consisting of a long lower caudal hair and a shorter upper caudal tuft of four to six branches; ventral brush well developed, restricted to the barred area; gills two, short, bulbous.

DISTRIBUTION. This species has been found in Florida (274, 562), Puerto Rico (561), the Bahamas, and the Virgin Islands. It seems to be found most commonly on the smaller islands of the Antilles.

BIONOMICS. The larvae often occur in large numbers in temporary rain-filled pools on the Florida Keys. Larvae were taken by Fisk (274) from various types of aquatic habitats, including a cistern containing brackish water at Key West, Florida. Hill and Hill (361) found the larvae in shallow open pools in mangrove swamps and in rockholes along the seashore in Jamaica. The adults are rarely encountered and little is known of their habits.

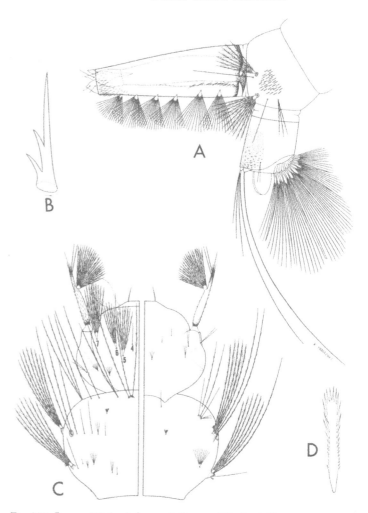

Fig. 230. Larva of *Culex bahamensis* Dyar and Knab. *A*, Terminal segments. *B*, Pecten tooth. *C*, Head and thorax (dorsal and ventral). *D*, Comb scale.

Culex (Culex) chidesteri Dyar

Culex chidesteri Dyar, 1921, Ins. Ins. Mens., *9*:117.
Culex chidesteri, Joyce, 1948, Mosquito News, *8*:102.

ADULT FEMALE (pl. 102). Medium-sized species. *Head:* Proboscis dark-scaled, with a broad median area of pale scales, usually more distinct ventrally; palpi short, dark. Occiput clothed with narrow curved pale bronze-brown scales and dark erect forked scales dorsally, with a patch of broad white scales laterally. *Thorax:* Integument of scutum brown; scutum clothed with fine bronze-brown scales; pale golden on anterior and lateral margins, at the scutal angle and around the prescutellar space. Pleura brown, with small patches of broad white scales. *Abdomen:* First tergite with a median patch of dark bronze-brown scales; remaining tergites dark-scaled, with bronze to metallic blue-green reflection, each with a narrow basal pale band joining basolateral white patches. Venter mostly white-scaled, usually a few black scales scattered along apical margin of the sternites. *Legs:* Legs dark-scaled, with metallic blue-green reflection; posterior surface of femora and tibiae pale, with the pale scales extending over most of the first tarsal segment; femora and tibiae white-tipped. Tarsi dark, with faint indications of pale rings on some specimens. *Wing:* Length about 3.5 mm. Scales narrow, dark.

ADULT MALE. Coloration similar to that of the female, but with the pale markings on the proboscis more distinct. *Terminalia* (fig. 231). Lobes of ninth tergite (IXT-L) slightly raised, each bearing several slender setae. Tenth sternite (X-S) crowned with many spines, the inner ones pointed, the outer ones blunt; basal arm (XS-BA) stout, strongly curved, sclerotized. Phallosome (Ph) formed of two large sclerotized plates connected at base. Each plate

with a long curved dorsal arm (D-A), a small thumb-shaped ventral arm (V-A), and a row of short teeth (Ph-T). Claspette absent. Basistyle (Bs) a little more than twice as long as the mid-width, clothed with many setae, longer on outer aspect. Subapical lobe (S-L) prominent, undivided, bearing three rods somewhat hooked at tip, a broad leaf, and a slender seta. Dististyle (Ds) nearly half as long as the basistyle, curved, tapered apically; claw (Ds-C) short, blunt.

LARVA (fig. 232). Antenna about as long as the head, curved, constricted beyond antennal tuft, part before constriction dark near base, pale to near insertion of tuft, spiculate, part beyond constriction darker and with fewer spicules; antennal tuft large, multiple, barbed, inserted at outer third of shaft. Head hairs: Postclypeal 4 short, single; upper frontal 5 about 6- to 8-branched, barbed, shorter than hair 6; lower frontal 6 long, barbed, usually triple; preantennal 7 multiple, barbed, reaching near insertion of antennal tuft. Thorax and abdomen spiculate. Prothoracic hairs: 1–3 long, single; 4 medium, usually double; 5–6 long, single; 7 medium, 3- or 4-branched. Comb of eighth segment of numerous scales in a patch; individual scale rounded apically and fringed with subequal spinules. Siphonal index more than 8.0; siphon slender and only slightly tapered; pecten of about 9 to 12 teeth on basal third of siphon; individual tooth with about 3 to 5 coarse teeth on one side; seven to nine (usually eight) pairs of 2- to 4-branched siphonal tufts inserted in a straight line, the proximal tuft inserted before the end of the pecten, individual hairs of proximal tuft shorter than the basal diameter of the siphon. Anal segment completely ringed by the saddle, entire surface spiculate; lateral hair double, shorter than the saddle; dorsal brush bilaterally consisting of a long lower caudal hair and a shorter upper caudal tuft of three hairs; ventral brush well developed, confined to the barred area; gills shorter than the saddle, tapered.

DISTRIBUTION. This species is apparently restricted in the United States (277, 408) to the Lower Rio Grande Valley of Texas. It also occurs in Mexico, Central America, and northern South America.

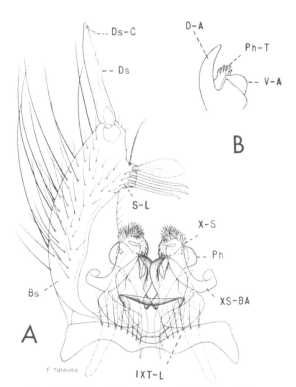

Fig. 231. *Culex chidesteri* Dyar. *A*, Male terminalia. *B*, Plate of phallosome.

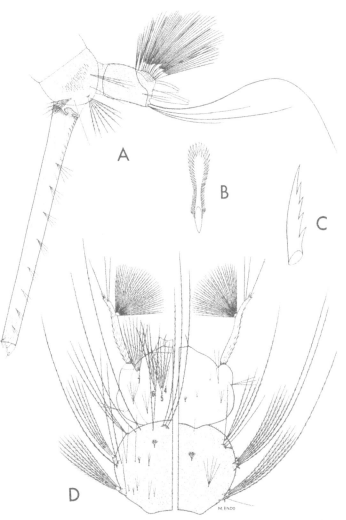

Fig. 232. Larva of *Culex chidesteri* Dyar. *A,* Terminal segments. *B,* Comb scale. *C,* Pecten tooth. *D,* Head and thorax (dorsal and ventral).

BIONOMICS. Joyce (408) collected the larvae of this species throughout most of the year in ponds overgrown with hyacinths and cattails at Brownsville, San Benito, and Harlingen, Texas. Adults, both males and females, were taken in a light trap operated at Brownsville.

Culex (Culex) coronator Dyar and Knab

Culex coronator Dyar and Knab, 1906, Jour. N.Y. Ent. Soc., *14*:215.
Culex ousqua Dyar, 1918, Ins. Ins. Mens., *6*:99.
Culex usquatus Dyar, 1918, Ins. Ins., Mens., *6*:122.
Culex usquatissimus Dyar, 1922, Ins. Ins., Mens., *10*: 19.

Culex camposi Dyar, 1925, Ins. Ins. Mens., *13*:21.
Culex cingulatus Theobald (not Fabricius), 1901, Mon. Culic., *2*:5.

ADULT FEMALE (pl. 103). Small- to medium-sized species. *Head:* Proboscis dark-scaled, broadly pale ventrally near middle; palpi short, dark. Occiput clothed with narrow golden scales and dark erect forked scales dorsally, with a patch of broad white scales laterally. *Thorax:* Integument of scutum brown; scutum clothed with narrow curved golden-brown scales; scales on the anterior and lateral margins and the prescutellar space light golden, often with two submedian pale-golden spots near middle. Scutellum with light-golden scales and dark-brown setae on the lobes. Pleura with small patches of white scales. *Abdomen:* First tergite with a median patch of dark bronze-brown scales; remaining tergites bronze-brown, each with a basal white band joined with white basolateral patches. Venter largely pale-scaled. *Legs:* Legs dark with metallic reflection. Femora and tibiae pale on posterior surface and tipped with white. Tarsi ringed with white basally and apically, rings usually less distinct on front and middle legs. *Wing:* Length about 3.0 to 3.3 mm. Scales narrow, dark.

ADULT MALE. Proboscis with a narrow white band beyond middle. Abdominal tergites with basal white bands broader than in the female. Otherwise the coloration is similar to that of the female. *Terminalia* (fig. 233). Lobes of ninth tergite (IXT-L) widely separated, only slightly raised, each bearing several slender setae of medium length. Tenth sternite (X-S) pointed, clothed with fine setae on apical part; basal arm (XS-BA) long, strongly curved. Phallosome (Ph) formed of two large heavily sclerotized plates connected near base. Each plate with a long stout curved pointed dorsal arm (D-A), a beaklike curved pointed ventral arm (V-A) and a group of about 5 or 6 closely set pointed teeth (Ph-T). Claspette absent. Basistyle (Bs) about two and a half times as long as the mid-width, clothed with many setae, much longer on outer aspect. Subapical lobe (S-L) prominent, undivided, bearing about ten spines (about half of which are rodlike), spines arranged as shown in figure 233, *A.* Dististyle (Ds) nearly half as long as the basistyle, stout

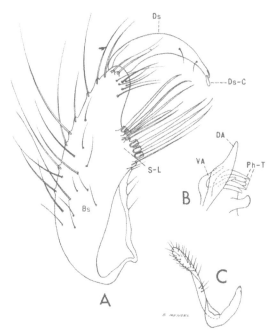

Fig. 233. Male terminalia of *Culex coronator* Dyar and Knab. *A*, Basistyle. *B*, Plate of phallosome. *C*, Tenth sternite.

inserted beyond the pecten, the two middle tufts inserted a little laterally out of line, each tuft about as long as the diameter of the siphon at the point of insertion of the tuft. Anal segment with prominent spicules, completely ringed by the saddle; lateral hair short, single or double; dorsal brush bilaterally consisting of a long lower caudal hair and usually two upper caudal hairs, one long and one short; ventral brush well developed, confined to the barred area; gills usually a little longer than the saddle, each tapered to a point.

DISTRIBUTION. This species is found in Texas (249, 571, 634), and south through Mexico (537a), Central America, and South America to Argentina.

on basal half, curved and tapered beyond middle; claw (Ds-C) short, broad.

LARVA (fig. 234). Antenna nearly as long as the head, curved, constricted beyond antennal tuft, with part before constriction pale and spiculate, part beyond constriction more slender, darker, and with fewer spicules; antennal tuft large, multiple, barbed, inserted at outer third of shart, reaching beyond tip. Head hairs: Postclypeal 4 short, single; upper frontal 5 and lower frontal 6 long, usually 3- or 4-branched, barbed; preantennal 7 multiple, barbed, reaching near insertion of antennal tuft. Prothoracic hairs: 1–3 long, single; 4 medium, usually double; 5–6 long, usually single; 7 medium, 2- to 4-branched. Thorax spicular. Lateral abdominal hair 6 triple on segments I and II, double or triple on III, double on IV–VI. Comb of eighth segment of numerous scales in a triangular patch; individual scale long, rounded apically, and fringed with subequal spinules. Siphonal index about 8.0 to 9.0; siphon slender and straight, bearing a crown of spines before apex; pecten of about 8 to 14 teeth on basal fourth of siphon; individual tooth with 2 to 5 coarse teeth on one side; four siphonal tufts usually double and

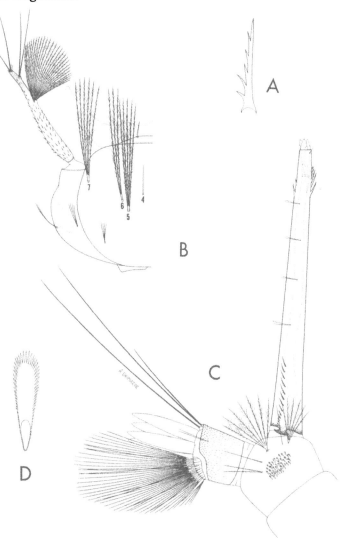

Fig. 234. Larva of *Culex coronator* Dyar and Knab. *A*, Pecten tooth. *B*, Head. *C*, Terminal segments. *D*, Comb scale.

BIONOMICS. *Culex coronator* occurs in large numbers in the southwestern counties of Texas, especially in the Rio Grande Valley, where the larvae are found in temporary rain-filled pools and artificial water containers. In Panama this is a common mosquito found in a variety of habitats. The females are occasionally captured in horse-baited traps in Panama but apparently do not feed on man.

Culex (Culex) erythrothorax Dyar

Culex erythrothorax Dyar, 1907, Proc. U.S. Nat. Mus., *32*:124.
Culex federalis Dyar, 1923, Ins. Ins. Mens., *11*:186.
Culex badgeri Dyar, 1924, Ins. Ins. Mens., *12*:125.

ADULT FEMALE (pl. 104). Medium-sized species. *Head:* Proboscis dark-scaled, with some pale scales ventrally; palpi short, dark. Occiput with narrow curved golden scales and dark-brown erect forked scales on broad dorsal region, with appressed yellowish-white scales laterally. *Thorax:* Integument of scutum reddish-brown; scutum clothed with fine hairlike golden scales, those on the prescutellar space pale golden. Scutellum with narrow pale-golden scales and golden setae on the lobes. Pleura reddish brown, with small patches of a few pale scales. *Abdomen:* First tergite with a small median patch of reddish-brown scales; tergites II–VII clothed with brownish-black scales and with narrow irregular and rather indistinct yellowish basal bands; eighth tergite and sometimes most of the seventh pale-scaled. Venter clothed with yellowish scales. *Legs:* Legs brown with bronze reflection; posterior surface of femora and tibiae yellowish; knee spots yellowish. *Wing:* Length about 3.5 to 4.0 mm. Scales narrow, dark brown, darker on costa.

ADULT MALE. Coloration similar to that of the female, but with the abdominal pale scaling usually more extensive. *Terminalia* (fig. 235). Lobes of ninth tergite (IXT-L) slightly raised, separated by about the width of one lobe, each bearing several setae. Tenth sternite (X-S) crowned with numerous spines; basal arm (XS-BA) stout, strongly curved, sclerotized. Phallosome (Ph) formed of two large heavily sclerotized plates connected near base. Each plate with a long stout curved pointed dorsal arm (D-A), a small slender pointed ventral

arm (V-A), and a group of 7 to 12 prominent, closely set, pointed, dark teeth (Ph-T). Claspette absent. Basistyle (Bs) a little more than twice as long as the mid-width, clothed with numerous setae, longer on outer aspect. Subapical lobe (S-L) prominent, undivided, bearing three long stout rods (one pointed, two with recurved tips), a slender hooked spine, a small broadly expanded striated leaf and a slender seta. Dististyle (Ds) about half as long as the basistyle, curved and tapered apically, bearing two short setae on distal half; claw (Ds-C) short, blunt.

LARVA (fig. 236). Antenna a little shorter than the head, constricted beyond the antennal tuft, part before constriction spiculate, dark near base then pale to near insertion of tuft, part beyond constriction darker, slender, and with few spicules; antennal tuft large, multiple, inserted at outer third of shaft, reaching much beyond tip. Head hairs: Postclypeal 4 short, single; upper frontal 5 long, usually 4- to 7-branched, barbed; lower frontal 6 long, usually 3- or 4-branched, barbed; preantennal 7 long, multiple, barbed. Prothoracic hairs: 1–3 long, single; 4 long, usually double; 5–6 long, single; 7 long, usually double or triple. Lateral abdominal hair 6 usually multiple on

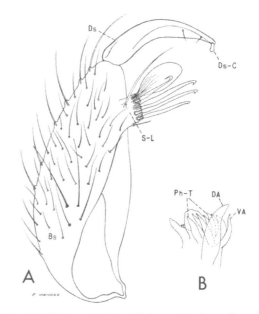

Fig. 235. Male terminalia of *Culex erythrothorax* Dyar. *A*, Basistyle. *B*, Plate of phallosome.

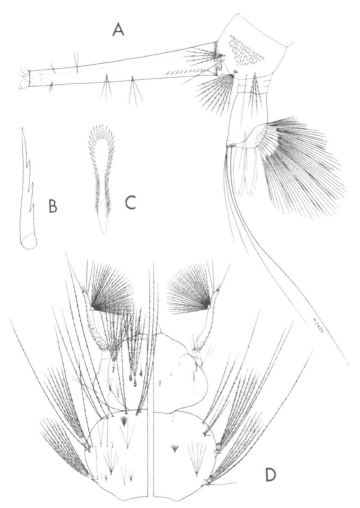

Fig. 236. Larva of *Culex erythrothorax* Dyar. *A*, Terminal segments. *B*, Pecten tooth. *C*, Comb scale. *D*, Head and thorax (dorsal and ventral).

segments I and II, double on III–VI. Comb of eighth segment with many scales in a patch; individual scale long, broad, and fringed apically with subequal spinules. Siphonal index about 6.0 to 7.0; pecten of about 11 to 20 teeth on basal fourth of siphon; individual pecten tooth with 3 or 4 coarse teeth on one side; usually five pairs of small 2- to 4-branched siphonal tufts inserted beyond the pecten, next to the last one or two pairs inserted laterally. Anal segment completely ringed by the saddle; lateral hair usually double (occasionally triple), a little shorter than the saddle; dorsal brush bilaterally consisting of a long lower caudal hair and about three upper caudal hairs

(one long and two short); ventral brush well developed, confined to the barred area; gills usually about as long as the saddle, bluntly pointed.

DISTRIBUTION. Known from Mexico (537a) and the southwestern United States. *United States:* California (287); Idaho (344); Utah (575).

BIONOMICS. The larvae of *Culex erythrothorax* are found in California in large shallow ponds containing heavy growths of vegetation. Larval production is continuous throughout the year in southern California. Rees (575) says that the larvae appear sporadically throughout the season in Utah in shallow ponds containing an abundance of vegetation and that they may become quite numerous in some places late in the season.

Rees (575) says that he has encountered the females in considerable numbers in limited areas in Utah and that they attack viciously, their bites being very painful. He further states that they were able to bite through light clothing and were not easily disturbed while feeding. The senior author encountered the adults of this species at a pond in Marin County, California, during the summer of 1952, in such numbers that as many as 40 to 50 were observed on the clothing at one time. They generally attack below the waist, but occasionally attack near one's shoulders or head. The adults will approach the collector to feed in the sunshine as well as on cloudy days or in shade. The males rest during the day among the vegetation along the margins of the pond. The females also rest during the daylight hours in a similar manner but readily leave the vegetation to feed.

Culex (Culex) interrogator
Dyar and Knab

Culex interrogator Dyar and Knab, 1906, Jour. N.Y. Ent. Soc., *14*:209.

Culex reflector Dyar and Knab, 1909, Smithson. Misc. Coll., *52*:256.

ADULT FEMALE (pl. 105). Small species. *Head:* Proboscis dark-scaled, with a median pale area ventrally; palpi short, dark. Occiput with narrow curved pale-golden scales and dark-brown erect forked scales dorsally, with broad white appressed scales laterally. *Thorax:* Integument of scutum brown; scutum clothed with very small narrow curved golden-brown scales; scales on prescutellar space light golden. Scutellum with narrow curved light-golden scales and dark-brown setae on the lobes. Pleura light brown, often with a greenish tinge, small patches of white scales present. *Abdomen:* First tergite with a median patch of dark scales; remaining tergites dark-scaled, with basal bands of white scales widening at the sides. Venter pale-scaled, with a few dark scales intermixed. *Legs:* Legs dark-scaled; posterior surface of femora and tibiae pale; femora and tibiae tipped apically with white. Tarsi often with very narrow brownish rings involving both ends of the joints. *Wing:* Length about 2.5 to 2.8 mm. Scales narrow, dark.

ADULT MALE. Coloration similar to that of the female, but with dorsal abdominal pale bands broader. *Terminalia* (fig. 237). Lobes of ninth tergite (IXT-L) rather prominent, rounded, separated by a little more than the width of one lobe, each bearing several setae. Tenth sternite (X-S) crowned with numerous spines; basal arm (XS-BA) long, stout, strongly curved, chitinized. Phallosome (Ph) formed of two large sclerotized plates. Each plate bears a series of denticles apically, a broad winglike

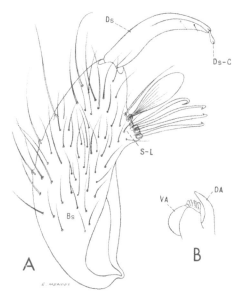

Fig. 237. Male terminalia of *Culex interrogator* Dyar and Knab. *A*, Basistyle. *B*, Plate of phallosome.

ventral arm (V-A) curved outwardly, and a large pointed dorsal arm (D-A). Claspette absent. Basistyle (Bs) about two and a half times as long as mid-width, clothed with numerous setae, longer on outer aspect. Subapical lobe (S-L) prominent, undivided, bearing three rods of about equal length, a shorter hooked spine, a small leaf, and a seta. Dististyle (Ds) about half as long as the basistyle, curved, gradually tapered beyond middle; claw (Ds-C) short, blunt.

LARVA (fig. 238). Head broader than long. Antenna shorter than the head, constricted beyond antennal tuft, with part before constriction dark at base, pale to near insertion of tuft, spiculate, part beyond constriction darker and with fewer spicules; antennal tuft large, multiple, barbed, inserted at outer fourth of shaft, reaching much beyond tip. Head hairs: Postclypeal 4 short, single; upper frontal 5 and lower frontal 6 long, usually triple, barbed; preantennal 7 long, multiple, barbed. Thorax sparsely spicular. Prothoracic hairs: 1–3 long, single; 4 moderate, double; 5–7 long, single. Lateral abdominal hair 6 double on segments I–IV, single or double on V and VI. Comb of eighth segment composed of numerous scales in a patch; individual scale long, narrow, with apex expanded and fringed with subequal spinules. Siphonal index 3.0 to 3.5; pecten consists of numerous teeth extending to distal third of siphon, the distal 3 to 6 teeth large, spinelike, and without lateral teeth, others with 2 to 5 coarse teeth on one side; four pairs of 2- or 3-branched siphonal tufts present, the first three tufts inserted within the pecten, the apical tuft inserted near end of pecten, the subapical tuft inserted laterally out of line. Anal segment a little longer than wide, completely ringed by the saddle; lateral hair single, shorter than the saddle; dorsal brush bilaterally consisting of a long lower caudal hair and three upper caudal hairs, one long and two short; ventral brush well developed, confined to the barred area; gills usually about two to three times as long as the saddle, bluntly pointed.

DISTRIBUTION. This species has been found at Harlingen and Edinburg, Texas (634, 695). Its range extends south through Mexico (537a) and Central America.

BIONOMICS. The larvae occur in foul water

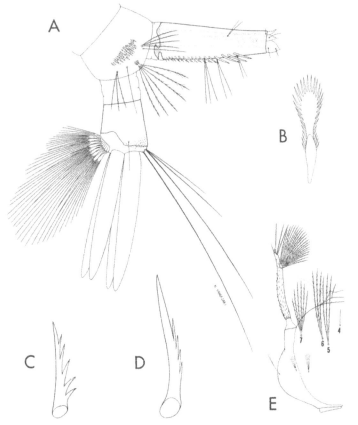

Fig. 238. Larva of *Culex interrogator* Dyar and Knab. *A*, Terminal segments. *B*, Comb scale. *C*, and *D*, Pecten teeth. *E*, Head.

in ground pools, wheel ruts, treeholes, rain barrels, and other similar aquatic habitats. Thurman *et al.* (695) found the larvae in irrigation seepage areas in the Rio Grande Valley (Cameron and Hidalgo counties), Texas. The adults are seldom encountered, and therefore little is known of their habits.

Culex (Culex) nigripalpus Theobald

Culex nigripalpus Theobald, 1901, Mon. Culic., *2*: 322.
Culex palus Theobald, 1903, Mon. Culic., *3*:194.
Culex similis Theobald, 1903, Mon. Culic., *3*:207.
Culex microsquamosus Grabham, 1905, Can. Ent., *37*:407.
Culex carmodyae Dyar and Knab, 1906, Jour. N.Y. Ent. Soc., *14*:210.
Culex mortificator Dyar and Knab, 1906, Jour. N.Y. Ent. Soc., *14*:210.
Culex factor Dyar and Knab, 1906, Jour. N.Y. Ent. Soc., *14*:212.

Culex regulator Dyar and Knab, 1906, Jour. N.Y. Ent. Soc., *14*:213.
Trichopronomyia microannulata Theobald, 1907, Mon. Culic., *4*:481.
Culex proximus Dyar and Knab, 1909, Proc. Ent. Soc. Wash., *11*:38.
Culex prasinopleurus Martini, 1914, Ins. Ins. Mens., *2*:68.

ADULT FEMALE (pl. 106). Medium-sized species. *Head:* Proboscis dark-scaled, usually paler underneath on basal half; palpi short, dark. Occiput with narrow curved pale golden-brown scales and dark erect forked scales dorsally, with a patch of broad dingy-white scales laterally. *Thorax:* Integument of scutum brown; scutum clothed with fine dark bronze-brown scales. Scutellum with brown setae and fine dark bronze-brown scales on the lobes. Pleura with few or no scales, rarely more than 5 or 6 scales in any single group. *Abdomen:* Tergites clothed with dark-brown to black scales with bronze to metallic blue-green reflection; narrow white basal bands occasionally present on some segments; basolateral white-scaled patches present. Venter pale-scaled. *Legs:* Legs dark-scaled with bronze to metallic blue-green reflection; posterior surface of femora and tibiae pale. *Wing:* Length 3.0 to 3.5 mm. Scales narrow, dark.

ADULT MALE. Coloration similar to that of the female. *Terminalia* (fig. 239). Eighth tergite (VIII-T) bearing many short stout setae. Lobes of ninth tergite (IXT-L) broadly rounded, separated by a deep emargination about the width of one lobe, each lobe bearing many slender setae. Tenth sternite (X-S) densely crowned with short spines, the apical ones pointed, the outer ones blunt; basal arm (XS-BA) long, stout, strongly curved, sclerotized. Phallosome (Ph) consists of two large sclerotized plates. Each plate bears a long pointed basal dorsal arm (DA), not bent at a right angle as in *C. salinarius;* a long pointed basal process nearly as large as the dorsal arm; and a stout curved ventral arm (VA), finely denticulate on outer surface beyond middle; between the dorsal and ventral arms arise 4 strong teeth (Ph-T). Claspette absent. Basistyle (Bs) about two and a half times as long as the mid-width, clothed with many setae, much longer on outer aspect. Subapical lobe (S-L) undi-

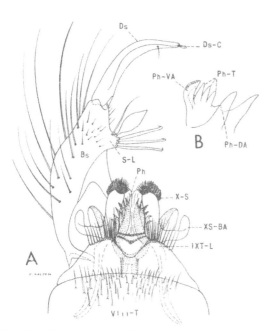

Fig. 239. *Culex nigripalpus* Theobald. *A*, Male terminalia.
B, Plate of phallosome.

usually four paired siphonal tufts inserted beyond pecten; proximal tuft usually double, occasionally single, as long or longer than the basal diameter of the siphon; second and third tufts usually double or triple and inserted somewhat laterally; distal tuft small, single to triple. Anal segment completely ringed by the saddle, with coarse spicules present on dorso-apical surface; lateral hair usually single, sometimes double, usually a little shorter than the saddle; dorsal brush bilaterally consisting of a long lower caudal hair and an upper caudal tuft of three hairs (one long and two short); ventral brush well developed, confined to the barred area; gills one to three times as long as the saddle, bluntly pointed.

DISTRIBUTION. This species is found in the southern United States, Mexico (537a), Central America, and in northern South America. *United States:* Alabama, Florida and Georgia (135, 419); Louisiana (419); Mississippi

vided, bearing three long strong rods hooked at tips, a large broad leaflike filament, and a strong seta. Dististyle (Ds) about half as long as the basistyle, bearing one or two small setae on inner surface before apex; claw (Ds-C) short, blunt.

LARVA (fig. 240). Antenna shorter than the head, constricted beyond antennal tuft, with part before constriction pale and spiculate, part beyond constriction darker and with few spicules; antennal tuft large, multiple, barbed, inserted at outer third of shaft, reaching well beyond tip. Head hairs: Postclypeal 4 short, single; upper frontal 5 and lower frontal 6 usually 3-branched, barbed, extending beyond preclypeus; preantennal 7 long, multiple, barbed. Prothoracic hairs: 1–3 long, single; 4 long, usually double; 5–6 long, single; 7 long, 2- to 4-branched. Thorax densely clothed with fine spicules. Lateral abdominal hair 6 usually 3- or 4-branched on segments I and II, usually double on III–VI. Comb of eighth segment with many scales in a patch; individual scale rounded apically and fringed with subequal spinules. Siphonal index 6.0 to 7.0; pecten of about 9 to 15 teeth on basal fourth of siphon; individual tooth with 2 to 6 coarse teeth on one side;

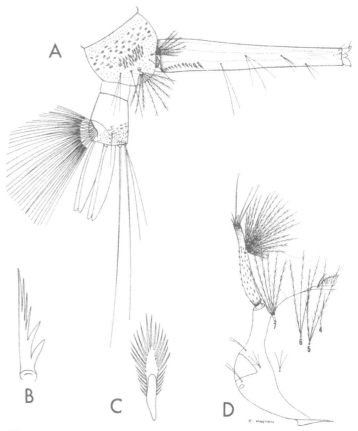

Fig. 240. Larva of *Culex nigripalpus* Theobald. *A*, Terminal segments. *B*, Pecten tooth. *C*, Comb scale. *D*, Head.

(135, 546); North Carolina (136); South Carolina (135, 742); Tennessee (135, 514); Texas (463, 634).

BIONOMICS. The larvae are found in ditches, grassy pools, and marshes of a semipermanent or permanent nature. They are occasionally found in water in wheel ruts, leaf axils of plants, and artificial containers. The females are sometimes taken in the southern United States while biting but seem to be less inclined to bite man than is *C. salinarius*. It is generally regarded as an outdoor species, but where adults are numerous they have occasionally been taken inside houses. Males and females are frequently captured in New Jersey light traps. Both the adults and larvae may be found throughout the year in the extreme southern United States.

Culex (Culex) pipiens Linnaeus[21]

Culex pipiens Linnaeus, 1758, Syst. Nat. Ed., *10*:602.

The *Culex pipiens* complex has recently been reviewed by Mattingly *et al.* (501). Rozeboom (629) states that there are four possible *C. pipiens* complex populations in North America north of Mexico: typical *pipiens* in the north, *quinquefasciatus* in the south autogenous *molestus* with an undetermined but apparently widespread distribution from New York to California, and *dispeticus* (? = *comitatus*) in southern California and western Mexico. The two widespread forms of the complex, *C. pipiens* and *C. quinquefasciatus*, are described in this study. Typical northern *pipiens* can be separated from typical southern *quinquefasciatus* by the relative positions of the arms of the phallosome. The extent of the distribution and prevalence of autogenous *molestus* in North America has not been fully explored; however, Galindo (300) believes that it is *molestus* rather than *pipiens* which is present in California. Studies of Farid (264), Sundararaman (679), and Barr and Kartman (32) show that *C. pipiens* and *quinquefasciatus* interbreed in captivity in the United States and that hybridization, giving rise to intermediate forms, takes place in nature. These writers thus concluded that the two forms should be treated as sub-

species. They are treated as full species in this study as a matter of convenience.

ADULT FEMALE (pl. 107). Very similar to *C. quinquefasciatus*, but usually has the pale basal bands of the abdominal tergites more broadly joined to the lateral white patches. Medium-sized species. *Head:* Proboscis dark-scaled; palpi short, dark. Occiput with narrow golden scales and erect forked scales dorsally (forked scales of central part usually pale, others dark brown), with broad white scales laterally. *Thorax:* Integument of scutum brown; scutum clothed with narrow curved golden-brown scales (coarser than on *C. restuans, salinarius*, and *nigripalpus*), paler on prescutellar space. Scutellum with narrow golden scales and brown setae on the lobes. Pleura with small patches of white scales. *Abdomen:* First tergite with a median patch of dark-bronze scales; remaining tergites dark-scaled with bronze to metallic blue-green reflection, with conspicuous basal bands and lateral patches of white scales; each band broadly rounded on posterior margin and narrowed at the sides where it joins the lateral patches. Venter predominantly whitish-scaled, usually speckled with a few brown scales. *Legs:* Legs dark-scaled with bronze to metallic blue-green reflection; posterior surface of femora and tibiae pale; femora and tibiae tipped apically with pale scales. *Wing:* Length 3.5 to 4.0 mm. Scales narrow, dark.

ADULT MALE. Coloration similar to that of the female, but with basal bands of the abdominal tergites more broadly joined to the lateral patches and not rounded on posterior margin. *Terminalia* (fig. 241). Lobes of ninth tergite (IXT-L) widely separated, slightly elevated, each bearing several slender setae. Tenth sternite (X-S) crowned with numerous short pointed spines; basal arm (XS-BA) variable in length, but usually represented by a short protuberance. Phallosome (Ph) formed of two large sclerotized plates connected at base. Each plate with the ventral arm (Ph-VA) large, winglike (usually narrower and more heavily sclerotized than in *C. quinquefasciatus*), curved outwardly, tapered to a point (a much smaller pointed process, similarly curved, present laterally near middle of plate on same plane as ventral arm); the dorsal arm (Ph-

---

[21]Consult Dyar (1928) and Edwards (1932) for synonyms.

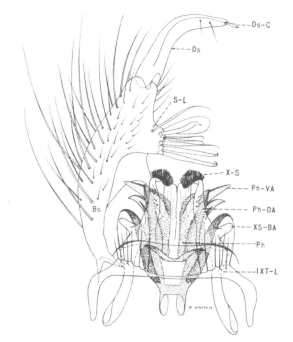

Fig. 241. Male terminalia of *Culex pipiens* Linnaeus.

DA) long, slender, straight, truncate or bluntly rounded at tip (rarely pointed), directed posterolaterally, and obliquely crossing over the winglike lateral extension of the ventral arm. Sundararaman (679) recommends using the ratio of the distance between the tips of the dorsal and ventral arms to that between the two dorsal arms for separating *C. pipiens* and *C. quinquefasciatus*. In *C. pipiens* the distance between the tips of the dorsal and ventral arms is very short owing to the inclination of the dorsal arm. It is much greater in *C. quinquefasciatus*. Intermediate forms occur which cannot be determined with certainty. Claspette absent. Basistyle (Bs) nearly two and a half times as long as mid-width, clothed with numerous setae, longer on outer aspect. Subapical lobe (S-L) prominent, undivided, and bearing the following appendages: two long stout rods and one long more slender rod, each pointed and usually slightly hooked at tip; two stout setae with tips somewhat recurved; a stout rod, about two-thirds as long as the first three rods, often with tip minutely hooked; a large broad leaflike filament; and a stout straight seta. Dististyle (Ds) about half as long as basistyle, curved, bearing two small setae before apex; claw (Ds-C) short, blunt.

LARVA (fig. 242). Antenna shorter than the head, constricted beyond antennal tuft, with part before constriction spiculate, part beyond constriction darker and sparsely spiculate; antennal tuft large, multiple, barbed, inserted at outer third of shaft, reaching beyond tip. Head hairs: Postclypeal 4 short, single; upper frontal 5 and lower frontal 6 usually long, with five or more branches, barbed; preantennal 7 long, multiple, barbed. Prothoracic hairs: 1–3 long, single; 4 medium, double; 5–6 long, single; 7 long, double. Body glabrous. Lateral abdominal hair 6 usually multiple on segments I and II, double on III–VI. Comb of eighth segment with many scales in a patch; individual scale rounded apically and fringed with subequal spinules. Siphonal index about 4.0; siphon gradually tapered to apex; pecten of about 6 to 13 teeth on basal third of siphon; individual tooth with 1 to 5 coarse teeth on one side; usually four pairs of siphonal tufts inserted beyond the pecten (the subapical tuft inserted laterally, the apical and subapical tufts usually double or triple, proximal tufts multiple and weakly barbed). Anal segment completely ringed by the saddle; lateral hair single, a little shorter than the saddle; dorsal brush bilaterally consisting of a long lower caudal hair and two shorter upper caudal hairs; ventral brush well developed, confined to the barred area; gills usually one and a half times as long as the saddle, pointed.

DISTRIBUTION. This species occurs in the United States from the Atlantic to the Pacific except in the extreme South, Canada, northern Europe and Asia, the southern part of South America, and in East and South Africa. *United States:* Alabama (738); Arkansas (679); California (287); Colorado (343); Connecticut (270, 494); Delaware (120, 167); District of Columbia (213, 318); Georgia (738); Idaho (344, 672); Illinois (615); Indiana (152, 345); Iowa (625, 626); Kansas (14, 360); Kentucky (570); Maine (45, 270); Maryland (64); Massachusetts (270); Michigan (391); Minnesota (533); Mississippi (135, 738); Missouri (6, 324); Montana (14); Nebraska (14, 680); New Hampshire (270); New Jersey (350); New York (31); North Carolina (646, 738); North Dakota (553); Ohio (482, 731a); Oklahoma (321);

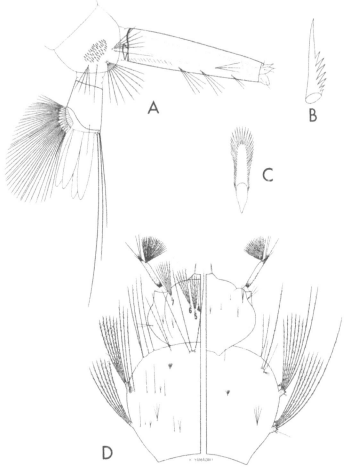

Fig. 242. Larva of *Culex pipiens* Linnaeus. *A*, Terminal segments. *B*, Pecten tooth. *C*, Comb scale. *D*, Head and thorax (dorsal and ventral).

Oregon (672); Pennsylvania (298, 756); Rhode Island (35, 270); South Carolina (738, 742); South Dakota (14); Tennessee (658, 738); Utah (575); Vermont (648); Virginia (176, 213); Washington (67, 672); Wisconsin (171); Wyoming (578). *Canada:* Alberta (598); British Columbia (212, 352); Manitoba (708); New Brunswick and Nova Scotia (536, 708); Ontario (212, 708); Quebec (536, 708).

BIONOMICS. The larvae are found in foul water in rain barrels, tubs, catch basins, faulty cesspools, ditches, and other similar habitats. Water containing vegetable wastes from food-processing plants often provides favorable conditions for larval development. It is a domesticated species developing in association with man. It is generally known as the northern house mosquito or rain-barrel mosquito and commonly infests houses and bites at night.

Adult females pass the winter hibernating in cellars, basements, outbuildings, caves, and other warm places which afford protection from the cold. Owen (533) describes a large mushroom cave in St. Paul, Minnesota, where millions of adults were observed in hibernation during early March.

MEDICAL IMPORTANCE. *Culex pipiens* is a known intermediate host of *Wuchereria bancrofti*. Western equine and St. Louis encephalitis viruses have been isolated from wild-caught specimens of *C. pipiens* in the Yakima Valley, Washington (332). It is known to transmit the organisms causing bird malaria and heartworm of dogs (*Dirofilaria immitis*). The virus of fowl pox is also known to be transmitted by this species.

Culex (Culex) quinquefasciatus Say[22]

Culex quinquefasciatus Say, 1823, Jour. Acad. Nat. Sci. Phila., *3*:10.

Culex fatigans Wiedemann, 1828, Ausser. Zweifl. Ins., *1*:10.

ADULT FEMALE (pl. 108). Similar to *C. pipiens* but has the pale basal bands of the abdominal tergites narrowly joining or entirely disconnected from the lateral patches.

ADULT MALE. Coloration similar to that of the female, but has the pale basal bands of the abdominal tergites broadly joined to the lateral patches and not prominently rounded on the posterior margin. *Terminalia* (fig. 243). Lobes of ninth tergite (IXT-L) widely separated, slightly elevated, each bearing several slender setae. Tenth sternite (X-S) densely crowned with short pointed spines; basal arm (XS-BA) variable in length, but usually represented by a short protuberance. Phallosome (Ph) consists of two large sclerotized plates connected near base. Each plate with the ventral arm (Ph-VA) large, winglike (usually broader and more weakly sclerotized than in *C. pipiens*), curved outwardly, tapered to a point (a much smaller pointed process, similarly curved,

---

[22]Consult Dyar (1928) and Edwards (1932) for additional synonyms.

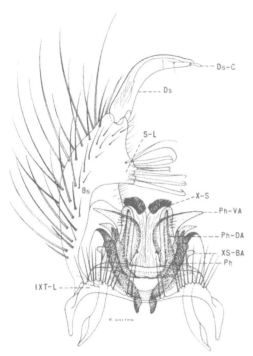

Fig. 243. Male terminalia of *Culex quinquefasciatus* Say.

LARVA (fig. 244). The larva is similar to that of *C. pipiens* except that the siphon is rather stout on the basal half and tapered distally. The siphon in *C. pipiens* is usually more slender. Antenna shorter than the head, constricted beyond the antennal tuft, part before constriction spiculate, part beyond constriction darker and with fewer spicules; antennal tuft large, multiple, barbed, inserted at outer third of shaft, reaching beyond tip. Head hairs: Postclypeal 4 short, single; upper frontal 5 and lower frontal 6 multiple (five or more branches), barbed; preantennal 7 long, multiple, barbed. Prothoracic hairs: 1–3 long, single; 4 medium, double; 5–6 long, single; 7 long, double. Body glabrous. Lateral abdominal hair 6 usually multiple on segments I and II, single or double on III–VI. Comb of eighth segment with numerous scales in a patch; individual scale rounded apically and fringed with subequal spinules. Siphonal index usually about 3.8 to 4.0; siphon with distal half tapered; pecten of about 8 to 12 teeth on basal

present laterally near middle of plate on same plane as ventral arm); the dorsal arm (Ph-DA) long, slender, straight, pointed or narrowly rounded at tip, directed posteriorly and crossing over the ventral arm nearly at a right angle to its winglike outward extension. The distances between the tips of the dorsal and ventral arms and the dorsal arms are about equal to one another in *C. quinquefasciatus*. In *C. pipiens* the distance between the tips of the dorsal and ventral arms is short. Intermediate forms occur which cannot be determined with certainty. Claspette absent. Basistyle (Bs) nearly two and a half times as long as mid-width, clothed with numerous setae, longer on outer aspect. Subapical lobe (S-L) undivided, bearing the following appendages: two long strong rods and one long more slender rod, each pointed and usually slightly hooked at tip; two stout setae with tips somewhat recurved; one stout rod, about two-thirds as long as the first three rods, often with tip minutely hooked; one large broad leaflike filament; and one strong straight seta. Dististyle (Ds) about half as long as the basistyle, bearing two small setae before apex; claw (Ds-C) short, blunt.

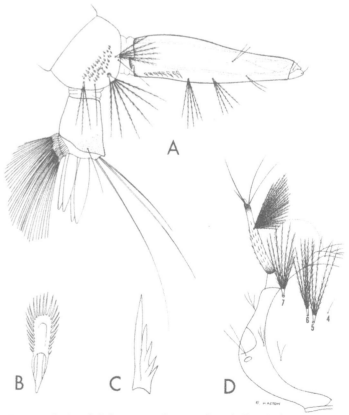

Fig. 244. Larva of *Culex quinquefasciatus* Say. *A*, Terminal segments. *B*, Comb scale. *C*, Pecten tooth. *D*, Head.

third of siphon; individual tooth with 1 to 5 coarse teeth on one side; usually four pairs of siphonal tufts inserted beyond the pecten (the subapical tuft inserted laterally, the apical and subapical tufts usually double or triple, proximal tufts multiple and weakly barbed). Anal segment completely ringed by the saddle; lateral hair single, a little shorter than the saddle; dorsal brush bilaterally consisting of a long lower caudal hair and two upper caudal hairs, one long and one short; ventral brush large, confined to the barred area; gills about one to one and a half times as long as the saddle, pointed.

DISTRIBUTION. This species occurs throughout the tropical and subtropical regions of the world. *United States:* Alabama (135, 419); Arizona (213); Arkansas (88, 121); California (287); District of Columbia (213, 318); Florida and Georgia (135, 419); Illinois (615); Iowa (531); Kansas (14, 360); Kentucky (213, 570); Louisiana (213, 759); Mississippi (135, 546); Missouri (6, 324); Nebraska (14, 680); New Mexico (20, 273); North Carolina (135, 419); Ohio (213); Oklahoma (628); South Carolina (135, 742); Tennessee (135, 658); Texas (249, 463); Utah (580); Virginia (176, 178).

BIONOMICS. The larvae are found in foul water in rain barrels, tubs, catch basins, cesspools, ditches, ground pools, and other similar habitats. It is commonly known as the southern house mosquito and is a troublesome domesticated species that commonly infests houses and bites at night. It is said to display a preference for avian blood. The adults do not occur in large numbers until July or August on the northern edge of its North American range, and they soon disappear, except for hibernating females, after the first cool weather in September or October. Both the larvae and adults can be found throughout the year in the extreme South.

MEDICAL IMPORTANCE. This species is an important vector of *Wuchereria bancrofti* in the tropical and subtropical regions of the world. It is also a known vector of the parasites causing bird malaria. Western equine and St. Louis encephalitis viruses have been isolated also from *Culex quinquefasciatus* in California.

Culex (Culex) restuans Theobald

Culex restuans Theobald, 1901, Mon. Culic., 2:142.
Culex brehmei Knab, 1916, Proc. Biol. Soc. Wash., 29:161.

ADULT FEMALE (pl. 109). Medium-sized species. *Head:* Proboscis dark-scaled, with some pale scales on ventral surface; palpi short, dark. Occiput clothed with narrow curved yellowish-white to golden-brown scales and dark-brown erect forked scales dorsally, with broad yellowish-white scales laterally. *Thorax:* Integument of scutum light brown to reddish brown; scutum clothed with fine narrow curved golden-brown scales, paler on anterior and lateral margins and on the prescutellar space; a pair of small pale-scaled submedian spots usually present near middle of scutum. Scutellum with narrow pale-golden scales and brown setae on the lobes. Pleura with small patches of broad pale scales. *Abdomen:* First tergite with a median patch of dark bronze-brown scales; remaining tergites dark bronze-brown, each with a basal band of white to yellowish-white scales (the bands are usually not as evenly rounded as in *C. quinquefasciatus* and are usually broadly joined with the basolateral patches of pale scales). Venter mostly pale-scaled. *Legs:* Legs dark-scaled with bronze to metallic blue-green reflection, posterior surface of femora and tibiae pale; femora and tibiae tipped apically with pale scales. Tarsi entirely dark or with faint brownish-yellow bands. *Wing:* Length about 4.0 to 4.4 mm. Scales narrow, dark.

ADULT MALE. Coloration similar to that of the female. *Terminalia* (fig. 245). Lobes of ninth tergite (IXT-L) large, rounded, each bearing several small setae. Tenth sternite (X-S) densely crowned with pointed spines; basal arm (XS-BA) long, stout, usually curved, weakly sclerotized. Phallosome (Ph) consists of two large sclerotized plates. Each plate with a long slender apical ventral arm (Ph-VA), a short blunt basal dorsal arm (Ph-DA) curved outwardly, and a short stout triangular tooth (Ph-T). Claspette absent. Basistyle (Bs) about two to two and a half times as long as the midwidth, clothed with numerous setae, longer on outer aspect. Subapical lobe (S-L) undivided, bearing three long strong rods with tips hooked,

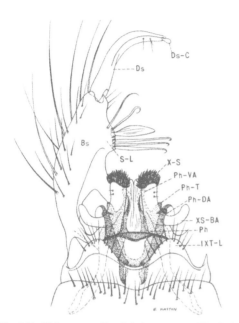

Fig. 245. Male terminalia of *Culex restuans* Theobald.

a large broad leaflike filament, and two stout setae, one of which is hooked. Dististyle (Ds) about half as long as the basistyle, bearing two small setae before apex; claw (Ds-C) short, blunt.

LARVA (fig. 246). Antenna shorter than the head, spiculate, slightly narrowed and darker beyond antennal tuft; antennal tuft multiple, barbed, inserted near middle of shaft, not reaching tip. Head hairs: Postclypeal 4 short, double or triple; upper frontal 5 and lower frontal 6 with four to eight branches, barbed; preantennal 7 multiple, barbed. Prothoracic hairs: 1–2 long, single; 3 long, single or double; 4 long, double; 5–6 long, single; 7 long, double or triple. Body glabrous. Lateral abdominal hair 6 double on segments I and II, long and single on III–VI. Comb of eighth segment with many scales in a patch; individual scale rounded and fringed apically with subequal spinules. Siphonal index 4.0 to 4.5; pecten of about 12 to 20 teeth on basal third of siphon; individual tooth with 1 to 5 coarse teeth on one side; siphonal tufts represented by three pairs of long single hairs irregularly placed and a pair of small subapical tufts of two or three branches each, inserted beyond the pecten. Anal segment completely ringed by

the saddle; saddle spiculate on dorsoapical surface; lateral hair single or double; dorsal brush bilaterally consisting of a long single upper and a long single lower caudal hair; ventral brush well developed, confined to the barred area; gills about two to three times as long as the saddle, tapered.

DISTRIBUTION. This species is widely distributed in North America from the Gulf of Mexico to southern Canada. It also occurs in Mexico (537a). *United States:* Alabama (135, 419); Arkansas (88, 121); California (287); Colorado (343); Connecticut (270, 494); Delaware (120, 167); District of Columbia (318); Florida and Georgia (135, 419); Idaho (344); Illinois (615); Indiana (152, 345); Iowa (625, 626); Kansas (14, 360); Kentucky (134, 570); Louisiana (419, 759);

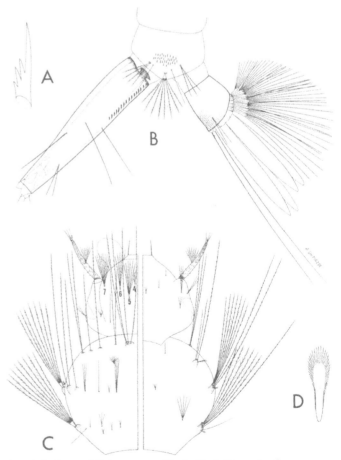

Fig. 246. Larva of *Culex restuans* Theobald. *A*, Pecten tooth. *B*, Terminal segments. *C*, Head and thorax (dorsal and ventral). *D*, Comb scale.

Maine (45, 270); Maryland (64, 161); Massachusetts (270); Michigan (391, 529); Minnesota (533); Mississippi (135, 546); Missouri (6, 324); Montana (475); Nebraska (14, 680); New Hampshire (66, 270); New Jersey (213, 350); New Mexico (273); New York (31); North Carolina (135); North Dakota (553); Ohio (213, 731a); Oklahoma (628); Pennsylvania (298, 756); Rhode Island (35, 270); South Carolina (135, 742); South Dakota (14); Tennessee (135, 658); Texas (463, 634); Utah (575); Vermont (270); Virginia (176, 178); West Virginia (213); Wisconsin (171); Wyoming (578). *Canada:* British Columbia (707); Manitoba (466, 598); New Brunswick (708); Ontario (212, 708); Quebec (289, 708); Saskatchewan (598, 708).

BIONOMICS. The larvae are found in a wide variety of aquatic habitats, such as ditches, pools in streams, woodland pools, and artificial containers. The species reaches its greatest abundance in the spring and early summer throughout most of its range, and occurs in lesser numbers during late summer and autumn. The larvae and adults may be found throughout the year in the extreme South. The females are regarded as troublesome biters by some observers, although others say that they rarely bite man.

MEDICAL IMPORTANCE. Norris (528) recovered a virus believed to be that of western equine encephalitis from wild-caught *C. restuans* in Manitoba during the summer of 1944.

Culex (Culex) salinarius Coquillett

Culex salinarius Coquillet, 1904, Ent. News, *15*:73.

ADULT FEMALE (pl. 110). Medium-sized species. *Head:* Proboscis dark-scaled, usually paler on ventral side; palpi short, dark. Occiput clothed dorsally with narrow curved pale golden-brown scales and dark erect forked scales, those along the margins of the eyes yellowish white, with a patch of broad dingy-white scales laterally. *Thorax:* Integument of scutum light brown to dark brown; scutum clothed with fine narrow curved golden-brown scales, somewhat paler on the anterior and lateral margins and on the prescutellar space. Scu-

tellum with narrow yellowish-white to light-golden scales and brown setae on the lobes. Pleura with several small patches of broad dingy-white scales, each patch usually comprised of more than six scales. *Abdomen:* First tergite with a median area of dark-brown scales; remaining tergites primarily dark-brown-scaled with bronze to metallic blue-green reflection, often with narrow to moderately broad basal bands of dingy-yellow scales contiguous with basolateral patches of pale scales; seventh and eighth tergites often entirely covered with dingy-yellow scales. Venter yellowish white. *Legs:* Legs dark-scaled with bronze reflection, posterior surface of femora and tibiae pale. *Wing:* Length 3.5 to 4.0 mm. Scales narrow, dark.

ADULT MALE. Coloration similar to that of the female, but with pale bands on abdominal tergites much broader. *Terminalia* (fig. 247). Eighth tergite (VIII-T) bearing numerous apical setae, longer and stronger than those on the lobes of the ninth tergite. Lobes of ninth tergite (IXT-L) widely separated, slightly raised, each bearing a row of several slender setae.

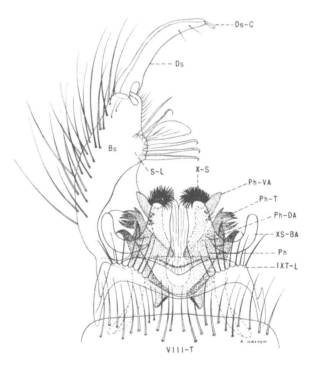

Fig. 247. Male terminalia of *Culex salinarius* Coquillett.

Tenth sternite (X-S) densely crowned with short pointed spines; basal arm (XS-BA) long, stout, strongly curved, sclerotized. Phallosome (Ph) consists of two sclerotized plates connected near base. Each plate bears a stout pointed dorsal arm (Ph-DA) bent medially at a right angle, a stout bluntly pointed ventral arm (Ph-VA) bearing a small projection on its inner margin, and a group of strong pointed teeth (Ph-T). Claspette absent. Basistyle (Bs) about two and a half times as long as the mid-width, slender, clothed with numerous setae, longer on outer aspect. Subapical lobe (S-L) undivided, bearing three long strong rods usually hooked at tips, a large broad leaflike filament, and two strong setae, one of which is hooked. Dististyle (Ds) about half as long as the basistyle, bearing two small setae on inner surface before apex; claw (Ds-C) short, blunt.

LARVA (fig. 248). Antenna shorter than the head, constricted beyond insertion of antennal tuft, with part before constriction pale and spiculate, part beyond constriction darker and with few spicules; antennal tuft large, multiple, barbed, inserted at outer third of shaft, reaching well beyond tip. Head hairs: Postclypeal 4 short, single; upper frontal 5 long, 3- to 6-branched, barbed; lower frontal 6 long, 3- or 4-branched, barbed; preantennal 7 long, multiple, barbed. Prothoracic hairs: 1–2 long, single; 3 long, single or double; 4 medium, double; 5–6 long, single; 7 long, double or triple. Thorax glabrous, although occasionally minute papillated hairs may be present. Lateral abdominal hair 6 triple on segments I and II, usually double or triple on III–VI. Comb of eighth segment with many scales in a patch; individual scale rounded apically and fringed with subequal spinules. Siphonal index about 6.0 to 7.0; pecten of about 10 to 16 teeth on basal fourth of siphon; individual tooth with two to five coarse teeth on one side; usually four (occasionally five) paired 2- to 4-branched siphonal tufts inserted beyond pecten (the proximal tuft as long or longer than basal diameter of the siphon), the subapical tuft inserted laterally. Anal segment completely ringed by the saddle; lateral hair double, sometimes single, a little shorter than the saddle; dorsal brush bilaterally consisting of a long

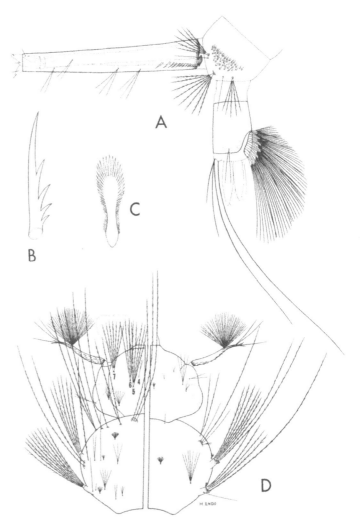

Fig. 248. Larva of *Culex salinarius* Coquillett. *A*, Terminal segments. *B*, Pecten tooth. *C*, Comb scale. *D*, Head and thorax (dorsal and ventral).

lower caudal hair and an upper caudal tuft of one long and two shorter hairs; ventral brush well developed, confined to the barred area; gills varying from about one to one and a half times as long as the saddle, bluntly pointed.

DISTRIBUTION. This species occurs in the eastern United States and southeastern Canada. Its range is known to extend west to Utah. It is found also in Mexico (537a). *United States:* Alabama (135, 419); Arkansas (88, 121); Colorado (343); Connecticut (270, 494); Delaware (120, 167); District of Columbia (213, 318); Florida and Georgia (135, 419); Idaho (344); Illinois (615); Indiana (133); Iowa (625, 626); Kansas (14, 530); Kentucky (134, 570); Louisiana (419, 759); Maine (45, 270); Maryland (64, 161); Massachusetts (270); Michigan (391, 529); Minnesota (533); Mississippi (135, 546); Missouri (6, 324); Nebraska (14, 680); New Hampshire (66, 270); New Jersey (213, 350); New Mexico (20, 273); New York (31); North Carolina (135, 213); North Dakota (553); Ohio (482, 731a); Oklahoma (628); Pennsylvania (298, 756); Rhode Island (35, 270); South Carolina (135, 742); South Dakota (14); Tennessee (135, 420); Texas (463, 634); Utah (575); Vermont (270); Virginia (176, 178); Wisconsin (171); Wyoming (578). *Canada:* Nova Scotia (707, 708).

BIONOMICS. The larvae are found either in fresh or foul water in grassy pools, ditches, ponds, occasionally in rain barrels, bilge water in boats, cattle tracks, and sometimes in stump holes. Larval development of this species begins early in the season and continues at a rather uniform rate during the summer and early fall throughout most of its range. Larvae and adults can be found any time during the year in the extreme South, but the females overwinter in hibernation farther north.

The adults are frequently found resting during the daytime in outbuildings and other similar shelters. The males and females are often captured in large numbers in New Jersey light traps in the Atlantic and Gulf coastal regions where the species probably reaches its greatest abundance.

The females bite readily outdoors and occasionally enter dwellings to feed on man. Rozeboom (628) took this mosquito while feeding on man more often than any other *Culex* in Oklahoma. Wallis and Spielman (736) describe the successful rearing of *C. salinarius* to the third generation in the laboratory. They observed the mating flight which was swift and brief, lasting only a few seconds.

Culex (Culex) stigmatosoma Dyar

Culex stigmatosoma Dyar, 1907, Proc. U.S. Nat. Mus., *32*:123.
Culex eumimetes Dyar and Knab, 1908, Proc. U.S. Nat Mus., *35*:61.

ADULT FEMALE (pl. 111). Similar to *C. tar-*salis but lacks white scales in a line or row of spots on outer surface of femora and tibiae, and has dark scales on each sternite of the abdomen in the form of an oval median spot, not in the form of an inverted V. Medium-sized species. *Head:* Proboscis dark-scaled, with a rather broad white median band; palpi short, dark except for a few white scales at tip. Occiput with narrow curved yellowish scales and dark erect forked scales dorsally, with broad white appressed scales laterally. *Thorax:* Integument of scutum dark brown; scutum clothed with narrow curved golden-brown scales; the anterior and lateral margins and the prescutellar space with yellowish-white scales; an indistinct narrow submedian yellowish-white line on either side of the prescutellar space reaching near the middle of the scutum and extending toward the scutal angle. Scutellum with narrow yellowish-white scales and dark-brown setae on the lobes. Pleura with small patches of white scales. *Abdomen:* First tergite with a median patch of dark scales; remaining tergites dark-scaled, with basal bands of yellowish-white scales widening at the sides. Venter pale-scaled, with an oval median spot of black scales on each sternite, generally distinct but sometimes suffused. *Legs:* Femora and tibiae dark, posterior surface pale, tipped with white. Hind tarsi dark, with broad white bands involving both ends of the joints; middle and front tarsi similarly marked but with the white bands narrower, much reduced or lacking on segments 4 and 5. *Wing:* Length 4.0 to 4.5 mm. Scales narrow, dark, a few white scales often present on costa.

ADULT MALE. Coloration similar to that of the female, but with white scaling on the abdominal tergites usually more extensive. *Terminalia* (fig. 249). Similar to that of *C. thriambus* but constant differences are found in the appendages of the subapical lobe. Lobes of ninth tergite (IXT-L) slightly elevated, widely separated, each bearing several slender setae. Tenth sternite (X-S) densely crowned with stout spines; basal arm (XS-BA) long, stout, strongly curved, sclerotized. Phallosome (Ph) formed of two large sclerotized plates. Each plate with a long pointed dorsal arm (D-A), a bluntly rounded ventral arm (V-A), and about 4 stout pointed teeth (Ph-T) bordered by a

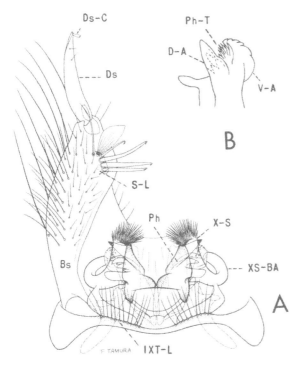

Fig. 249. *Culex stigmatosoma* Dyar. *A*, Male terminalia.
B, Plate of phallosome.

frontal 5 and lower frontal 6 long, usually 4-
to 7-branched, barbed; preantennal 7 multiple,
barbed, reaching near insertion of antennal
tuft. Prothoracic hairs: 1–3 long, single; 4 medium, double; 5–6 long, single; 7 long, double
or triple. Thorax glabrous. Lateral abdominal
hair 6 usually triple on segments I–IV and
double or triple on V and VI. Comb of eighth
segment of numerous scales in a patch; individual scale rounded apically and fringed with
subequal spinules. Siphonal index 4.0 to 5.0;
pecten consists of about 9 to 15 teeth on basal
third of siphon; individual pecten tooth with
1 to 5 coarse teeth on one side; five pairs of
siphonal tufts present, the subapical pair inserted laterally out of line. Anal segment completely ringed by the saddle; spicules conspicuously enlarged and dark on dorsoapical
part of the saddle; lateral hair single or double,
shorter than the saddle; dorsal brush bilaterally consisting of a long lower caudal hair and

group of compact slender teeth arising from
the inner side of the ventral arm. Claspette
absent. Basistyle (Bs) a little more than twice
as long as the mid-width, clothed with numerous setae, longer on outer aspect. Subapical
lobe (S-L) prominent, undivided, bearing
three stout rods, a short slender hooked spine,
a rather broad leaflike appendage and a long
slender spine. Dististyle (Ds) about half as
long as the basistyle, gradually tapered; claw
(Ds-C) short, blunt.

LARVA (fig. 250). Similar to *C. tarsalis* but
has conspicuous dark spicules on dorsoapical
part of the saddle of the anal segment. These
are lacking in *C. tarsalis*. The subapical tuft of
the siphon is usually lateral or sublateral in
position in *C. stigmatosoma* and is usually
ventral in position in *C. tarsalis*. Head broader
than long. Antenna shorter than the head, constricted beyond the antennal tuft, part before
constriction dark near base, pale to near insertion of tuft, spiculate, part beyond constriction darker and with fewer spicules:
antennal tuft large, multiple, barbed, inserted
at outer third of shaft, reaching beyond tip.
Head hairs: Postclypeal 4 short, single; upper

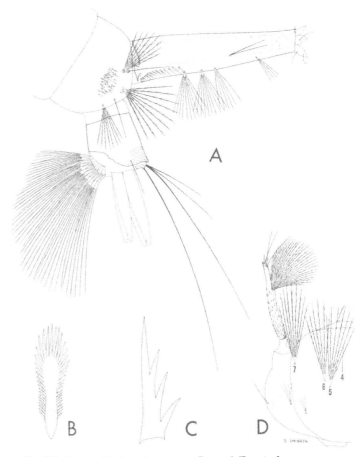

Fig. 250. Larva of *Culex stigmatosoma* Dyar. *A*, Terminal
segments. *B*, Comb scale. *C*, Pecten tooth. *D*, Head.

an upper caudal tuft of three hairs, one nearly as long as the lower caudal; ventral brush well developed, confined to the barred area; gills usually a little longer than the saddle, bluntly pointed.

DISTRIBUTION. This species is known to occur in the western United States from Washington south to Mexico, Central America, and northern South America. *United States:* California (287); Oklahoma (321); Oregon (672); Texas (249); Washington (672). Dyar (237) and Matheson (492) list the species as also occurring in Utah; however, Rees (575) did not find it but states that it may occur there but not abundantly.

BIONOMICS. The larvae of *Culex stigmatosoma* are generally found in stagnant or foul water at sewage plants, in street drains, in polluted water on farms especially around dairies, and occasionally in rather clean water. The larvae are sometimes found in artificial containers. Freeborn (284) says this species finds its optimum breeding conditions in the pools of dry arroyos in southern California. The females rarely feed on man but can be induced to feed on chickens and guinea pigs in the laboratory.

MEDICAL IMPORTANCE. Western equine encephalitis virus has been isolated from wild-caught specimens of *C. stigmatosoma* in Kern County, California (337).

Culex (Culex) tarsalis Coquillett

Culex tarsalis Coquillett, 1896, Can. Ent., *28*:43.
Culex willistoni Giles, 1900, Handbook Gnats or Mosq., p. 281.
Culex affinis Adams, 1903, Kans. Univ. Sci. Bull., *20*:25.
Culex kelloggi Theobald, 1903, Can. Ent., *25*:211.
Culex peus Speiser, 1904, Insektenbörse, *21*:148.

ADULT FEMALE (pl. 112). Medium-sized species. *Head:* Proboscis dark-scaled, with a rather broad median white band; palpi short, dark except for a few white scales at tip and at apex of third segment. Occiput with narrow white scales in a median triangular patch, broad posteriorly, narrow anteriorly; eyes margined with narrow white scales; erect forked scales on dorsal surface dark, a few pale ones on the median area; a submedian

patch of narrow golden-brown scales behind the eye margin; broad white scales on lateral region of occiput. *Thorax:* Integument of scutum dark brown to black; scutum clothed with fine narrow curved golden-brown scales dorsally, narrowly margined anteriorly and laterally with narrow white scales. Prescutellar space with white scales; a pair of narrow submedian white lines extending forward to near middle of scutum and each terminating in a small white submedian spot which is often separated from the white line by brown scales. Scutellum with narrow whitish scales and brown setae on the lobes. Pleura with small patches of broad dingy-white scales. *Abdomen:* First tergite with a median patch of dark bronze-brown scales, usually with a few pale ones intermixed; second tergite dark-scaled, with a median basal triangular patch of pale scales; remaining tergites dark, with prominent basal bands of white or yellowish-white scales; the terminal segments often with apical pale scaling as well as basal, the eighth segment often entirely pale-scaled. Venter pale-scaled, with a V-shaped marking of dark scales on each sternite, the base of the V at the median anterior margin. *Legs:* Hind legs dark-scaled except for the following markings: posterior surface of femora and tibiae mostly yellowish-white-scaled; a narrow line of white scales or a row of spots on anterior surface of femora, tibiae, and often on first segment of tarsi; femora and tibiae tipped apically with white; wide basal and apical white bands on tarsal segments. Front and middle legs similarly marked but with tarsal bands narrower on segments 1-3, reduced or absent on segments 4 and 5. *Wing:* Length about 4.0 to 4.4 mm. Scales narrow, dark, a few white scales on costa and subcosta.

ADULT MALE. Coloration similar to that of the female, but with the white scaling on the abdominal tergites usually more extensive. *Terminalia* (fig. 251). Lobes of ninth tergite (IXT-L) slightly elevated, widely separated, each bearing several slender setae. Tenth sternite (X-S) densely crowned with short spines, the apical ones pointed, the outer ones blunt; basal arm (XS-BA) long, stout, strongly curved, sclerotized. Phallosome (Ph) consists of two sclerotized plates. Each plate with a long dorsal arm (D-A) directed posteriorly, a long

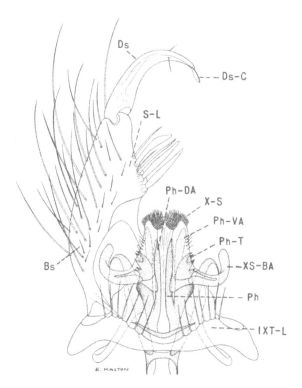

Fig. 251. Male terminalia of *Culex tarsalis* Coquillett.

hair 6 usually triple on segments I–V, double or triple on VI. Comb of eighth segment with many scales in a patch; individual scale rounded apically and fringed with subequal spinules. Siphonal index about 4.5 to 5.5; pecten of about 10 to 15 teeth on basal third of siphon; individual tooth with 1 to 5 coarse teeth on one side; five pairs of multiple siphonal tufts present, all inserted in a straight line, the proximal pair often inserted near or slightly before end of pecten. Anal segment completely ringed by the saddle; lateral hair usually double or triple, occasionally single, shorter than the saddle; dorsal brush bilaterally consisting of a long lower caudal hair and an upper caudal tuft of three hairs, one nearly as long as the lower caudal; ventral brush well developed, confined to the barred area; gills varying in length, usually one to one and a half times as long as the saddle, tapered.

DISTRIBUTION. The western, central, and

ventral arm (V-A), and several stout teeth (Ph-T). Claspette absent. Basistyle (Bs) about two and a half times as long as the mid-width, clothed with numerous setae, longer on outer aspect. Subapical lobe (S-L) undivided, bearing appendages as follows: two stout rods; a stout spine; a weak spine with tip slightly hooked; a narrow leaflike filament; and a long, slender spine. Dististyle (Ds) about half as long as the basistyle, bearing two small setae before apex; claw (Ds-C) short, blunt.

LARVA (fig. 252). Antenna shorter than the head, constricted beyond the antennal tuft, with part before constriction spiculate, dark near base, then pale to near insertion of tuft, part beyond constriction darker and with fewer spicules; antennal tuft large, multiple, barbed, inserted at outer third of shaft, reaching beyond tip. Head hairs: Postclypeal 4 small, single; upper frontal 5 and lower frontal 6 multiple, barbed; preantennal 7 multiple, barbed. Prothoracic hairs: 1–2 long, single; 3 long, single or double; 4 medium, usually double; 5–6 long, single; 7 long, double or triple. Thorax glabrous. Lateral abdominal

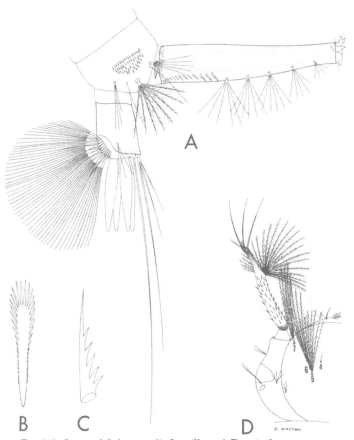

Fig. 252. Larva of *Culex tarsalis* Coquillett. *A*, Terminal segments. *B*, Comb scale. *C*, Pecten tooth. *D*, Head.

southern United States and southwestern Canada. Its range also extends south to Mexico (537a). *United States:* Alabama (124, 135); Arizona (213); Arkansas (88, 121); California (287); Colorado (343); Florida (135, 514); Georgia (124, 135); Idaho (344, 672); Illinois (615); Indiana (152, 345); Iowa (625, 626); Kansas (14, 360); Kentucky (134, 570); Louisiana (419, 759); Michigan (391); Minnesota (533); Mississippi (124, 135); Missouri (6, 324); Montana (475); Nebraska (14, 680); Nevada (213); New Mexico (20, 273); North Dakota (553); Oklahoma (628); Oregon (213, 672); South Carolina (124, 742); South Dakota (14, 530); Tennessee (124, 658); Texas (213, 463); Utah (575); Virginia (237); Washington (67, 672); Wisconsin (171); Wyoming (578). *Canada:* Alberta (707, 708); British Columbia (212, 352); Manitoba (536, 708); Northwest Territories (289); Saskatchewan (212, 464).

BIONOMICS. The larvae are found in clear or foul water in a variety of habitats including ditches, irrigation systems, ground pools, marshes, pools in stream beds, rain barrels, hoofprints, and ornamental pools. Foul water in corrals and around slaughter yards appear to be favorite larval habitats in many localities. The species has been found to 7,000 feet in the Sierra in California and to 9,000 feet in the Rockies in Utah. Larval production commences during the late spring and continues until early autumn throughout most of its range, with several generations being produced. The maximum adult population is generally reached during August or September.

The females are painful and usually persistent biters, attacking at dusk and after dark, and readily entering dwellings for blood meals. Domestic and wild birds seem to be the preferred hosts, and man, cows, and horses are generally incidental hosts. The adults hide in sheltered places during the day. Adult females pass the winter in colder climates in hibernation in basements, cellars, caves, and outbuildings where protection is afforded from cold. In the temperate climate of the West, production may continue throughout the year. Brennan and Harwood (103) report the successful

establishment of a laboratory colony of this species.

MEDICAL IMPORTANCE. *Culex tarsalis* is believed to be the chief vector of western equine encephalitis virus under natural conditions. The virus has been isolated from wild-caught *C. tarsalis* on several occasions in areas in which the disease was both epidemic and epizoötic. The viruses of both St. Louis and California encephalitis have been isolated from this mosquito.

Culex (Culex) thriambus Dyar

Culex (Culex) thriambus Dyar, 1921, Ins. Ins. Mens., *9*:33.

Culex (Culex) stigmatosoma, Dyar, 1928 (in part), Carnegie Inst. Wash. Pub., *387*:368.

Culex (Culex) stigmatosoma var. *thriambus*, Edwards, 1932, Gen. Insect. Fasc., *194*:206.

Culex stigmatosoma, Matheson, 1929 (in part), Handbook Mosq. No. Amer., p. 176.

Culex (Culex) thriambus, Galindo and Kelley, 1943, Pan-Pac. Ent., *19*:87.

ADULT FEMALE (pl. 113). Medium-sized species. *Head:* Proboscis dark-scaled, with a broad median white area on ventral side, extending around the sides in some specimens but never completely ringed; palpi short, dark. Occiput clothed dorsally with narrow yellowish-white scales and brown erect forked scales, with a patch of broad appressed grayish-white scales laterally. *Thorax:* Integument brown; scutum clothed with narrow curved golden-brown scales, with pale-golden scales on anterior and lateral margins and on the prescutellar space. Scutellum with narrow pale-golden scales and brown setae on the lobes. Pleura with small patches of grayish-white scales. *Abdomen:* First tergite with a median patch of dark scales; remaining tergites dark, with white basal bands joining with white basolateral patches. Venter pale, with triangular dark patches posteriorly on segments II–VI. *Legs:* Femora and tibiae dark, posterior surface pale; knee spots white. Hind tarsi dark, with narrow brownish to rather broad white bands involving both ends of joints, segment 5 occasionally all white; front and middle tarsi dark, except for medium to narrow pale bands on segments 1 to 3, usually indistinct or obsolete on 4 and 5. *Wing:* Length about 3.6 to 4.0 mm. Scales narrow, dark.

ADULT MALE. Coloration similar to that of the female, but with abdominal white scaling usually more extensive and proboscis with a complete ring. *Terminalia* (fig. 253). Very similar to that of *C. stigmatosoma* but constant differences are found in the subapical lobe. The subapical lobe in *C. thriambus* has three stout rods, a rather broad leaflike appendage, and a long slender spine. The subapical lobe in *C. stigmatosoma* has three stout rods, a short slender hooked spine, a leaflike appendage, and a long slender spine. Lobes of ninth tergite (IXT-L) slightly raised, widely separated, each bearing several slender setae. Tenth sternite (X-S) densely crowned with stout spines; basal arm (XS-BA) long, stout, strongly curved. Phallosome (Ph) composed of two sclerotized plates. Each plate bears a long pointed dorsal arm (D-A), a short bluntly rounded ventral arm (V-A), and about 4 stout teeth (Ph-T) adjacent to a group of slender teeth arising from the inner side of the ventral arm. Claspette absent. Basistyle (Bs) a little

more than twice as long as mid-width, clothed with many setae, longer on outer aspect. Subapical lobe (S-L) prominent, undivided, bearing three stout rods, a rather broad leaflike appendage and a long slender spine. Dististyle (Ds) about half as long as the basistyle, curved, and gradually tapered distally; claw (Ds-C) short, blunt.

LARVA (fig. 254). Antenna shorter than the head, constricted beyond antennal tuft, with part before constriction spiculate, part beyond constriction darker and with fewer spicules; antennal tuft large, multiple, barbed, inserted at outer third of shaft, extending considerably beyond tip. Head hairs: Postclypeal 4 short, single; upper frontal 5 and lower frontal 6 long, multiple, barbed; preantennal 7 long, multiple, barbed. Prothoracic hairs: 1–3 long, single; 4 long, double; 5–6 long, single; 7 long, usually double. Thorax glabrous. Lateral abdominal hair 6 usually triple on segments I and II, double on III–V and single on VI. Comb of eighth segment of numerous scales in a patch; individual scale rounded and fringed apically with subequal spinules. Siphonal index about 6.0; pecten of about 10 to 14 teeth on basal third of siphon; individual pecten tooth with 1 to 4 coarse teeth on one side; siphonal tufts represented by three pairs of long, single (rarely double) hairs irregularly placed and a pair of small 2- or 3-branched subapical tufts (these subapical tufts are sometimes represented by single hairs). Anal segment completely ringed by the saddle, with coarse dark spicules on dorsoapical surface; lateral hair single, a little shorter than the saddle; dorsal brush bilaterally consisting of a long single lower caudal hair and two upper caudal hairs, one long and one short; ventral brush well developed, confined to the barred area; gills usually a little longer than the saddle, pointed.

DISTRIBUTION. Mexico (537a) and the southwestern United States. *United States:* California (287, 301); Oklahoma (321, 472); Texas (208, 463, 634).

BIONOMICS. Dyar (208) found the larvae of *Culex thriambus* in a small dirty pool beside the river at Kerrville, Texas. Galindo and Kelley (301) refer to one collection of this species made in California at an elevation of 5,000 feet. Seaman (654) found the larvae in marshes

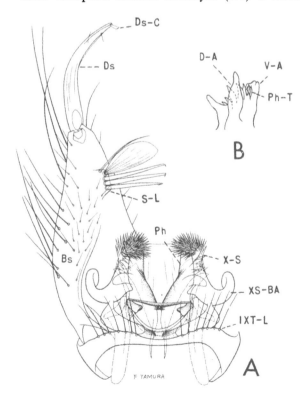

Fig. 253. *Culex thriambus* Dyar. *A*, Male terminalia. *B*, Plate of phallosome.

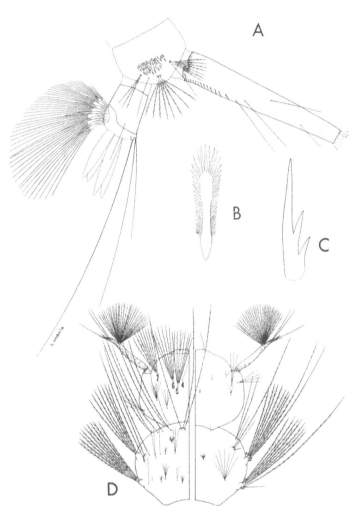

Fig. 254. Larva of *Culex thriambus* Dyar. *A*, Terminal segments. *B*, Comb scale. *C*, Pecten tooth. *D*, Head and thorax (dorsal and ventral).

and marshy ponds in southern California. The larvae have been found by Freeborn and Bohart (287) in leaf-filled rock pools beside streams. The adults are apparently not known to feed on man.

Culex (Culex) virgultus Theobald

Culex virgultus Theobald, 1901, Mon. Culic., *2*:123.

Culex declarator Dyar and Knab, 1906, Jour. N.Y. Ent. Soc., *14*:211.

Culex inquisitor Dyar and Knab, 1906, Jour. N.Y. Ent. Soc., *14*:211.

Culex proclamator Dyar and Knab, 1906, Jour. N.Y. Ent. Soc., *14*:211.

Culex jubilator Dyar and Knab, 1907, Jour. N.Y. Ent. Soc., *15*:201.

Culex dictator Dyar and Knab, 1909, Smithson. Misc. Coll., *52*:199.

Culex vindicator Dyar and Knab, 1909, Smithson. Misc. Coll., *52*:255.

Culex revelator Theobald, 1910, Mon. Culic., *5*:614.

Culex bidens Dyar, 1922, Ins. Ins. Mens., *10*:190.

ADULT FEMALE (pl. 114). Medium-sized to small species. *Head:* Proboscis dark-scaled, with a median pale area ventrally; palpi dark scaled. Occiput with narrow curved pale-golden scales and brown erect forked scales dorsally, with broad appressed white scales laterally. *Thorax:* Integument brown, scutum clothed with fine narrow bronze-brown curved scales, somewhat paler on the anterior and lateral margins and on the prescutellar space. Scutellum with pale-golden scales and dark brown setae on the lobes. Pleura light brown (often with dark-brown or greenish spots present), with small patches of white scales. *Abdomen:* First tergite with a median patch of dark scales; remaining tergites dark-scaled with bronze to metallic blue-green reflection, with narrow basal bands of white scales widening at the sides to form lateral spots (bands may be lacking or represented by a few pale scales on each segment). Venter largely pale-scaled. *Legs:* Legs dark with bronze reflection; posterior surface of femora and tibiae pale-scaled; femora and tibiae tipped with white. Hind tarsi with narrow bands of pale scales, often obsolete on front and middle tarsi. *Wing:* Length about 3.0 to 3.3 mm. Scales narrow, dark.

ADULT MALE. Coloration similar to that of the female, but with dorsal abdominal pale bands much broader. *Terminalia* (fig. 255). Lobes of ninth tergite (IXT-L) widely separated, only slightly raised, each bearing several setae. Tenth sternite (X-S) crowned with numerous coarse spines; basal arm (XS-BA) long, stout, curved, weakly sclerotized. Phallosome (Ph) formed of two sclerotized plates. Each plate with a group of 3 or more large teeth followed by 2 to 5 smaller teeth (Ph-T). Claspette absent. Basistyle (Bs) a little more than twice as long as mid-width, clothed with numerous setae, longer on outer aspect. Subapical lobe (S-L) prominent, undivided, bear-

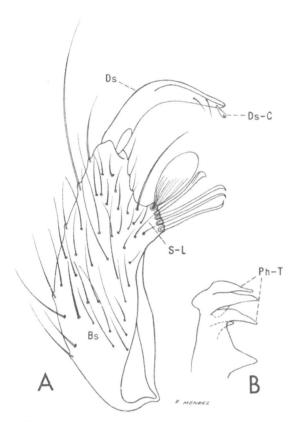

Fig. 255. Male terminalia of *Culex virgultus* Theobald.
A, Basistyle. *B*, Plate of phallosome.

on III–V. Comb of eighth segment of numerous scales in a patch; individual scale long, broadly expanded apically and fringed with subequal spinules. Siphonal index 5.0 to 6.0; pecten of numerous teeth on basal one-third to two-fifths of siphon, spirally twisted distally; individual pecten tooth with 2 to 6 coarse side teeth; three pairs of siphonal tufts present, the proximal tuft long, usually 2-branched and inserted beyond the pecten, the middle tuft usually 2- or 3-branched and inserted laterally, the distal tuft small, usually 2- to 4-branched. Anal segment completely ringed by the saddle; lateral hair single, nearly as long as the saddle; dorsal brush bilaterally consisting of a long lower caudal hair and two upper caudal hairs, one long and one short; ventral brush well developed, confined to the barred area; gills usually one and a half to two times as long as the saddle, each tapered to a point.

DISTRIBUTION. This species is known to occur in the United States in the vicinity of Brownsville, Texas (237, 277, 634). It has recently

ing three long rods of about equal length and with curved tips, a filament flattened apically, a rather broad leaf, and a long slender seta. Dististyle (Ds) a little less than half as long as the basistyle, curved, tapered beyond middle, bearing two small setae before apex; claw (Ds-C) short, blunt.

LARVA (fig. 256). Head broader than long. Antenna nearly as long as the head, constricted beyond antennal tuft, part before constriction dark at base, spiculate, part beyond constriction darker and with fewer spicules; antennal tuft large, multiple, barbed, inserted at outer third, extending considerably beyond tip of antenna. Head hairs: Postclypeal 4 short, single; upper frontal 5 and lower frontal 6 long, usually triple, barbed; preantennal 7 large, multiple, barbed. Prothoracic hairs: 1–3 long, single; 4 moderate, single or double; 5–6 long, single; 7 long, double or triple. Thorax spiculate. Lateral abdominal hair 6 usually double or triple on segments I and II, double

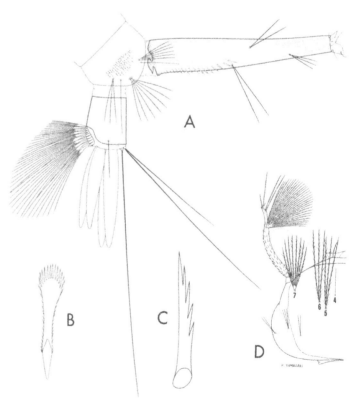

Fig. 256. Larva of *Culex virgultus* Theobald. *A*, Terminal segments. *B*, Comb scale. *C*, Pecten tooth. *D*, Head.

been taken as far north as Luling, Texas, by Dr. O. P. Breland. Its range extends south through Mexico (537a), Central America, the Lesser Antilles, and South America.

BIONOMICS. This species appears to be rare in Texas but is rather common in Panama and in many other areas in the tropics. The larvae are found in a variety of habitats in Panama, including rock pools, swamps, cement drains, rot cavities in trees, and coconut husks. The adults are seldom encountered; therefore, little is known of their habits.

Culex (Melanoconion) abominator
Dyar and Knab

Culex abominator Dyar and Knab, 1909, Smithson. Misc. Coll., *52*:257.

Culex abominator, Howard, Dyar, and Knab (in part), 1915, Carnegie Inst. Wash. Pub., *159*, 3:378.

Culex abominator, King and Bradley, 1937, Ann. Ent. Soc. Amer., *30*:352.

Culex abominator, Eads, 1943, Jour. Econ. Ent., *36*: 336.

ADULT FEMALE (pl. 115). Small species. *Head:* Proboscis long, dark, slightly swollen at tip; palpi very short, dark. Occiput with narrow curved light-golden scales and brown erect forked scales on posterodorsal region; anterodorsal and lateral regions with broad appressed dingy-white scales with metallic blue-green reflection. *Thorax:* Integument of scutum dark brown; scutum clothed with narrow curved golden-brown scales. Scutellum with narrow golden scales and brown setae on the lobes. *Abdomen:* Tergites dark-brown-scaled, with bronze to metallic blue-green reflection; narrow white basal bands occasionally present on some segments. Sternites pale-scaled basally, dark apically. *Legs:* Legs dark-scaled with bronze to metallic blue-green reflection, except for pale posterior surface of femora. *Wing:* Length about 2.5 mm. Scales all dark. Plume scales narrow, squame scales broader.

ADULT MALE. Coloration similar to that of the female, but with more pale scales on the abdomen. *Terminalia* (fig. 257). Lobes of ninth tergite (IXT-L) small, triangular, bearing a prominent patch of hairs on inner surface of basal half and a prominent large terminal hair. Tenth sternite (X-S) with an apical comb-

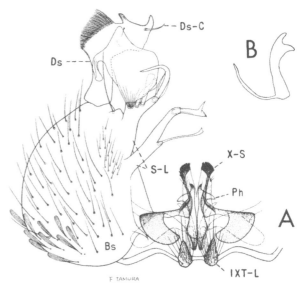

Fig. 257. *Culex abominator* Dyar and Knab. *A*, Male terminalia. *B*, Inner plate of phallosome.

like row of spines. Phallosome (Ph) with inner plate bearing a stout curved tooth at apex, a strong tooth on outer surface at distal three-fourths, and a curved basal hook. Claspette absent. Basistyle (Bs) large, nearly spherical, clothed with a few scales and numerous rather long setae on outer aspect. Basistyle with an angular projection between the base of the sub-apical lobe and the dististyle. Subapical lobe (S-L) divided into two distinct divisions: basal division bearing a stout hooked apical filament, and a slender capitate filament arising from a tubercle near middle; apical division bearing a large irregular fan-shaped leaf, two short flattened filaments, a slender seta, and a long strong curved filament. Dististyle (Ds) about half as long as the basistyle, greatly expanded beyond middle; the crest bears a row of long fine reflexed setae and is followed by a stout horn; claw (Ds-C) long, inserted before apex, reaching beyond curved tip of the dististyle.

LARVA (fig. 258). Antenna nearly as long as the head, constricted beyond antennal tuft, with part before constriction spiculate, dark near base, part beyond constriction darker and with fewer spicules; antennal tuft large, multiple, barbed, inserted at outer third of shaft, reaching beyond tip. Head hairs: Postclypeal 4 small, single or double; upper frontal 5 double (rarely single or triple), about half as long as hair 6; lower frontal 6 long, single; preantennal 7 long, multiple, barbed. Prothoracic hairs: 1-2 long, single; 3 short, multiple; 4-6 long, single; 7 medium, 2- to 4-branched. Thorax spiculate. Lateral abdominal hair 6 usually double on segments I and II, triple on III-VI. Comb of eighth segment with many scales in a patch; individual scale short, rounded apically, and fringed with subequal spinules. Siphonal index about 5.0; pecten of about 16 to 20 teeth on basal third of siphon; individual pecten tooth fringed on one side to near tip; five pairs of long multiple barbed siphonal tufts inserted beyond pecten, proximal tuft about twice the diameter of the siphon at point of insertion of the tuft, two small 3-branched subdorsal tufts present. Anal segment completely ringed by the saddle; lateral hair small, multiple; dorsal brush bilaterally consisting of a long upper caudal hair and a lower caudal tuft of one long and two short hairs; ventral brush well developed, confined to the barred area; gills about half as long as the saddle, pointed.

DISTRIBUTION. Known only from Texas (246, 463, 571, 634).

BIONOMICS. Eads (246) describes the larva of this species from collections made during August, September, and October, 1942, in grass overhanging a permanent pool near New Braunfels, Texas. Larvae have been taken also at Kelley Field, San Antonio, Texas. Rueger and Druce (634) report larval collections from Camp Bowie in Brown County and Camp Maxey in Lamar County. Nothing is known of the habits of the adults.

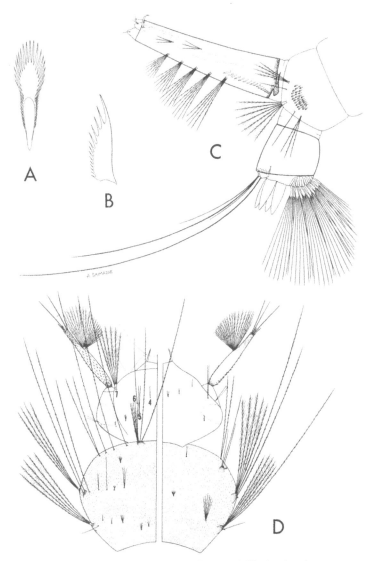

Fig. 258. Larva of *Culex abominator* Dyar and Knab. *A.* Comb scale. *B*, Pecten tooth. *C*, Terminal segments. *D*, Head and thorax (dorsal and ventral).

Culex (Melanoconion) anips Dyar

Culex anips Dyar, 1916, Ins. Ins. Mens., 4:48.
Culex anips, Brookman and Reeves, 1953, Ann. Ent. Soc. Amer., 46:231 (larva described).

ADULT FEMALE. *Head:* Proboscis long, dark, slightly swollen at tip; palpi very short, dark. Occiput with appressed dark brownish scales and brownish erect forked scales dorsally, with broad appressed whitish scales laterally and behind the eyes. *Thorax:* Integument of scutum dark brown; scutum clothed with narrow curved bronze-brown scales. Scutellum with narrow bronze-brown scales and dark setae on the lobes. Pleura with small patches of a few dingy-white scales. *Abdomen:* Tergites dark-brown-scaled, with bronze to metallic blue-green reflection; white basolateral patches distinct on segments V to VII. Sternites pale-scaled basally and apically (not sharply banded). *Legs:* Legs dark-scaled with bronze to metallic blue-green reflection, except for pale posterior surface of femora. *Wing:* Length about 2.5 to 3.0 mm. Scales all dark. Plume scales narrow, squame scales broader.

ADULT MALE. Coloration similar to that of

the female. *Terminalia* (fig. 259). Lobes of ninth tergite (IXT-L) widely separated, somewhat ovate, each lobe bearing about ten to twelve setae on apical half. Tenth sternite (X-S) with an apical comblike row of blunt spines. Phallosome (Ph) with inner plate bearing a pointed apical tooth, a stout subapical tooth on outer margin, and a taillike basal hook directed ventrally. Claspette absent. Basistyle (Bs) less than twice as long as mid-width, outer margin strongly curved, clothed with scales and numerous setae, with long setae on outer aspect. Subapical lobe (S-L) deeply divided into two distinct divisions: basal division bearing a slender pointed lateral filament and a strong apical rod arising from a strong arm retrorsely spined at tip; apical division bearing three irregularly placed filaments, a short spine, and a large greatly expanded fan-shaped leaf arising from a stout branch; arising from the basal part of this stout branch is a slender seta. Dististyle (Ds) about half as long as the basistyle, constricted before middle, outer margin near tip expanded caplike and bearing a patch of long hairs and a subterminal patch of short appressed spines; claw (Ds-C) short, inserted before apex.

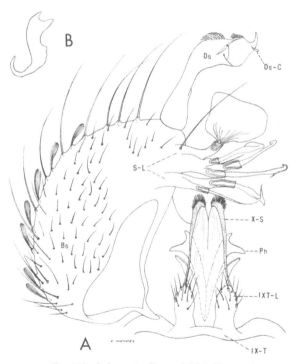

Fig. 259. *Culex anips* Dyar. *A*, Male Terminalia. *B*, Inner plate of phallosome.

LARVA (fig. 260). Antenna about as long as the head, constricted beyond the antennal tuft, with part before constriction spiculate, part beyond constriction darker and sparsely spiculate; antennal tuft large, multiple, barbed, inserted at outer third of shaft, reaching beyond tip. Head hairs: Postclypeal 4 short, single; upper frontal 5 weak, single or double (rarely triple), about half as long as hair 6; lower frontal 6 long, single; preantennal 7 large, multiple, barbed. Mentum with 4 closely placed teeth and one additional basal tooth on either side of the central tooth. Prothoracic hairs: 1–2 long, single; 3 short, multiple; 4–6 long, single; 7 medium, usually triple (occasionally 2- or 4-branched). Body spiculate. Lateral abdominal hair 6 double on segments I and II, triple on III–VI. Comb of eighth segment with many scales in a patch; individual scale rounded apically and fringed with subequal spinules. Siphonal index about 5.0 to 7.0; pecten of numerous teeth on basal third of siphon; individual tooth slender and fringed with fine lateral teeth on one side nearly to tip; about five pairs of multiple barbed siphonal tufts inserted beyond pecten, the proximal tuft about two and a half times as long as the basal diameter of the siphon but less than half its length; two small 2- or 3-branched subdorsal tufts present; dorsal preapical spine as long or longer than the apical pecten tooth, curved. Anal segment completely ringed by the saddle, saddle spiculate on dorsoapical surface and along posterior margin; lateral hair short, double or triple; dorsal brush bilaterally consisting of a long lower caudal hair and one long and two short upper caudal hairs; ventral brush well developed, confined to the barred area; gills about one-third as long as the saddle, each tapering to a blunt point.

DISTRIBUTION. Known only from Baja California, Mexico (106), and from near San Diego, California (184, 284).

BIONOMICS. Pupae of this species were collected by Dyar from a large deep isolated pool near the mouth of the San Diego River in California, and the male and female descriptions are based on the reared specimens. The surface of the pool had mats of *Lemna* and was margined with cattails. Brookman and Reeves

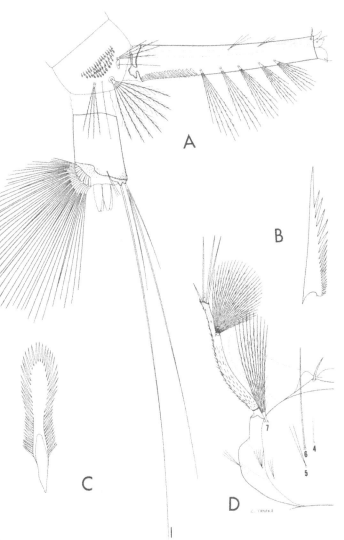

Fig. 260. Larva of *Culex anips* Dyar. *A*, Terminal segments. *B*, Pecten tooth. *C*, Comb scale. *D*, Head.

Thorax: Integument of scutum dark brown; scutum clothed with narrow reddish-brown scales. Scutellum with narrow reddish-brown scales and brown setae. Pleura brown, greenish in some areas, with a small patch of broad white scales on sternopleuron. *Abdomen:* Tergites dark bronze-brown-scaled, with basolateral patches of white scales. Sternites white basally, dark apically. *Legs:* All legs dark-scaled with bronze reflection, except for pale posterior surface of femora and small pale knee spots. *Wing:* Length about 2.5 mm. Scales dark brown, mostly rather narrow, but with ovate scales intermixed on vein 3 and branches of veins 2 and 4.

ADULT MALE. Coloration similar to that of the female. *Terminalia* (fig. 261). Lobes of ninth tergite (IXT-L) each broad at base and tapered and pointed apically, bearing several strong, barbed and smooth setae on basal half, apical half bare. Tenth sternite (X-S) with an apical comblike row of blunt spines. Phallosome (Ph) formed of two simple curved

(106) found the larvae in Baja California, Mexico, in a similar aquatic habitat.

Culex (Melanoconion) atratus Theobald

Culex atratus Theobald, 1901, Mon. Culic., *2*:55.
Culex falsificator Dyar and Knab, 1909, Smithson. Misc. Coll., *52*:257.

ADULT FEMALE (pl. 116). Small species. *Head:* Proboscis long, dark, slightly swollen at tip; palpi very short, dark. Occiput with broad brown appressed scales, light-brown lanceolate scales and dark erect forked scales dorsally, with broad pale appressed scales laterally.

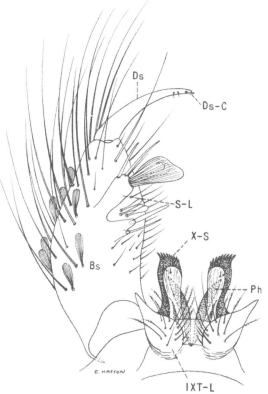

Fig. 261. Male terminalia of *Culex atratus* Theobald.

divergent bladelike plates, connected at bases, pointed apically. Claspette absent. Basistyle (Bs) about twice as long as mid-width, clothed with scales and long setae on outer aspect; three or four short greatly flattened setae present between the divisions of the subapical lobe. Subapical lobe (S-L) widely separated into two distinct divisions: basal division bearing a stout rod at apex and a slender subapical spine; apical division bearing a large broad striated leaflike filament, a flattened recurved spine, two straight spines, and one or two smaller apical setae. Dististyle (Ds) about half as long as the basistyle, broad at base, tapered and curved distally; two small setae on inner surface near tip; claw (Ds-C) small, blunt, spinelike, inserted a little before apex.

LARVA (fig. 262). Antenna nearly as long as the head, constricted beyond antennal tuft, the part before constriction spiculate, part beyond constriction darker and with few spicules; antennal tuft large, multiple, barbed, inserted at outer third of shaft, reaching much beyond tip. Head hairs: Postclypeal 4 small, single; upper frontal 5 small, 5- to 7-branched; lower frontal 6 long, single; preantennal 7 long, multiple, barbed. Prothoracic hairs: 1–2 long, single; 3 short, multiple; 4 long, single or double; 5–6 long, single; 7 long, double or triple. Body sparsely spiculate. Lateral abdominal hair 6 long, single or double on segments I and II, shorter and usually multiple on III–VI. Comb of eighth segment with many scales in a patch; individual scale rounded apically and fringed with subequal spinules. Siphonal index about 7.0 to 8.0; pecten of about 12 to 21 teeth on basal two-fifths of siphon; individual tooth fringed on one side nearly to tip; four or five pairs of multiple siphonal tufts inserted beyond pecten, the proximal tuft a little longer than the basal diameter of the siphon; usually three small tufts inserted dorsolaterally on siphon. Anal segment completely ringed by the saddle; a few long slender pale spicules on apical margin of the saddle; lateral hair small, double or multiple; dorsal brush bilaterally consisting of a long lower caudal hair and an upper caudal tuft of one long and two to four short hairs; ventral brush large, confined to the barred area; gills shorter than the saddle, bluntly pointed.

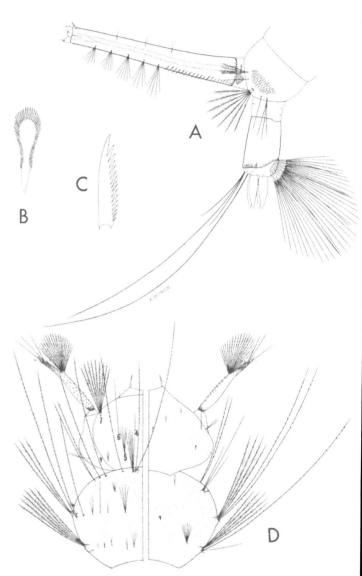

Fig. 262. Larva of *Culex atratus* Theobald. *A*, Terminal segments. *B*, Comb scale. *C*, Pecten tooth. *D*, Head and thorax (dorsal and ventral).

DISTRIBUTION. The Florida Keys (562, 624), the Greater Antilles, and the Virgin Islands. Males were taken at Cortez and Copeland on the Florida mainland in 1946 by Pritchard *et al.* (562).

BIONOMICS. The larvae have been found in ground pools. Pritchard *et al.* (562) state that this is a common species throughout the Florida Keys. They list approximately fifteen different keys on which larval collections have been made. Males have been taken in light-trap collections, but apparently nothing is known of the habits of the females.

Culex (Melanoconion) erraticus
(Dyar and Knab)

Melanoconion atratus Dyar (not *Culex atratus* Theobald, 1901), 1905, Jour. N.Y. Ent. Soc., *13*:26.

Mochlostyrax erraticus Dyar and Knab, 1906, Jour. N.Y. Ent. Soc., *14*:224 (larva described).

Culex leprincei Dyar and Knab, 1907, Jour. N.Y. Ent. Soc., *15*:202.

Culex egberti Dyar and Knab, 1907, Jour. N.Y. Ent. Soc., *15*:214.

Culex trachycampa Dyar and Knab, 1909, Can. Ent., *41*:101.

Culex peribleptus Dyar and Knab, 1917, Ins. Ins. Mens., *5*:181.

Culex pose Dyar and Knab, 1917, Ins. Ins. Mens., *5*:182.

Culex moorei Dyar, 1918, Ins. Ins. Mens., *6*:108.

Culex peccator Dyar and Barrett, 1918 (not Dyar and Knab), Ins. Ins. Mens., *6*:119.

Culex degustator Dyar and Knab, 1921, Ins. Ins. Mens., *9*:39.

Culex borinqueni Root, 1922, Amer. Jour. Hyg., *2*:400.

Culex erraticus, King and Bradley, 1937, Ann. Ent. Soc. Amer., *30*:345 (additional synonymy given).

ADULT FEMALE (pl. 117). Small species. *Head:* Proboscis long, dark-scaled, slightly swollen at tip; palpi very short, dark. Occiput with narrow curved light-golden to brown scales and dark erect forked scales dorsally; anterodorsal and lateral regions with broad appressed scales, the anterodorsal ones predominantly brown with bronze or metallic blue-green reflection, the lateral ones mostly dingy white. Cibarial armature of 7 or 8 teeth. *Thorax:* Integument of scutum dark brown; scutum clothed with narrow curved golden-brown scales (scales usually paler and coarser than those of *C. peccator* and *C. pilosus*). Scutellum with narrow golden scales and brown setae on the lobes. Pleura with small patches of dingy-white scales. *Abdomen:* Tergites dark-brown-scaled, usually with bronze to metallic blue-green reflection; narrow white basal bands usually present, sometimes absent; white basolateral patches present. Sternites each white banded basally, dark apically. *Legs:* Legs

dark-scaled with bronze to metallic blue-green reflection, except for pale posterior surface of femora and small pale knee spots. *Wing:* Length about 2.5 to 3.0 mm. Scales all dark. Plume scales narrow, squame scales broader.

ADULT MALE. Coloration similar to that of the female. *Terminalia* (fig. 263). Lobes of ninth tergite (IXT-L) large, ovoid, narrowly separated basally, clothed with long setae arising from distinct tubercles. Tenth sternite (X-S) with an apical comblike row of blunt spines. Phallosome (Ph) with inner plate bearing a short stout apical tooth, a strong slender subapical tooth on outer margin, a stout ventrolateral tooth at basal third, and a long taillike basal hook directed ventrally. Claspette absent. Basistyle (Bs) about twice as long as midwidth, outer margin strongly curved, clothed with scales and numerous setae, longer on outer aspect. Subapical lobe (S-L) deeply divided into two distinct divisions: basal division cleft nearly to base into two stout branches, each bearing a long strong rodlike filament with tip flattened and hooked; apical division bearing a long flattened pointed strongly hooked rodlike filament, a short curved rodlike filament, two blunt rodlike filaments, and a large broadly

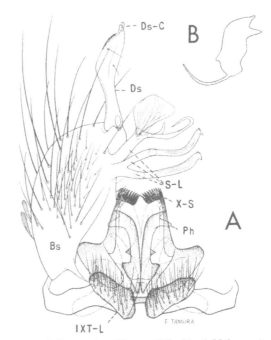

Fig. 263. *Culex erraticus* (Dyar and Knab). *A*, Male terminalia. *B*, Inner plate of phallosome.

expanded leaflike appendage. Dististyle (Ds) about two-thirds as long as the basistyle, broader on apical half, bearing a row of very short reflexed setae on outer margin near tip, tapered and curved apically; claw (Ds-C) short, curved, blunt, inserted before apex.

LARVA (fig. 264). Antenna about as long as the head, constricted beyond the antennal tuft, with part before constriction spiculate, part beyond constriction darker and sparsely spiculate; antennal tuft large, multiple, barbed, inserted at outer third of shaft, reaching beyond tip. Head hairs: Postclypeal 4 small, single; upper frontal 5 short, weak, 4- to 8-branched, barbed; lower frontal 6 long, single, barbed; preantennal 7 large, multiple, barbed. Prothoracic hairs: 1–2 long, single; 3 short,

multiple; 4–6 long, single; 7 medium, usually 2- to 4-branched. Body densely spiculate. Lateral abdominal hair 6 usually double or triple on segments I and II, usually triple on III–VI. Comb of eighth segment with many scales in an irregular single or double row; individual scale thorn-shaped, with short lateral spinules. Siphonal index 6.0 to 7.0; pecten of about 11 to 17 teeth on basal third of siphon; individual tooth fringed on one side nearly to tip; about five pairs of multiple barbed siphonal tufts inserted beyond the pecten, the proximal tuft much longer than the basal diameter of the siphon but less than half its length; two small 2- or 3-branched subdorsal tufts present; dorsal preapical spine as long or longer than apical pecten tooth, curved. Anal segment completely ringed by the saddle; lateral hair short, single or double; dorsal brush bilaterally consisting of a long lower caudal hair and one long and one short upper caudal hair; ventral brush well developed, confined to the barred area; gills varying in length from about two-thirds to one and a half times as long as the saddle, each tapering to a blunt point.

DISTRIBUTION. Known from the southern United States north to Michigan and west to North Dakota and Texas, Mexico (537a), Jamaica (685), and Panama. *United States:* Alabama (135, 419); Arkansas (88, 121); Delaware (167); District of Columbia (318); Florida and Georgia (135, 419); Illinois (615); Indiana (152, 345); Iowa (625, 626); Kansas (14, 530); Kentucky (134, 570); Louisiana (419, 759); Maryland (64, 161); Michigan (529); Mississippi (135, 546); Missouri (6, 324); Nebraska (14, 680); North Carolina (135); North Dakota (14); Ohio (482, 731a); Oklahoma (628); South Carolina (135, 742); South Dakota (14); Tennessee (135, 658); Texas (249, 463); Virginia (176, 178).

BIONOMICS. The larvae are found in the grassy shallow margins of ponds, lakes, marshes, and streams where they are frequently associated with the larvae of *Anopheles*, particularly with *A. quadrimaculatus* Say in the southern United States and *A. albimanus* Wiedemann in the Caribbean region. The larvae are often very abundant in Najas beds in the Chagres River and Upper Gatun Lake in

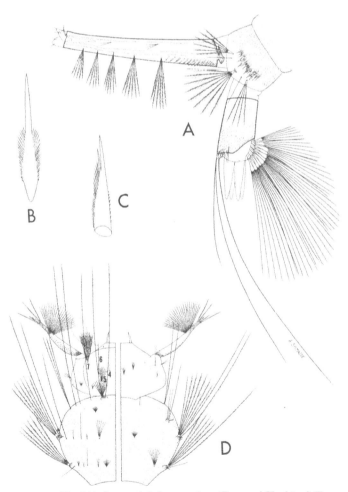

Fig. 264. Larva of *Culex erraticus* (Dyar and Knab). *A,* Terminal segments. *B,* Comb scale. *C,* Pecten tooth. *D,* Head and thorax (dorsal and ventral).

Panama during the dry season where they are associated with the larvae of *A. albimanus*. The immature stages can be found throughout the year in the extreme southern United States and from about May to October farther north.

The adults are occasionally seen in large numbers in the southern United States but are not troublesome biters. Observations made by King *et al.* (419) in Louisiana indicate that the females may attack man at night in outdoor situations but that they seem to prefer the blood of fowl.

Culex (Melanoconion) iolambdis Dyar

Culex (Choeroporpa) iolambdis Dyar, 1918, Ins. Ins. Mens., *6*:106.

Culex (Mochlostyrax) iolambdis, Dyar, 1928, Carnegie Inst. Wash. Pub., *387*:329.

Culex (Melanoconion) iolambdis, Pratt, 1952, Proc. Ent. Soc. Wash., *54*:27.

ADULT FEMALE (pl. 118). Small species. *Head:* Proboscis long, dark, slightly swollen at tip; palpi very short, dark. Occiput clothed dorsally with broad appressed pale to dark-brown scales with metallic reflection and dark erect forked scales; lateral region of occiput with a patch of broad appressed pale scales. *Thorax:* Integument of scutum dark brown to black; scutum clothed with fine narrow curved bronze-brown scales. Scutellum with narrow brown scales and dark-brown setae on the lobes. Pleura brown with a greenish tinge. *Abdomen:* Tergites clothed dorsally with bronze-brown scales; small basolateral patches of pale scales on posterior segments. Sternites each pale-scaled basally and broadly dark-scaled apically. *Legs:* Legs dark-scaled with bronzy reflection, except for pale posterior surface of femora. *Wing:* Length about 2.5 mm. Scales all dark. Plume scales narrow, squame scales broader.

ADULT MALE. Coloration similar to that of the female. *Terminalia* (fig. 265). Lobes of ninth tergite (IXT-L) ovoid and cushionlike, each bearing about ten to fifteen slender setae arising from distinct tubercles. Tenth sternite (X-S) with an apical comblike row of blunt spines. Phallosome (Ph) with inner plate broad distally and bearing two or three blunt

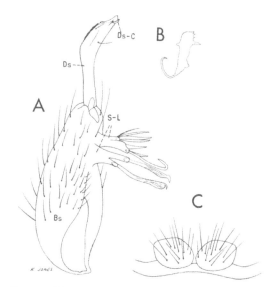

Fig. 265. Male terminalia of *Culex iolambdis* Dyar. *A*, Basistyle. *B*, Inner plate of phallosome. *C*, Lobes of ninth tergite.

apical teeth, a strong subapical lateral tooth and a long curved basal hook. Claspette absent. Basistyle (Bs) about twice as long as mid-width, outer margin curved. Subapical lobe (S-L) deeply divided into two distinct divisions: basal division consisting of two stout unequal branches, each bearing a long strong rod flattened distally, pointed and curved at apex; apical division bearing a long hooked filament, a short filament, one or two large middle filaments, and a group of two or three flattened filaments. Dististyle (Ds) about half as long as the basistyle, broader beyond middle, and snoutlike apically, bearing a row of short reflexed setae on outer margin before tip; claw (Ds-C) inserted before apex, curved, blunt.

LARVA (fig. 266). Antenna nearly as long as the head, constricted beyond the antennal tuft, part before constriction spiculate, part beyond constriction darker and sparsely spiculate; antennal tuft large, multiple, barbed, inserted at outer third of shaft, reaching considerably beyond tip. Head hairs: Postclypeal 4 small, single or branched distally; upper frontal 5 usually double, sparsely barbed, about three-fourths as long as hair 6; lower frontal 6 long, single, sparsely barbed; preantennal 7 multiple, barbed, reaching insertion of antennal tuft. Prothoracic hairs: 1–2 long, single; 3 medium, multiple; 4 long, single or double;

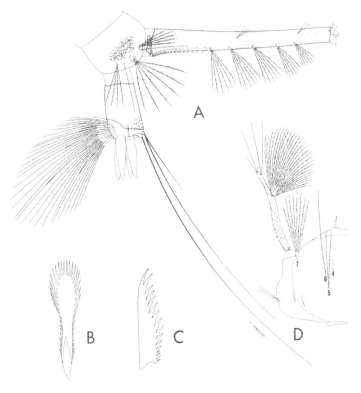

Fig. 266. Larva of *Culex iolambdis* Dyar. *A*, Terminal segments. *B*, Comb scale. *C*, Pecten tooth. *D*, Head.

5–6 long, single; 7 medium, double or triple. Thorax sparsely spiculate. Lateral abdominal hair 6 usually double on segments I and II, usually triple on III–VI. Comb of eighth segment with many scales in a patch; individual scale long, rounded apically, and fringed with subequal spinules. Siphonal index about 5.0 to 6.0; siphon with a rather conspicuous dark band a little beyond middle; pecten of numerous evenly spaced teeth on basal third of siphon; individual pecten tooth fringed on one side nearly to tip; five pairs of rather long barbed siphonal tufts inserted beyond pecten, proximal tuft about twice as long as diameter of siphon at point of insertion of tuft; two pairs of small subdorsal tufts present. Anal segment completely ringed by the saddle, with a dorso-apical patch of prominent spicules; lateral hair double or triple, shorter than the saddle; dorsal brush bilaterally consisting of a long lower caudal hair and an upper caudal tuft of one long and one or two short hairs; ventral brush well developed, confined to the barred area; gills shorter than the saddle, pointed.

DISTRIBUTION. Mexico (537a), Panama (235), Puerto Rico and Florida (559). Florida records listed by Pratt and Seabrook (559) include larvae and pupae collected at Key Largo, Dade County, on April 23 and 28 and July 23, 1944 [reported as *C. elevator* by Wirth (758)] and also later collections obtained by other workers in Martin and Palm Beach counties.

BIONOMICS. The larval collections of *Culex iolambdis* from Florida came from the edges of small ponds or from standing water around the aerial roots of the black mangrove. Pratt found the larvae in similar densely shaded brackish water in mangrove swamps in Puerto Rico. According to Dyar (235) the larvae have been found along the edges of streams in Panama. Adults were collected by Pratt and Seabrook in Florida in aspirators, after they were disturbed in their resting places on tree trunks and decaying branches.

Culex (Melanoconion) mulrennani
Basham

Culex mulrennani Basham, 1948, Ann. Ent. Soc. Amer., *41*:1.

ADULT FEMALE. Small species. *Head:* Proboscis long, dark-scaled, slightly swollen at tip; palpi very short, dark. Occiput with bluish-white scales and dark-brown erect forked scales dorsally, with broad appressed silver-gray scales laterally. *Thorax:* Integument of scutum brown; scutum clothed with fine narrow curved bronze-brown scales. Scutellum with bronze-brown scales and brown setae on the lobes. *Abdomen:* Tergites dark-scaled with bronze to metallic blue-green reflection, with or without small basolateral patches of pale scales on posterior segments. Sternites each dark with a few pale scales basally. *Legs:* All legs dark with bronze to metallic blue-green reflection, except for pale posterior surface of femora and small pale knee spots. *Wing:* Length about 2.0 to 2.5 mm. Scales all dark. Plume scales narrow, squame scales broader.

ADULT MALE. Coloration similar to that of the female, but with abdominal pale scaling more prominent. *Terminalia* (fig. 267). Lobes of ninth tergite (IXT-L) large and ovate, approximate at bases, each with an apical pro-

reflexed setae; claw (Ds-C) inserted before apex and extending beyond tip of the dististyle.

LARVA (fig. 268). Antenna about as long as the head, constricted beyond antennal tuft, with part before constriction spiculate, part beyond constriction darker and with fewer spicules; antennal tuft large, multiple, barbed, inserted at outer third of shaft, reaching beyond tip. Head hairs: Postclypeal 4 short, double or triple; upper frontal 5 usually triple (occasionally 2- or 4-branched), one-half to three-fourths as long as hair 6; lower frontal 6 long, single; preantennal 7 multiple, barbed, extending to insertion of antennal tuft. Prothoracic hairs; 1–2 long, single; 3 short, multiple; 4–6 long, single; 7 medium, double or triple.

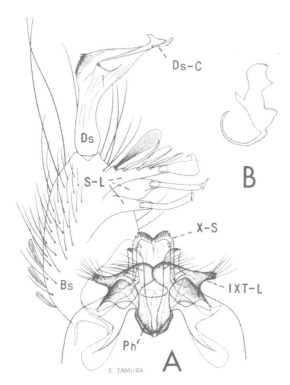

Fig. 267. *Culex mulrennani* Basham. *A*, Male terminalia. *B*, Inner plate of phallosome.

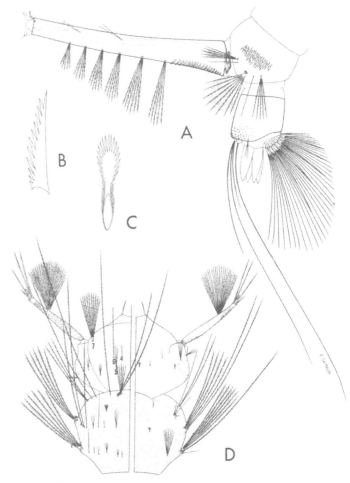

Fig. 268. Larva of *Culex mulrennani* Basham. *A*, Terminal segments. *B*, Pecten tooth. *C*, Comb scale. *D*, Head and thorax (dorsal and ventral).

jection extending laterally, basal and apical parts clothed with setae arising from distinct tubercles. Tenth sternite (X-S) with an apical comblike row of blunt spines. Phallosome (Ph) with inner plate bearing four or five blunt apical points, a tooth on outer margin, and a longe taillike basal hook. Claspette absent. Basistyle (Bs) large, nearly spherical, clothed with scales and numerous setae, longer on outer aspect. Subapical lobe (S-L) divided into two distinct divisions: basal division cleft nearly to base into two stout branches, each bearing a long strong rod with tip flattened, curved, and pointed, with a small apical retrorse barb; apical division bearing a broad striated leaf arising near base, followed by a rodlike curved filament, a lanceolate blade, a short acuminate seta, and a narrow strongly curved leaf. Dististyle (Ds) about three-fourths as long as the basistyle, basal half narrow, outer margin of distal half expanded caplike and bearing a row of short reflexed setae, constricted preapically, and expanded apically; a long spine arises from a groove underneath the row of

Thorax and abdomen spiculate. Lateral abdominal hair 6 usually double on segments I and II, triple on III–VI. Comb of eighth segment with many scales in a triangular patch; individual scale expanded and rounded apically, fringed with subequal spinules. Siphonal index about 5.5 to 7.0; median dark-pigmented band variable; pecten of about 13 to 20 teeth on basal third of siphon; individual pecten tooth fringed on one side to tip; five or six pairs of siphonal tufts inserted beyond pecten, proximal tuft about one and a half to two times as long as basal diameter of siphon; two 2- or 3-branched subdorsal tufts present; dorsal preapical spine longer than apical pecten tooth, curved. Anal segment completely ringed by the saddle; saddle with prominent dorsoapical spicules present; lateral hair 3- or 4-branched from near base, shorter than the saddle; dorsal brush bilaterally consisting of a long lower caudal hair and an upper caudal tuft of one long and two short hairs; ventral brush well developed, confined to the barred area; gills a little shorter than the saddle, tapered.

DISTRIBUTION. This species was described by Basham (36) from specimens taken on Big Pine, Cudjoe, and Ramrod keys, Monroe County, Florida.

BIONOMICS. Larvae were collected by Basham from a man-made well, from limestone solution holes, and from rockholes. Males and females were found resting on the sides of the depressions in a few instances. Very little is known of the habits of the adults.

Culex (Melanoconion) opisthopus Komp

Culex (Choeroporpa) opisthopus Komp, 1926, Ins. Ins. Mens., *14*:44 (male and female described).

Culex (Mochlostyrax) mychonde Komp, 1928, *in* Dyar, Carnegie Inst. Wash. Pub., *387*:295.

Culex (Melanoconion) opisthopus, Pratt, Wirth, and Denning, 1945, Proc. Ent. Soc. Wash., *47*:245 (larva described).

ADULT FEMALE (pl. 119). Small species. *Head:* Proboscis long, dark; palpi very short, dark. Occiput clothed dorsally with long slender curved silvery scales, black erect forked scales, and prominent setae directed forward; occiput with patches of whitish scales laterally.

Thorax: Integument of scutum dark brown; scutum clothed with fine narrow curved bronze-brown scales and prominent brown setae. Scales light golden around the prescutellar space and in a small patch on the anterior margin of the scutum. Scutellum with narrow golden scales and brown setae on the lobes. Pleura with small patches of grayish-white scales. *Abdomen:* Tergites bronze-brown, with purplish reflection in some lights; a row of pale hairs at apex of segments; small basolateral white patches on some of the posterior segments. Sternites pale basally and dark apically. *Legs:* Legs dark, except for pale posterior surface of femora and narrow basal and apical white rings on hind tarsal segments 1 to 4 and segment 5 which is almost entirely white. *Wing:* Length about 2.5 mm. Scales all dark. Plume scales narrow, squame scales broader.

ADULT MALE. Coloration similar to that of the female. *Terminalia* (fig. 269). Lobes of ninth tergite (IXT-L) large, finger-shaped, ovate, and wrinkled apically, bearing many slender setae. Tenth sternite (X-S) with an apical comblike row of blunt teeth. Phallosome (Ph) consists of two plates connected basally; each plate broad distally and with a blunt point at the dorsal and ventral angles, basal hook

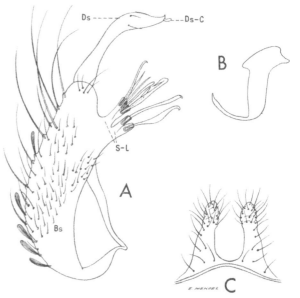

Fig. 269. Male terminalia of *Culex opisthopus* Komp. *A,* Basistyle. *B,* Inner plate of phallosome. *C,* Lobes of ninth tergite.

long and slender. Claspette absent. Basistyle (Bs) about twice as long as the mid-width, outer margin curved, clothed with scales and many setae. Subapical lobe (S-L) deeply divided into two distinct divisions: basal division bearing two long crooked subequal filaments with recurved tips, the basal filament flattened and more curved; apical division bearing four to six filaments of various types and lengths inserted apically. Dististyle (Ds) a little more than half as long as the basistyle, broader at apical third, and tapered snout-shaped apically; claw (Ds-C) short, blunt, inserted before apex.

LARVA (fig. 270). Antenna nearly as long as the head, uniformly dark brown, spiculate, constricted beyond antennal tuft, part beyond constriction more sparsely spiculate; antennal tuft large, multiple, barbed, inserted at outer third of shaft, reaching beyond tip. Head hairs: Postclypeal 4 small, single or double; upper frontal 5, 5- or 6-branched, barbed, about one-half to two-thirds as long as hair 6; lower frontal 6 long, single, barbed; preantennal 7 multiple, barbed, reaching near insertion of antennal tuft. Prothoracic hairs: 1–2 long, single; 3 short, multiple; 4 long, single or double; 5–6 long, single; 7 medium, 4- or 5-branched. Lateral abdominal hair 6 usually double on segments I and II, short and multiple on III–VI. Comb of eighth segment with many scales in a patch; individual scale rounded apically and fringed with subequal spinules. Siphonal index 8 to 10; pecten of 9 to 12 teeth on basal fourth of siphon; individual tooth fringed on one side; four short 2- to 4-branched siphonal tufts inserted beyond the pecten in a straight line; one small subdorsal tuft inserted before apex of siphon; the tufts slightly longer than the diameter of the siphon at the point of insertion of the tufts; dorsal preapical spine about as long as the apical pecten tooth, curved. Anal segment completely ringed by the saddle; lateral hair multiple, shorter than the saddle; dorsal brush bilaterally consisting of a long lower caudal hair and an upper caudal tuft of one long and several short hairs; ventral brush well developed, confined to the barred area; gills short, nearly as long as the saddle, tapered.

DISTRIBUTION. Known from southern Florida

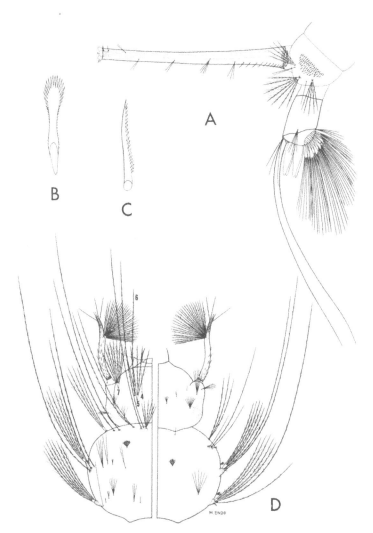

Fig. 270. Larva of *Culex opisthopus* Komp. *A*, Terminal segments. *B*, Comb scale. *C*, Pecten tooth. *D*, Head and thorax (dorsal and ventral).

and Puerto Rico (560), Honduras (437), Panama (237), and Mexico (537a).

BIONOMICS. The larva of this species was described by Pratt *et al.* (560) from specimens collected in crabholes near Fort Lauderdale, Florida, and in a sluggish stream and in pools beside the stream in Puerto Rico. The adults were found by Komp (437) resting on the walls of a hospital in Honduras. Adults have been collected also in calf-baited and horse-baited traps and in light traps in Puerto Rico (560). They also have been captured in light traps at Fort Lauderdale, Jupiter, Cape Sable, and Charlotte Harbor, Florida.

Culex (Melanoconion) peccator
Dyar and Knab

Culex peccator Dyar and Knab, 1909, Smithson. Misc. Coll., *52*:256.

Culex incriminator Dyar and Knab, 1909, Smithson. Misc. Coll., *52*:257.

Culex peccator, King and Bradley, 1937, Ann. Ent. Soc. Amer., *30*:350.

ADULT FEMALE (pl. 120). Small species. *Head:* Proboscis long, dark, slightly swollen at tip; palpi very short, dark. Occiput clothed with broad appressed dark-brown scales with bronze or metallic blue-green reflection and dark erect forked scales dorsally; scales on lateral region dingy white; a small postero-median area of brown lanceolate scales present. Cibarial armature consisting of 7 or 8 teeth. *Thorax:* Integument of scutum dark brown; scutum clothed with fine narrow curved brown scales. Scutellum with fine brown scales and brown setae on the lobes. Pleura brownish, with small patches of grayish-white scales. *Abdomen:* Tergites clothed with dark scales with bronze to metallic blue-green reflection; small basolateral white patches present. Sternites primarily pale-scaled, becoming darker on apical margin. *Legs:* Legs dark-scaled with bronze to metallic blue-green reflection, except for pale posterior surface of femora and small pale knee spots. *Wing:* Length about 2.5 mm. Scales all dark. Plume scales narrow, squame scales broader.

ADULT MALE. Coloration similar to that of the female. *Terminalia* (fig. 271). Lobes of ninth tergite (IXT-L) set close together, triangular, each bearing many long setae arising from distinct tubercles. Tenth sternite (X-S) with an apical comblike row of blunt spines. Phallosome (Ph) with inner plate bearing a short stout curved tooth at apex, a large blunt subapical lateral process and a long basal tail-like hook directed ventrally. Claspette absent. Basistyle (Bs) nearly spherical, clothed with scales and numerous rather fine to medium-sized setae. (*Culex peccator* can be recognized macroscopically by the greatly swollen basistyles.) Subapical lobe (S-L) deeply divided into two distinct divisions: basal division bearing a slender lateral rod somewhat flattened and angulate near tip, and a stronger apical rod arising from a strong arm and retrorsely

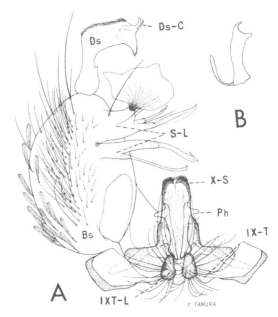

Fig. 271. *Culex peccator* Dyar and Knab. *A,* Male terminalia. *B,* Inner plate of phallosome.

spined at tip; apical division bearing three large flattened bladelike filaments arising from small branches and a large greatly expanded fan-shaped appendage arising from a stout branch; arising from the basal part of this stout branch is a long slender seta. Dististyle (Ds) about half as long as the basistyle, constricted at basal third, quadrately expanded on distal half, hirsute on crest; claw short, curved, inserted before apex and reaching beyond curved tip of the dististyle.

LARVA (fig. 272). Antenna about as long as the head, constricted beyond antennal tuft, with part before constriction spiculate, dark at base and pale to near base of tuft, part beyond constriction darker and sparsely spiculate; antennal tuft large, multiple, barbed, inserted at outer third of shaft, reaching beyond tip. Head hairs: Postclypeal 4 small, single; upper frontal 5 double or triple (occasionally single), less than half as long as hair 6; lower frontal 6 long, single, sparsely barbed; preantennal 7 long, multiple, barbed. Prothoracic hairs: 1–2 long, single; 3 short, multiple; 4–6 long, single; 7 medium, usually triple. Body spiculate. Lateral abdominal hair 6 usually double on segments I and II and triple on III–VI. Comb of eighth segment with many scales in a patch; individual scale rounded apically and fringed with subequal spinules. Siphonal index about 6.0; siphon with a darkly pigmented band just beyond middle; pecten of numerous teeth on basal third or two-fifths of siphon; individual pecten tooth fringed on one side to tip; about five multiple barbed siphonal tufts inserted beyond pecten, the proximal tuft much longer than the basal diameter of the siphon; two 2- or 3-branched subdorsal tufts present; dorsal pre-apical spine as long or longer than apical pecten tooth, curved. Anal segment completely ringed by the saddle, saddle spiculate posteriorly; lateral hair small, usually triple; dorsal brush bilaterally consisting of a long lower caudal hair and one long and two short upper caudal hairs; ventral brush well developed, confined to the barred area; gills rarely longer than the saddle, each tapered to a blunt point.

DISTRIBUTION. Known from the southeastern United States, north to Michigan and west to Kansas and Texas. *United States:* Alabama (135, 419); Arkansas (88, 121); Delaware (120, 167); Florida and Georgia (135, 419); Illinois (615); Kansas (14, 531); Kentucky (570); Louisiana (419, 759); Michigan (391); Mississippi (135, 510); Missouri (6, 14); North Carolina (135, 646); Oklahoma (628); South Carolina (135, 742); Tennessee (135, 658); Texas (634); Virginia (176, 178).

BIONOMICS. The larvae have been found in pools in streams and in marshy areas. The adults occasionally hide in diurnal shelters, particularly in damp situations, in the southern United States. The senior author has collected adults of this species on several occasions from underneath damp rock ledges near streams. They are occasionally captured in New Jersey light traps. The females cannot be separated with certainty from some other species of the subgenus *Melanoconion*, and therefore nothing is known of their feeding habits.

Culex (Melanoconion) pilosus
(Dyar and Knab)

Mochlostyrax pilosus Dyar and Knab, 1906, Jour. N.Y. Ent. Soc., *14*:223.
Mochlostyrax cubensis Dyar and Knab (not *Culex cubensis* Bigot), 1906, Jour. N.Y. Ent. Soc., *14*: 223.

Mochlostyrax floridanus Dyar and Knab, 1906, Proc. Biol. Soc. Wash., *19*:171.
Mochlostyrax jamaicensis Grabham (not *Culex jamaicensis* Theob.), 1906, Can. Ent., *38*:318.
Culex agitator Dyar and Knab, 1907, Jour. N.Y. Ent. Soc., *15*:100.
Culex deceptor Dyar and Knab, 1909, Smithson. Misc. Coll., *52*:257.
Culex reductor Dyar and Knab, 1909, Smithson. Misc. Coll., *52*:257.
Culex ignobilis Dyar and Knab, 1909, Proc. Ent. Soc. Wash., *11*:39.
Culex mastigia Howard, Dyar, and Knab, 1915, Carnegie Inst. Wash. Pub., *159, 3*:426.
Culex (Mochlostyrax) curopinensis Bonne-Wepster and Bonne, 1920, Ins. Ins. Mens., *7*:177.
Culex pilosus, King and Bradley, 1937, Ann. Ent. Soc. Amer., *30*:353.

ADULT FEMALE (pl. 121). Small species.

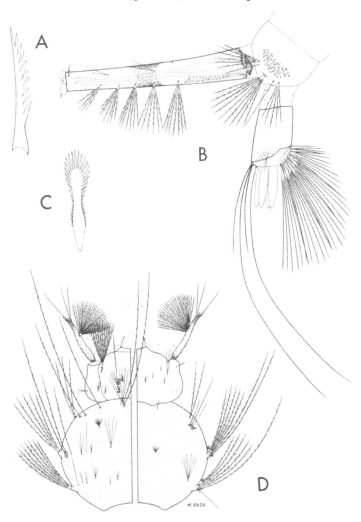

Fig. 272. Larva of *Culex peccator* Dyar and Knab. *A*, Pecten tooth. *B*, Terminal segments. C, Comb scale. *D*, Head and thorax (dorsal and ventral).

Head: Proboscis long, dark, slightly swollen at tip; palpi very short, dark. Occiput clothed dorsally with broad appressed pale to dark-brown scales with bronze to metallic blue-green reflection and dark erect forked scales, a small posteromedian area of narrow curved yellowish-white to golden scales present; lateral region of occiput with broad appressed dingy-white scales. Cibarial armature consisting of three teeth. *Thorax:* Integument of scutum dark brown; scutum clothed with fine narrow curved dark bronze-brown scales. Scutellum with fine brown scales and brown setae on the lobes. Pleura brownish, with small patches of broad grayish-white scales. *Abdomen:* Tergites clothed dorsally with dark-brown or black scales with bronze to metallic blue-green reflection; narrow, more or less complete, white basal bands occasionally present on some segments; small basolateral patches of white scales present. Sternites each with a white basal band, dark apically. *Legs:* Legs dark-scaled with bronze to metallic blue-green reflection, except for pale posterior surface of femora and small pale knee spots. *Wing:* Length about 2.5 mm. Scales all dark. Plume scales narrow, squame scales broader.

ADULT MALE. Coloration similar to that of the female. *Terminalia* (fig. 273). Lobes of ninth tergite (IXT-L) widely separated, long and slender, bearing two or three setae arising from distinct tubercles, each lobe arises from a broad plate bearing many setae distally. Tenth sternite (X-S) with an apical comblike row of blunt spines. Phallosome (Ph) with inner plate bearing a stout apical tooth, a stout subapical tooth on outer margin, a strong tooth pointed ventrally, a slender tooth pointed ventrally near base, and a long taillike basal hook directed ventrally. Claspette absent. Basistyle (Bs) nearly twice as long as mid-width, outer margin strongly curved, clothed with scales and numerous setae, long setae on outer aspect. Subapical lobe (S-L) deeply divided into two distinct divisions: basal division cleft nearly to base into two stout branches, each bearing a long strong rod with tip flattened and curved; apical division bearing a long flattened strongly hooked pointed rod, three shorter slightly flattened rods, a moderately broad weakly striated leaflike appendage, and two narrower leaflike

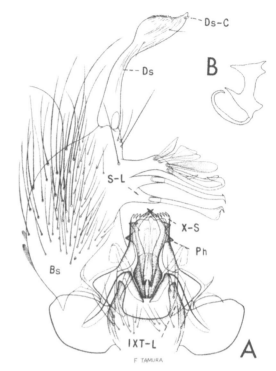

Fig. 273. *Culex pilosus* (Dyar and Knab). *A*, Male terminalia. *B*, Inner plate of phallosome.

filaments. Dististyle (Ds) nearly three-fourths as long as the basistyle, curved, narrow medially, outer margin expanded caplike near tip and bearing a row of very short reflexed setae, tip pointed and slightly reflexed; claw (Ds-C) short, curved, inserted a little before apex.

LARVA (fig. 274). An ovoid gill inserted on ventral side of head at base of antenna. Antenna as long or longer than the head, constricted beyond antennal tuft, with part before constriction spiculate, part beyond constriction darker and with few spicules; antennal tuft large, multiple, barbed, inserted at outer third of shaft, reaching beyond tip. Head hairs: Postclypeal 4 small, single; upper frontal 5 single (occasionally double), barbed, about half as long as hair 6; lower frontal 6 single, barbed; preantennal 7 long, multiple, barbed. Prothoracic hairs: 1–2 long, single; 3 short, double or multiple; 4 long, single or double; 5–6 long, single; 7 long, usually double. Body spiculate. Lateral abdominal hair 6 usually triple on segments I and II and double on III–VI. Comb of eighth segment with about 8 to 12 scales in a single curved or irregular row; individual scale thorn-shaped, with minute spinules on basal part. Siphonal index 3.0 to 4.0; siphon distinctly upturned; pecten of about 6 to 11 teeth on basal third of siphon; individual pecten tooth fringed on one side to tip; eight pairs of long multiple barbed siphonal tufts present, the basal two tufts inserted within the pecten and about as long as the siphon; two small 2- or 3-branched tufts inserted laterally; dorsal preapical spine as long or longer than the apical pecten tooth, curved. Anal segment completely ringed by the saddle; lateral hair, small, double or triple; dorsal brush bilaterally consisting of a long lower caudal hair and two upper caudal hairs, one long and one short; ventral brush well developed, confined to the barred area; gills one and a half to two times as long as the saddle, the dorsal pair shorter than the ventral pair, each tapered to a blunt point.

DISTRIBUTION. The southeastern United States, Mexico, Central America, the West Indies, and northern South America. *United States:* Alabama, Florida, and Georgia (135, 419); Kentucky (570); Louisiana (419); Mississippi (135, 546); North Carolina (135, 213); South Carolina (135, 742).

BIONOMICS. The larvae have been found in semipermanent and permanent pools including ditches, floodwater areas, grassy pools, streams, and occasionally in bilge water of boats and other artificial collections of water. King *et al.* (419) state that the eggs are able to withstand drying—an unusual trait in this genus. The females cannot at this time be separated with certainty from some other species of the subgenus *Melanoconion;* therefore nothing is known of their feeding habits.

Culex (Neoculex) apicalis Adams

Culex apicalis Adams, 1903, Kans. Univ. Sci. Bull., 2:26.
Culex apicalis, Bohart, 1948, Ann. Ent. Soc. Amer., 41:336.

ADULT FEMALE (pl. 122). Rather small species. *Head:* Proboscis long, dark-scaled; palpi short, dark, usually a few pale scales at apex of subterminal segment and at base of terminal segment. Occiput clothed dorsally with narrow curved ash-white scales and brown erect forked scales, with broad dingy-

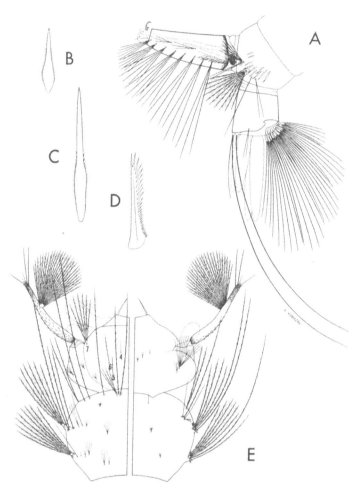

Fig. 274. Larva of *Culex pilosus* (Dyar and Knab). *A,* Terminal segments. *B,* and *C,* Comb scales. *D,* Pecten tooth. *E,* Head and thorax (dorsal and ventral).

white scales laterally. *Thorax:* Integument of scutum brown to dark brown; scutum clothed with narrow grayish to light-brown scales, often forming a median line and a curved lateral line, scales on anterior and lateral margins and on the prescutellar space paler. Scutal bristles strong and dark. Scutellum clothed with grayish scales and brown setae. Pleura with patches of broad grayish-white scales. *Abdomen:* First tergite with a median patch of dark scales; remaining tergites dark-brown to black-scaled with bronze to metallic blue-green reflection, each with an apical band of white scales joining a triangular patch of white scales on either side. Venter clothed with grayish-white scales. *Legs:* Legs dark, with bronze to metallic blue-green reflection except for pale knee spots and pale posterior surface of femora (pale stripe on posterior surface of hind femur not quite reaching apex). *Wing:* Length about 3.0 to 3.2 mm. Scales narrow, dark.

ADULT MALE. Coloration similar to that of the female. Palpi long and with moderately dense hairs toward apex; long hairs at apex of long segment limited to an area not more than one-half as long as the terminal segment. *Terminalia* (fig. 275). Lobes of ninth tergite (IXT-L) rounded, about as long as broad, each bearing several long setae. Tenth sternite (X-S) crowned with short blunt spines, arranged in a comblike row. Phallosome (Ph) formed of a pair of stout columnar lateral plates strongly joined at base; apex of each column blunt, crowned with many small denticles. Claspette absent. Basistyle (Bs) about two and a half times as long as mid-width, with long strong setae on outer aspect. Subapical lobe (S-L) prominent, undivided, bearing two

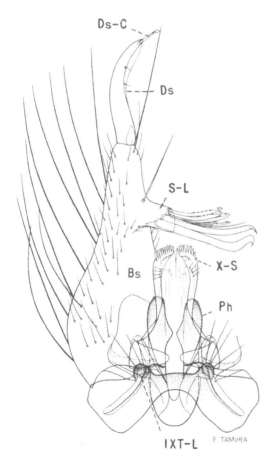

Fig. 275. Male terminalia of *Culex apicalis* Adams.

strong rods with tips pointed and recurved, and about six smaller spines, the three most apical of which are flattened and serrate on one side. Dististyle (Ds) about half as long as the basistyle, curved, distal half bearing two small setae; claw (Ds-C) about one-fifth as long as the dististyle, with tip blunt and flattened.

LARVA. (fig. 276). Antenna about as long as the head, constricted beyond antennal tuft, with part before constriction spiculate, dark near base, part beyond constriction dark and with few spicules; antennal tuft large, multiple, barbed, inserted at outer third of shaft, reaching well beyond tip. Head hairs: Postclypeal 4 short, single; upper frontal 5 double or triple, barbed, about half as long as hair 6; lower frontal 6 long, double (rarely single), barbed; preantennal 7 long, multiple, barbed. Prothoracic hairs; 1–2 very long, single; 3 medium, single or double; 4 long, double; 5–6 very long, single; 7 long, triple (occasionally 2- or 4-branched). Body spiculate. Lateral abdominal hair 6 usually multiple on segments I and II, double on III–VI. Abdominal segments rather evenly pigmented. Comb of eighth abdominal segment of many scales in a patch; individual scale rounded apically and fringed with subequal spinules. Siphonal index about 7.0 to 9.0; basal diameter of siphon nearly twice the apical diameter; pecten of about 10 to 14 teeth on basal fourth to one-third of siphon; individual tooth with 1 to 4 coarse teeth on one side; about five or six pairs of siphonal tufts inserted beyond the pecten; individual hairs of proximal tuft rarely more than one to one and a half times as long as the basal diameter of the siphon. Anal segment completely ringed by the saddle, dorsoapical part of saddle spiculate; lateral hair usually double (rarely single or triple), shorter than the saddle; dorsal brush bilaterally consisting of a long lower caudal hair and a shorter multiple upper caudal tuft of several branches; ventral brush well developed; gills usually shorter than the saddle.

DISTRIBUTION. This species is known from California (70, 287), Arizona (70), and Mexico (537a).

BIONOMICS. According to Freeborn and Bohart (287) this species is a common inhabitant of woodland pools in California at elevations less than 5,000 feet. It has been found also in

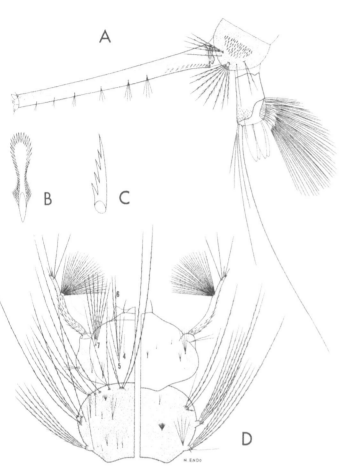

Fig. 276. Larva of *Culex apicalis* Adams. *A*, Terminal segments. *B*, Comb scale. *C*, Pecten tooth. *D*, Head and thorax (dorsal and ventral).

canyon streams in central Arizona. The females have not been observed feeding on man, and it is believed that they confine their feeding to cold-blooded animals, possibly to frogs alone.

Culex (Neoculex) arizonensis Bohart

Culex arizonensis Bohart, 1948, Ann. Ent. Soc. Amer., *41*:341.

ADULT FEMALE (pl. 123). Rather small species. *Head:* Proboscis long, dark-scaled, palpi short, dark. Occiput with narrow curved white scales and dark-brown erect forked scales dorsally, with broad dingy-white scales laterally. *Thorax:* Integument of scutum dark brown; scutum clothed with narrow brown scales, a median stripe of whitish scales reach-

ing the prescutellar space, scutal angle with whitish scales which extend posteriorly in two submedian lines, one on either side of the prescutellar space; anterior and lateral margins and prescutellar space with whitish scales. Scutal bristles strong and dark. Scutellum with narrow whitish scales and dark setae on the lobes. Pleura with patches of broad white scales. *Abdomen:* First tergite with a median patch of dark and pale scales; remaining tergites dark-brown to black-scaled with bronze to metallic blue-green reflection, each with a narrow apical band of white scales. Venter clothed with grayish-white scales. *Legs:* Legs dark-scaled with bronze to metallic blue-green reflection except for small pale knee spots and pale posterior surface of femora (pale posterior stripe on basal five-sixths of hind femur). *Wing:* Length about 3.0 mm. Scales narrow, dark.

ADULT MALE. Coloration similar to that of the female. Palpi long and with sparse and relatively short hairs, a few erect hairs near apex of long segment, last segment with three or four moderate hairs on inner surface and one or two toward base of outer side. *Terminalia* (fig. 277). Lobes of ninth tergite (IXT-L) rounded, nearly as long as broad, each bearing many long setae. Tenth sternite (X-S) crowned with short blunt spines arranged in a comblike row. Phallosome (Ph) formed of a pair of stout columnar lateral plates strongly joined at base and with a narrow transverse bridge near apical third, apex of each column broad and with many denticles. Claspette absent. Basistyle (Bs) about two and a half times as long as mid-width, with long setae on outer aspect, with rather long setae on inner margin below the subapical lobe. Subapical lobe (S-L) prominent, undivided, bearing two strong rods and about six or seven spines, some of which are serrate on one side. Dististyle (Ds) about half as long as the basistyle, curved, with two small setae on distal half; claw (Ds-C) about one-fifth as long as the dististyle, with tip blunt and flattened.

LARVA (fig. 278). Antenna about as long as the head, constricted beyond antennal tuft, with part before constriction spiculate, dark near base, part beyond constriction darker and with fewer spicules; antennal tuft large, mul-

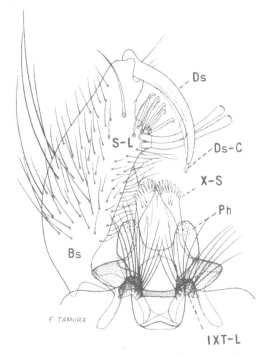

Fig. 277. Male terminalia of *Culex arizonensis* Bohart.

tiple, barbed, inserted at outer third of shaft, reaching beyond tip. Head hairs: Postclypeal 4 short, single; upper frontal 5 and lower frontal 6 long, double, of about equal length, barbed; preantennal 7 long, multiple, barbed. Prothoracic hairs: 1–2 long, single; 3 short, 2- to 4-branched; 4 medium, single or double; 5–6 long, single; 7 medium, double or triple. Thorax spiculate. Lateral abdominal hair 6 usually multiple on segments I and II, double on III–VI. Abdominal segments rather evenly pigmented. Comb of eighth segment with numerous scales in a patch; individual scale rounded apically and fringed with subequal spinules. Siphonal index about 7.0; siphon only slightly expanded at apex, with a broad basal darkly pigmented band present; pecten of about 15 teeth on basal fourth of siphon; individual tooth with 1 to 4 long coarse side teeth; six pairs of 2- to 4-branched siphonal tufts inserted beyond the pecten (proximal tuft sometimes inserted within the pecten); individual hairs of proximal tuft about one-third to two-fifths as long as the siphon. Anal segment ringed by the saddle; dorsoapical part of saddle spiculate; lateral hair usually double, shorter than the saddle; dorsal brush bilaterally consisting of a long lower caudal hair and a shorter upper

caudal tuft of about four branches; ventral brush well developed, confined to the barred area; gills slightly longer than the saddle, pointed.

DISTRIBUTION. Known from Arizona (70) and Mexico (537a).

BIONOMICS. Bohart (70) states that the type series was taken from pools in a creek bed, shaded by oak trees and by a bridge near Prescott, Arizona. The larvae were numerous in all stages and were associated with the larvae of *C. apicalis*. Bohart further states that adult females were numerous about the pools but did not attempt to bite.

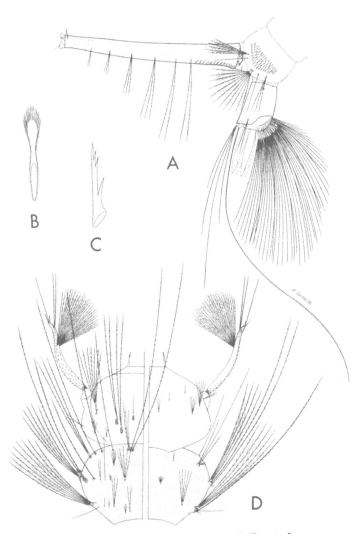

Fig. 278. Larva of *Culex arizonensis* Bohart. *A*, Terminal segments. *B*, Comb scale. *C*, Pecten tooth. *D*, Head and thorax (dorsal and ventral).

Culex (Neoculex) boharti
Brookman and Reeves

Culex reevesi Bohart, 1948, Ann. Ent. Soc. Amer., *41*:342.

Culex boharti Brookman and Reeves, 1950, Pan-Pac. Ent., *26*:159.

This species was originally named *Culex reevesi* Bohart (70), but the name was antedated while still in press by *C. reevesi* Wirth. The resulting homonymy was explained, and a new name was given the species by Brookman and Reeves (105).

ADULT FEMALE (pl. 124). Rather small species. *Head:* Proboscis long, dark-scaled; palpi short, dark. Occiput with narrow curved yellowish-brown scales and brown erect forked scales dorsally, with broad dingy-white scales laterally. *Thorax:* Integument of scutum light brown; scutum clothed with narrow brown scales, paler on the anterior and lateral margins and on the prescutellar space; a pair of more or less distinct submedian spots of light-golden scales near middle of scutum. Scutal bristles strong and dark. Scutellum clothed with light-golden scales and dark setae. Pleura with patches of broad white scales. *Abdomen:* First tergite with a median patch of dark scales; remaining tergites dark-brown to black-scaled with bronze to metallic blue-green reflection, each with an apical band of white scales joining a triangular patch of white scales on either side. Venter clothed with grayish-white scales. *Legs:* Legs dark with bronze to metallic blue-green reflection except for small pale knee spots and pale posterior surface of femora and tibiae. *Wing:* Length about 3.0 mm. Scales narrow, dark.

ADULT MALE. Coloration similar to that of the female. Palpi with rather short hairs toward apex, long segment with erect hairs at tip only, last segment with a few erect hairs near base on inner side. *Terminalia* (fig. 279). Lobes of ninth tergite (IXT-L) low and broadly rounded, each bearing several setae. Tenth sternite (X-S) crowned with short blunt spines arranged in a comblike row. Phallosome (Ph) formed of a pair of stout columnar lateral plates joined at base and with a transverse bridge near apex; apex of each column strongly narrowed and heavily sclerotized apically,

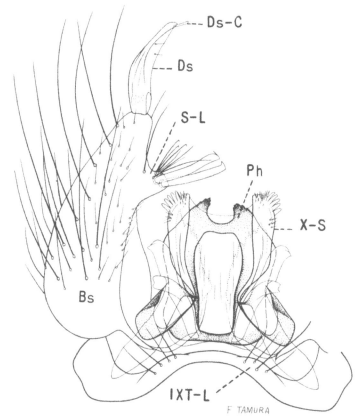

Fig. 279. Male terminalia of *Culex boharti* Brookman and Reeves.

crowned with a few denticles. Claspette absent. Basistyle (Bs) about twice as long as mid-width, with long strong setae on outer aspect. Subapical lobe (S-L) undivided, bearing two strong rods with tips pointed and recurved and about six small spines, the three most apical of which are flattened and serrate on one side. Dististyle (Ds) about half as long as the basistyle, curved, distal half with two small setae; claw (Ds-C) about one-fifth as long as the dististyle, with tip blunt and flattened.

LARVA (fig. 280). Similar to *C. reevesi* and *C. territans*, but these three species can usually be separated by the characters given in the key. Antenna about as long as the head, constricted beyond antennal tuft, with part before constriction spiculate, dark near base, part beyond constriction darker and with fewer spicules; antennal tuft multiple, barbed, inserted at outer third of shaft, reaching beyond tip. Head hairs: Postclypeal 4 short, single; upper frontal 5 double (rarely triple), barbed, about two-

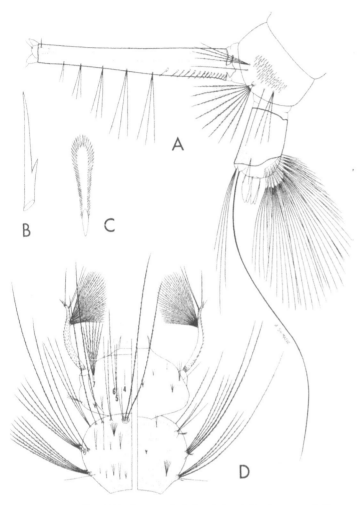

Fig. 280. Larva of *Culex boharti* Brookman and Reeves.
A, Terminal segments. *B*, Pecten tooth. *C*, Comb scale. *D*, Head and thorax (dorsal and ventral).

the proximal tuft often inserted within the pecten; individual hairs of proximal tuft about one-third as long as the siphon. Anal segment completely ringed by the saddle; dorsoapical part of plate spiculate; lateral hair single or double, a little shorter than the saddle; dorsal brush bilaterally consisting of a long lower caudal hair and a shorter 3- to 5-branched upper caudal tuft; ventral brush large, with one to three precratal tufts; gills varying from about one-third to one and a half times as long as the saddle.

DISTRIBUTION. Known from the lowlands and foothills of California (287).

BIONOMICS. Freeborn and Bohart (287) state that the larvae prefer partly sunlit stream pools with an abundant growth of aquatic vegetation. The adults are said to be shy and are not known to bite man.

Culex (Neoculex) reevesi Wirth

Culex reevesi Wirth, 1948, Univ. Calif. Syllabus Series, Ent., *133*:230.

Culex reevesi, Brookman and Reeves, 1950, Pan-Pac. Ent., *26*:159.

Culex reevesi, Brookman and Reeves, 1953, Ann. Ent. Soc. Amer., *46*:226.

This mosquito was originally described by Galindo (300) from specimens collected at Monterey, California. The first use of the name *Culex reevesi* was by Wirth, who presented characters sufficient for separating it from other *Culex* in California.

ADULT FEMALE. Rather small species. *Head:* Proboscis long, dark-scaled; palpi short, dark, about two and a half times as long as flagellar segment 4 of antenna. Occiput clothed dorsally with narrow curved grayish to light-brown scales and brown erect forked scales, with broad dingy-white scales laterally. *Thorax:* Integument of scutum light brown; scutum clothed with narrow yellowish or light-brown scales, a little paler on the anterior and lateral margins and on the prescutellar space. Scutal bristles strong and dark. Scutellum clothed with narrow grayish scales and dark setae. Pleura with patches of broad white scales. *Abdomen:* First tergite with a median patch of bronze-brown scales; remaining tergites dark-

thirds as long as hair 6; lower frontal 6 long, single, barbed; preantennal 7 long, multiple, barbed. Prothoracic hairs: 1–2 long, single; 3 short, double or multiple; 4 medium, single or double; 5–6 long, single; 7 medium, double or triple. Thorax spiculate. Lateral abdominal hair 6 usually multiple on segments I and II, double on III to VI. Abdominal segments unevenly pigmented (segment IV much paler than III or V). Comb of eighth segment with numerous scales in a patch; individual scale rounded apically and fringed with subequal spinules. Siphonal index about 6.0; pecten of about 10 to 17 teeth on basal third of siphon; individual tooth with 1 to 4 long coarse side teeth; usually five pairs of 2- to 4-branched siphonal tufts, brown to black-scaled with bronze to metallic blue-green reflection, with lateral triangular patches of white scales apically but not joined by narrow dorsoapical bands; tergites V to VII with scattered grayish scales along apical margins. Venter clothed with grayish-white scales. Abdominal segment V of dry unengorged specimens longer than broad. *Legs:* Legs dark-scaled with bronze to metallic blue-green reflection except for small pale knee spots, and pale posterior surface of femora and tibiae. *Wing:* Length about 3.0 mm. Scales narrow, dark.

ADULT MALE. Coloration similar to that of the female. Palpi with moderate to long hairs along the entire length of the terminal segment, long segment with three or four erect hairs toward tip only, palpi a little longer than the proboscis. *Terminalia* (fig. 281). Lobes of ninth tergite (IXT-L) rounded, nearly as long as broad, each bearing several setae. Tenth sternite (X-S) with a comblike apical row of short spines. Phallosome (Ph) formed of a pair of stout columnar lateral plates strongly joined at base and at distal third; apex of each column broadly rounded and without denticles or teeth. Claspette absent. Basistyle (Bs) about two and a half times as long as mid-width, with long setae on outer aspect. Subapical lobe (S-L) prominent, undivided, bearing two strong rods with tips pointed and recurved, about four flattened spines (two of which are serrated along one edge), and a few simple setae. Dististyle (Ds) nearly half as long as the basistyle, strongly narrowed on apical third, distal half bearing two small setae; claw (Ds-C) short, with tip blunt and flattened.

LARVA (fig. 282). Antenna about as long as the head, constricted beyond antennal tuft, with part before constriction spiculate, dark near base, part beyond constriction darker and with fewer spicules; antennal tuft large, multiple, barbed, inserted at outer third of shaft, reaching beyond tip. Head hairs: Postclypeal 4 short, single; upper frontal 5 triple, barbed, about two-thirds as long as hair 6; lower frontal 6 long, double, barbed; preantennal 7 long, multiple, barbed. Prothoracic hairs: 1–2 long, single; 3 short, double; 4 medium, double; 5–6 long, single; 7 medium, triple. Thorax spiculate. Lateral abdominal hair 6 usually multiple on segments I and II, double on III–VI. Abdominal segments unevenly pigmented (segment IV much paler than segments III or V). Comb of eighth segment with many scales in a patch; individual scale rounded apically and fringed with subequal spinules. Siphonal index about 5.5 to 6.0; pecten of about 13 to 20 teeth on basal two-fifths of siphon; individual tooth with 1 to 4 coarse side teeth; five pairs of 2- to 4-branched siphonal tufts, individual hairs of proximal tuft about one-third as long as the siphon. Anal segment completely ringed by the saddle; dorsoapical part of saddle spiculate; lateral hair double, shorter than the saddle; dorsal brush bilaterally consisting of a long lower caudal hair and a shorter upper caudal tuft of several hairs; ventral brush well developed; gills shorter than the saddle, pointed.

DISTRIBUTION. *Culex reevesi* has been reported by Galindo (300) as occurring along the California coast from Point Reyes, Marin County, south to Monterey and San Luis Obispo County, and possibly San Diego. It has also been reported from Baja California, Mexico (106). Brookman and Reeves (106) state that the female specimens reported from Point Reyes and San Luis Obispo may have been *C. boharti* Brookman and Reeves; however, Dr.

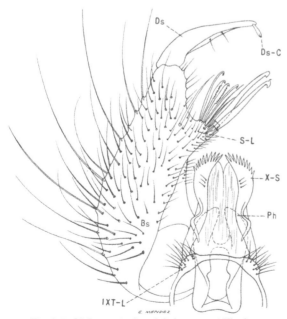

Fig. 281. Male terminalia of *Culex reevesi* Wirth.

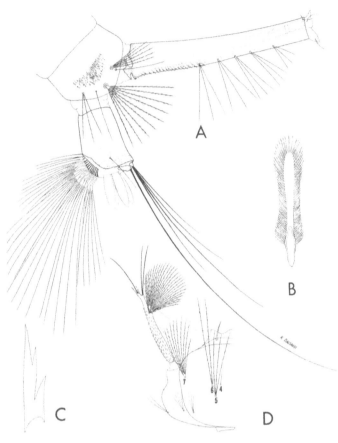

Fig. 282. Larva of *Culex reevesi* Wirth. *A*, Terminal segments. *B*, Comb scale. *C*, Pecten tooth. *D*, Head.

Richard M. Bohart has recently collected *C. reevesi* in San Luis Obispo County.

BIONOMICS. Brookman and Reeves (106) discovered the larvae in Baja California, Mexico, in a pond formed by a sandbar across the mouth of a stream. The larvae were found along the fringe of the pond among thick growths of tule, algae, and duckweed.

Culex (Neoculex) territans Walker

Culex territans Walker, 1856, Ins. Saund., Dipt., *1*: 428.

Culex saxatilis Grossbeck, 1905, Can. Ent., *37*:360.

Culex frickii Ludlow, 1906, Can. Ent., *38*:132.

Culex apicalis Adams (of Dyar and other authors, not of Adams).

Culex territans, Bohart, 1948, Ann. Ent. Soc. Amer., *41*:332.

ADULT FEMALE (pl. 125). Rather small

species. *Head:* Proboscis long, dark-scaled; palpi short, dark. Occiput with narrow curved ash-white to golden scales and brown erect forked scales dorsally, with broad dingy-white scales laterally. *Thorax:* Integument of scutum usually light brown; scutum clothed with narrow light-brown scales, light gray to dark brown on some specimens, scales on anterior and lateral margins and the prescutellar space paler; a pair of indefinite submedian spots of pale scales often present near middle of scutum. Scutellum clothed with grayish scales and brown setae. Pleura with patches of broad white scales. *Abdomen:* First tergite with a median patch of dark scales; remaining tergites dark-brown to black-scaled with bronze to metallic blue-green reflection, each with a narrow apical band of white scales joining a triangular patch on either side. Venter clothed with grayish-white scales. *Legs:* Legs dark-scaled with bronze to metallic blue-green reflection except for small pale knee spots, pale posterior surface of femora and tibiae and a pale streak usually present on segment 1 of tarsi (pale posterior stripe on hind femur complete). *Wing:* Length about 3.0 to 3.3 mm. Scales narrow, dark.

ADULT MALE. Coloration similar to that of the female. Palpi with long and dense hairs toward apex; long hairs at apex of long segment occupying an area about as long as the terminal segment. *Terminalia* (fig. 283). Lobes of ninth tergite (IXT-L) rounded, about as long as broad, each bearing several setae. Tenth sternite (X-S) with a comblike apical row of short blunt spines. Phallosome (Ph) formed of a pair of stout columnar lateral plates strongly joined at base and with a narrow transverse sclerotized bridge near apical fourth; apex of each column broad and blunt, crowned with numerous small denticles. Claspette absent. Basistyle (Bs) about two and a half times as long as mid-width, with long strong setae on outer aspect. Subapical lobe (S-L) prominent, undivided, bearing two strong rods with tips pointed and recurved, and five smaller spines, the three most apical of which are flattened and serrate on one side. Dististyle (Ds) about half as long as the basistyle, curved, distal half with many short setae, two longer setae, and a row of membranous lobes seen at high

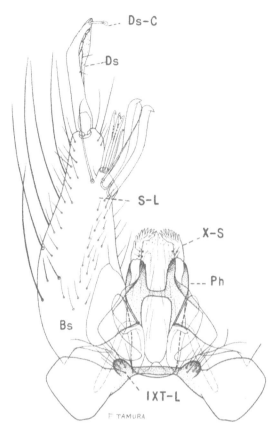

Fig. 283. Male terminalia of *Culex territans* Walker.

merous scales in a patch; individual scale rounded apically and fringed with subequal spinules. Siphonal index 6.0 to 7.0; siphon slightly expanded at the apex; pecten of about 12 to 16 teeth on basal third of siphon; individual tooth with 1 to 4 long coarse side teeth; four or five pairs of siphonal tufts inserted beyond the pecten, the individual hairs of proximal tuft about one to two times as long as basal diameter of the siphon; apical tuft inserted somewhat laterally. Anal segment completely ringed by the saddle; dorsoapical part of saddle spiculate; lateral hair usually double (rarely 3- or 4-branched), shorter than the saddle; dorsal brush bilaterally consisting of

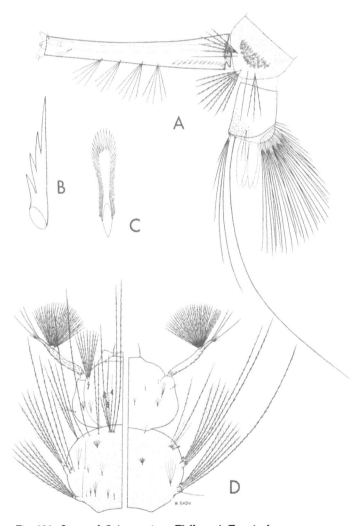

Fig. 284. Larva of *Culex territans* Walker. *A*, Terminal segments. *B*, Pecten tooth. *C*, Comb scale. *D*, Head and thorax (dorsal and ventral).

magnification; claw (Ds-C) about one-fifth as long as the dististyle, with tip blunt and flattened.

LARVA (fig. 284). Antenna about as long as the head, constricted beyond antennal tuft, with part before constriction spiculate, dark near base, part beyond constriction darker and with fewer spicules; antennal tuft large, multiple, barbed, inserted at outer third of shaft, reaching much beyond tip. Head hairs: Postclypeal 4 small, single; upper frontal 5 single (occasionally double, rarely triple), barbed, about one-half to two-thirds as long as hair 6; lower frontal 6 long, single (rarely double), barbed; preantennal 7 long, multiple, barbed. Prothoracic hairs: 1–2 long, single; 3 short, double or multiple; 4 medium, usually double; 5–6 long, single; 7 medium, double or triple. Thorax spiculate. Lateral abdominal hair 6 usually multiple on segments I and II, double on III–VI. Abdominal segments rather evenly pigmented. Comb of eighth segment with nua long lower caudal hair and a shorter upper caudal tuft of about four branches; ventral brush large, with two or three precratal tufts; gills variable in length but usually as long or a little longer than the saddle, tapered.

DISTRIBUTION. The subgenus *Neoculex* in North America north of Mexico was recently revised by Bohart (70). *Culex territans* is now known to occur in eastern Europe, Alaska, Canada, and throughout most of the United States. Earlier records of *C. apicalis,* east of the Rocky Mountains, undoubtedly refer to *C. territans.* The following records are those given by Bohart (70) and subsequent workers. *United States:* California (287) ; Florida and Georgia (70); Idaho (672); Iowa, Louisiana, Maryland, Massachusetts, Michigan, Minnesota, Mississippi, Missouri, Montana, New York and North Carolina (70) ; Ohio (731a) ; Oklahoma (70) ; Oregon (70, 672) ; Rhode Island (70) ; Texas (70, 249) ; Vermont and Virginia (70) ; Washington (70, 672). *Canada:* British Columbia (70). *Alaska:* (70).

BIONOMICS. The larvae are found in semipermanent and permanent pools in streams, swamps, and ponds. They do not seem to favor foul water. The larvae may be collected any time during the year in the southern United States but farther north, breeding is limited mostly to the spring, summer, and fall seasons.

The females are not known to bite man but have been observed feeding on cold-blooded vertebrates, particularly frogs. The adults are seldom taken inside buildings but are encountered occasionally among vegetation and in shelters near their larval habitats. The females apparently pass the winter in hibernation in the colder areas.

Genus DEINOCERITES Theobald

Deinocerites Theobald, 1901, Mon. Culic., 2:215.

Brachomyia Theobald, 1901, Mon. Culic., 2:343.

Dinomimetes Knab, 1907, Jour. N.Y. Ent. Soc., 15: 120.

Dinanamesus Dyar and Knab, 1909, Smithson. Misc. Coll., 52:259.

ADULT. Medium-sized, dull-colored mosquitoes. Many characters of *Deinocerites* are the same as in the genus *Culex,* from which it differs chiefly in characteristics of the antennae. The first flagellar segment of the antenna is elongate in both sexes; usually several of the following segments are elongate in the male. The antenna of the male is not plumose. Palpi short in both sexes. Abdomen of the female more pointed than in *Culex* and the cerci are prominent. *Male terminalia.* Ninth tergite with two long fingerlike lobes. Tenth sternite with an apical transverse comb of strong close-set teeth. Phallosome strongly toothed. Basistyle stout; basal lobe small, not divided, bearing a few stout blunt spines. Dististyle rather short, pubescent, with two claws.

PUPA. Unmodified. Paddle with smooth margin as in *Culex,* but with one apical hair.

LARVA. Head almost circular in outline and with a prominent triangular pouch on either side. Antenna with tuft inserted near middle of shaft. Comb of eight segment consisting of a large patch of scales. Siphon moderate; pecten present; one pair of large siphonal tufts inserted near middle, two pairs of shorter tufts distally. Anal segment with a dorsal plate and a smaller ventral plate, the two widely separated laterally. Gills two, short, bulbous, or rudimentary.

BIONOMICS. The larvae of the members of this genus are found in salt water in crabholes. The adults are frequently found resting in the upper part of the hole above the water level. The females of most species are known to bite, but not readily.

CLASSIFICATION. Edwards (251) divides the genus into three groups as follows:

Group A (Deinocerites) *D. cancer* and *D. pseudes*
Group B (Dinomimetes) *D. epitedeus*
Group C (Dinanamesus) *D. spanius*

Two species, *Deinocerites cancer* and *D. spanius,* are known to occur in this region. A single male specimen of *Deinocerites epitedeus* (Knab) was reported by Rueger and Druce (634) from a light-trap collection at Harlingen Army Air Field, Texas, during March 1945. This is believed to represent either an intrusion from Central America or a misidentification, and it is not likely that *D. epitedeus* is established in the area. However, it has been collected at San Jose del Cabo, Lower California, Mexico.

DISTRIBUTION. The genus *Deinocerites* contains a few species which are confined to the Caribbean area, the Gulf of Mexico, and adjacent areas.

KEYS TO THE SPECIES

ADULT FEMALES

1. Small species; antenna with the first flagellar segment about twice as long as the second
spanius D. and K., p. 327
Medium-sized species; antenna with the first flagellar segment three times as long as the second
cancer Theob., p. 325

MALE TERMINALIA

1. Dististyle with claws equal; phallosome with a row of spinelike filaments laterally
cancer Theob., p. 326
Dististyle with claws unequal; phallosome with two strong spines laterally
spanius D. and K., p. 327

LARVAE (FOURTH INSTAR)

1. Upper frontal head hair 5 and lower frontal 6 multiple..................................*spanius* D. and K., p. 328
Upper frontal head hair 5, 2- to 4-branched, lower frontal 6 single or double..*cancer* Theob., p. 326

Deinocerites cancer Theobald

Deinocerites cancer Theobald, 1901, Mon. Culic., 2: 215.

Brachomyia magna Theobald, 1901, Mon. Culic., 2: 344.

Deinocerites melanophylum Dyar and Knab, 1907, Jour. N.Y. Ent. Soc., 15:200.

Deinocerites tetraspathus Dyar and Knab, 1909, Smithson. Misc. Coll., 52:260.

Deinocerites troglodytus Dyar and Knab, 1909, Smithson. Misc. Coll., 52:260.

ADULT FEMALE (pl. 126). Medium-sized

species. *Head:* Proboscis long, slender, dark-scaled; palpi short, dark; antennae much longer than the proboscis, the first flagellar segment three times as long as the second and equal in length to the second, third, and fourth flagellar segments combined. Occiput with copper-brown narrow curved scales and erect forked scales dorsally (the erect forked scales are dense posteriorly), with a patch of broad flat light copper-brown scales laterally. Tori light brown. *Thorax:* Integument of scutum brown; scutum clothed with narrow brown scales and dark setae with a coppery sheen. Scutellum with narrow curved light-brown scales and dark setae on the lobes. Sternopleura densely shingled with rather broad, coppery scales. Spiracular and postspiracular bristles absent. *Abdomen:* Tergites with dull brown scales with metallic bronze and blue luster. Venter tan. Abdomen with the terminal segments somewhat compressed, truncate, with prominent cerci. *Legs:* Legs with bronze-brown scales with blue reflection in some lights; femora yellowish on posterior surface. *Wing:* Length about 3.0 to 3.5 mm. Scales narrow (some moderately broad), brown.

ADULT MALE. Coloration similar to that of the female. Palpi similar to those of the female. First segment of antenna less than twice as long as the second, and not equal in length to the second, third, and fourth segments combined. *Terminalia* (fig. 285). Ninth tergite (IX-T) with dorsal band narrow; lobes (IXT-L) widely separated, heavily sclerotized, long, fingerlike and reaching apical fourth of basistyle, bearing numerous small setae on rounded inner surface near base. Tenth sternite (X-S) stout, bearing an apical transverse row of dark blunt teeth and many small setae on inner surface subapically. Phallosome (Ph) composed of two sclerotized plates, open dorsally and ventrally; each plate has two arms, one of which is slender and curved apically and bears a row of many long slender spinelike filaments laterally. Claspette absent. Basistyle (Bs) about two to two and a half times as long as midwidth, clothed with many long setae on the outer aspect and short setae on the inner aspect; a fleshy, fingerlike projection, bearing many setae, arises from the inner margin near apex of the basistyle and probably represents the

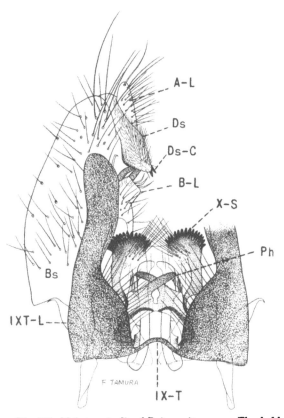

Fig. 285. Male terminalia of *Deinocerites cancer* Theobald.

apical lobe; basal lobe (B-L) arises near the middle of the basistyle, and bears one long slender spine, two short blunt dark rods, and about three smaller spines. Dististyle (Ds) short, dark, densely clothed with fine setae on outer surface, roundly expanded on inner margin at apical third, tapered apically; claws two, stout, curved and pointed, nearly equal in size, inserted a little before apex.

LARVA (fig. 286). Head almost circular in outline, with a prominent triangular pouch on either side. Antenna about half as long as the head, sparsely spiculate on basal third; antennal tuft multiple, sparsely barbed, inserted near middle of shaft. Eyes small, rounded. Head hairs: Postclypeal 4 small, single, inserted near anterior margin of the clypeus; upper frontal 5, 3- or 4-branched, occasionally double, barbed, shorter than hair 6; lower frontal 6 long, single, sparsely barbed; preantennal 7 multiple, barbed. Prothoracic hairs: 1–2 medium, single; 3 short, double; 4 medium 3- or 4-branched; 5–6 long, single; 7 medium, 2- to 4-branched. Lateral abdominal hair 6 double or triple on segment I, double on II and III, single or double on IV and V, single on VI. Comb of eighth segment with many scales in a patch; individual scale broadened and rounded apically, fringed with subequal spinules. Siphonal index 4.0 to 5.0; pecten of 6 to 8 teeth on basal third of siphon, teeth progressively more widely spaced distally; siphonal tuft long, 2- or 3-branched, inserted near middle of siphon; two small subapical single or double tufts present, one of which is inserted subdorsally. Anal segment with saddle sclerotized dorsally and ventrally, indistinct laterally; lateral hair short, single; dorsal brush bilaterally consisting of a long lower caudal hair and a shorter multiple upper caudal tuft; ventral brush large, confined to the barred area; gills two, short, bulbous.

DISTRIBUTION. Southern Florida (135, 213, 419, 562), the Antilles, Mexico, and Central America.

BIONOMICS. The larvae of *Deinocerites cancer* develop in water in the holes of land crabs in the tidal marshes along the coast. They are occasionally found in artificial receptacles such as tin cans. The adults generally rest during the daytime in the upper part of the crabhole. The senior author has frequently observed the males and females in copula in crabholes. The females occasionally bite humans. This species is readily attracted to horse-baited traps in Panama.

Deinocerites spanius (Dyar and Knab)

Dinanamesus spanius Dyar and Knab, 1909, Smithson. Misc. Coll., *52*:259.

Deinocerites spanius, Fisk, 1941, Ann. Ent. Soc. Amer., *34*:543.

ADULT FEMALE (pl. 127). Small species. *Head:* Proboscis long, slender, dark-scaled; palpi short, dark; antennae longer than the proboscis, the first flagellar segment about twice as long as the second and about equal in length to the second and third flagellar segments combined. Occiput with copper-brown narrow curved scales and erect forked scales dorsally, the erect forked scales dense posteriorly, with

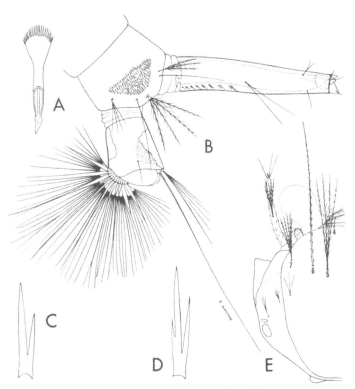

Fig. 286. Larva of *Deinocerites cancer* Theobald. *A,* Comb scale. *B,* Terminal segments. *C,* and *D,* Pecten teeth. *E,* Head.

a patch of broad flat light copper-brown scales laterally. Tori mottled, light brown to dark brown. *Thorax:* Integument of scutum brown; scutum sparsely clothed with narrow curved coppery scales and dark setae. Scutellum with narrow curved light-brown scales and dark setae on the lobes. Sternopleura with rather broad coppery scales. Spiracular and postspiracular bristles absent. *Abdomen:* Tergites with dull brown scales with metallic bronze and blue luster. Venter tan. Terminal segments of abdomen somewhat compressed, truncate, with large cerci. *Legs:* Legs with bronze-brown scales, with blue reflection in some lights; femora yellowish on posterior surface. *Wing:* Length about 2.0 mm. Scales narrow (some moderately broad), brown.

ADULT MALE. Coloration similar to that of the female. Palpi similar to those of female. *Terminalia* (fig. 287). Ninth tergite (IX-T) with dorsal band rather narrow, somewhat expanded medially; lobes (IXT-L) widely separated, each broad at base, long, slender and

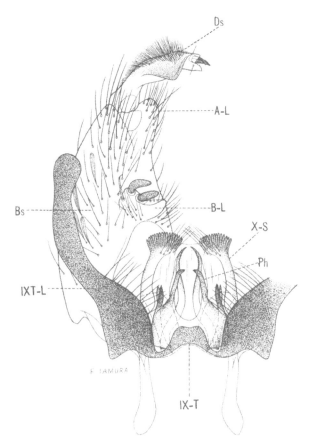

Fig. 287. Male terminalia of *Deinocerites spanius* (Dyar and Knab).

projection of the lobe. Dististyle (Ds) short, dark, densely clothed with fine setae on outer surface, angularly expanded on inner surface before apex; claws two, stout, curved, pointed, unequal in size, inserted a little before apex.

LARVA (fig. 288). Head almost circular in outline, with a triangular pouch on either side. Antenna about half as long as the head, sparsely spiculate on basal half; antennal tuft short, multiple, inserted near outer third. Head hairs: Postclypeal 4 small, single, inserted near anterior margin of clypeus; upper frontal 5 and lower frontal 6, 3- to 5-branched, barbed; preantennal 7 multiple, barbed, reaching near insertion of antennal tuft. Prothoracic hairs: 1-2

fingerlike (more slender than in *D. cancer*), with numerous setae on rounded inner surface near base. Tenth sternite (X-S) bearing an apical transverse row of dark blunt teeth and small setae subapically. Phallosome (Ph) composed of two sclerotized plates, open dorsally and ventrally; each plate has two arms, one of which is slender and curved apically and bears strong spines laterally. Claspette absent. Basistyle (Bs) about twice as long as mid-width, clothed with scales and long setae on outer aspect and short setae on inner aspect; a fleshy fingerlike projection bearing many setae arises from the inner margin near apex of the basistyle and probably represents the apical lobe; basal lobe (B-L) arises near the middle of the basistyle and is expanded apically with a prominent blunt projection on inner surface, the basal lobe bears one long, slender spine, two short blunt dark rods and two or three smaller spines which arise from the inner blunt

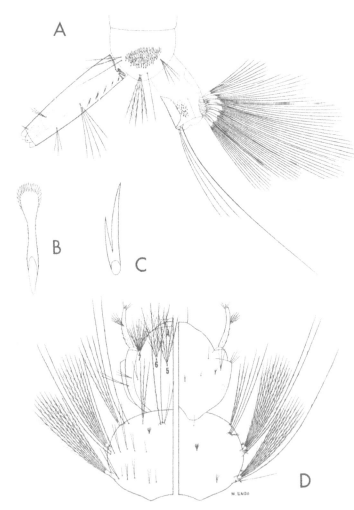

Fig. 288. Larva of *Deinocerites spanius* (Dyar and Knab). *A*, Terminal segments. *B*, Comb scale. *C*, Pecten tooth. *D*, Head and thorax (dorsal and ventral).

medium, single; 3 small, single or double; 4 medium, 3- or 5-branched; 5–6 long, single; 7 medium, usually 3- to 5-branched. Lateral abdominal hair 6 double on segment I, single (sometimes double) on II–V, single on VI. Comb of eighth segment of many scales in a large patch; individual scale broadened and rounded apically, fringed with subequal spinules. Siphonal index about 4.0; pecten of usually 5 teeth on basal third to two-fifths of siphon, teeth progressively more widely spaced distally; siphonal tuft usually 5-branched, inserted near middle of siphon; two pairs of small multiple tufts subapically, one of which is inserted subdorsally. Anal segment with the saddle sclerotized dorsally and ventrally, indistinct laterally; lateral hair small, single, sometimes double; dorsal brush bilaterally consisting of a long lower caudal hair and a short multiple upper caudal tuft; ventral brush well developed, confined to the barred area; gills two, short, blunt and indistinct.

DISTRIBUTION. Known from Texas (249, 275) and Panama.

BIONOMICS. Fisk (275) collected the adults of this species during November, 1939, in a light trap operated near Brownsville, Texas. He found the larvae in the holes of the fiddler crab (*Uca pugilator*) in 1940. Observations made by Fisk indicate that the adults are nocturnal and that the females feed by preference on cold-blooded animals. The species has been since collected by Dr. E. S. Ross at Brownsville in September, 1942, and in Kleberg County (249).

Bibliography

1. Aarons, T., J. R. Walker, H. F. Gray, and E. G. Mezger. 1951. Studies of the Flight Range of *Aedes squamiger* (Coquillett). Calif. Mosquito Control Assoc. Ann. Confr. Proc. and Papers, *19*:65-69.
2. Abdel-Malek, Albert. 1948. The Biology of *Aedes trivittatus*. Jour. Econ. Ent., *41*:951-954.
3. Abdel-Malek, Albert. 1949. A Study of the Morphology of the Immature Stages of *Aedes trivittatus* (Coquillett) (Diptera, Culicidae). Ann. Ent. Soc. Amer., *42*:19-37.
4. Adams, C. F. 1934. A Preliminary Report on the Culicidae (Mosquitoes) of Missouri. Proc. Mo. Acad. Sci., *1*:77-80.
5. Adams, C. F. 1942. *Aedes aegypti* Linn. The Yellow Fever Mosquito in Central Missouri. Science, *95*:19.
6. Adams, C. F., and Wm. M. Gordon. 1943. Notes on the Mosquitoes of Missouri (Diptera, Culicidae). Ent. News, *54*:232-235.
7. Aitken, T. H. G. 1939. Two *Aedes* Records for California. Pan-Pac. Ent., *15*:13-14.
8. Aitken, T. H. G. 1939. The *Anopheles maculipennis* Complex of Western North America. Pan-Pac. Ent., *15*:191-192.
9. Aitken, T. H. G. 1940. The Genus *Psorophora* in California (Diptera, Culicidae). Rev. de Ent., *11*:672-682.
10. Aitken, T. H. G. 1941. A New American Subgenus and Species of *Aedes*. Pan-Pac. Ent., *17*:81-84.
11. Aitken, T. H. G. 1942. Contributions Toward a Knowledge of the Insect Fauna of Lower California (Diptera: Culicidae). Proc. Calif. Acad. Sci., *24*:161-170.
12. Aitken, T. H. G. 1945. Studies on The Anopheline Complex of Western North America. Univ. Calif. Publ. Ent., *7*:273-364.
13. Aitken, T. H. G. 1948. Recovery of Anopheline Eggs from Natural Habitats, An Aid to Rapid Survey Work. Ann. Ent. Soc. Amer., *40*:327-329.
14. Anonymous. 1951. Mosquito Records from the Missouri River Basin States. Survey Section, CDC, Public Health Service, Federal Security Agency, 93 pp. (mimeo.).
15. Arnett, R. H. 1948. Notes on the Distribution, Habits and Habitats of Some Panama Culicines (Diptera: Culicidae). Part II, *Uranotaenia* through *Deinocerites*. Jour. N.Y. Ent. Soc., *56*:175-193.
16. Baker, F. C. 1936. A New Species of *Orthopodomyia*, *O. alba*, sp.n. Proc. Ent. Soc. Wash., *38*:1-7.
17. Bang, F. B., G. E. Quinby, and T. W. Simpson. 1940. *Anopheles walkeri* Theobald: A Wild-Caught Specimen Harboring Malaria Plasmodia. U.S. Pub. Hlth. Rpts., *55*:119-120.
18. Bang, F. B., G. E. Quinby, and T. W. Simpson. 1943. Studies on *Anopheles walkeri* Theobald Conducted at Reelfoot Lake, Tennessee, 1935-1941. Amer. Jour. Trop. Med., *23*:247-273.
19. Banks, H. W. 1937. New and Significant Experiences in Mosquito Control in Delaware County, Pennsylvania. N.J. Mosq. Exterm. Assoc. Proc., *24*:143-147.
20. Barber, M. A. 1939. Further Observations (1938) on the *Anopheles* of New Mexico. Amer. Jour. Trop. Med., *19*:345-356.
21. Barber, M. A., and L. R. Forbrich. 1933. Malaria in the Irrigated Regions of New Mexico. U.S. Pub. Hlth. Rpts., *48*:610-623.
22. Barber, M. A., and T. B. Hayne. 1924. Some Notes on the Relation of Domestic Animals to *Anopheles*. U.S. Pub. Hlth. Rpts., *39*:139-144.
23. Barber, M. A., and T. B. Hayne. 1924. Some Observations on the Dispersal of Adult *Anopheles*. U.S. Pub. Hlth. Rpts., *39*:195-203.
24. Barber, M. A., and W. H. W. Komp. 1930. Breeding Places of *Anopheles* in the Yazoo-Mississippi Delta. U.S. Pub. Hlth. Rpts., *44*:2457-2462.
25. Barber, M. A., W. H. W. Komp, and T. B. Hayne.

1924. Some Observations on the Winter Activities of *Anopheles* in Southern United States. U.S. Pub. Hlth. Rpts., *39*:231-246.

26. Barber, M. A., W. H. W. Komp, and T. B. Hayne. 1929. Malaria and the Malaria Danger in Certain Irrigated Regions of the Southwestern United States. U.S. Pub. Hlth. Rpts., *44*:1300-1315.

27. Barber, M. A., and J. B. Rice. 1936. Methods of Dissecting and Making Permanent Preparations of the Salivary Glands and Stomachs of *Anopheles*. Amer. Jour. Hyg., *24*:32-40.

28. Barnes, R. C. 1945. *Anopheles walkeri* in Diurnal Shelters in Massachusetts. Jour. Econ. Ent. *38*:114.

29. Barnes, R. C. 1946. The Occurrence of *A. quadrimaculatus*, *A. occidentalis* and *A. walkeri* in Vermont, Mosquito News, *6*:32-33.

30. Barnes, R. C., and H. L. Fellton. 1944. *Anopheles quadrimaculatus* in Northeastern United States. N.J. Mosq. Exterm. Assoc. Proc., *31*:48-51.

31. Barnes, R. C., H. L. Fellton, and C. A. Wilson. 1950. An Annotated List of the Mosquitoes of New York. Mosquito News, *10*:69-84.

32. Barr, A. R., and L. Kartman. 1951. Biometrical Notes on the Hybridization of *Culex pipiens* L. and *C. quinquefasciatus* Say. Jour. Parasitol., *37*:419-420.

33. Barraud, P. J. 1934. Fauna of British India Including Ceylon and Burma. Diptera, Vol. V (Culicidae, Megarhinini and Culicini). London: Taylor and Francis, 463 pp.

34. Barret, H. P. 1919. Observations on the Life History of *Aedes bimaculatus* Coq. (Diptera, Culicidae). Ins. Ins. Mens., *7*:63-64.

35. Bartosewitz, W. V. 1945. Notes on Mosquitoes in Rhode Island. Mosquito News, *5*:19-20.

36. Basham, E. H. 1948. *Culex (Melanoconion) mulrennani*, A New Species from Florida (Diptera, Culicidae). Ann. Ent. Soc. Amer., *41*:1-7.

37. Basham, E. H., and J. S. Haeger. 1948. Records of *Anopheles quadrimaculatus* Say for the Florida Keys. Mosquito News, *8*:72.

38. Basham, E. H., J. A. Mulrennan, and A. J. Obermuller. 1947. The Biology and Distribution of *Megarhinus* Robineau-Desvoidy in Florida. Mosquito News, *7*:64-66.

39. Bates, Marston. 1940. The Nomenclature and Taxonomic Status of the Mosquitoes of the *Anopheles maculipennis* Complex. Ann. Ent. Soc. Amer., *33*:343-356.

40. Bates, Marston. 1943. Mosquitoes as Vectors of Dermatobia in Eastern Colombia. Ann. Ent. Soc. Amer., *36*:21-24.

41. Bates, Marston. 1944. Notes on the Construction and Use of Stable Traps for Mosquito Studies. Jour. Nat. Mal. Soc., *3*:135-145.

42. Bates, Marston. 1949. Ecology of Anopheline Mosquitoes. *In* Boyd, Malariology, W. B. Saunders Co., Philadelphia. 2 vols., 1,643 pp.

43. Bates, Marston. 1949. The Natural History of Mosquitoes. New York: Macmillan, 379 pp.

44. Beadle, L. D. 1952. Eastern Equine Encephalitis in the United States. Mosquito News, *12*:102-107.

45. Bean, J. L. 1946. A Preliminary List of the Mosquitoes of Maine (Culicidae, Diptera). Can. Ent., *78*:25-28.

46. Beaudette, F. R. 1946. Common Mosquito-borne Diseases of Birds. N.J. Mosq. Exterm. Assoc. Proc., *33*:31-36.

47. Belding, G. L. 1942. Textbook of Clinical Parasitology. New York: Appleton-Century Co., 888 pp.

48. Belkin, J. N. 1950. A Revised Nomenclature for the Chaetotaxy of the Mosquito Larva (Diptera: Culicidae). Amer. Mid. Nat., *44*:678-698.

49. Belkin, J. N. 1952. The Homology of the Chaetotaxy of Immature Mosquitoes and a Revised Nomenclature for the Chaetotaxy of the Pupa (Diptera, Culicidae). Proc. Ent. Soc. Wash., *54*:115-130.

50. Belkin, J. N., N. Ehmann, and G. Heid. 1951. Preliminary Field Observations on the Behavior of the Adults of *Anopheles franciscanus* McCracken in Southern California. Mosquito News, *11*:23-31.

51. Bellamy, R. E. 1939. An Anopheline From Inland Georgia Resembling the Brackish-Water Race of *Anopheles crucians*. Jour. Parasitol., *25*:186.

52. Bellamy, R. E. 1950. An Unusual Winter Population of *Anopheles quadrimaculatus* Say. Jour. Nat. Mal. Soc., *9*:80-83.

53. Bellamy, R. E., and J. Andrews. 1938. The Occurrence of *Anopheles walkeri* Theobald in Georgia. South. Med. Jour., *31*:797.

54. Bellamy, R. E., and W. C. Reeves. 1952. A Portable Mosquito Bait-Trap. Mosquito News, *12*:256-258.

55. Bellamy, R. E., and R. P. Repass. 1950. Notes on the Ova of *Anopheles georgianus* King. Jour. Nat. Mal. Soc., *9*:84-88.

56. Bennett, B. L., F. C. Baker, and A. W. Sellards. 1939. Susceptibility of the Mosquito *Aedes triseriatus* to the Virus of Yellow Fever Under Experimental Conditions. Ann. Trop. Med. and Parasitol., *33*:101.

57. Bequaert, J. 1946. *Aedes aegypti* (Linnaeus), the Yellow Fever Mosquito in Arizona (Diptera). Brooklyn Ent. Soc. Bull., *41*:157.

58. Berg, M., and S. Lang. 1948. Observations of Hibernating Mosquitoes in Massachusetts. Mosquito News, *8*:70-71.

59. Berner, L. 1947. Notes on the Breeding Habits of *Aedes (Stegomyia) aegypti* (Linnaeus). Ann. Ent. Soc. Amer., *40*:528-529.

60. Beyer, G. E. 1923. Mosquitoes of Louisiana. La. State Bd. Hlth. Quart. Bull., *14*:54-84.

61. Beyer, G. E. 1923. A New Species of *Anopheles* in Louisiana, *Anopheles atropos* Dyar and Knab. Amer. Jour. Trop. Med., *3*:351-363.

62. Bick, G. H. 1946. Collections of Mosquitoes on Parris Island during 1945. Jour. Econ. Ent., *39*:89-91.

63. Bickley, W. E. 1945. The Anal Gills of Mosquito Larvae. Mosquito News, *5*:18.

64. Bishopp, F. C., E. N. Cory, and Alan Stone. 1933. Preliminary Results of a Mosquito Survey in the Chesapeake Bay Section. Proc. Ent. Soc. Wash., *35*:1-6.

65. Blakeslee, T. E., and G. S. Payne. 1953. *Aedes (O.) sollicitans* (Walker) and *Culiseta (C.) morsitans* (Theobald) in Kentucky. Mosquito News, *13*:210.

66. Blickle, R. L. 1952. Notes on the Mosquitoes (Culicinae) of New Hampshire. N.J. Mosq. Exterm. Assoc. Proc., *39*:198-202.

67. Boddy, D. W. 1948. An Annotated List of the Culicidae of Washington. Pan-Pac. Ent., *24*:85-94.

68. Bodman, M. T., and N. Gannon. 1950. Some Habitats of Eggs of *Aedes vexans*. Jour. Econ. Ent., *43*:547-548.

69. Bohart, R. M. 1948. Differentiation of Larvae and Pupae of *Aedes dorsalis* and *Aedes squamiger* (Diptera, Culicidae). Proc. Ent. Soc. Wash., *50*:216-218.

70. Bohart, R. M. 1948. The Subgenus *Neoculex* in America North of Mexico (Diptera, Culicidae). Ann. Ent. Soc. Amer., *41*:330-345.

71. Bohart, R. M. 1950. Observations on Snow Mosquitoes in California (Diptera: Culicidae). Pan-Pac. Ent., *26*:111-118.

72. Bohart, R. M. 1950. A new Species of *Orthopodomyia* from California (Diptera, Culicidae). Ann. Ent. Soc. Amer., *43*:399-404.

73. Bonne, C., and J. Bonne-Wepster. 1925. Mosquitoes of Surinam. Amsterdam: Het Koloniaal Institut te Amsterdam, 558 pp.

74. Bordas, E., and W. G. Downs. 1951. Control of *Anopheles pseudopunctipennis* in Mexico with DDT Residual Sprays Applied in Buildings. Part IV. Activity Pattern of Adult *A. pseudopunctipennis* Theobald. Amer. Jour. Hyg., *53*:217-223.

75. Botsford, R. C. 1936. Progress of Mosquito Control in Connecticut, 1935. N.J. Mosq. Exterm. Assoc. Proc., *23*:151-153.

76. Boyd, M. F. 1927. Studies on the Bionomics of North American Anophelines. I. The Number of Annual Generations of *Anopheles quadrimaculatus*. Amer. Jour. Hyg., *7*:264-275.

77. Boyd, M. F. 1930. An Introduction to Malariology. Cambridge, Mass.: Harvard Univ. Press, 437 pp.

78. Boyd, M. F. 1949. Epidemiology: Factors Related to the Definitive Host. *In* Boyd, Malariology, *1*:608-697.

79. Boyd, M. F., and F. W. Aris. 1929. A Malaria Survey of the Island of Jamaica, B.W.I. Amer. Jour. Trop. Med., *9*:309-399.

80. Boyd, M. F., S. F. Kitchen, and J. A. Mulrennan. 1936. On the Relative Susceptibility of the Inland and Coastal Varieties of *A. crucians* Wied. to *P. falciparum* Welch. Amer. Jour. Trop. Med., *16*:159-161.

81. Bradley, G. H. 1925. The Larva of *Aedes thibaulti* Dyar and Knab (Diptera: Culicidae). Ins. Ins. Mens., *13*:89-91.

82. Bradley, G. H. 1932. Some Factors Associated with the Breeding of *Anopheles* Mosquitoes. Jour. Agr. Res., *44*:381-399.

83. Bradley, G. H. 1936. On the Identification of Mosquito Larvae of the Genus *Anopheles* Occurring in the United States (Diptera: Culicidae). South. Med. Jour., *29*:859-861.

84. Bradley, G. H., R. F. Fritz, and L. E. Perry. 1944. Additional Mosquito Records for the Southeastern States. Jour. Econ. Ent., *37*:109.

85. Bradley, G. H., M. H. Goodwin, Jr., and Alan Stone. 1949. Entomologic Technics as Applied to Anophelines. *In* Boyd, Malariology, *1*:331-378.

86. Bradley, G. H., and W. V. King. 1944. Bionomics and Ecology of Nearctic *Anopheles*. Amer. Assoc. Adv. Sci., *15*:79-87.

87. Bradley, G. H., and T. E. McNeel. 1935. Mosquito Collections in Florida with the New Jersey Light Trap. Jour. Econ. Ent., *28*:780-786.

88. Brandenburg, J. F., and R. D. Murrill. 1947. Occurrence and Distribution of Mosquitoes in Arkansas. Ark. Hlth. Bull., *4*:4-6.

89. Breeland, S. G. 1951. The Identification of the Early Larval Instars of Three Common Anophelines of Southern Georgia, U.S.A. Jour. Nat. Mal. Soc., *10*:224-232.

90. Breland, O. P. 1947. Notes on Pennsylvania Mosquitoes. Mosquito News, *7*:76-77.

91. Breland, O. P. 1947. Variations in the Larvae of the Mosquito *Orthopodomyia alba* Baker (Diptera, Culicidae). Bull. Brooklyn Ent. Soc., *42*:81-86.

92. Breland, O. P. 1947. *Orthopodomyia alba* Baker in Texas with Notes on Biology (Diptera: Culicidae). Proc. Ent. Soc. Wash., *49*:185-187.

93. Breland, O. P. 1948. Notes on the Mosquito, *Uranotaenia syntheta* Dyar and Shannon (Diptera: Culicidae). Mosquito News, *8*:108-109.

94. Breland, O. P. 1948. Some Bicolored Mosquito Larvae. Jour. Kans. Ent. Soc., *21*:120-121.

95. Breland, O. P. 1948. Notes on Some Carnivorous Mosquito Larvae. Ent. News, *59*:156-157.

96. Breland, O. P. 1949. The Biology and the Immature Stages of the Mosquito, *Megarhinus septentrionalis* Dyar and Knab. Ann. Ent. Soc. Amer., *42*:38-47.

97. Breland, O. P. 1949. Distinctive Features of the Larvae of *Aedes alleni* Turner (Diptera: Culicidae). Jour. N.Y. Ent. Soc., *57*:93-100.

98. Breland, O. P. 1950. Additional Studies on the Larvae of *Uranotaenia syntheta* Dyar and Shannon (Diptera: Culicidae). Mosquito News, *10*:25-28.

99. Breland, O. P. 1952. Keys to the Larvae of Texas Mosquitoes with Notes on Recent Synonymy. I. Key to Genera and to the Species of the Genus *Aedes*. Tex. Jour. Sci., *4*:65-72.

100. Breland, O. P. 1952. The Stirrup-Shaped Piece as an Aid in the Taxonomic Study of Mosquito Larvae. Mosquito News, *12*:253-255.

101. Breland, O. P. 1953. Keys to the Larvae of Texas Mosquitoes with Notes on Recent Synonymy. II. The Genus *Culex* Linnaeus. Tex. Jour. Sci., *5*:114-119.

102. Brennan, J. M. 1951. The Occurrence of *Anopheles crucians* in Guatemala. Amer. Jour. Trop. Med., *31*:138.

103. Brennan, J. M., and R. F. Harwood. 1953. A Preliminary Report on the Laboratory Colonization of the Mosquito, *Culex tarsalis* Coquillett. Mosquito News, *13*:153-157.

104. Brookman, B. 1950. The Ecology of *Culex* Mosquitoes. Calif. Mosquito Control Assoc. Ann. Confr. Proc. and Papers, *18*:74-76.

105. Brookman, B., and W. C. Reeves. 1950. A New Name for a California Mosquito (Diptera, Culicidae). Pan-Pac. Ent., *26*:159-160.

106. Brookman, B., and W. C. Reeves. 1953. New Records of Mosquitoes from Lower California, Mexico, with Notes and Descriptions (Diptera: Culicidae). Ann. Ent. Soc. Amer., *46*:225-236.

107. Brooks, I. C. 1947. Tree-hole Mosquitoes in Tippecanoe County, Indiana. Ind. Acad. Sci. Proc., *56*:154-156.

108. Brown, F. R., and J. W. Pearson. 1938. Some Culicidae of the Reelfoot Lake Region. Jour. Tenn. Acad. Sci., *13*:126-132.

109. Brown, W. L. 1948. Results of the Pennsylvania Mosquito Survey for 1947. Jour. N.Y. Ent. Soc., *56*:219-232.

110. Buren, W. F. 1944. An Anomalous *Anopheles quadrimaculatus* Larva from Louisiana. Jour. Econ. Ent., *37*:555.

111. Buren, W. F. 1946. Some Observations on the Larval Habitat of *Psorophora varipes* (Coq.). Mosquito News, *6*:120-121.

112. Buren, W. F. 1946. *Psorophora pygmaea* (Theob.), An Exotic Mosquito Now Established in Florida. Mosquito News, *6*:185.

113. Buren, W. F. 1947. A New *Aedes* from the Florida Keys (Diptera, Culicidae). Proc. Ent. Soc. Wash., *49*:228-229.

114. Burgess, R. W. 1948. The Experimental Hybridization of *Anopheles quadrimaculatus* Say and *Anopheles maculipennis freeborni* Aitken. Amer. Jour. Hyg., *48*:171-172.

115. Burgess, R. W. 1950. Colonization of *Anopheles albimanus* from the Florida Keys. Jour. Econ. Ent., *43*:108.

116. Burgess, R. W., and M. D. Young. 1946. Experimental Transmission of *P. falciparum* by *Anopheles freeborni*. Jour. Nat. Mal. Soc., *5*: 151-152.

117. Burgess, R. W., and M. D. Young. 1950. The Comparative Susceptibility of *Anopheles quadrimaculatus* and *Anopheles freeborni* to Infection by *Plasmodium vivax* (St. Elizabeth Strain). Jour. Nat. Mal. Soc., *9*:218-221.

118. Buxton, J. A., and O. P. Breland. 1952. Some Species of Mosquitoes Reared from Dry Materials. Mosquito News, *12*:209-214.

119. Buxton, P. A., and H. S. Leeson. 1949. Anopheline Mosquitoes: Life History. *In* Boyd, Malariology, *1*:257-283.

120. Cairns, D. W. 1936. Mosquitoes of Fort Dupont, Delaware. Military Surgeon, *78*:369-386.

121. Carpenter, S. J. 1941. The Mosquitoes of Arkansas. Ark. State Bd. of Hlth., Little Rock, 87 pp., rev. ed.

122. Carpenter, S. J. 1942. Mosquito Studies in Military Establishments in the Seventh Corps Area during 1941. Jour. Econ. Ent., *35*:558-561.

123. Carpenter, S. J. 1945. Anopheline Surveys in the Fourth Service Command. Jour. Nat. Mal. Soc., *4*:115-121.

124. Carpenter, S. J. 1945. Collection Records of *Culex tarsalis* in Army Camps in the Southeastern United States During 1942, 1943 and 1944. Jour. Econ. Ent., *38*:404-406.

125. Carpenter, S. J. 1945. Packing Mosquito Larvae for Storage or Shipment. Jour. Econ. Ent., *38*:501.

126. Carpenter, S. J. 1945. Artificial Shelters for Measuring Densities of *Anopheles quadrimaculatus*. Jour. Econ. Ent., *38*:721-722.

127. Carpenter, S. J. 1948. Gynandromorphism in *Aedes canadensis*. Jour. Econ. Ent., *41*:522-523

128. Carpenter, S. J. 1949. Notes on Mosquito Collections in Pennsylvania and Canada During 1948. Mosquito News, *9*:172-173.

129. Carpenter, S. J. 1949. Collection of a Fourth Instar Larva of *Anopheles albimanus* at Boca Raton Field in 1944. Jour. Econ. Ent., *42*: 834.

130. Carpenter, S. J. 1950. Notes on Mosquitoes in North America: I—New Distribution Records for Eastern United States During 1946 and 1947. Mosquito News, *10*:64-65.

131. Carpenter, S. J. 1952. Further Observations on Sexual Dimorphism in Mosquito Pupae (Diptera, Culicidae). Mosquito News, *12*:7-8.

132. Carpenter, S. J. 1952. *Mansonia indubitans* Dyar and Shannon in Panama. Mosquito News, *12*: 27-28.

133. Carpenter, S. J. 1952. Notes on Mosquitoes in North America: II—Collections at Military Installations in Indiana During 1944 and 1945. Mosquito News, *12*:251-252.

134. Carpenter, S. J. 1952. Notes on Mosquitoes in North America: III—Collections at Military Installations in Kentucky During 1944 and 1945. Mosquito News, *12*:252-253.

135. Carpenter, S. J., and R. W. Chamberlain. 1946. Mosquito Collections at Army Installations in the Fourth Service Command, 1943. Jour. Econ. Ent., *39*:82-88.

136. Carpenter, S. J., R. W. Chamberlain, and J. F. Wanamaker. 1945. New Distribution Records for the Mosquitoes of the Southeastern States in 1944. Jour. Econ Ent., *38*:401-402.

137. Carpenter, S. J., and D. W. Jenkins. 1945. A New State Record of *Megarhinus rutilus* Coquillett in South Carolina. Mosquito News, *5*: 88.

138. Carpenter, S. J., and W. W. Middlekauff. 1944. Inland Records of Salt Marsh Mosquitoes. Jour. Econ. Ent., *37*:108.

139. Carpenter, S. J., W. W. Middlekauff, and R. W. Chamberlain. 1946. The Mosquitoes of the Southern United States East of Oklahoma and Texas. Amer. Mid. Nat. Mon. No. 3, 292 pp.

140. Carpenter, S. J., and E. L. Peyton. 1952. Mosquito Studies in the Panama Canal Zone During 1949 and 1950 (Diptera, Culicidae). Amer. Mid. Nat., *48*:673-682.

141. Carr, H. P., and J. F. Melendez. 1942. Malaria Reconnaissance of the Province of Pinar Del Rio in Cuba. Amer. Jour. Trop. Med., *22*:51-61.

142. Carr, H. P., J. F. Melendez, and A. F. Melendez. 1941. Malaria Reconnaissance of the Province of Camaguey in Cuba. Amer. Jour. Trop. Med., *21*:739-750.

143. Carr, H. P., J. F. Melendez, and A. Ros. 1940. Malaria Reconnaissance of the Province of Oriente in Cuba. Amer. Jour. Trop Med., *20*: 81-97.

144. Causey, O. R. 1943. A Method for the Collection, Transportation and Study of Anopheline Eggs and Adults. Amer. Jour. Trop. Med., *23*: 133-137.

145. Causey, O. R., and H. W. Kumm. 1948. Dispersal of Forest Mosquitoes in Brazil. Amer. Jour. Trop. Med., *28*:469-480.

146. Chamberlain, R. W., and T. E. Duffey. 1945. Collection Records of *Mansonia titillans* (Walker) and *Mansonia indubitans* Dyar and Shannon in Florida, with Keys to the Species of *Mansonia* in the United States (Diptera, Culicidae). Mosquito News, *5*:96-97.

147. Chamberlain, R. W., H. Rubin, R. E. Kissling, and M. E. Eidson. 1951. Recovery of Virus of Eastern Equine Encephalomyelitis from a Mosquito, *Culiseta melanura* (Coquillett). Proc. Soc. Expm. Biol. and Med., *77*:396-397.

148. Chamberlain, R. W., and R. K. Sikes. 1950. A Safe Method of Handling Mosquitoes for Virus Transmission Experiments. Jour. Parasitol., *36*:499-500.

149. Chamberlin, J. C., and F. R. Lawson. 1945. A Mechanical Trap for the Sampling of Aerial Insect Populations. Mosquito News, *5*:4-7.

150. Chang, H. T. 1946. Studies of the Use of Fluorescent Dyes for Marking *Anopheles quadrimaculatus* Say. Mosquito News, *6*:122-125.

151. Chang, S. L., and F. E. Richart. 1951. Studies on Anopheline Larvae. I. The Anatomy and Function of the so-called "Notched Organs" of Nuttall and Shipley on the Thorax of Larvae of *Anopheles quadrimaculatus*. Jour. Nat. Mal. Soc., *10*:287-292.

152. Christensen, G. R., and F. C. Harmston. 1944. A Preliminary List of the Mosquitoes of Indiana. Jour. Econ. Ent., *37*:110-111.

153. Christophers, S. R. 1933. The Fauna of British India. Diptera, Vol. IV (Culicidae, Anophelini). London: Taylor and Francis, 371 pp.

154. Christophers, S. R. 1952. The recorded Parasites of Mosquitoes. Riv. di Parasitol., *13*: 21-28.

155. Christophers, S. R., J. A. Sinton, and G. Covell. 1939. How to do a Malaria Survey. Calcutta Hlth. Bull No. 14, 208 pp.

156. Clark, H. C. 1926. *Anopheles* Mosquitoes in Our Tropical Divisions. United Fruit Company, Med. Dept., 15th Ann. Rpt., 45-47.

157. Cockerell, T. D. A. 1918. The Mosquitoes of Colorado. Jour. Econ. Ent., *11*:195-200.

158. Cockerell, T. D. A., and J. T. Scott. 1918. Culicidae of Colorado. Jour. Econ. Ent., *11*:387-388.

159. Coher, E. I. 1948. A Study of the Female Genitalia of Culicidae: With Particular Reference to Characters of Generic Value. Ent. Amer., *28*:75-112.

160. Cole, F. R., and A. L. Lovett. 1921. An Annotated List of the Diptera (Flies) of Oregon. Proc. Calif. Acad. Sci., *11*:197-344.

161. Cory, E. N., G. S. Langford, S. L. Crosthwait, and C. Graham. 1934. Anti-Mosquito Work in Maryland. Univ. Md. Bull., *73*:3-31.

162. Crampton, G. C. 1942. External Morphology of the Diptera, Guide to the insects of Connecticut, Part VI, Fasc. I. Conn. State Geo. and Nat. Hist. Surv. Bull., *64*:10-165.

162a. Curtis, L. C. 1953. Observations on Mosquitoes at Whitehorse, Yukon Territory (Culicidae: Diptera). Can. Ent., *85*:353-370.

163. Daggy, R. H., O. J. Muegge, and W. A. Riley. 1940. A Preliminary Survey of the Anopheline Mosquito Fauna of Southeastern Minnesota and Adjacent Wisconsin Areas. U.S. Pub. Hlth. Rpts., *56*:883-895.

164. Dampf, A. 1943. Distribucion y Ciclo Anual de *Uranotaenia syntheta* Dyar and Shannon en Mexico y Descripcion del Hipopigio Masculino (Insecta, Diptera). Rev. Soc. Mexico Hist. Nat., *4*:147-169.

165. Darsie, R. F. 1949. Pupae of the Anopheline Mosquitoes of the Northeastern United States (Diptera, Culicidae). Rev. de Entomologia, *20*:509-530.

166. Darsie, R. F. 1951. Pupae of the Culicine Mosquitoes of the Northeastern United States (Diptera, Culicidae, Culicini). Cornell Agr. Exp. Sta. Mem., *304*:3-67.

167. Darsie, R. F., Donald MacCreary, and L. A. Stearns. 1951. An Annotated List of the Mosquitoes of Delaware. N.J. Mosq. Exterm. Assoc. Proc., *38*:137-146.

168. Davis, W. A. 1940. A Study of Birds and Mosquitoes as Hosts for the Virus of Eastern Equine Encephalomyelitis. Amer. Jour. Hyg., *32*:45-59.

169. Day, M. F. 1943. Report on Mosquitoes Collected in St. Louis County During 1942. Trans. St. Louis Acad. Sci., *31*:29-45.

170. Dethier, V. G., and F. H. Whitley. 1944. Population Studies of Florida Mosquitoes. Jour. Econ. Ent., *37*:480-484.

171. Dickinson, W. E. 1944. The Mosquitoes of Wisconsin. Milwaukee Pub. Mus. Bull., *8*:269-365.

172. Dodge, H. R. 1945. Notes on the Morphology of Mosquito Larvae. Ann. Ent. Soc. Amer., *38*:163-167.

173. Dodge, H. R. 1946. Studies Upon Mosquito Larvae. Abs. Doctoral Diss., Ohio State Univ., *50*:17-22.

174. Dodge, H. R. 1946. The Identification of *Anopheles bradleyi* Larvae. Mosquito News, *6*:125-126.

175. Dodge, H. R. 1947. A new Species of *Wyeomyia* from the Pitcher Plant. Proc. Ent. Soc. Wash., *49*:117-122.

176. Dorer, R. E., W. E. Bickley, and H. P. Nicholson. 1944. An Annotated List of the Mosquitoes of Virginia. Mosquito News, *4*:48-50.

177. Dorer, R. E., R. G. Carter, and W. E. Bickley. 1950. Observations on the Pupae of *Mansonia perturbans* (Walk.) in Virginia. N.J. Mosq. Exterm. Assoc. Proc., *37*:110-112.

178. Dorsey, C. K. 1944. Mosquito Survey Activities at Camp Peary, Va. Ann. Ent. Soc. Amer., *37*:376-387.

179. Downs, W. G. 1943. Polyvinol Alcohol: A Medium for Mounting and Clearing Biological Specimens. Science, *97*:539-540.

180. Downs, W. G., and E. Bordas. 1949. A Malaria Survey of the Southern Territory of Lower California. Amer. Jour. Trop. Med., *29*:695-699.

181. Dutilly, A. 1946. A List of Insects of the Mackenzie River Basin. Can. Field Nat., *60*:35-44.

182. Dyar, H. G. 1906. The Larvae of Culicidae Classified as Independent Organisms. Jour. N.Y. Ent. Soc., *14*:169-230.

183. Dyar, H. G. 1907. Report on the Mosquitoes of the Coast Region of California, with Descriptions of New Species. Proc. U.S. Nat. Mus., *32*:121-129.

184. Dyar, H. G. 1916. Mosquitoes at San Diego, California. Ins. Ins. Mens., *4*:46-51.

185. Dyar, H. G. 1916. The Eggs and Oviposition in Certain Species of *Mansonia* (Diptera, Culicidae). Ins. Ins. Mens., *4*:61-68.

186. Dyar, H. G. 1916. New *Aedes* from the Mountains of California (Diptera, Culicidae). Ins. Ins. Mens., *4*:80-90.

187. Dyar, H. G. 1917. The Mosquitoes of the Mountains of California (Diptera, Culicidae). Ins. Ins. Mens., *5*:11-21.

188. Dyar, H. G. 1917. The Mosquitoes of the Pacific Northwest (Diptera, Culicidae). Ins. Ins. Mens., *5*:97-102.

189. Dyar, H. G. 1917. Notes on *Aedes* at Lake Pend D'Oreille, Idaho (Diptera, Culicidae). Ins. Ins. Mens., *5*:102-104.

190. Dyar, H. G. 1917. Notes on the *Aedes* of Montana (Diptera, Culicidae). Ins. Ins. Mens., *5*:104-121.

191. Dyar, H. G. 1917. New American Mosquitoes (Diptera, Culicidae). Ins. Ins. Mens., *5*:165-169.

192. Dyar, H. G. 1917. The Genus *Culex* in the United States (Diptera, Culicidae). Ins. Ins. Mens., *5*:170-183.

193. Dyar, H. G. 1917. The Larva of *Aedes idahoensis* (Diptera, Culicidae). Ins. Ins. Mens., *5*:187-188.

194. Dyar, H. G. 1918. The Male Genitalia of *Aedes* as Indicative of Natural Affinities (Diptera, Culicidae). Ins. Ins. Mens., *6*:71-86.

195. Dyar, H. G. 1918. A Revision of the American Species of *Culex* on the Male Terminalia (Diptera, Culicidae). Ins. Ins. Mens., *6*:86-111.

196. Dyar, H. G. 1918. New American Mosquitoes (Diptera, Culicidae). Ins. Ins. Mens., *6*:120-129.

197. Dyar, H. G. 1919. Westward Extension of the Canadian Mosquito Fauna (Diptera, Culicidae). Ins. Ins. Mens., *7*:11-39.

198. Dyar, H. G. 1919. The Mosquitoes Collected by the Canadian Arctic Expedition 1913–1918. Rpt. Can. Arct. Exp., *3*(C):32.

199. Dyar, H. G. 1920. The Mosquitoes of British Columbia and Yukon Territory, Canada (Diptera, Culicidae). Ins. Ins. Mens., *8*:1-27.

200. Dyar, H. G. 1920. Notes on European Mosquitoes (Diptera, Culicidae). Ins. Ins. Mens., *8*:51-54.

201. Dyar, H. G. 1920. A New Mosquito from Mexico (Diptera, Culicidae). Ins. Ins. Mens., *8*:81-82.

202. Dyar, H. G. 1920. The Classification of American *Aedes* (Diptera, Culicidae). Ins. Ins. Mens., *8*:103-106.

203. Dyar, H. G. 1920. The American *Aedes* of the *stimulans* Group (Diptera, Culicidae). Ins. Ins. Mens., *8*:106-120.

204. Dyar, H. G. 1920. The Larva of *Aedes campestris* Dyar and Knab (Diptera, Culicidae). Ins. Ins. Mens., *8*:120.

205. Dyar, H. G. 1920. A Note on *Aedes niphadopsis* Dyar and Knab (Diptera, Culicidae). Ins. Ins. Mens., *8*:138-139.

206. Dyar, H. G. 1920. The *Aedes* of the Mountains of California and Oregon (Diptera, Culicidae). Ins. Ins. Mens., *8*:165-173.

207. Dyar, H. G. 1921. The Swarming of *Culex quinquefasciatus* Say (Diptera, Culicidae). Ins. Ins. Mens., *9*:32.

208. Dyar, H. G. 1921. Ring-Legged *Culex* in Texas (Diptera, Culicidae). Ins. Ins. Mens., *9*:32-34.

209. Dyar, H. G. 1921. Three New Mosquitoes from Costa Rica (Diptera, Culicidae). Ins. Ins. Mens., *9*:34-36.

210. Dyar, H. G. 1921. The American *Aedes* of the *punctor* Group (Diptera, Culicidae). Ins. Ins. Mens., *9*:69-80.

211. Dyar, H. G. 1921. The Mosquitoes of Argentina (Diptera, Culicidae). Ins. Ins. Mens., *9*:148-150.

212. Dyar, H. G. 1921. The Mosquitoes of Canada (Diptera, Culicidae). Royal Can. Inst., Trans. *13*:71-120.

213. Dyar, H. G. 1922. The Mosquitoes of the United States. Proc. U.S. Nat. Mus., *62*:1-119.

214. Dyar, H. G. 1922. New Mosquitoes from Alaska (Diptera, Culicidae). Ins. Ins. Mens., *10*:1-3.

215. Dyar, H. G. 1922. The American *Aedes* of the *impiger (Decticus)* Group (Diptera, Culicidae). Ins. Ins. Mens., *10*:3-8.

216. Dyar, H. G. 1922. The American *Aedes* of the *scapularis* Group (Diptera, Culicidae). Ins. Ins. Mens., *10*:51-60.

217. Dyar, H. G. 1922. Two Mosquitoes New to the Mountains of California (Diptera, Culicidae). Ins. Ins. Mens., *10*:60-61.

218. Dyar, H. G. 1922. The Mosquitoes of the Palaearctic and Nearctic Regions (Diptera, Culicidae). Ins. Ins. Mens., *10*:65-75.

219. Dyar, H. G. 1922. The Mosquitoes of the Glacier National Park, Montana (Diptera, Culicidae). Ins. Ins. Mens., *10*:80-88.

220. Dyar, H. G. 1922. The American *Aedes* of the *serratus* Group (Diptera, Culicidae). Ins. Ins. Mens., *10*:157-166.

221. Dyar, H. G. 1923. The Mosquitoes of the Yellowstone National Park (Diptera, Culicidae). Ins. Ins. Mens., *11*:36-46.

222. Dyar, H. G. 1923. Note on the Swarming of *Aedes cinereoborealis* Felt and Young (Diptera, Culicidae). Ins. Ins. Mens., *11*:56-57.

223. Dyar, H. G. 1923. On *Aedes riparius* Dyar and Knab (Diptera, Culicidae). Ins. Ins. Mens., *11*:88-92.

224. Dyar, H. G. 1923. Note on the Habits and Distribution of *Aedes flavescens* (Müller) in America (Diptera, Culicidae). Ins. Ins. Mens., *11*:92-94.

225. Dyar, H. G. 1924. Notes on *Aedes punctor* Kirby (Diptera, Culicidae). Ins. Ins. Mens., *12*:24-26.

226. Dyar, H. G. 1924. The Mosquitoes of Colorado (Diptera, Culicidae). Ins. Ins. Mens., *12*:39-46.

227. Dyar, H. G. 1924. Note on *Culex tarsalis* Coq. (Diptera, Culicidae). Ins. Ins. Mens., *12*:95-96.

228. Dyar, H. G. 1924. Two New Mosquitoes from California (Diptera, Culicidae). Ins. Ins. Mens., *12*:125-127.

229. Dyar, H. G. 1924. The Larva of *Aedes alleni* Turner (Diptera, Culicidae). Ins. Ins. Mens., *12*:131-132.

230. Dyar, H. G. 1924. The Larva of *Aedes thelcter* Dyar (Diptera, Culicidae). Ins. Ins. Mens., *12*:132.

231. Dyar, H. G. 1924. The Larva of *Culiseta maccrackenae* Dyar and Knab (Diptera, Culicidae). Ins. Ins. Mens., *12*:144.

232. Dyar, H. G. 1924. Note on *Aedes aloponotum* and other Species of its Region (Diptera, Culicidae). Ins. Ins. Mens., *12*:176-179.

233. Dyar, H. G. 1924. The American Forms of *Aedes cinereus* Meigen (Diptera, Culicidae). Ins. Ins. Mens., *12*:179-180.

234. Dyar, H. G. 1924. Notes on *Aedes ventrovittis* Dyar (Diptera, Culicidae). Ins. Ins. Mens., *12*:181-182.

235. Dyar, H. G. 1925. The Mosquitoes of Panama (Diptera, Culicidae). Ins. Ins. Mens., *13*:101-195.

236. Dyar, H. G. 1926. Titles of Papers on Mosquitoes Published in Insecutor Insecitiae Menstruus, 1913-1927. Ins. Ins. Mens., *14*:191-197.

237. Dyar, H. G. 1928. The Mosquitoes of the Americas. Carnegie Inst. Wash. Pub., *387*:616 pp.

238. Dyar, H. G. 1929. A New Species of Mosquito from Montana with Annotated List of the Species Known from the State. Proc. U.S. Nat. Mus., 75, Art., 23:1-8.

239. Dyar, H. G., and F. Knab. 1907. On the Classification of Mosquitoes. Can. Ent., *39*:47-50.

240. Dyar, H. G., and F. Knab. 1909. On the Identity of *Culex pipiens* Linnaeus. Proc. Ent. Soc. Wash., *11*:30-34.

241. Dyar, H. G., and F. Knab. 1916. Eggs and Oviposition in Certain Species of *Mansonia* (Diptera, Culicidae). Ins. Ins. Mens., *4*:61-68.

242. Dyar, H. G., and F. Knab. 1917. Notes on *Aedes curriei* (Coq.). Ins. Ins. Mens., *5*:122-125.

243. Dyar, H. G., and F. Knab. 1917. New American

Mosquitoes (Diptera, Culicidae). Ins. Ins. Mens., 5:165-169.

244. Dyar, H. G., and R. C. Shannon. 1924. The American Species of *Uranotaenia* (Diptera, Culicidae). Ins. Ins. Mens., *12*:187-192.

245. Dyar, H. G., and R. C. Shannon. 1924. The American Chaoborinae (Diptera, Culicidae). Ins. Ins. Mens., *12*:201-216.

246. Eads, R. B. 1943. The Larva of *Culex abominator* Dyar and Knab. Jour. Econ. Ent., *36*:336-338.

247. Eads, R. B. 1946. A New Record of *Anopheles albimanus* in Texas. Jour. Econ. Ent., *39*:420.

248. Eads, R. B., and G. C. Menzies. 1950. Distribution Records of *Mansonia* Blanchard (Diptera, Culicidae) in Texas. Mosquito News, *10*:3-5.

249. Eads, R. B., G. C. Menzies, and L. J. Ogden. 1951. Distribution Records of West Texas Mosquitoes. Mosquito News, *11*:41-47.

250. Earle, W. C. 1949. Trapping and Deflection of Anopheline Mosquitoes. *In* Boyd, Malariology, *2*:1221-1231.

251. Edwards, F. W. 1932. Diptera, Fam. Culicidae. *In* P. Wytsman, *Genera insectorum*, Fas. 194. Brussels: V. Verteneuil and L. Desmet, 258 pp.

252. Edwards, F. W. 1941. Mosquitoes of the Ethiopian Region (Vol. III, Culicine, Adults and Pupae). London: British Museum of Natural History, 499 pp.

253. Edwards, G. A., and E. V. Porter. 1945. Mosquitoes of Northwest Florida. Jour. Econ. Ent., *38*:613-615.

254. Ellis, J. M. 1944. Notes on the Collection and Oviposition of *Anopheles walkeri*. Jour. Tenn. Acad. Sci., *19*:29-30.

255. Evans, A. M. 1938. Mosquitoes of the Ethiopian Region (Vol. II, Anophelini, Adults and Early Stages). London: British Museum of Natural History, 417 pp.

256. Eyles, D. E. 1943. A Method for Catching, Marking and Re-examining Large Numbers of *Anopheles quadrimaculatus* Say. Jour. Nat. Mal. Soc., *2*:85-91.

257. Eyles, D. E. 1944. A Critical Review of the Literature Relating to the Flight and Dispersion Habits of Anopheline Mosquitoes. U.S. Pub. Hlth. Bull., *287*:1-39.

258. Eyles, D. E., and L. K. Bishop. 1942. An Experiment on the Range of Dispersion of *Anopheles quadrimaculatus*. Amer. Jour. Hyg., *37*:239-245.

259. Eyles, D. E., and L. K. Bishop. 1943. The Microclimate of Diurnal Resting Places of *Anopheles quadrimaculatus* Say in the Vicinity of Reelfoot Lake. U.S. Pub Hlth. Rpts., *58*:217-230.

260. Eyles, D. E., and R. W. Burgess. 1945. *Anopheles walkeri* in South Carolina. Jour. Econ. Ent., *38*:115.

261. Eyles, D. E., and W. W. Cox. 1943. The Measurement of a Population of *Anopheles quadrimaculatus* Say. Jour. Nat. Mal. Soc., *2*:71-83.

262. Eyles, D. E., C. W. Sabrosky, and J. C. Russell. 1945. Long-range Dispersal of *Anopheles quadrimaculatus*. U.S. Pub. Hlth. Rpts., *60*:1265-1273.

263. Fair, G. M., S. L. Chang, and F. E. Richart. 1951. Studies on Anopheline Larvae. II. The Mechanism Involved in the Flotation of Larvae of *A. quadrimaculatus* on a Water Surface. Jour. Nat. Mal. Soc., *10*:293-305.

264. Farid, M. A. 1949. Relationships between Certain Populations of *Culex pipiens* Linnaeus and *Culex quinquefasciatus* Say in the United States. Amer. Jour. Hyg., *49*:83-100.

265. Farnsworth, M. W. 1947. The Morphology and Musculature of the Larval Head of *Anopheles quadrimaculatus* Say. Ann. Ent. Soc. Amer., *40*:137-151.

266. Farrar, W. W. 1945. Occurrence of Pacific Coast *Anopheles* in Brackish Water. Jour. Econ. Ent., *38*:722.

267. Feemster, R. F., and V. A. Getting. 1941. Special Report of the Department of Public Health Relative to Varieties and Prevalence of Mosquitoes in the Commonwealth. Massachusetts, House Bill 2260, 70 pp.

268. Fellton, H. L. 1944. The Breeding of the Salt-marsh Mosquito in Midwestern States. Jour. Econ. Ent., *37*:245-247.

269. Fellton, H. L., R. C. Barnes, and C. A. Wilson. 1945. Malaria Control in a Nonendemic Area. Jour. Nat. Mal. Soc., *4*:201-208.

270. Fellton, H. L., R. C. Barnes, and C. A. Wilson. 1950. New Distribution Records for the Mosquitoes of New England. Mosquito News, *10*:84-91.

271. Fellton, H. L., R. C. Barnes, C. A. Wilson, and M. Boyers. 1945. Malaria Control in the Northeastern States. N.J. Mosq. Exterm. Assoc. Proc., *32*:223-230.

272. Felt, E. P. 1904. Mosquitoes or Culicidae of New York State. N.Y. State Mus. Bull., *79*:241-400.

273. Ferguson, F. F., and T. E. McNeel. 1954. The Mosquitoes of New Mexico. Mosquito News, *14*:30-31.

274. Fisk, F. W. 1939. New Mosquito Records from Key West, Florida. Jour. Econ. Ent., *32*:469.

275. Fisk, F. W. 1941. *Deinocerites spanius* at Brownsville, Texas, with Notes on its Biology and a Description of the Larva. Ann. Ent. Soc. Amer., *34*:543-550.

276. Fisk, F. W., and J. H. LeVan. 1940. Mosquito Collections at Charleston, South Carolina, Using the New Jersey Light Trap. Jour. Econ. Ent., *33*:578-579.

277. Fisk, F. W., and J. H. LeVan. 1941. Mosquito Collections at Brownsville, Texas. Jour. Econ. Ent., *33*:944-945.

278. Fletcher, O. K. 1946. Apparent Anomalies in the Behavior of Anopheline Mosquitoes in Southwestern Georgia. Jour. Nat. Mal. Soc., 5:142-150.

279. Foote, R. H. 1952. The Larval Morphology and Chaetotaxy of the *Culex* Subgenus *Melanoconion* (Diptera, Culicidae). Ann. Ent. Soc. Amer., 45:445-472.

280. Foote, R. H. 1952. A Method of Making Whole Mounts of Mosquito Larvae for Special Study. Jour. Parasitol., 38:494-495.

281. Foote, R. H. 1953. The Pupal Morphology and Chaetotaxy of the *Culex* Subgenera *Melanoconion* and *Mochlostyrax* (Diptera, Culicidae). Proc. Ent. Soc. Wash., 55:89-100.

282. Fox, Irving and J. M. Capriles. 1953. Light Trap Studies on Mosquitoes and *Culicoides* in Western Puerto Rico. Mosquito News, 13:165-166.

283. Freeborn, S. B. 1924. The Terminal Abdominal Structures of Male Mosquitoes. Amer. Jour. Hyg., 4:188-212.

284. Freeborn, S. B. 1926. The Mosquitoes of California. Univ. Calif. Publ. Entom., 3:333-460.

285. Freeborn, S. B. 1932. The Seasonal Life History of *Anopheles maculipennis* with Reference to Humidity Requirements and "Hibernation." Amer. Jour. Hyg., 16:215-253.

286. Freeborn, S. B. 1949. Anophelines of the Nearctic Region. *In* Boyd, Malariology, 1:379-398.

287. Freeborn, S. B., and R. M. Bohart. 1951. The Mosquitoes of California. Bull. Calif. Insect Survey, 1:25-78.

288. Freeborn, S. B., and B. Brookman. 1943. Identification Guide to the Mosquitoes of the Pacific Coast States. U.S. Pub. Hlth. Serv., MCWA, Atlanta, Georgia, 23 pp.

289. Freeman, T. N. 1952. Interim Report of the Distribution of the Mosquitoes Obtained in the Northern Insect Survey. Defense Research Board of Ottawa, Environmental Protection, Tech. Rpt. 1, 2 pp., plus 43 maps.

290. Frohne, W. C. 1951. Seasonal Incidence of Mosquitoes in the Upper Cook Inlet, Alaska. Mosquito News, 11:213-216.

291. Frohne, W. C. 1952. Mosquito News from Alaska. *Culiseta incidens* (Thomson) Present in Southeastern Alaska. Mosquito News, 12:263.

292. Frohne, W. C. 1952. Mosquito News from Alaska. Alaskan Mosquitoes Disturb Snowshoers. Mosquito News, 12:263.

293. Frohne, W. C. 1953. Mosquito Breeding in Alaskan Salt Marshes, with Especial Reference to *Aedes punctodes* Dyar. Mosquito News, 13:96-103.

294. Frohne, W. C., and R. G. Frohne. 1952. Mating Swarms of Males of the Mosquito, *Aedes punctor* (Kirby), in Alaska. Mosquito News, 12:248-251.

295. Frohne, W. C., and J. W. Hart. 1949. Over-wintering of *Anopheles crucians* Wied. in South Carolina. Jour. Nat. Mal. Soc., 8:171-173.

296. Frohne, W. C., and D. A. Sleeper. 1951. Reconnaissance of Mosquitoes, Punkies and Blackflies in Southeast Alaska. Mosquito News, 11:209-213.

297. Frost, S. W. 1950. Mosquitoes of 26 Kinds Pester Pennsylvanians. Pa. Agr. Expm. Sta. Bull., 515(Sup. 1):1.

298. Frost, S. W. 1950. Results of the Pennsylvania Mosquito Survey for 1948. Mosquito News, 10:65-68.

299. Gabaldon, A. 1939. A Method for Mounting Anopheline Eggs. Jour. Parasitol., 25:281.

300. Galindo, Pedro. 1943. Contribution to our Knowledge of the Genus *Culex* in California. Unpublished M.S. thesis, University of California, Berkeley.

301. Galindo, Pedro, and T. F. Kelley. 1943. *Culex thriambus* Dyar, A New Mosquito Record for California (Diptera, Culicidae). Pan-Pac. Ent., 19:87-90.

302. Galindo, Pedro, S. J. Carpenter, and Harold Trapido. 1951. Ecological Observations on Forest Mosquitoes of an Endemic Yellow Fever Area in Panama. Amer. Jour. Trop. Med., 31:98-137.

303. Gerhardt, R. W. 1951. The Mosquitoes of Wyoming. Univ. Wyoming Pubs., 16:126-127.

304. Gibson, Arthur. 1934. Mosquito Suppression Work in Canada in 1933. N.J. Mosq. Exterm. Assoc. Proc., 21:102-112.

305. Gibson, Arthur. 1935. Mosquito Suppression Work in Canada in 1934. N.J. Mosq. Exterm. Assoc. Proc., 22:77-92.

306. Gibson, Arthur. 1936. Mosquito Suppression Work in Canada in 1935. N.J. Mosq. Exterm. Assoc. Proc., 23:88-98.

307. Gibson, Arthur. 1937. Mosquito Suppression Work in Canada in 1936. N.J. Mosq. Exterm. Assoc. Proc., 24:96-108.

308. Gibson, Arthur. 1938. Mosquito Suppression Work in Canada in 1937. N.J. Mosq. Exterm. Assoc. Proc., 25:176-185.

309. Gibson, Arthur. 1939. Mosquito Suppression Work in Canada in 1938. N.J. Mosq. Exterm. Assoc. Proc., 26:109-116.

310. Gibson, Arthur. 1940. Mosquito Suppression Work in Canada in 1939. N.J. Mosq. Exterm. Assoc. Proc., 27:44-53.

311. Gibson, Arthur. 1941. Mosquito Suppression Work in Canada in 1940. N.J. Mosq. Exterm. Assoc. Proc., 28:167-176.

312. Gilyard, R. T. 1944. Mosquito Transmission of Venezuelan Virus Equine Encephalomyelitis in Trinidad. Bull. U.S. Army Med. Dept., 75:96-107.

313. Gjullin, C. M. 1937. The Female Genitalia of *Aedes* Mosquitoes of the Pacific Coast States. Proc. Ent. Soc. Wash., 39:252-266.

314. Gjullin, C. M. 1946. A Key to the *Aedes* Females of America, North of Mexico (Diptera: Culicidae). Proc. Ent. Soc. Wash., *48*:215-236.

315. Gjullin, C. M., and W. W. Yates. 1945. *Anopheles* and Malaria in the Northwestern States. Mosquito News, *5*:121-127.

316. Gjullin, C. M., W. W. Yates, and H. H. Stage. 1939. The Effect of Certain Chemicals on the Hatching of Mosquito Eggs. Science, *89*:539-540.

317. Gjullin, C. M., W. W. Yates, and H. H. Stage. 1950. Studies on *Aedes vexans* (Meig.) and *Aedes sticticus* (Meig.), Flood-water Mosquitoes in the Lower Columbia River Valley. Ann. Ent. Soc. Amer., *43*:262-275.

318. Good, N. E. 1945. A List of the Mosquitoes of the District of Columbia. Proc. Ent. Soc. Wash., *47*:168-179.

319. Goodwin, M. H. 1942. Studies on Artificial Resting Places of *Anopheles quadrimaculatus* Say. Jour. Nat. Mal. Soc., *1*:93-99.

320. Goodwin, M. H. 1949. Observations on Dispersal of *Anopheles quadrimaculatus* Say from a Breeding Area. Jour. Nat. Mal. Soc., *8*:192-197.

321. Griffith, M. E. 1952. Additional Species of Mosquitoes in Oklahoma. Mosquito News, *12*:10-14.

322. Griffitts, T. H. D. 1927. *Anopheles atropos* Dyar and Knab, A Note on its Breeding and Other Habits. U.S. Pub. Hlth. Rpts., *42*:1903-1905.

323. Griffitts, T. H. D., and J. J. Griffitts. 1931. Mosquitoes Transported by Airplanes: Staining Methods Used in Determining Their Importation. U.S. Pub. Hlth. Rpts., *46*:2775-2782.

324. Gurney, A. B. 1943. A Mosquito Survey of Camp Crowder, Missouri, During 1942. Jour. Econ. Ent., *36*:927-935.

325. Guyton, F. E. 1935. Pest Mosquito Control in Alabama under C.W.A. Jour. Econ. Ent., *28*:786-790.

326. Hackett, L. W., and A. Missiroli. 1935. The Varieties of *Anopheles maculipennis* and Their Relation to the Distribution of Malaria in Europe. Riv. di Malariol., *14*:45-109.

327. Haeger, J. S. 1949. A Review of the *Anopheles albimanus* Breeding on the Florida Keys. Proc. Fla. Anti-Mosquito Assoc. 20th Ann. Meeting, 118-121.

328. Hammon, W. McD. 1948. The Arthropod-borne Virus Encephalitides. Amer. Jour. Trop. Med., *28*:515-525.

329. Hammon, W. McD., and W. C. Reeves. 1942. *Culex tarsalis* Coq. as a Proven Vector of St. Louis Encephalitis. Proc. Soc. Expm. Biol. and Med., *51*:142-143.

330. Hammon, W. McD., and W. C. Reeves. 1943. Laboratory Transmission of St. Louis Encephalitis Virus by Three Genera of Mosquitoes. Jour. Expm. Med., *78*:241-253.

331. Hammon, W. McD., and W. C. Reeves. 1943. Laboratory Transmission of Western Equine Encephalomyelitis Virus by Mosquitoes of the Genera *Culex* and *Culiseta*. Jour. Expm. Med., *78*:425-434.

332. Hammon, W. McD., W. C. Reeves, S. R. Benner, and B. Brookman. 1945. Human Encephalitis in the Yakima Valley, Washington, 1942, with Forty-nine Virus Isolations (Western Equine and St. Louis Types) from Mosquitoes. Jour. Amer. Med. Assoc., *128*:1133-1139.

333. Hammon, W. McD., W. C. Reeves, B. Brookman, and C. M. Gjullin. 1942. Mosquitoes and Encephalitis in the Yakima Valley, Washington. V. Summary of Case Against *Culex tarsalis* Coquillett as a Vector of the St. Louis and Western Equine Viruses. Jour. Inf. Dis., *70*:278-283.

334. Hammon, W. McD., W. C. Reeves, B. Brookman, and E. M. Izumi. 1942. Mosquitoes and Encephalitis in the Yakima Valley, Washington. I. Arthropods Tested and Recovery of Western Equine and St. Louis Viruses from *Culex tarsalis* Coq. Jour. Inf. Dis., *70*:263-266.

335. Hammon, W. McD., W. C. Reeves, B. Brookman, E. M. Izumi, and C. M. Gjullin. 1941. Isolation of the Viruses of Western Equine and St. Louis Encephalitis from *Culex tarsalis* Mosquitoes. Science, *94*:328-330.

336. Hammon, W. McD., W. C. Reeves, and P. Galindo. 1945. Epizoology of Western Equine Encephalomyelitis. Eastern Nebraska Field Survey of 1943 with Isolations of the Virus from Mosquitoes. Amer. Jour. Vet. Res., *6*:145-148.

337. Hammon, W. McD., W. C. Reeves, and P. Galindo. 1945. Epidemiologic Studies of Encephalitis in the San Joaquin Valley of California, 1943, with the Isolation of Viruses from Mosquitoes. Amer. Jour. Hyg., *42*:299-306.

338. Hammon, W. McD., W. C. Reeves, and M. Gray. 1943. Mosquito Vectors and Inapparent Animal Reservoirs of St. Louis and Western Equine Encephalitis Viruses. Amer. Jour. Pub. Hlth., *33*:201-207.

339. Hammon, W. McD., W. C. Reeves, and J. V. Irons. 1944. Survey of the Arthropod-borne Virus Encephalitides in Texas with Particular Reference to the Lower Rio Grande Valley in 1942. Texas Rpts. on Biol. and Med., *2*:366-375.

340. Hammon, W. McD., W. C. Reeves, and G. Sather. 1952. California Encephalitis Virus, A Newly Described Agent. II. Isolations and Attempts to Identify and Characterize the Agent. Jour. Immunology, *69*:493-510.

341. Harden, P. H. 1945. The Occurrence of *Ortho-*

podomyia alba Baker at New Orleans, Louisiana. Mosquito News, *5*:131.

342. Hardman, N. F. 1947. Studies on Imported Malarias: 3. Laboratory Rearing of Western Anophelines. Jour. Nat. Mal. Soc., *6*:165-172.

343. Harmston, F. C. 1949. An Annotated List of the Mosquito Records from Colorado. Great Basin Nat., *9*:65-75.

344. Harmston, F. C., and D. M. Rees. 1946. Mosquito Records from Idaho. Pan-Pac. Ent., *22*:148-156.

345. Hart, J. W. 1944. A Preliminary List of the Mosquitoes of Indiana. Amer. Mid. Nat., *31*:414-416.

346. Hassett, C. C., and D. W. Jenkins. 1949. Production of Radioactive Mosquitoes. Science, *110*:109-110.

347. Hatch, M. H. 1938. A Bibliographical Catalogue of the Injurious Arachnids and Insects of Washington. Univ. Wash. Pub. Biol., *1*:163-224.

348. Haufe, W. O. 1952. Observations on the Biology of Mosquitoes (Diptera, Culicidae) at Goose Bay, Labrador. Can. Ent., *84*:254-263.

349. Headlee, T. J. 1921. The Mosquitoes of New Jersey and Their Control. N.J. Agr. Expm. Sta. Bull., *348*:1-229.

350. Headlee, T. J. 1945. The Mosquitoes of New Jersey and Their Control. New Brunswick, N.J.: Rutgers Univ. Press, 326 pp.

351. Hearle, E. 1926. The Mosquitoes of the Lower Fraser Valley, British Columbia and Their Control. Nat. Res. Council Rpt., *17*:1-94.

352. Hearle, E. 1927. List of Mosquitoes of British Columbia. Proc. Ent. Soc. British Columbia, *24*:11-19.

353. Hearle, E. 1927. Notes on the Occurrence of *Aedes (Ochlerotatus) nearcticus* Dyar in the Rocky Mountain Park, Alberta (Culicidae, Diptera). Can. Ent., *59*:61-63.

354. Hearle, E. 1929. The Life History of *Aedes flavescens* Müller. Trans. Royal Soc. Can., Third Series, *23*:85-101.

355. Hearle, E. 1932. Notes on the More Important Mosquitoes of Western Canada. N.J. Mosq. Exterm. Assoc. Proc., *19*:7-15.

356. Hedeen, R. A. 1953. The Biology of the Mosquito *Aedes atropalpus* (Coquillett). Jour. Kans. Ent. Soc., *26*:1-10.

357. Hedeen, R. A., and H. L. Keegan. 1952. The Use of Lindane and Dieldrin as Mosquito Adulticides in Alaska. Mosquito News, *12*:242-243.

358. Herms, W. B., and S. B. Freeborn. 1920. The Egg-laying Habits of California Anophelines. Jour. Parasitol., 7:69-79.

359. Hess, A. D., and R. L. Crowell. 1949. Seasonal History of *Anopheles quadrimaculatus* in the Tennessee Valley. Jour. Nat. Mal. Soc., *8*:159-170.

360. Hill, N. DeM. 1939. Biological and Taxonomic Observations on the Mosquitoes of Kansas. Trans. Kans. Acad. Sci., *42*:255-265.

361. Hill, R. B., and C. McD. Hill. 1948. The Mosquitoes of Jamaica. Bull. Inst. Jamaica, Science Series, *4*:1-60.

362. Hinman, E. H. 1930. A Study of the Food of Mosquito Larvae (Culicidae). Amer. Jour. Hyg., *12*:238-270.

363. Hinman, E. H. 1932. A Description of the Larva of *Anopheles atropos* Dyar and Knab, with Biological Notes on the Species. Proc. Ent. Soc. Wash., *34*:138-142.

364. Hinman, E. H. 1934. Preliminary Observations on the Hibernation of *Anopheles quadrimaculatus* in Southern Louisiana. South. Med. Jour., *27*:461-464.

365. Hinman, E. H. 1935. Biological Notes on *Uranotaenia* spp. in Louisiana. Ann. Ent. Soc. Amer., *28*:404-407.

366. Hinman, E. H., and H. S. Hurlbut. 1940. A Study of Winter Activities and Hibernation of *Anopheles quadrimaculatus* in the Tennessee Valley. Amer. Jour. Trop. Med., *20*:431-446.

367. Hinman, E. H., and H. S. Hurlbut. 1942. A Collection of Anopheline Mosquitoes from Southern Ontario. Can. Ent., *74*:20.

368. Hinman, E. J. 1950. A Preliminary Study of Mosquitoes at Lake Texoma, Oklahoma. Proc. Okla. Acad. Sci., *31*:51-52.

369. Hocking, B. 1952. Autolysis of Flight Muscles in Mosquitoes. Nature (London), *169*:1101.

370. Hocking, B., W. R. Richards, and C. R. Twinn. 1950. Observations on the Bionomics of Some Northern Mosquito Species (Culicidae: Diptera). Can. Jour. Res. D., *28*:58-80.

371. Hodgen, B. B. 1943. *Aedes aegypti* (Linnaeus), the Yellow Fever Mosquito in Oklahoma. Jour. Kans. Ent. Soc., *16*:154.

372. Hollingsworth, W. P. 1948. Mosquitoes in the Vicinity of Nacogdoches. Tenn. Acad. Sci. Proc. and Trans., *30*:125-129.

373. Hopkins, G. H. E. 1952. Mosquitoes of the Ethiopian Region (Vol. I. Larval Bionomics of Mosquitoes and Taxonomy of Culicine Larvae). London: British Museum of Natural History, 2d ed., 355 pp.

374. Horsfall, W. R. 1936. Occurrence and Sequence of Mosquitoes in Southeastern Arkansas in 1935. Jour. Econ. Ent., *29*:676-679.

375. Horsfall, W. R. 1937. Mosquitoes of Southeastern Arkansas. Jour. Econ. Ent., *30*:743-748.

376. Horsfall, W. R. 1939. Habits of *Aedes thibaulti* Dyar and Knab (Diptera, Culicidae). Jour. Kans. Ent. Soc., *12*:70-71.

377. Horsfall, W. R. 1942. Breeding Habits of a Rice Field Mosquito. Jour. Econ. Ent., *35*:478-482.

378. Horsfall, W. R. 1942. Biology and Control of Mosquitoes in the Rice Area. Ark. Agr. Expm. Sta. Bull., *427*:1-46.

379. Horsfall, W. R. 1943. Some Responses of the Malaria Mosquito to Light. Ann. Ent. Soc. Amer., 36:41-45.

380. Horsfall, W. R. 1946. Area Sampling of Population of Larval Mosquitoes in Rice Fields. Ent. News, 57:242-244.

381. Horsfall, W. R., R. C. Miles, and J. T. Sokatch. 1952. Eggs of Floodwater Mosquitoes. I. Species of *Psorophora*. Ann. Ent. Soc. Amer., 45:618-624.

382. Howard, L. O., H. G. Dyar, and F. Knab. 1912-1917. The Mosquitoes of North and Central America and the West Indies. Carnegie Inst. Wash. Pub., 4 vols., 159:1,064 pp.

383. Howitt, B. F., H. R. Dodge, L. K. Bishop, and R. H. Gorrie. 1949. Recovery of the Virus of Eastern Equine Encephalomyelitis from Mosquitoes *(Mansonia perturbans)* Collected in Georgia. Science, 110:141-142.

384. Hubert, A. A. 1953. Observations on the Continuous Rearing of *Culiseta incidens* (Thomson). Mosquito News, 13:207-208.

385. Huffaker, C. B. 1942. Observations on the Overwintering of Mosquitoes near Fort duPont, Delaware. Mosquito News, 2:37-40.

386. Huffaker, C. B., and R. C. Back. 1945. A Study of the Effective Flight Range, Density and Seasonal Fluctuation in the Abundance of *Anopheles quadrimaculatus* Say in Delaware. Amer. Jour. Trop. Med., 25:155-161.

387. Hurlbut, H. S. 1938. Further Notes on the Overwintering of the Eggs of *Anopheles walkeri* Theobald with a Description of the Eggs. Jour. Parasitol., 24:521-526.

388. Hurlbut, H. S. 1938. A Study of the Larval Chaetotaxy of *Anopheles walkeri* Theobald. Amer. Jour. Hgy., 28:149-173.

389. Hurlbut, H. S. 1941. First Instar Characters for Distinguishing the Common Inland Species of Anophelines of Eastern United States. Amer. Jour. Hyg., 34:47-48.

390. Hurlbut, H. S. 1943. The Rate of Growth of *Anopheles quadrimaculatus* in Relation to Temperature. Jour. Parasitol., 29:107-113.

391. Irwin, W. H. 1941. A Preliminary List of the Culicidae of Michigan. Part I. Culicinae (Diptera). Ent. News, 52:101-105.

392. Irwin, W. H. 1942. The Mosquitoes of Three Selected Areas in Cheboygan County, Michigan. Papers Mich. Acad. Sci. Arts and Letters, 28:379-396.

393. Irwin, W. H. 1942. The Role of Certain North Michigan Bog Mats in Mosquito Production. Ecology, 23:466-477.

394. Jachowski, L. A., and C. Schultz. 1948. Notes on the Biology and Control of Mosquitoes at Umiat, Alaska. Mosquito News, 8:155-165.

395. James, M. T. 1942. A Two-Season Light Trap Study of Mosquitoes in Colorado. Jour. Econ. Ent., 35:945.

396. Jenkins, D. W. 1948. Ecological Observations on the Mosquitoes of Central Alaska. Mosquito News, 8:140-147.

397. Jenkins, D. W. 1949. *Toxorhynchites* Mosquitoes of the United States. Proc. Ent. Soc. Wash., 51:225-229.

398. Jenkins, D. W. 1949. A Field Method of Marking Arctic Mosquitoes with Radiophosphorus. Jour. Econ. Ent., 42:988-989.

399. Jenkins, D. W. 1950. Bionomics of *Culex tarsalis* in Relation to Western Equine Encephalomyelitis. Amer. Jour. Trop. Med., 30:909-916.

400. Jenkins, D. W., and S. J. Carpenter. 1946. Ecology of the Tree Hole Breeding Mosquitoes of Nearctic North America. Ecol. Monog., 16:33-48.

401. Jenkins, D. W., and C. C. Hassett. 1951. Dispersal and Flight Range of Subarctic Mosquitoes Marked with Radiophosphorus. Can. Jour. Zool., 29:178-187.

402. Jenkins, D. W., and K. L. Knight. 1950. Ecological Survey of the Mosquitoes of Great Whale River, Quebec (Diptera, Culicidae). Proc. Ent. Soc. Wash., 52:209-223.

403. Jenkins, D. W., and K. L. Knight. 1952. Ecological Survey of the Mosquitoes of Southern James Bay. Amer. Mid. Nat., 47:456-468.

404. Johnson, C. W. 1925. List of Diptera, Fauna of New England, 15. Boston Soc. Nat. Hist., 7:1-326.

405. Johnson, H. A. 1937. Notes on the Continuous Rearing of *Aedes aegypti* in the Laboratory. U.S. Pub. Hlth. Rpts., 52:1177-1179.

406. Johnson, P. T., and E. B. Thurman. 1950. The Occurrence of *Aedes (Ochlerotatus) pullatus* (Coq.) in California (Diptera, Culicidae). Pan-Pac. Ent., 26:107-110.

407. Joyce, C. R. 1945. The Occurrence of *P. mexicana* (Bellardi) in the United States. Mosquito News, 5:86.

408. Joyce, C. R. 1948. *Culex chidesteri* Dyar (Diptera, Culicidae) at Brownsville, Texas. Mosquito News, 8:102-105.

409. Judd, W. W. 1950. Mosquitoes Collected in the Vicinity of Hamilton, Ontario during the Summer of 1948. Mosquito News, 10:57-59.

410. Kartman, L., and R. P. Repass. 1952. The Effects of Desiccation on the Eggs of *Anopheles quadrimaculatus* Say. Mosquito News, 12:107-110.

411. Keener, G. G. 1945. Detailed Observations on the Life History of *Anopheles quadrimaculatus*. Jour. Nat. Mal. Soc., 4:263-270.

412. Keener, G. G. 1952. Observations on Overwintering of *Culex tarsalis* Coquillett (Diptera, Culicidae) in Western Nebraska. Mosquito News, 12:205-209.

413. Kelser, R. A. 1933. Mosquitoes as Vectors of the Virus of Equine Encephalomyelitis. Jour. Amer. Vet. Med. Assoc., 82:767-771.

414. King, W. V. 1937. On the Distribution of

Anopheles albimanus and its Occurrence in the United States. South. Med. Jour., *30*:943-946.

415. King, W. V. 1939. Varieties of *Anopheles crucians* Wied. Amer. Jour. Trop. Med., *19*:461-471.

416. King, W. V., and G. H. Bradley. 1937. Notes on *Culex erraticus* and Related Species in the United States (Diptera, Culicidae). Ann. Ent. Soc. Amer., *30*:345-357.

417. King, W. V., and G. H. Bradley. 1941. General Morphology of *Anopheles* and Classification of the Nearctic Species. Amer. Assoc. Adv. Sci. Pub., *15*:63-70.

418. King, W. V., and G. H. Bradley. 1941. Distribution of the Nearctic Species of *Anopheles*. Amer. Assoc. Adv. Sci. Pub., *15*:71-78.

419. King, W. V., G. H. Bradley, and T. E. McNeel. 1944. The Mosquitoes of the Southeastern States (rev. ed.). U.S. Dept. Agr., Misc. Pub., *336*:1-96.

420. King, W. V., L. Roth, J. Toffaleti, and W. W. Middlekauff. 1943. New Distribution Records for the Mosquitoes of the Southeastern United States during 1942. Jour. Econ. Ent., *36*:573-577.

421. King, W. V., and C. G. Bull. 1923. The Blood Feeding Habits of Malaria Carrying Mosquitoes. Amer. Jour. Hyg., *3*:497-513.

422. Kitchen, S. F., and G. H. Bradley. 1936. *Anopheles walkeri* Theobald as a Vector of *Plasmodium falciparum* (Welch). Amer. Jour. Trop. Med., *16*:579-581.

423. Kitzmiller, J. B. 1945. *Orthopodomyia alba* in Kentucky. Jour. Econ. Ent., *38*:409.

424. Kitzmiller, J. B. 1952. Inbred Strains of *Culex* Mosquitoes. Science, *116*:66-67.

425. Kligler, I. J., and M. Aschner. 1931. Demonstration of Presence of Fowl-pox Virus in Wild-Caught Mosquitoes (*Culex pipiens*). Proc. Soc. Expm. Biol. and Med., *28*:463-465.

426. Kligler, I. J., R. S. Muckenfuss, and T. M. Rivers. 1928. Transmission of Fowl-pox by Mosquitoes. Proc. Soc. Expm. Biol. and Med., *26*:128-129.

427. Knab, F. 1906. The Swarming of *Culex pipiens*. Psyche, pp. 123-133.

428. Knab, F. 1908. Observations on the Mosquitoes of Saskatchewan. Smithson. Misc. Coll., *50*:540-547.

429. Knight, K. L. 1948. A Taxonomic Treatment of the Mosquitoes of Umiat, Alaska. U.S. Nav. Med. Res. Inst. Proj. NM005-017, Rept. 2, 13 pp.

430. Knight, K. L. 1951. The *Aedes (Ochlerotatus) punctor* Subgroup in North America (Diptera, Culicidae). Ann. Ent. Soc. Amer., *44*:87-99.

431. Knight, K. L. 1953. Hybridization Experiments with *Culex pipiens* and *C. quinquefasciatus*

(Diptera, Culicidae). Mosquito News, *13*:110-115.

432. Knight, K. L., and R. W. Chamberlain. 1948. A New Nomenclature for the Chaetotaxy of the Mosquito Pupa based on a Comparative Study of the Genera (Diptera, Culicidae). Proc. Helminthological Soc. Wash., *15*:1-10.

433. Knowlton, G. F., and J. A. Rowe. 1935. Notes on Utah Mosquitoes. Utah Acad. Sci. Arts and Letters, *12*:245-247.

434. Knowlton, G. F., and J. A. Rowe. 1935. Handling Mosquitoes on Equine Encephalomyelitis Investigation. Jour. Econ. Ent., *28*:824-829.

435. Knowlton, G. F., and J. A. Rowe. 1936. Mosquito Studies. Utah Acad. Sci. Arts and Letters, *13*:283-287.

436. Knutson, H. 1943. The Status of the Mosquitoes of the Great Swamp in Rhode Island during 1942. Jour. Econ Ent., *36*:311-319.

437. Komp, W. H. W. 1926. A New *Culex* from Honduras. Ins. Ins. Mens., *14*:44-45.

438. Komp, W. H. W. 1926. Observations on *Anopheles walkeri* and *Anopheles atropos* (Diptera, Culicidae). Ins. Ins. Mens., *14*:168-176.

439. Komp, W. H. W. 1936. A Guide to the Identification of the Common Mosquitoes of the Southeastern United States. U.S. Pub. Hlth. Rpts., *38*:1061-1080.

440. Komp, W. H. W. 1937. The Nomenclature of the Thoracic Sclerites in the Culicidae, and their Setae. Proc. Ent. Soc. Wash., *39*:241-252.

441. Komp, W. H. W. 1942. A Technique for Staining, Dissecting and Mounting the Male Terminalia of Mosquitoes. U.S. Pub. Hlth. Rpts., *57*:1327-1333.

442. Komp, W. H. W. 1942. The Anopheline Mosquitoes of the Caribbean Region. Nat. Inst. Hlth. Bull., *179*:1-195.

443. Kumm, H. W. 1929. Geographical Distribution of Malaria Carrying Mosquitoes. Amer. Jour. Hyg. Mon. Ser., *10*:1-178.

444. Kumm, H. W. 1942. *Anopheles crucians* found in Northern Nicaragua. Amer. Jour. Trop. Med., *22*:511-512.

445. Kumm, H. W., and M. Frobisher. 1932. Attempts to Transmit Yellow Fever with Certain Brazilian Mosquitoes (Culicidae) and with Bedbugs (*Cimex hemipterus*). Amer. Jour. Trop. Med., *12*:349-361.

446. Kumm, H. W., and L. M. Ram. 1941. Observations on the *Anopheles* of British Honduras. Amer. Jour. Trop. Med., *21*:559-566.

447. Kumm, H. W., and H. Zuniga. 1942. The Mosquitoes of El Salvador. Amer. Jour. Trop. Med., *22*:399-415.

448. Kuyp, Edwin Van Der. 1948. Mosquito Records of Aruba and Bonaire. Amer. Jour. Trop. Med., *28*:895-897.

448a. Lake, R. W. 1953. New Mosquito Distribution

Records for New Jersey. N.J. Mosq. Exterm. Assoc. Proc., *40*:152-155.

449. Lane, John. 1939. Catalogo dos Mosquitos Neotropicos. Bol. Biol. Sao Paulo. Ser. Mon. I., XIV, 218 pp.

450. Lane, John. 1951. Synonymy of Neotropical Culicidae (Diptera). Proc. Ent. Soc. Wash., *53*:333-336.

451. Lane, John, and N. L. Cerqueira. 1942. Os Sabetineos Da America (Diptera, Culicidae). Arq. Zool. Estad. Sao Paulo, 7:473-849.

452. Lasky, W. R. 1946. Report of Mosquitoes Collected at Fitzsimmons General Hospital, Denver, Colorado, during the Seasons of 1944-1945. Ent. News, *57*:222-228.

453. Lathrop, F. H. 1939. Mosquitoes Collected in Maine. Maine Agr. Expm. Sta. Bull., *397*: 828-829.

454. Lee, Dorothy. 1947. The Mosquito Control Program in Oregon. N.J. Mosq. Exterm. Assoc. Proc., *34*:69-72.

455. Leeson, H. S., and P. A. Buxton. 1949. Anopheline Mosquitoes: Morphology in Different Stages and Instars. *In* Boyd, Malariology, *1*:235-256.

456. LeVan, J. H. 1940. Viability of *Aedes aegypti* Eggs. U.S. Pub. Hlth. Rpts., *55*:900.

457. Lewis, D. J. 1942. A Method of Transporting Living Mosquito Larvae. Bull. Ent. Res., *33*: 227-228.

458. Lowry, P. R. 1929. Mosquitoes of New Hampshire, A Preliminary Report. N.H. Agr. Expm. Sta. Bull., *243*:1-23.

459. Ludlow, C. S. 1906. The Distribution of Mosquitoes in the United States. Med. Record, *69*:95-98.

460. Macan, T. T. 1930. *Culicella morsitans* in a Tree Hole. Entomologist, *63*:164.

461. McClesky, O. L. 1951. The Bionomics of the Culicidae of the Dallas Area. Field and Lab., *19*:5-14.

462. MacCreary, D., and L. A. Stearns. 1937. Mosquito Migration Across the Delaware Bay. N.J. Mosq. Exterm. Assoc. Proc., *24*:188-197.

463. McGregor, T., and R. B. Eads. 1943. Mosquitoes of Texas. Jour. Econ. Ent., *36*:938-940.

464. McLaine, L. S. 1943. Mosquito Suppression Work in Canada in 1942. N.J. Mosq. Exterm. Assoc. Proc., *30*:51-56.

465. McLeod, J. A., and J. McLintock. 1947. Anophelism and Climate in Relation to Malaria in Manitoba. Can. Jour. Res., E., *25*:33-42.

466. McLintock, J. 1944. The Mosquitoes of the Greater Winnipeg Area. Can. Ent. 76:89-104.

467. McLintock, J. 1946. Collecting and Handling Mosquitoes on Western Equine Encephalitis Investigations in Manitoba. Can. Jour. Res., E., *24*:55-62.

468. McLintock, J. 1952. Continuous Laboratory Rearing of *Culiseta inornata* (Will.) (Diptera: Culicidae). Mosquito News, *12*:195-201.

469. McNeel, T. E. 1931. A Method for Locating the Larvae of the Mosquito *Mansonia*. Science, 74:155.

470. McNeel, T. E. 1932. Observations on the Biology of *Mansonia perturbans* (Walker) (Diptera, Culicidae). N.J. Mosq. Exterm. Assoc. Proc., *19*:91-96.

471. McNeel, T. E., and F. F. Freguson. 1952. *Psorophora cyanescens* (Coquillett) New to the Mosquito Fauna of New Mexico. Mosquito News, *12*:241.

472. McNeel, T. E., and F. F. Ferguson. 1954. Mosquito Distribution and Abundance in the Arkansas-White-Red River Basins. U. S. Pub. Hlth. Rpts., *69*:385-390.

473. Magoon, E. H. 1935. A Portable Stable Trap for Capturing Mosquitoes. Bull. Ent. Res., *26*: 363-372.

474. Mail, G. A. 1930. Viability of Eggs of *Aedes campestris* D. and K. (Culicidae). Science, 72:170.

475. Mail, G. A. 1934. The Mosquitoes of Montana. Mont. Agr. Expm. Sta., Bull., *288*:1-72.

476. Majid, S. A. 1937. An Improved Technique for Marking and Catching Mosquitoes. Rec. Mal. Surv. India, 7:105-107.

477. Manzelli, M. A. 1948. A Survey of the Arthopod Vectors of Equine Encephalomyelitis and Encephalitis. Jour. N.Y. Ent. Soc., *56*:79-107.

478. Markos, B. G. 1950. The Ecology of California *Anopheles*. Calif. Mosq. Control Assoc. Ann. Confr. Proc. and Papers, *18*:69-73.

479. Markos, B. G. 1951. Distribution and Control of Mosquitoes in Rice Fields in Stanislaus County, California. Jour. Nat. Mal. Soc., *10*:233-247.

480. Marshall, J. F. 1938. The British Mosquitoes. London: British Museum of Natural History, 341 pp.

481. Martini, E. 1935. Los Mosquitos de Mexico. Dept. Salub. Pub., Bol. Tec. Ser. A: Entom. Med. y Parasitol., *1*:1-65.

482. Masters, C. O. 1949. A Study of the Adult Mosquito Population of a Northern Ohio Woods. Ohio Jour. Sci., *49*:12-14.

483. Masters, C. O. 1951. Storing Adult Mosquitoes. Mosquito News, *11*:74.

484. Masters, C. O. 1953. The Effect of Contamination Upon Mosquito Larvae in Rain-water Barrels. Mosquito News, *13*:152.

485. Masters, C. O. 1953. Season Succession of Mosquito Species and the Relationship Existing Between Dissolved Minerals in Mosquito-Breeding Waters and Species. Mosquito News, *13*:159-161.

486. Matheson, Robert. 1924. Notes on Culicidae *(Aedes)*. Ins. Ins. Mens., *12*:22-24.

487. Matheson, Robert. 1924. The Genera of Culicidae of North America. Can. Ent., *56*:151-161.

488. Matheson, Robert. 1924. The Culicidae of the Douglas Lake Region (Michigan). Can. Ent., 56:289-290.

489. Matheson, Robert. 1930. Distribution Notes on Culicidae (Mosquitoes). Bull. Brooklyn Ent. Soc., 25:291-294.

490. Matheson, Robert. 1933. A New Species of Mosquito from Colorado (Diptera, Culicidae). Proc. Ent. Soc. Wash., 35:69-71.

491. Matheson, Robert. 1934. Notes on *Psorophora (Janthinosoma) horridus* (Dyar and Knab). Proc. Ent. Soc. Wash., 36:41-43.

492. Matheson, Robert. 1944. A Handbook of the Mosquitoes of North America. Ithaca, N.Y.: Comstock Pub. Assoc., 314 pp.

493. Matheson, Robert. 1945. Notes on *Anopheles occidentalis* D. and K. and *Anopheles quadrimaculatus* Say. Mosquito News, 5:1-3.

494. Matheson, Robert. 1945. Family Culicidae, The Mosquitoes, Guide to the Insects of Connecticut, Part VI, Fasc. 2. Conn. State Geol. and Nat. Hist. Surv. Bull., 68:1-48.

495. Matheson, Robert. 1948. The Status of *Anopheles earlei* Vargas (1943). N.J. Mosq. Exterm. Assoc. Proc., 35:69.

496. Matheson, Robert, and J. Belkin. 1943. Notes on *Anopheles occidentalis* in Central New York. N.J. Mosq. Exterm. Assoc. Proc., 30: 7-11.

497. Matheson, Robert, M. F. Boyd, and W. K. Stratman-Thomas. 1933. *Anopheles walkeri* Theobald as a Vector of *Plasmodium vivax* Grassi and Feletti. Amer. Jour. Hyg., 17:515-516.

498. Matheson, Robert, E. L. Brunett, and A. L. Brody. 1931. The Transmission of Fowl-pox by Mosquitoes. Preliminary Report. Poultry Science, 10:211-223.

499. Matheson, Robert, and H. S. Hurlbut. 1937. Notes on *Anopheles walkeri* Theobald. Amer. Jour. Trop. Med., 17:237-242.

500. Matheson, Robert, and R. C. Shannon. 1923. The Anophelines of Northeastern America (Diptera, Culicidae). Ins. Ins. Mens., 11: 57-64.

501. Mattingly, P. F., L. E. Rozeboom, K. L. Knight, H. Laven, F. H. Drummond, S. R. Christophers, and P. G. Shute. 1951. The *Culex pipiens* Complex. Trans. Royal Ent. Soc. London, 102:331-382.

502. Mayne, B. 1919. The Occurrence of Malaria Parasites in *Anopheles crucians* in Nature: Percentage of Infection of *Anopheles quadrimaculatus* and Latest Date Found Infected in Northern Louisiana. U.S. Pub. Hlth. Rpts., 34:1355-1357.

503. Mayne, B., and T. H. D. Griffitts. 1931. *Anopheles atropos* D. and K. A New Potential Carrier of Malaria Organisms. U.S. Pub. Hlth. Rpts., 46:3107-3115.

504. Meleney, H. E., E. L. Bishop, and F. L. Roberts. 1929. Observations on the Malaria Problem of West Tennessee. South. Med. Jour., 22: 382-394.

505. Menon, M. A. U. 1951. On Certain Little Known External Characters of Adult Mosquitoes and Their Taxonomic Significance. Royal Ent. Soc. London, Proc. Ser. B: Tax., 20:63-71.

506. Mentzner, R. L. 1938. New Developments in Mosquito Control in Delaware County, Pennsylvania. N.J. Mosq. Exterm. Assoc. Proc., 25:125-130.

507. Merrill, M. H., D. W. Lacaillade, and C. Tenbroeck. 1934. Mosquito Transmission of Equine Encephalomyelitis. Science, 80:251-252.

508. Michener, C. D. 1944. Differentiation of Females of Certain Species of *Culex* by the Cibarial Armature. Jour. N.Y. Ent. Soc., 52: 263-266.

509. Michener, C. D. 1945. Seasonal Variations in Certain Species of Mosquitoes (Diptera, Culicidae). Jour. N.Y. Ent. Soc., 53:293-300.

510. Michener, C. D. 1947. Mosquitoes of a Limited Area in Southern Mississippi. Amer. Mid. Nat., 37:325-374.

511. Middlekauff, W. W. 1944. A New Species of *Aedes* from Florida (Diptera: Culicidae). Proc. Ent. Soc. Wash., 46:42-44.

512. Middlekauff, W. W. 1944. A Rapid Method for Making Permanent Mounts of Mosquito Larvae. Science, 99:206.

513. Middlekauff, W. W. 1944. Gynandromorphism in Recently Collected Mosquitoes. Jour. Econ. Ent., 37:297.

514. Middlekauff, W. W., and S. J. Carpenter. 1944. New Distribution Records for the Mosquitoes of the Southeastern United States in 1943. Jour. Econ. Ent., 37:88-92.

515. Miles, V. I. 1945. Differentiating the Larvae of *Anopheles georgianus* King, *A. bradleyi* King, and *A. punctipennis* (Say). Jour. Nat. Mal. Soc., 4:235-242.

516. Miles, V. I., and R. W. Rings. 1946. Distribution Records for Mosquitoes of the Southeastern States in 1945. Jour. Econ, Ent., 39: 387-391.

517. Moore, J. A., M. D. Young, N. H. Hardman, and T. H. Stubbs. 1945. Studies on Imported Malarias: 2. Ability of California Anophelines to Transmit Malaria of Foreign Origin and Other Considerations. Jour. Nat. Mal. Soc., 4:307-329.

518. Moorefield, H. H. 1951. Sexual Dimorphism in Mosquito Pupae. Mosquito News, 11:175-177.

519. Mulhern, T. D. 1942. New Jersey Mechanical Trap for Mosquito Surveys. N.J. Agr. Expm. Sta. Cir., 421:1-8.

520. Mulrennan, J. A. 1946. Some New Records of Florida Mosquitoes. Fla. Anti-Mosquito Assoc. Ann. Rpt., 17:23-24.

521. Natvig, L. R. 1945. Notes on *Culex alpinus* Lin-

naeus and *Aedes nigripes* (Zett.). Norsk. Ent. Tidsskr., 7:99-106.

522. Natvig, L. R. 1946. Differential Characters of the Female *Aedes nigripes* (Zett.) and *A. nearcticus* Dyar (Diptera, Culicidae). Norsk. Ent. Tidsskr., 7:164-167.

523. Natvig, L. R. 1948. Contributions to the Knowledge of the Danish and Fennoscandian Mosquitoes: Culicini. Norsk. Ent. Tidsskr., Sup. 1, 567 pp.

524. Nelson, T. C., and M. L. Morris. 1938. Heartworm in Dogs; A Mosquito Menace in New Jersey. N.J. Mosq. Exterm. Assoc. Proc., 25:227-232.

525. Newton, W. L., W. H. Wright, and I. Pratt. 1945. Experiments to Determine Potential Mosquito Vectors of *Wuchereria bancrofti* in Continental U.S. Amer. Jour. Trop. Med., 25:253-261.

526. Nielsen, E. T., and H. Greve. 1950. Studies on the Swarming Habits of Mosquitoes and other Nematocera. Bull. Ent. Res. 41:227-258.

527. Nielsen, E. T., and A. T. Nielsen. 1953. Field Observations on the Habits of *Aedes taenioryhnchus*. Ecology, 34:141-156.

528. Norris, Marjorie. 1946. Recovery of a Strain of Western Equine Encephalitis Virus from *Culex restuans* (Theob.) (Diptera; Culicidae). Can. Jour. Res., E, 24:63-70.

529. Obrecht, C. B. 1949. Notes on the Distribution of Michigan Mosquitoes (Diptera, Nematocera). Amer. Mid. Nat., 41:168-173.

530. Olson, T. A., and H. L. Keegan. 1944. The Mosquito Collecting Program of the Seventh Service Command for 1942-1943. Jour. Econ. Ent., 37:780-785.

531. Olson, T. A., and H. L. Keegan. 1944. New Mosquito Distribution Records from the Seventh Service Command Area. Jour. Econ. Ent., 37:847-848.

532. O'Neill, K., L. J. Ogden, and D. E. Eyles. 1944. Additional Species of Mosquitoes found in Texas. Jour. Econ. Ent., 37:555.

533. Owen, W. B. 1937. The Mosquitoes of Minnesota, with Special Reference to Their Biologies. Univ. Minn. Agr. Expm. Sta., Tech. Bull., 126:1-75.

534. Owen, W. B. 1942. The Biology of *Theobaldia inornata* Williston, in Captive Colony. Jour. Econ. Ent., 35:903-907.

535. Owen, W. B. 1951. Important Species of Mosquitoes and Control Work in Wyoming. Mosquito News, 11:163-166.

536. Ozburn, R. H. 1945. Preliminary Report on Anopheline Mosquito Survey in Canada. Pt. 1. A Report on Light Trap Collections. Ent. Soc. Ontario Ann. Rpt. (1944), 75:37-44.

537. Palacios, A. M. 1950. Identification de Los Mosquitos Mexicanos Del Subgenera *Culex* (Diptera, Culicidae) Por La Genitalia Mas-

culina. Rev. Soc. Mexico Hist. Nat., 11:183-187.

537a. Palacios, A. M. 1952. Nota Sobre La Distribucion de Los Mosquitos *Culex* en Mexico (Diptera, Culicidae). Rev. Soc. Mexico Hist. Nat., 13:75-87.

538. Pelaez, D. 1946. *Anopheles walkeri* Theobald, 1901; *Anopheles darlingi* Root, 1926. Bol. Epidemiologias (Mexico), 1:13-16.

539. Penn, G. H. 1949. The Pupae of the Mosquitoes of New Guinea. Pacific Sci., 3:3-85

540. Penn, G. H. 1949. Pupae of Nearctic Anopheline Mosquitoes North of Mexico. Jour. Nat. Mal. Soc., 8:50-69.

541. Penn, G. H., and Coleman, S. A. 1949. An Analysis of the Pupal Chaetoxaxy of *Anopheles quadrimaculatus* Say. Mosquito News, 9:174-175.

542. Perez, M. 1930. An Anopheline Survey of the State of Mississippi. Amer. Jour. Hyg., 11:696-710.

543. Perez-Vigueras, I. 1948. Notes Sobre la *Psorophora johnstonii, Mansonia titillans, Anopheles atropos* y Sobre la Presencia en Cuba de la *Psorophora ciliata*. Univ. de la Habana, 73-75:293-302.

544. Peters, H. T. 1943. Studies on the Biology of *Anopheles walkeri* Theobald (Diptera: Culicidae). Jour. Parasitol., 29:117-122.

545. Peterson, Alvah. 1943. Some New Killing Fluids for Larvae of Insects. Jour. Econ. Ent., 36:115.

546. Peterson, A. G., and W. W. Smith. 1945. Occurrence and Distribution of Mosquitoes in Mississippi. Jour. Econ. Ent., 38:378-383.

547. Philip, C. B. 1943. Flowers as a Suggested Source of Mosquitoes during Encephalitis Studies and Incidental Mosquito Records in the Dakotas in 1941. Jour. Parasitol., 29:328-329.

548. Pierce, W., W. E. Duclus, and M. Y. Longacre. 1945. Mosquitoes of Los Angeles and Vicinity. Sanitarian (Los Angeles), 7:718-726.

549. Pletsch, D. J. 1946. *Anopheles* Mosquito Records and Observations in Montana. Great Basin Nat., 7:23-28.

550. Pletsch, D. J. 1946. Anopheline Mosquitoes in Montana. Jour. Lancet, 66:289.

551. Porter, J. E. 1946. The Larva of *Uranotaenia syntheta* (Diptera, Culicidae). Amer. Mid. Nat., 35:535-537.

552. Portman, R. W. 1943. New Mosquito Records for Colorado. Jour. Kans. Ent. Soc., 16:155.

553. Post, R. L., and J. A. Munro. 1949. Mosquitoes of North Dakota. N.D. Agr. Expm. Sta. Bimonthly Bull., 11:173-183.

554. Pratt, H. D. 1945. *Mansonia indubitans* Dyar and Shannon—A New Mosquito Addition to the United States Fauna. Jour. Kans. Ent. Soc., 18:121-129.

555. Pratt, H. D. 1946. The Larva of *Psorophora*

(Janthinosoma) coffini D. and K. and a Key to the *Psorophora* Larvae of the U.S. and Greater Antilles (Diptera, Culicidae). Proc. Ent. Soc. Wash., *48*:209-214.

556. Pratt, H. D. 1946. The Genus *Uranotaenia* L.-A. in Puerto Rico. Ann. Ent. Soc. Amer., *39*: 576-584.

557. Pratt, H. D. 1952. Notes on *Anopheles earlei* and other American Species of the *Anopheles maculipennis* Complex. Amer. Jour. Trop. Med. and Hyg., *1*:484-493.

558. Pratt, H. D. 1953. Notes on American *Mansonia* Mosquitoes (Diptera, Culicidae). Proc. Ent. Soc. Wash., *55*:9-19.

559. Pratt, H. D., and E. L. Seabrook. 1952. The Occurrence of *Culex iolambdis* Dyar in Florida and Puerto Rico, with a Description of the Larva (Diptera, Culicidae). Proc. Ent. Soc. Wash., *54*:27-32.

560. Pratt, H. D., W. W. Wirth, and D. G. Denning. 1945. The Occurrence of *Culex opisthopus* Komp in Puerto Rico and Florida with a Description of the Larva (Diptera, Culicidae). Proc. Ent. Soc. Wash., *47*:245-251.

561. Pritchard, A. E., and H. D. Pratt. 1944. I. A Comparison of Light Trap and Animal Bait Trap Anopheline Mosquito Collections in Puerto Rico. II. A List of the Mosquitoes of Puerto Rico. U.S. Pub. Hlth. Rpts., *59*:221-233.

562. Pritchard, A. E., E. L. Seabrook, and J. A. Mulrennan. 1947. The Mosquitoes of the Florida Keys. Fla. Ent., *30*:8-15.

563. Pritchard, A. E., E. L. Seabrook, and M. W. Provost. 1946. The Possible Endemicity of *A. albimanus* in Florida. Mosquito News, *6*:183-184.

564. Proctor, W. 1938. Biological Survey of the Mount Desert Region. Philadelphia: Wistar Inst. of Anat. and Biology (Insecta). 496 pp.

565. Provost, M. W. 1951. The Occurrences of Salt-marsh Mosquitoes in the Interior of Florida. Fla. Ent., *34*:48-53.

566. Provost, M. W. 1952. The Dispersal of *Aedes taeniorhynchus*. I. Preliminary Studies. Mosquito News, *12*:174-190.

567. Pryor, J. E. and R. W. Chamberlain. 1944. Differentiating the Larvae of *Uranotaenia* in the Southeast. Jour. Econ Ent., *37*:543-544.

568. Quayle, H. T. 1906. Mosquito Control. Univ. Calif. Agr. Expm. Sta. Bull., *178*:1-55.

569. Quinby, G. E. 1941. Additions to the Mosquitoes (Culicidae) of the Reelfoot Lake Region. Rpt. Reelfoot Lake Biol. Sta., *5*:17-21.

570. Quinby, G. E., R. E. Serfling, and J. K. Neel. 1944. Distribution and Prevalence of the Mosquitoes of Kentucky. Jour. Econ. Ent., *37*:547-550.

571. Randolph, N. M., and Kellie O'Neill. 1944. The Mosquitoes of Texas. Bull. Texas State Hlth. Dept., 100 pp.

572. Rees, D. M. 1934. Mosquito Records from Utah. Pan-Pac. Ent., *10*:161-165.

573. Rees, D. M. 1935. Observations on Mosquito Flight in Salt Lake City. Bull. Univ. Utah, *25*:1-6.

574. Rees, D. M. 1942. Overwintering Habits in Utah of *Anopheles maculipennis freeborni* Aitken (Diptera, Culicidae). Ent. News, *53*:282.

575. Rees, D. M. 1943. The Mosquitoes of Utah. Bull. Univ. Utah, *33*:1-99.

576. Rees, D. M. 1944. A New Mosquito Record from Utah (Diptera: Culicidae). Pan-Pac. Ent., *20*:19.

577. Rees, D. M., and F. C. Harmston. 1946. Observations on the Habits of *Anopheles freeborni* in Northern Utah and Southern Idaho. Mosquito News, *6*:73-75.

578. Rees, D. M., and F. C. Harmston. 1948. Mosquito Records from Wyoming and Yellowstone National Park (Diptera, Culicidae). Pan-Pac. Ent., *24*:181-188.

579. Rees, D. M., and L. T. Nielsen. 1947. On the Biology and Control of *Aedes dorsalis* (Meigen) in Utah. N.J. Mosq. Exterm. Assoc. Proc., *34*:160-165.

580. Rees, D. M., and L. T. Nielsen. 1951. Four New Mosquito Records from Utah (Diptera, Culicidae). Pan-Pac. Ent., *27*:11-12.

581. Rees, D. M., and K. Onishi. 1951. Morphology of the Terminalia and Internal Reproductive Organs and Copulation in the Mosquito, *Culiseta inornata* (Williston) (Diptera, Culicidae). Proc. Ent. Soc. Wash., *53*:233-246.

582. Reeves, W. C. 1940. Research with *Aedes varipalpus* (Coq.), the Pacific Coast Treehole Mosquito. Calif. Mosq. Cont. Assoc. Ann. Conf. Proc. and Papers, *11*:39-43.

583. Reeves, W. C. 1941. The Mosquito Genus *Mansonia* in California. Pan-Pac. Ent., *17*:28.

584. Reeves, W. C. 1941. The Genus *Orthopodomyia* Theobald in California (Diptera, Culicidae). Pan-Pac. Ent., *17*:69-72

585. Reeves, W. C. 1944. Preliminary Studies on the Feeding Habits of Pacific Coast Anophelines. Jour. Nat. Mal. Soc., *3*:261-266.

586. Reeves, W. C. 1951. Field Studies on Carbon Dioxide as a Possible Host Simulant to Mosquitoes. Proc. Soc. Expm. Biol. and Med., *77*:64-66.

587. Reeves, W. C. 1951. The Encephalitis Problem in the United States. Amer. Jour. Pub. Hlth., *41*:678-686.

588. Reeves, W. C. 1953. Quantitative Field Studies on a Carbon Dioxide Chemotropism of Mosquitoes. Amer. Jour. Trop. Med. and Hyg., *2*:325-331.

589. Reeves, W. C., B. Brookman, and W. McD. Hammon. 1948. Studies on the Flight Range of Certain *Culex* Mosquitoes, Using a Fluorescent Dye Marker, with Notes on *Culiseta* and *Anopheles*. Mosquito News, *8*:61-69.

590. Reeves, W. C., and W. McD. Hammon. 1942. Mosquitoes and Encephalitis in the Yakima Valley, Washington. IV. A Trap for Collecting Live Mosquitoes. Jour. Inf. Dis., 70: 275-277.

591. Reeves, W. C., and W. McD. Hammon. 1944. Feeding Habits of the Proven and Possible Mosquito Vectors of Western Equine and St. Louis Encephalitis in the Yakima Valley, Washington. Amer. Jour. Trop. Med., 24: 131-134.

592. Reeves, W. C., and W. McD. Hammon. 1946. Laboratory Transmission of Japanese B Encephalitis Virus by Seven Species (Three Genera) of North American Mosquitoes. Jour. Expm. Med., 83:185-194.

593. Reeves, W. C., and W. McD. Hammon. 1952. California Encephalitis Virus, A Newly Described Agent. III. Mosquito Infection and Transmission. Jour. Immunology, 69:511-514.

594. Reeves, W. C., W. McD. Hammon, and E. M. Izumi. 1942. Experimental Transmission of St. Louis Encephalitis Virus by Culex pipiens Linn. Proc. Soc. Expm. Biol. and Med., 50: 125-128.

595. Reeves, W. C., W. McD. Hammon, A. S. Lazarus, B. Brookman, H. E. McClure, and W. H. Doetschman. 1952. The Changing Picture of Encephalitis in the Yakima Valley, Washington. Jour. Inf. Dis., 90:291-301.

596. Reeves. W. C., W. N. Mack, and W. McD. Hammon. 1947. Epidemiological Studies on Western Equine Encephalomyelitis and St. Louis Encephalitis in Oklahoma, 1944. Jour. Inf. Dis., 81:191-196.

597. Remington, C. L. 1945. The Feeding Habits of Uranotaenia lowii Theobald (Diptera: Culicidae). Ent. News, 56:32-37; 64-68.

598. Rempel, J. G. 1950. A Guide to the Mosquito Larvae of Western Canada. Can. Jour. Res., D., 28:207-248.

599. Repass, R. P. 1952. Laboratory Colonization of Aedes triseriatus. Jour. Econ. Ent., 45:1076.

600. Ribbands, C. R. 1946. Man's Reaction to Mosquito Bites. Nature, 158:912-913.

601. Richards, A. G. 1941. A Stenogamic Autogenous Strain of Culex pipiens L. in North America (Diptera, Culicidae). Ent. News, 52:211-216.

602. Riley, W. A. 1938. The Role of Insects and Allied Forms in the Transmission of Diseases due to Filterable Viruses. Minnesota Med., 21:817-824.

603. Riley, W. A. 1939. Indigenous Malaria and Its Vectors in Minnesota. Jour. Lancet, 59:311.

604. Riley, W. A., and W. Chalgren. 1939. The Pest Mosquito Problem in the Minneapolis-St. Paul Metropolitan Area. Jour. Econ. Ent., 32:553-557.

605. Rings, R. W. 1946. Gynandromorphism in Culex nigripalpus. Jour. Econ. Ent., 39:415.

606. Rings, R. W., and S. O. Hill. 1946. The Larva of Aedes (Ochlerotatus) mathesoni Middlekauff (Diptera, Culicidae). Proc. Ent. Soc. Wash., 48:237-240.

607. Rings, R. W., and S. O. Hill. 1948. The Taxonomic Status of Aedes mathesoni (Diptera, Culicidae). Proc. Ent. Soc. Wash., 50:41-49.

608. Root, F. M. 1924. Notes on the Mosquitoes of Lee County, Georgia. Amer. Jour. Hyg., 4: 449-455.

609. Root, F. M. 1925. Rapid Development of Psorophora pygmaea. Jour. Parasitol., 21:59.

610. Root, F. M. 1932. The Pleural Hairs of American Anopheline Larvae. Amer. Jour. Hyg., 15:777-784.

611. Rosenstiel, R. G. 1947. Dispersion and Feeding Habits of Anopheles freeborni. Jour. Econ. Ent., 40:795-800.

612. Ross, E. S. 1943. New and Additional Lower California Mosquito Records (Diptera, Culicidae). Pan-Pac. Ent., 19:86.

613. Ross, E. S. 1943. The Identity of Aedes bimaculatus (Coquillett) and a New Subspecies of Aedes fulvus (Wiedemann) from the United States (Diptera: Culicidae). Proc. Ent. Soc. Wash., 45:143-151.

614. Ross, E. S., and H. R. Roberts. 1943. Mosquito Atlas. Part 1. The Nearctic Anopheles, Important Malaria Vectors of the Americas and Aedes aegypti and Culex quinquefasciatus. Amer. Ent. Soc., 44 pp.

615. Ross, H. H. 1947. The Mosquitoes of Illinois. Bull. Ill. Nat. Hist. Surv., 24:1-96.

616. Roth, L. M. 1943. A Key to the Culex (Diptera: Culicidae) of the Southeastern United States by Male Terminalia. Jour. Kans. Ent. Soc., 16:117-133.

617. Roth, L. M. 1944. A Key to the Anopheles of the Southeastern United States by Male Genitalia (Diptera, Culicidae). Amer. Mid. Nat., 31:96-110.

618. Roth, L. M. 1945. The Male and Larva of Psorophora (Janthinosoma) horrida (Dyar and Knab) and a New Species of Psorophora from the United States (Diptera, Culicidae). Proc. Ent. Soc. Wash., 47:1-23.

619. Roth, L. M. 1945. Aberrations and Variations in Anopheline Larvae of the Southeastern United States (Diptera: Culicidae). Proc. Ent. Soc. Wash., 47:257-278.

620. Roth, L. M. 1946. The Female Genitalia of the Wyeomyia of North America (Diptera: Culicidae). Ann. Ent. Soc. Amer., 39:292-297.

621. Roth, L. M. 1948. Mosquito Gynandromorphs. Mosquito News, 8:168-174.

622. Roth, L. M. 1948. A Study of Mosquito Behavior. An Experimental Laboratory Study of the Sexual Behavior of Aedes aegypti (Linnaeus). Amer. Mid. Nat. 40:265-352.

623. Roth, L. M., and E. R. Willis. 1952. Notes on Three Gynandromorphs of Aedes aegypti

(L.) (Diptera, Culicidae). Proc. Ent. Soc. Wash., 54:189-193.

624. Roth, L. M., and F. N. Young. 1944. *Culex (Melanoconion) atratus* Theobald in Florida; A New Continental North American Record, with Notes on the Other *Melanoconions* of the Southeastern United States. Ann. Ent. Soc. Amer., 37:84-88.

625. Rowe, J. A. 1942. Preliminary Report on Iowa Mosquitoes. Iowa State Coll. Jour. Sci., 16:211-225.

626. Rowe, J. A. 1942. Mosquito Light Trap Catches from Ten Iowa Cities, 1940. Iowa State Coll. Jour. Sci., 16:487-518.

627. Rozeboom, L. E. 1939. The Larva of *Psorophora (Janthinosoma) horrida* (Dyar and Knab) (Diptera: Culicidae). Jour. Parasitol., 25:145-147.

628. Rozeboom, L. E. 1942. The Mosquitoes of Oklahoma. Okla. Agr. Expm. Sta., Tech. Bull. T-16:1-56.

629. Rozeboom, L. E. 1951. The *Culex pipiens* Complex in North America. Royal Ent. Soc. London, Trans., 102:343-353.

630. Rozeboom, L. E. 1952. The Significance of *Anopheles* Species Complexes in Problems of Disease Transmission and Control. Jour. Econ. Ent., 45:222-226.

631. Rozeboom, L. E. 1952. *Anopheles (A.) earlei* Vargas, 1943, in Montana: Identity and Adaptation to Laboratory Conditions (Diptera, Culicidae). Amer. Jour. Trop. Med. and Hyg., 1:477-483.

632. Rozeboom, L. E., and W. H. W. Komp. 1950. A Review of the Species of *Culex* of the Subgenus *Melanoconion* (Diptera, Culicidae). Ann. Ent. Soc. Amer., 43:75-114.

633. Rudolphs, W. 1929. The Composition of Water and Mosquito Breeding. Amer. Jour. Hyg., 9:160-180.

634. Rueger, M. E., and S. Druce. 1950. New Mosquito Distribution Records for Texas. Mosquito News, 10:60-63.

635. Russell, P. F., L. E. Rozeboom, and A. Stone. 1943. Keys to the Anopheline Mosquitoes of the World. Amer. Ent. Soc., 152 pp.

636. Ryckman, R. E. 1952. Ecological Notes on Mosquitoes of Lafayette County, Wisconsin (Diptera, Culicidae). Amer. Mid. Nat., 47:469-470.

637. Ryckman, R. E., and K. Y. Arakawa. 1951. *Anopheles freeborni* Hibernating in Wood Rats' Nests (Diptera: Culicidae). Pan-Pac. Ent., 27:172.

638. Ryckman, R. E., and K. Y. Arakawa. 1952. Notes on the Ecology of *Culiseta maccrackenae* in San Bernardino and Riverside Counties, California (Diptera, Culicidae). Pan-Pac. Ent., 28:104-105.

639. Sabrosky, C. W. 1944. A Malaria Mosquito Survey of Southern Michigan. Jour. Econ. Ent., 37:312-313.

640. Sabrosky, C. W. 1946. The Occurrence of Malaria Mosquitoes in Southern Michigan. Mich. State Coll. Agr. Expm. Sta. Tech. Bull., 202:1-50.

641. Sabrosky, C. W., G. E. McDaniel, and R. F. Reider. 1946. A High Rate of Natural *Plasmodium* Infection in *Anopheles crucians*. Science, 104:247-248.

642. Sasse, B. E., and L. W. Hackett. 1950. Note on the Host Preferences of *Anopheles pseudopunctipennis*. Jour. Nat. Mal. Soc., 9:181-182.

643. Schmitt, J. B. 1942. Five Species of Mosquitoes New to New Jersey, Found in the Last Five Years. Mosquito News, 2:26-29.

644. Schmitt, J. B. 1945. Species Associations of Mosquito Larvae in New Jersey. N.J. Mosq. Exterm. Assoc. Proc., 32:202-210.

645. Schoof, H. F. 1944. Adult Observation Stations to Determine Effectiveness of the Control of *Anopheles quadrimaculatus*. Jour. Econ. Ent., 37:770-779.

646. Schoof, H. F., and D. F. Ashton. 1944. Notes and New Distribution Records on the Mosquitoes of North Carolina. Jour. Elisha Mitchell Sci. Soc., 60:1-10.

647. Schwardt, H. H. 1939. Biologies of Arkansas Rice Field Mosquitoes. Ark. Agr. Expm. Sta. Bull., 377:1-22.

648. Scott, J. W. 1948. Initial Mosquito Work in Vermont. N.J. Mosq. Exterm. Assoc. Proc., 35:102-105.

649. Scott, J. W. 1949. Mosquito Breeding Situation in Vermont for 1948. N.J. Mosq. Exterm. Assoc. Proc., 36:170-171.

650. Scott, J. W. 1950. Report on Mosquitoes in Vermont for 1949. N.J. Mosq. Exterm. Assoc. Proc., 37:187-189.

651. Seabrook, E. L. 1949. Observations on *Mansonia* Mosquitoes in S. E. Florida. Jour. Newell Ent. Soc., 4:27-30.

652. Seabrook, E. L., and T. E. Duffey. 1946. The Occurrence of *Megarhinus rutilus* Coq. in S. E. Florida. Mosquito News, 6:193-194.

653. Seaman, E. A. 1945. An Insect Light Trap for Use with Auto Vehicles in the Field. Mosquito News, 5:79-81.

654. Seaman, E. A. 1945. Ecological Observations and Recent Records on Mosquitoes of San Diego and Imperial Counties, California. Mosquito News, 5:89-95.

655. Shannon, R. C. 1934. The Genus *Mansonia* in the Amazon Valley. Proc. Ent. Soc. Wash., 36:99-110.

656. Shields, S. E. 1938. Tennessee Valley Mosquito Collections. Jour. Econ. Ent., 31:426-430.

657. Shields, S. E., and V. I. Miles. 1937. The Occurrence of *Orthopodomyia alba* in Alabama (Diptera: Culicidae). Proc. Ent. Soc. Wash., 39:237.

658. Shlaifer, A., and D. E. Harding. 1946. The Mosquitoes of Tennessee. Jour. Tenn. Acad. Sci., 21:241-256.

659. Simmons, J. S. 1941. The Transmission of Malaria by the *Anopheles* Mosquitoes of North America. Amer. Assoc. Adv. Sci. Pub., 15:113-130.

660. Simmons, J. S., and T. H. G. Aitken. 1942. The *Anopheles* Mosquitoes of the Northern Half of the Western Hemisphere and of the Philippine Islands. Army Med. Bull., 59:1-213.

661. Smith, G. E. 1942. The Keg Shelter as a Diurnal Resting Place of *Anopheles quadrimaculatus.* Amer. Jour. Trop. Med., 22:257-269.

662. Smith, J. B. 1904. Report of the New Jersey State Agricultural Experiment Station Upon the Mosquitoes Occurring within the State, Their Habits, Life History . . . New Brunswick, N.J., 482 pp.

663. Smith, M. E. 1952. A New Northern *Aedes* Mosquito, with Notes on its Close Ally, *Aedes diantaeus* H., D., and K. (Diptera, Culicidae). Bull. Brooklyn Ent. Soc., 47:19-28; 29-40.

664. Smithburn, K. C., A. J. Haddow, and W. H. R. Lumsden. 1949. Rift Valley Fever: Transmission of the Virus by Mosquitoes. Br. Jour. Expm. Path., 30:35-47.

665. Snow, W. E. 1949. Studies on Portable Resting Stations for *Anopheles quadrimaculatus* in the Tennessee Valley. Jour. Nat. Mal. Soc., 8:336-343.

666. Stabler, R. M. 1945. New Jersey Light Trap Versus Human Bait as a Mosquito Sampler. Ent. News, 56:93-99.

667. Stabler, R. M. 1948. Notes on Certain Species of Mosquitoes from Delaware County, Pennsylvania. Mosquito News, 8:17-19.

668. Stackelberg, A. A. 1937. Family *Culicidae* (Subfamily *Culicinae*). Faune de l'URSS Insectes Dipteres, 3:1-258.

669. Staebler, A. E., and W. F. Buren. 1946. *Aedes tortilis* (Theob.), A Mosquito New to the United States. U.S. Pub. Hlth. Rpts., 61:685.

670. Stage, H. H., and J. C. Chamberlin. 1945. Abundance and Flight Habits of Certain Alaskan Mosquitoes as Determined by Means of a Rotary-Type Trap. Mosquito News, 5:8-16.

671. Stage, H. H., C. M. Gjullin, and W. W. Yates. 1938. Flight Range and Longevity of Floodwater Mosquitoes in the Lower Columbia River Valley. Jour. Econ. Ent., 30:940-945.

672. Stage, H. H., C. M. Gjullin, and W. W. Yates. 1952. Mosquitoes of the Northwestern States. U.S. Dept. Agr. Handbook 46:1-95.

673. Stage, H. H., and E. A. McKinlay. 1946. A Preliminary List of Mosquitoes Occurring in the Vicinity of Nome, Alaska. Mosquito News, 6:131.

674. Stauber, L. A. 1942. Notes on Anophelines in Southern New Jersey. N.J. Mosq. Exterm. Assoc. Proc., 27:44.

675. Stone, Alan. 1948. A Change of Name in Mosquitoes (Diptera, Culicidae). Proc. Ent. Soc. Wash., 50:161.

676. Stone, Alan. 1953. The Halteres of *Anopheles walkeri* Theobald. Mosquito News, 13:209-210.

677. Stough, B. D., M. A. King, and D. E. Eyles. 1949. Extension of Known Range of *Aedes atropalpus* (Coquillett) Southeastward into Georgia. Mosquito News, 9:173.

678. Stratman-Thomas, W. K., and F. C. Baker. 1936. *Anopheles barberi* Coquillett as a Vector of *Plasmodium vivax* Grassi and Feletti. Amer. Jour. Hyg., 24:182-183.

679. Sundararaman, S. 1949. Biometrical Studies on Intergradation in the Genitalia of Certain Populations of *Culex pipiens* and *Culex quinquefasciatus* in the United States. Amer. Jour. Hyg., 50:307-314.

680. Tate, H. D., and D. B. Gates. 1944. The Mosquitoes of Nebraska. Univ. Neb. Agr. Expm. Sta. Res. Bull., 133:1-27.

681. Tate, H. D., and W. W. Wirth. 1942. Notes on the Mosquitoes in Nebraska (Diptera: Culicidae). Ent. News, 53:211-215.

682. Theobald, F. V. 1901-1910. A Monograph of the Culicidae or Mosquitoes. London: British Museum of Natural History, 5 vols., 2:1-459.

683. Thibault, J. K. 1910. Notes on the Mosquitoes of Arkansas. Proc. Ent. Soc. Wash., 12:13-26.

684. Thompson, G. A. 1946. Occurrence of *Anopheles atropos* in Jamaica, B.W.I. Mosquito News, 6:193.

685. Thompson, G. A. 1947. A List of the Mosquitoes of Jamaica, B.W.I. Mosquito News, 7:78-80.

686. Thompson, G. A. 1953. Observations of Early Spring Activity of *Culiseta inornata* (Williston) (Diptera, Culicidae) in South Central Nebraska. Mosquito News, 13:17.

687. Thompson, G. A., B. F. Howitt, R. Gorrie, and T. A. Cockburn. 1951. Encephalitis in Midwest. VI. Western Equine Encephalomyelitis Virus Isolated from *Aedes dorsalis* Meigen. Proc. Soc. Expm. Biol. and Med., 78:289-290.

688. Thurman, E. B., J. S. Haeger, and J. A. Mulrennan. 1949. The Occurrence of *Aedes (Ochlerotatus) thelcter* Dyar in the Florida Keys. Mosquito News, 9:171-172.

689. Thurman, E. B., J. S. Haeger, and J. A. Mulrennan. 1951. The Taxonomy and Biology of *Psorophora (Janthinosoma) johnstonii* (Grabham, 1905). Ann. Ent. Soc. Amer., 44:144-147.

690. Thurman, E. B., and P. T. Johnson. 1950. The Taxonomic Characters of the Larvae of the Genus *Culiseta* Felt, 1904 in California (Diptera, Culicidae). Pan-Pac. Ent., 26:179-187.

691. Thurman, E. B., and E. C. Winkler. 1950. A New Species of Mosquito in California, *Aedes*

(Ochlerotatus) bicristatus (Diptera, Culicidae). Proc. Ent. Soc. Wash., 52:237-250.

692. Thurman, D. C. 1948. A Far South Record of Anopheles quadrimaculatus Say in Florida. Mosquito News, 8:19.

693. Thurman, D. C. 1950. The Ecology of Aedes Mosquitoes in California. Calif. Mosquito Control Assoc. Ann. Confr. Proc. and Papers, 18:62-65.

694. Thurman, D. C., and E. W. Mortenson. 1950. A Method of Obtaining an Index to Aedes Densities in Irrigated Pastures. Mosquito News, 10:199-201.

695. Thurman, D. C., L. J. Ogden, and D. E. Eyles. 1945. A United States Record for Culex interrogator. Jour. Econ. Ent., 38:115.

696. Trembley, H. L. 1945. Laboratory Rearing of Aedes atropalpus. Jour. Econ. Ent., 38:408-409.

697. Trembley, H. L. 1946. Aedes atropalpus (Coq.), A New Mosquito Vector of Plasmodium gallinaceum Brumpt. Jour. Parasitol., 32:499-501.

698. Trembley, H. L. 1947. Biological Characteristics of Laboratory Reared Aedes atropalpus. Jour. Econ. Ent., 40:244-250.

699. Tulloch, G. S. 1934. Mosquito Investigations in Alaska. Psyche, 41:201-210.

700. Tulloch, G. S. 1937. The Mosquitoes of Puerto Rico. Jour. Agr. Univ. P.R., 21:137-168.

701. Tulloch, G. S. 1939. A Key to the Mosquitoes of Massachusetts. Psyche, 46:113-136.

702. Turner, R. L. 1924. A New Mosquito from Texas (Diptera, Culicidae). Ins. Ins. Mens., 12:84.

703. Twinn, C. R. 1925-26. Notes on the Mosquito Fauna of Quebec. Rpt. Soc. Protection Plants, 84-92.

704. Twinn, C. R. 1926. Notes on the Mosquitoes of the Ottawa District. Can. Ent., 58:108-111.

705. Twinn, C. R. 1927. Mosquitoes from Baffin Land. Can. Ent., 59:47-49.

706. Twinn, C. R. 1931. Notes on the Biology of Mosquitoes of Eastern Canada. N.J. Mosq. Exterm. Assoc. Proc., 18:10-22.

707. Twinn, C. R. 1945. Report of a Survey of Anopheline Mosquitoes in Canada in 1944. N.J. Mosq. Exterm. Assoc. Proc., 32:242-251.

708. Twinn, C. R. 1949. Mosquitoes and Mosquito Control in Canada. Mosquito News, 9:35-41.

709. Twinn, C. R. 1950. Studies of the Biology and Control of Biting Flies in Northern Canada. Jour. Arctic Inst. N.A., 3:14-26.

710. Twinn, C. R. 1953. Mosquitoes and Their Control in Prince Edward Island. Mosquito News, 13:185-190.

711. Twinn, C. R., B. Hocking, W. C. McDuffie, and H. F. Cross. 1948. A Preliminary Account of the Biting Flies at Churchill, Manitoba. Can. Jour. Res., D, 26:334-357.

712. Vargas, L. 1939. Datos Acerca del A. pseudopunctipennis y de un Anopheles nuevo de California. "Medicina," Rev. Mexico, 19:356-362.

713. Vargas, L. 1940. Clava para Indentificar las Larvas de Anopheles Mexicanos. Ciencia, 1:66-68.

714. Vargas, L. 1940. Anopheles (Anopheles) barberi en Mexico. Rev. Inst. Salub. Enf. Trop., 1:319-322.

715. Vargas, L. 1941. New Variety of Anopheles pseudopunctipennis (Diptera, Culicidae). Bull. Brooklyn Ent. Soc., 36:73-74.

716. Vargas, L. 1941. El Problema de Las Variedades de A. punctipennis. Descripcion de A. punctipennis stonei n. var. Rev. Soc. Mexico Hist. Nat., 2:175-186.

717. Vargas, L. 1942. El Huevo de Anopheles barberi Coquillett, 1903. Rev. Inst. Salub. Enf. Trop., 3:329-331.

718. Vargas, L. 1943. Anopheles earlei Vargas, 1942, n. sp. Norteamericana del Grupo Maculipennis. Bol. Of. Sanit. Panamer., 22:8-12.

719. Vargas, L. 1943. El "Grupo Maculipennis" del Nuevo Mendo y el Anopheles earlei. Rev. Inst. Salub. Enf. Trop., 4:279-286.

720. Vargas, L. 1948. Bionomia Del Genero Aedes. Notas Sobre Los Aedes Mexicanos. Rev. Soc. Mexico Hist. Nat., 9:91-119.

721. Vargas, L. 1949. Lista de Las Espacies Americanas de Aedes con Referencias a Los Subgeneros y a la Distribucion Geografica. Rev. Inst. Salub. Enf. Trop., 10:327-344.

722. Vargas, L. 1949. Lista del Sinonimos de los Aedes Americanos (Diptera, Culicidae). Rev. Soc. Mexico Hist. Nat., 10:219-224.

723. Vargas, L. 1949. Nueva Larva que Corresponde Probablemente a Aedes purpureipes Aitken, 1941. Rev. Inst. Salub. Enf. Trop., 10:261-265.

724. Vargas, L. 1950. Nota Sobre Aedes. Revision de Algunos Ochlerotatus Americanos, Kummyia n. Subgen. (Diptera, Culicidae). Rev. Soc. Mexico Hist. Nat., 11:149-159.

725. Vargas, L. 1950. Los Subgeneros de Aedes: Downsiomyia n. Subgen. (Diptera: Culicidae). Rev. Inst. Salub. Enf. Trop., 11:61-69.

726. Vargas, L. 1950. Malaria Along the Mexico-United States Border. Bull. World Hlth. Orgn., 2:611-620.

727. Vargas, L. 1952. Relacion de Las Especies de Culex de Las Americas con Los Sinonimos Mas Importantes (Insecta: Diptera). Bol. Inst. Est. Med. Biol. Mexico, 9:167-206.

728. Vargas, L. 1953. Megarhinus de Norte America (Diptera, Culicidae). Rev. Inst. Salub. Enf. Trop., 13:27-32.

729. Vargas, L., and Robert Matheson. 1948. Estado Actual del Anopheles earlei Vargas 1943, y Anopheles occidentalis Dyar and Knab, 1906, con Claves para Larvas, Pupas y Adultos del Llamado Complejo maculipennis de Norteamerica. Rev. Inst. Salub. Enf. Trop., 9:27-33.

730. Vargas, L., and A. M. Palacios. 1950. Estudio Taxonomico de los Mosquitos *Anofelinos* de Mexico. Mexico, D. F., 143 pp.

731a. Venard, C. E., and F. W. Mead. 1953. An Annotated List of Ohio Mosquitoes. Ohio Jour. Sci., *53*:327-331.

731. Viosca, Percy. 1924. Report of the Entomologist. Parish of New Orleans and City of New Orleans. Bd. Hlth. Ann. Rpt., 1923, 31-47.

732. Viosca, Percy. 1925. A Bionomical Study of the Mosquitoes of New Orleans and Southeastern Louisiana. N.J. Mosq. Exterm. Assoc. Proc., *12*:34-50.

733. Vockeroth, J. R. 1950. Specific Characters in Tarsal Claws of Some Species of *Aedes* (Diptera, Culicidae). Can. Ent., *82*:160-162.

734. Vockeroth, J. R. 1952. The Specific Status of *Aedes pionips* Dyar. Can. Ent., *84*:243-247.

734a. Vockeroth, J. R. 1954. Notes on Northern Species of *Aedes*, with Descriptions of Two New Species (Diptera, Culicidae). Can. Ent., *86*:109-116.

735. Vogt, G. G. 1947. Salinity Tolerance of *Anopheles quadrimaculatus* and Habitat Preference of *A. crucians bradleyi*. Jour. Econ. Ent., *40*: 320-325.

736. Wallis, R. C., and A. Spielman. 1953. Laboratory Rearing of *Culex salinarius* (Diptera, Culicidae). Proc. Ent. Soc. Wash., *55*:140-142.

737. Wanamaker, J. F. 1944. An Improved Method for Mounting Mosquito Larvae. Amer. Jour. Trop. Med., *24*:385-386.

738. Wanamaker, J. F., R. W. Chamberlain, and S. J. Carpenter. 1944. Distribution of *Culex pipiens* in the Southeastern United States. Jour. Econ. Ent., *37*:106-107.

739. Warren, M., and S. O. Hill. 1947. Gynandromorphism in Mosquitoes. Jour. Econ. Ent., *40*:139.

740. Weathersbee, A. A. 1944. A Note on the Mosquito Distribution Records of Puerto Rico and of the Virgin Islands. Puerto Rico Jour. Pub. Hlth. and Trop. Med., June 1944, 643-645.

741. Weathersbee, A. A. 1945. Mosquito Control at the Naval Base, Norfolk, Va. N.J. Mosq. Exterm. Assoc. Proc., *32*:96-99.

742. Weathersbee, A. A., and F. T. Arnold. 1947. A Resume of the Mosquitoes of South Carolina. Jour. Tenn. Acad. Sci., *22*:210-229.

743. Weathersbee, A. A., and F. T. Arnold. 1948. A Revision of the Genus *Wyeomyia* Theobald (Diptera, Culicidae) in South Carolina. Jour. Tenn. Acad. Sci., *23*:249-250.

744. Weathersbee, A. A., F. T. Arnold, and M. M. Askey. 1951. Additional Mosquito Distribution Records for South Carolina. Jour. Tenn. Acad. Sci., *26*:79-84.

745. Weber, N. A. 1950. A Survey of the Insects and Related Arthropods of Arctic Alaska. Part I. Amer. Ent. Soc. Trans., *76*:147-206.

746. Weidhaas, J. A. 1952. A First Record of *Aedes diantaeus* H., D., and K. for Massachusetts, with Notes on Associated Species. Mosquito News, *12*:8-9.

747. Welch, E. V. 1945. The Finding of *Anopheles albimanus* Wied. at West Palm Beach, Florida (Diptera, Culicidae). Mosquito News, *5*:145.

748. Wellington, W. G. 1944. The Effect of Ground Temperature Inversions Upon the Flight Activity of *Culex* sp. (Diptera, Culicidae). Can. Ent., *76*:223.

749. Wesenberg-Lund, C. 1920-1921. Contributions to the Biology of the Danish Culicidae, Copenhagen: A. F. Host & Son, 210 pp.

750. Whitehead, F. E. 1951. Host Preference of *Psorophora confinnis* and *P. discolor*. Jour. Econ. Ent., *44*:1019.

751. Wigglesworth, V. B. 1933. The Function of the Anal Gills of the Mosquito Larva. Jour. Expm. Biol., *10*:16-26.

752. Wigglesworth, V. B. 1933. The Adaptation of Mosquito Larvae to Salt Water. Jour. Expm. Biol., *10*:27-37.

753. Wigglesworth, V. B. 1949. The Physiology of Mosquitoes. *In* Boyd, Malariology, *1*:204-301.

754. Wilkins, O. P., and O. P. Breland. 1949. Recovery of the Mosquito *Culiseta inornata* (Williston) from Dry Material (Diptera, Culicidae). Proc. Ent. Soc. Wash., *51*:27-28.

755. Wilkins, O. P., and O. P. Breland. 1951. The Larval Stages and the Biology of the Mosquito, *Orthopodomyia alba* Baker (Diptera, Culicidae). Jour. N.Y. Ent. Soc., *59*:225-240.

756. Wilson, C. A., R. C. Barnes, and H. L. Fellton. 1946. A List of the Mosquitoes of Pennsylvania with Notes on their Distribution and Abundance. Mosquito News, *6*:78-84.

757. Wirth, W. W. 1944. Notes on the Occurrence of *Anopheles georgianus* King in Louisiana. Jour. Econ. Ent., *37*:446.

758. Wirth, W. W. 1945. The Occurrence of *Culex (Melanoconion) elevator* Dyar and Knab in Florida with Keys to the Melanoconions of the United States (Diptera, Culicidae). Proc. Ent. Soc. Wash., *47*:201-210.

759. Wirth, W. W. 1947. Notes on the Mosquitoes of Louisiana. Jour. Econ. Ent., *40*:742-744.

760. Wishart, G., and H. G. James. 1945. Notes on the Anopheline Mosquitoes of the Kingston, Trenton, and Peterborough, Ontario, Areas. Ann. Rept. Ent. Soc. Ontario, *76*:39-48.

761. Wray, F. C. 1946. Six Generations of *Culex pipiens* without a Blood Meal. Mosquito News, *6*:71-72.

762. Yamaguti, Satyu, and W. J. LaCasse. 1951. Mosquito Fauna of North America. Parts I-V,

629 pp. Office of the Surgeon, Hq. Japan Logistical Command.

763. Yates, W. W. 1943. Variations Noted in Anatomical Larval Structures of *Culex tarsalis* Coq. (Diptera: Culicidae). Proc. Ent. Soc. Wash., *45*:180-181.

764. Young, D. A. 1948. Notes on a *Mansonia perturbans* Problem at Edgewood Arsenal, Maryland. Mosquito News, *8*:57-60.

765. Young, F. N., and W. N. Christopher. 1944. Unusual Breeding Places of Mosquitoes in the Vicinity of Keesler Field, Mississippi. Amer. Jour. Trop. Med., *24*:379.

766. Young, M. D. 1945. Studies of Foreign Malarias. Field Bull., MCWA, Aug.-Sept., U.S. Pub. Hlth. Serv., Atlanta, Ga.

767. Young, M. D., and R. W. Burgess. 1947. The Transmission of *P. malariae* by *Anopheles m. freeborni*. Amer. Jour. Trop. Med., *27*:39-40.

768. Young, M. D., T. H. Stubbs, J. A. Moore, F. C. Ehrman, N. F. Hardman, J. M. Ellis, and R. W. Burgess. 1945. Studies of Imported Malarias: I. Ability of Domestic Mosquitoes to Transmit Vivax Malaria of Foreign Origin. Jour. Nat. Mal. Soc., *4*:127-132.

769. Zukel, J. W. 1945. Marking *Anopheles* Mosquitoes with Fluorescent Compounds. Science, *102*:157.

770. Zukel, J. W. 1949. A Winter Study of *Anopheles* Mosquitoes in Southwestern Georgia, with Notes on Some Culicine Species. Jour. Nat. Mal. Soc., *8*:224-233.

Systematic Index

(Families, subfamilies, and tribes shown in capitals; valid genera and subgenera in bold type; valid species in roman; all synonyms in italics; bold numerals refer to main references.)

Plates

The adult mosquitoes in these plates are shown enlarged to a maximum page size. There is, thus, no relationship in size of each species shown.

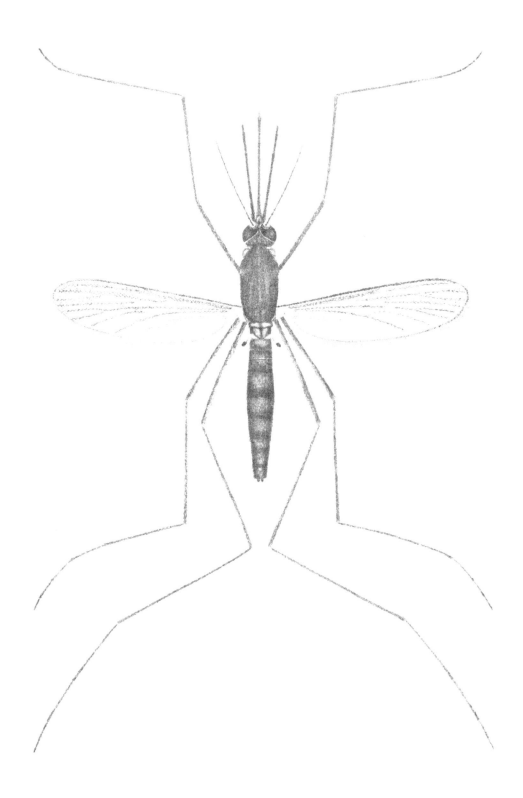

Plate 1. *Anopheles atropos* Dyar and Knab, female.

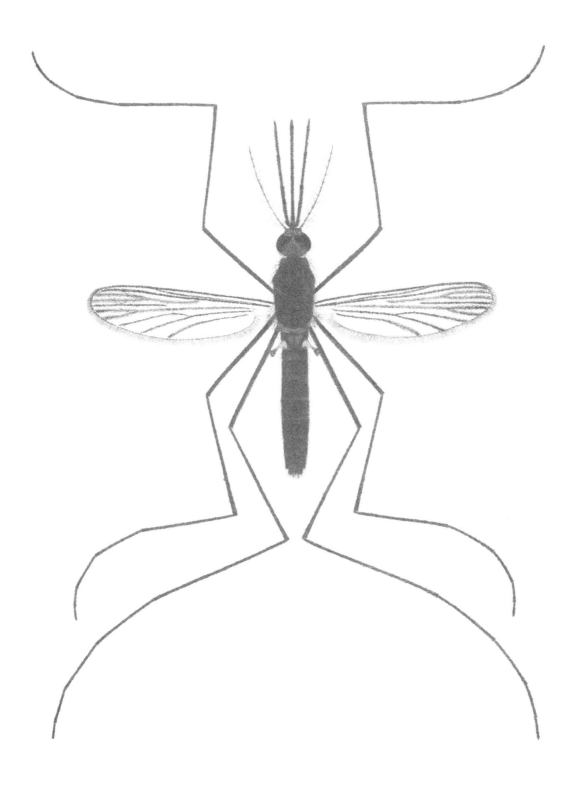

Plate 2. *Anopheles barberi* Coquillett, female.

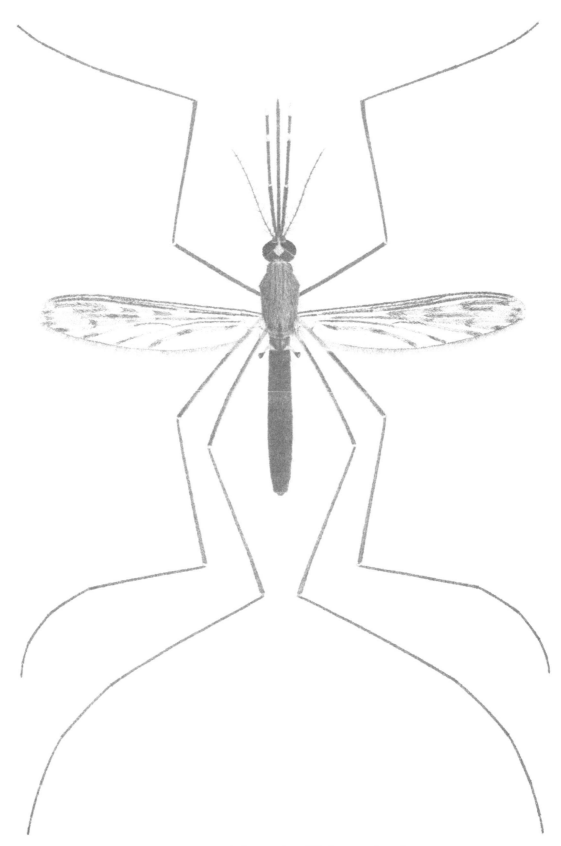

Plate 3. *Anopheles crucians* Wiedemann, female.

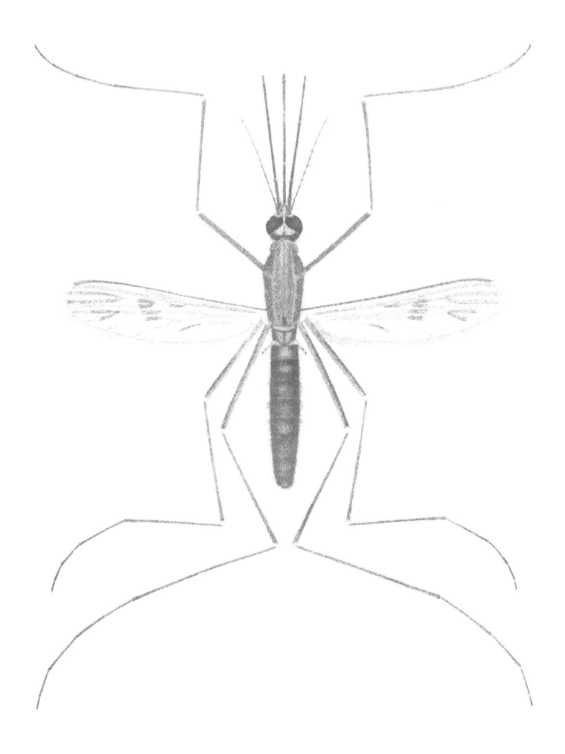

Plate 4. *Anopheles earlei* Vargas, female.

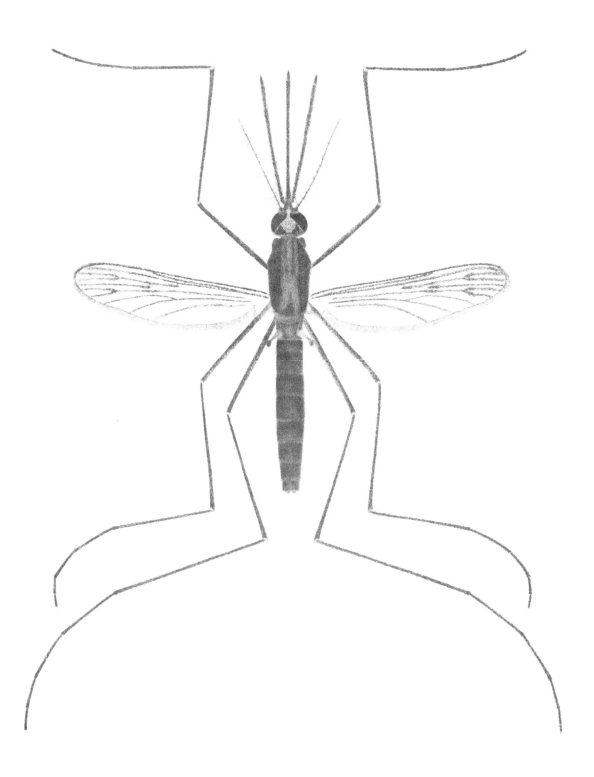

Plate 5. *Anopheles freeborni* Aitken, female.

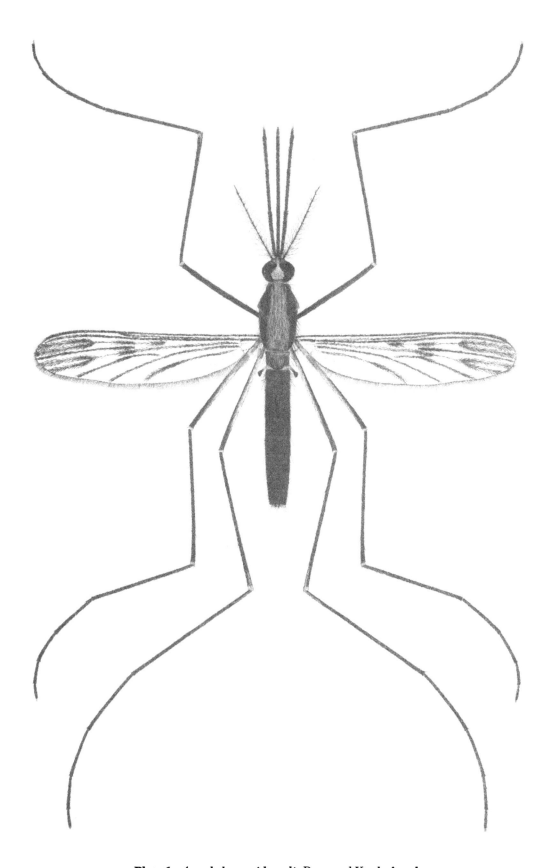

Plate 6. *Anopheles occidentalis* Dyar and Knab, female.

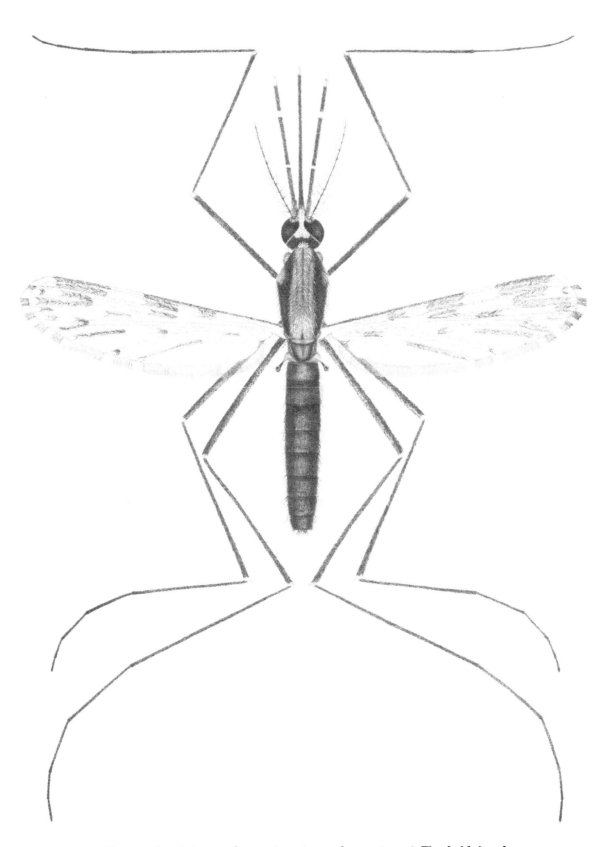

Plate 7. *Anopheles pseudopunctipennis pseudopunctipennis* Theobald, female.

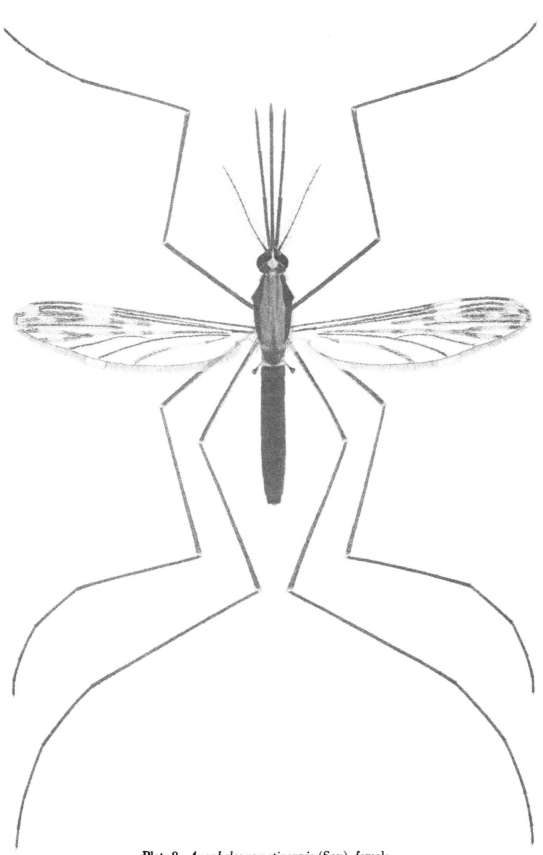

Plate 8. *Anopheles punctipennis* (Say), female.

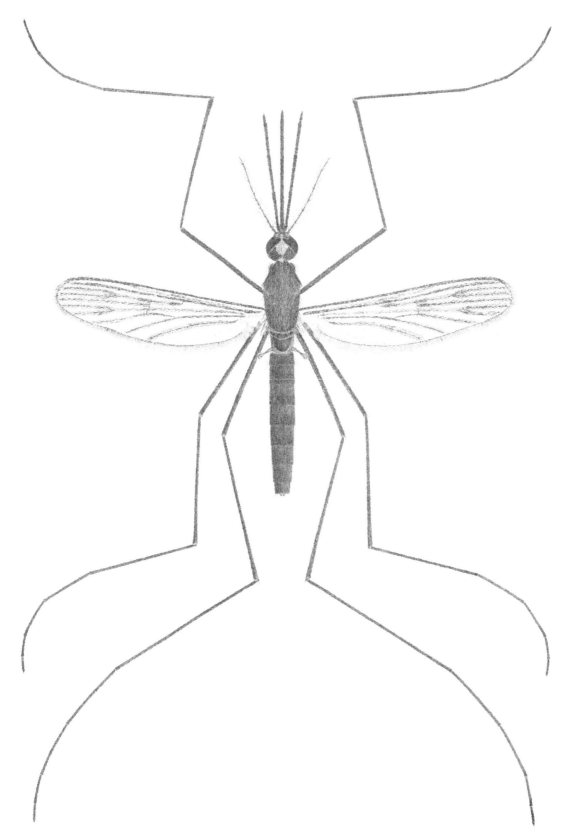

Plate 9. *Anopheles quadrimaculatus* Say, female.

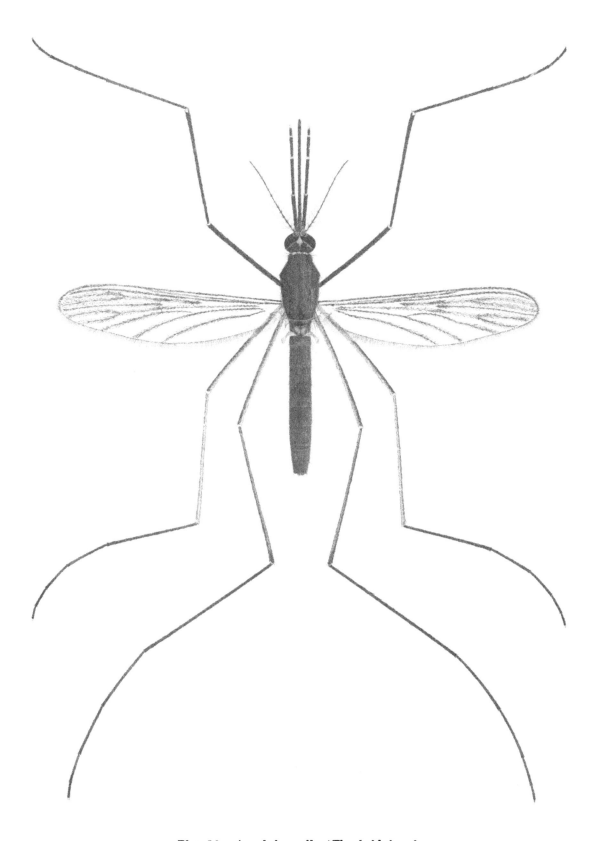

Plate 10. *Anopheles walkeri* Theobald, female.

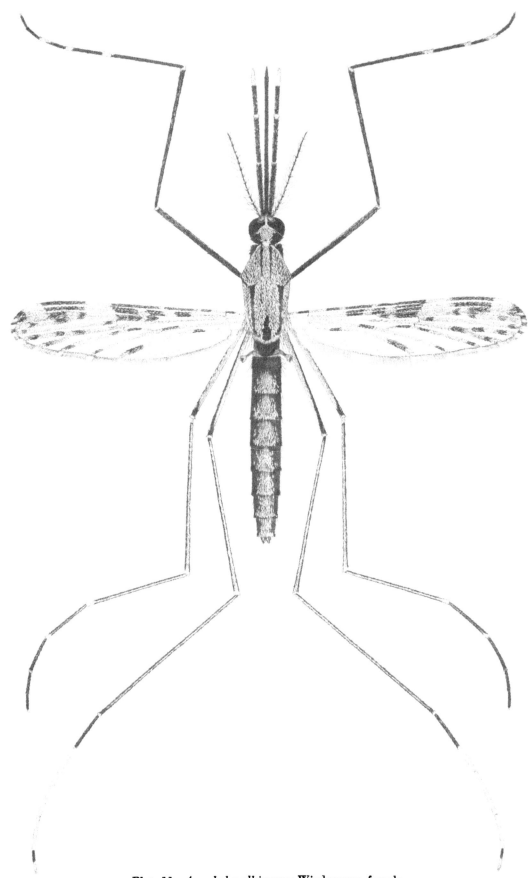

Plate 11. *Anopheles albimanus* Wiedemann, female.

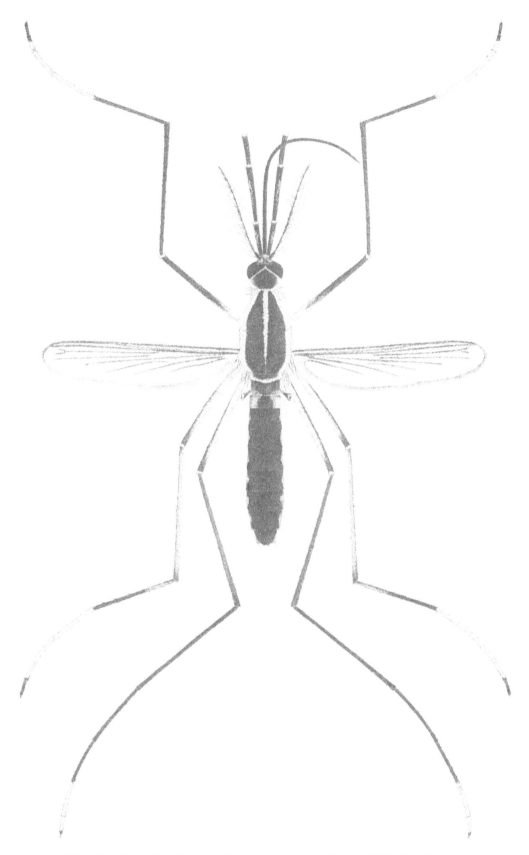

Plate 12. *Toxorhynchites rutilus septentrionalis* (Dyar and Knab), female.

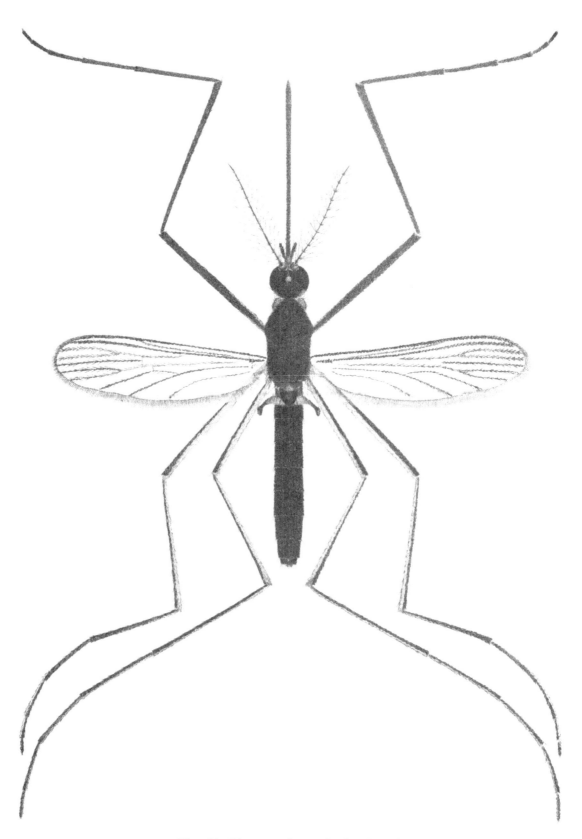

Plate 13. *Wyeomyia haynei* Dodge, female.

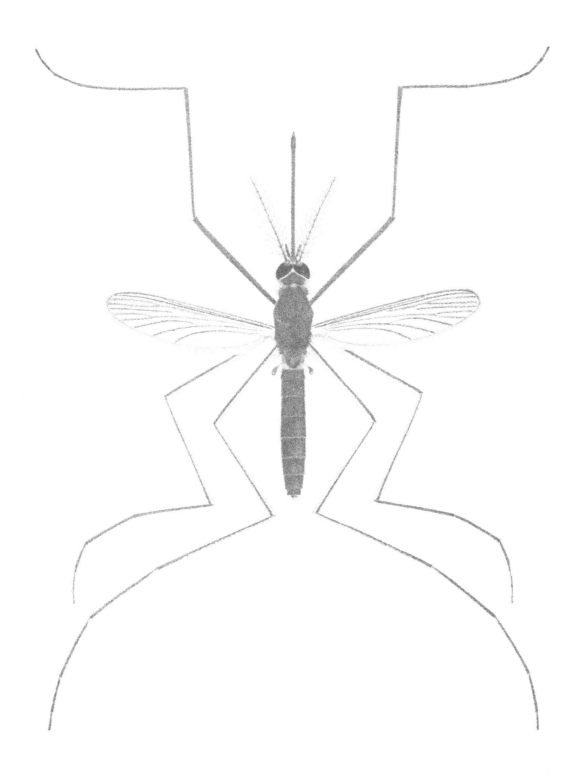

Plate 14. *Wyeomyia mitchellii* (Theobald), female.

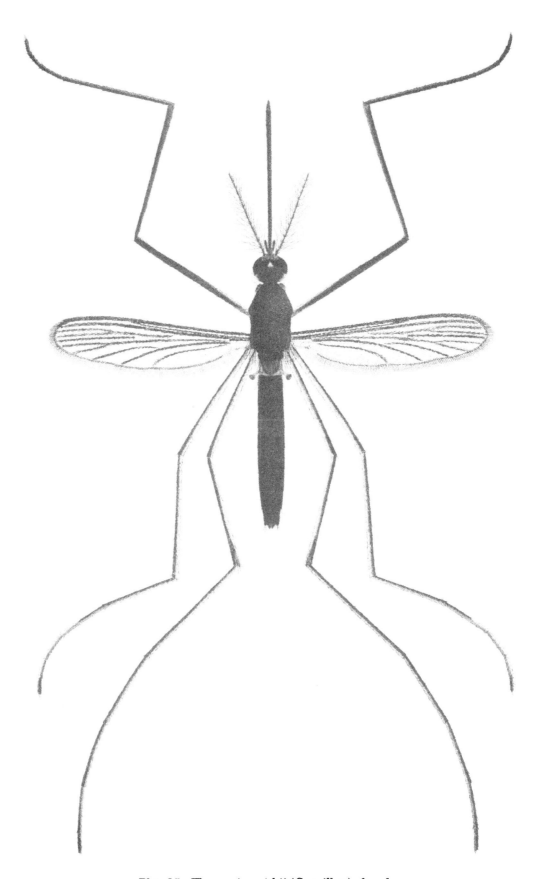

Plate 15. *Wyeomyia smithii* (Coquillett), female.

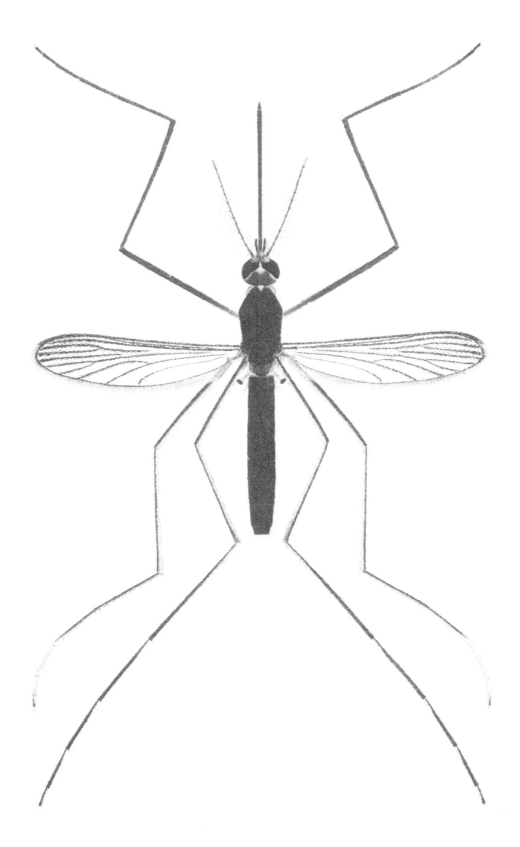

Plate 16. *Wyeomyia vanduzeei* Dyar and Knab, female.

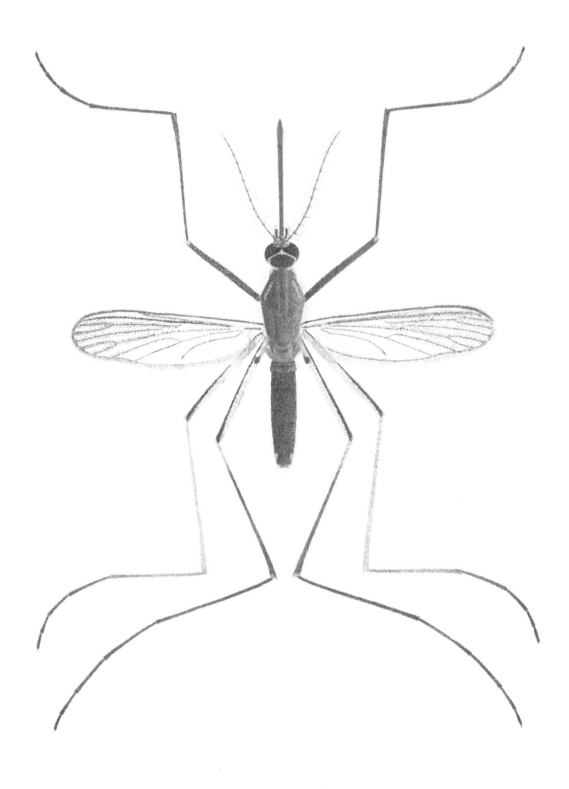

Plate 17. *Uranotaenia anhydor* Dyar, female.

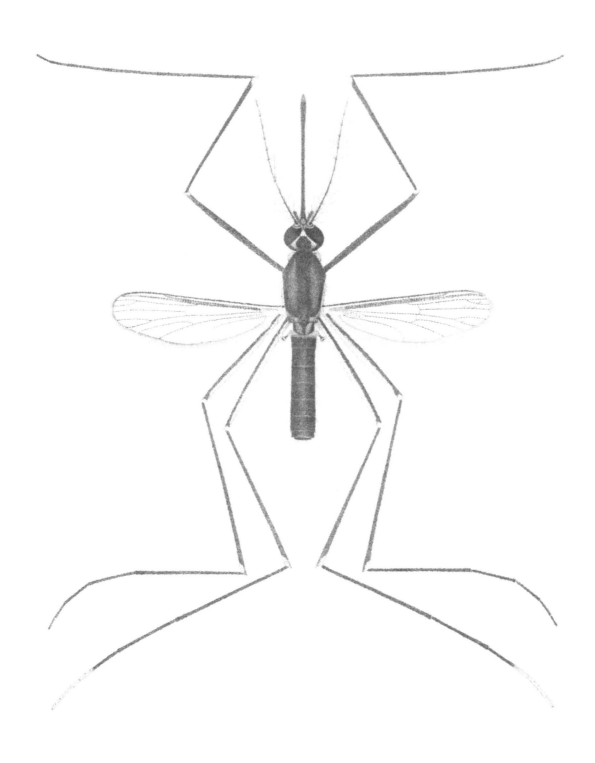

Plate 18. *Uranotaenia lowii* Theobald, female.

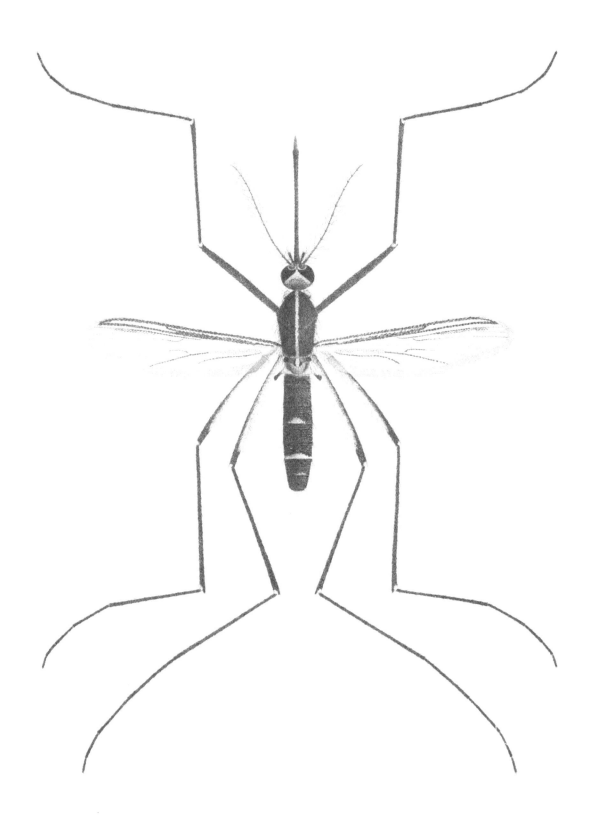

Plate 19. *Uranotaenia sapphirina* (Osten Sacken), female.

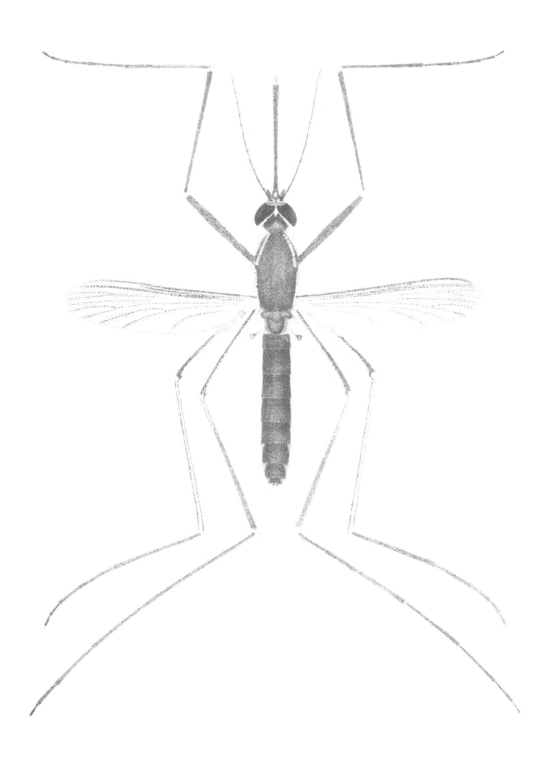

Plate 20. *Uranotaenia syntheta* Dyar and Shannon, female.

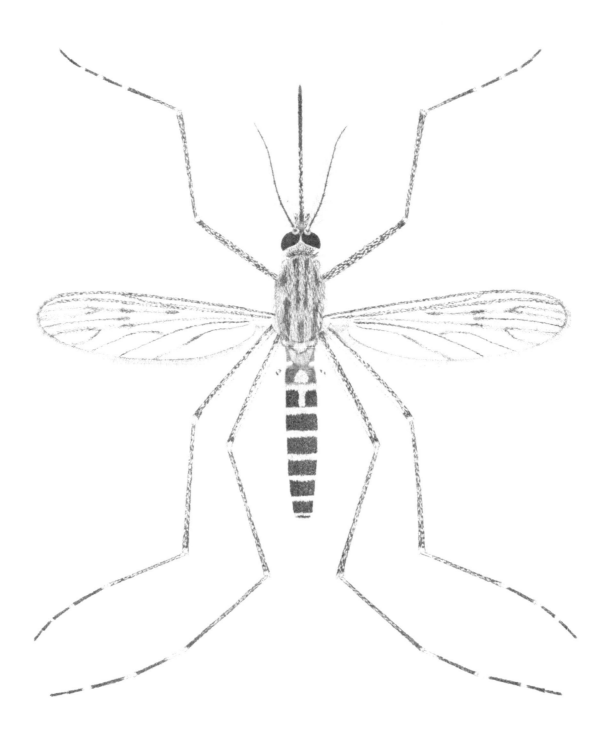

Plate 21. *Culiseta alaskaensis* (Ludlow), female.

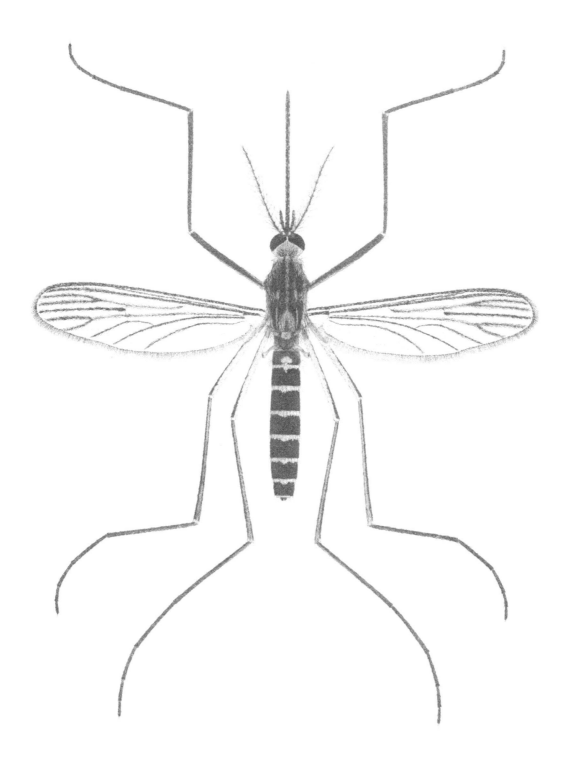

Plate 22. *Culiseta impatiens* (Walker), female.

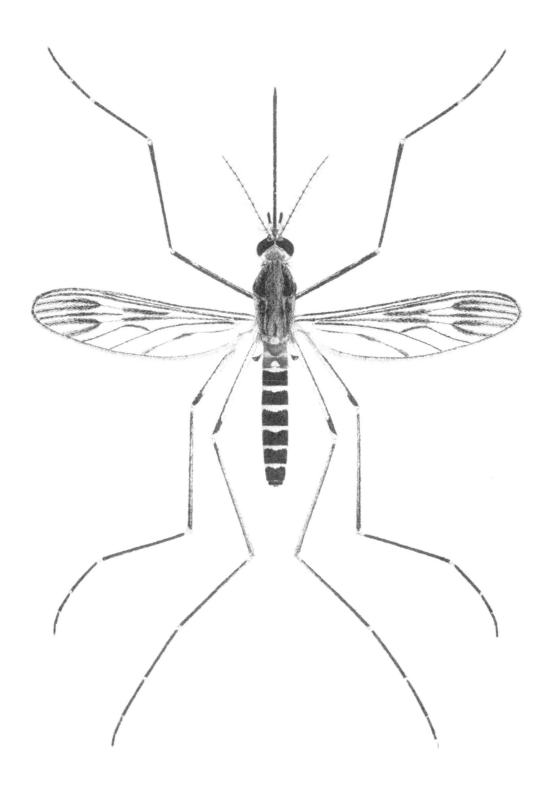

Plate 23. *Culiseta incidens* (Thomson), female.

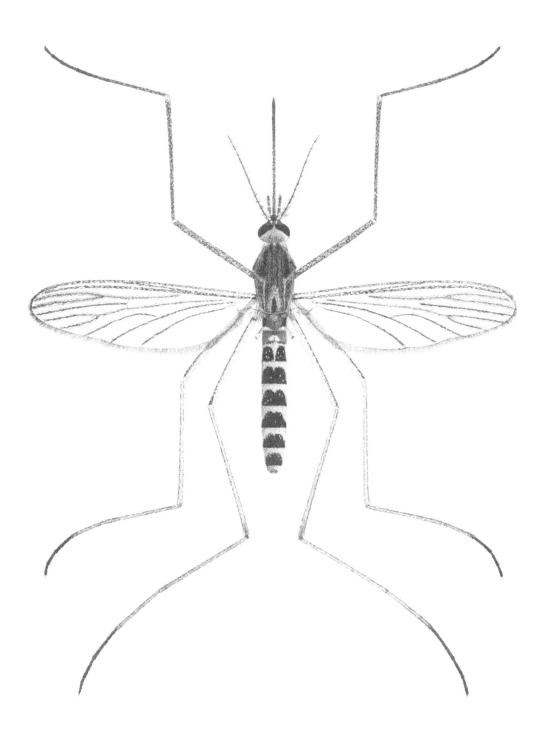

Plate 24. *Culiseta inornata* (Williston), female.

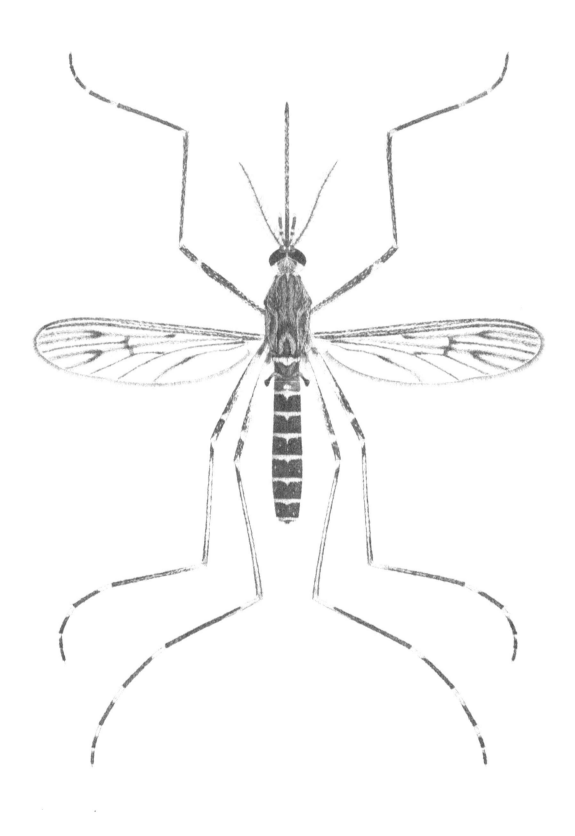

Plate 25. *Culiseta maccrackenae* Dyar and Knab, female.

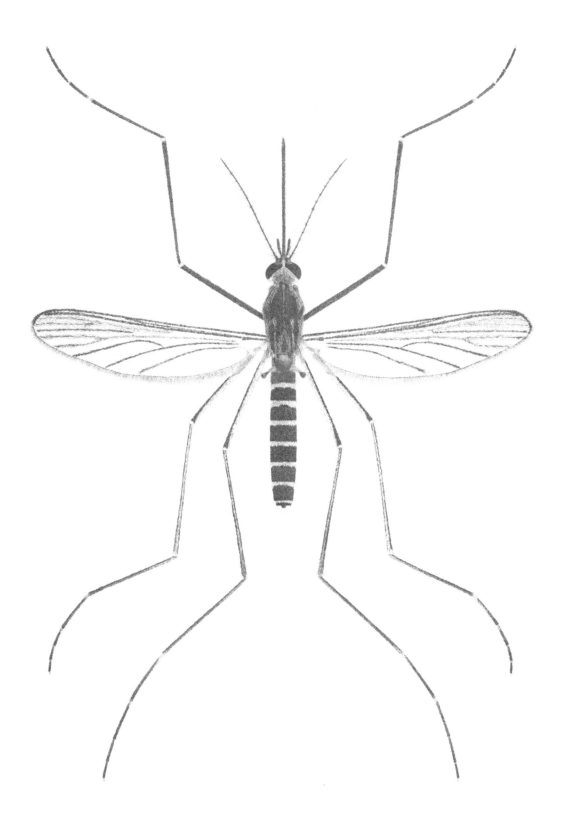

Plate 26. *Culiseta morsitans* (Theobald), female.

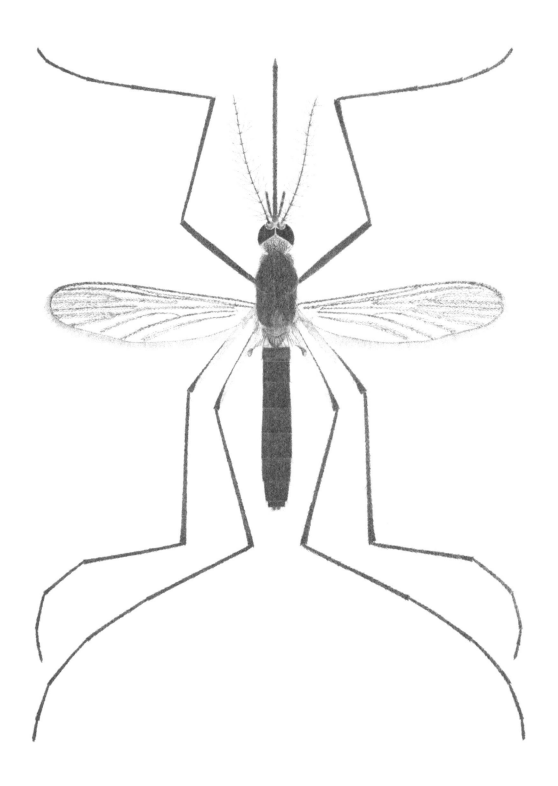

Plate 27. *Culiseta melanura* (Coquillett), female.

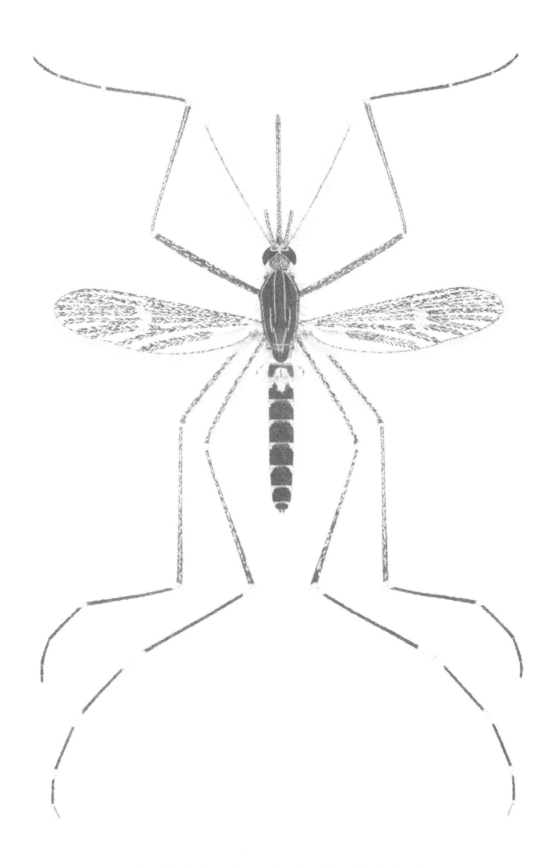

Plate 28. *Orthopodomyia signifera* (Coquillett), female.

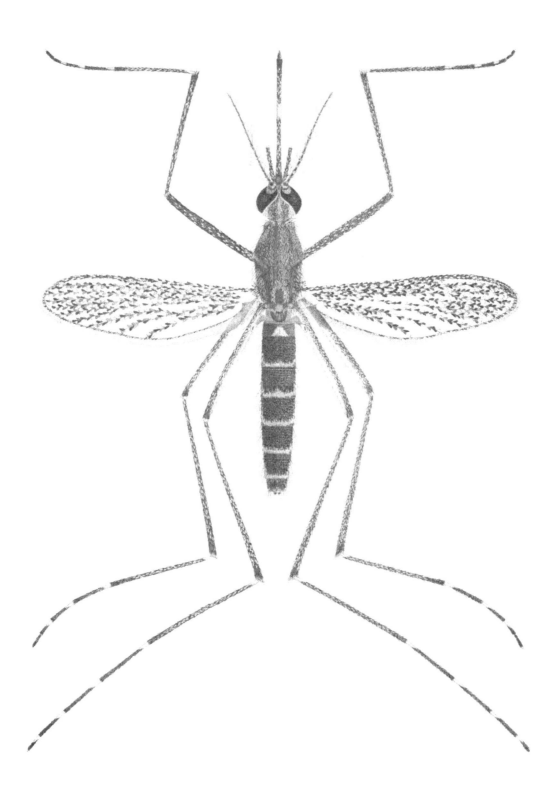

Plate 29. *Mansonia indubitans* Dyar and Shannon, female.

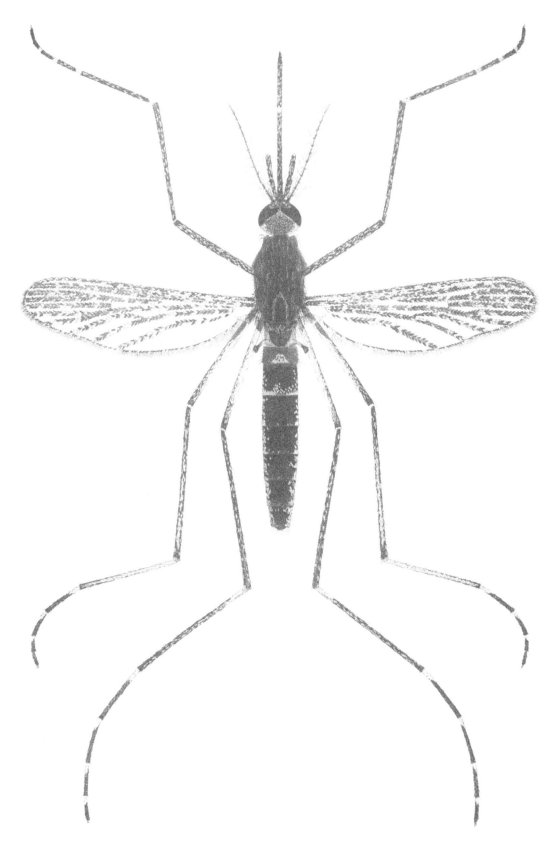

Plate 30. *Mansonia titillans* (Walker), female.

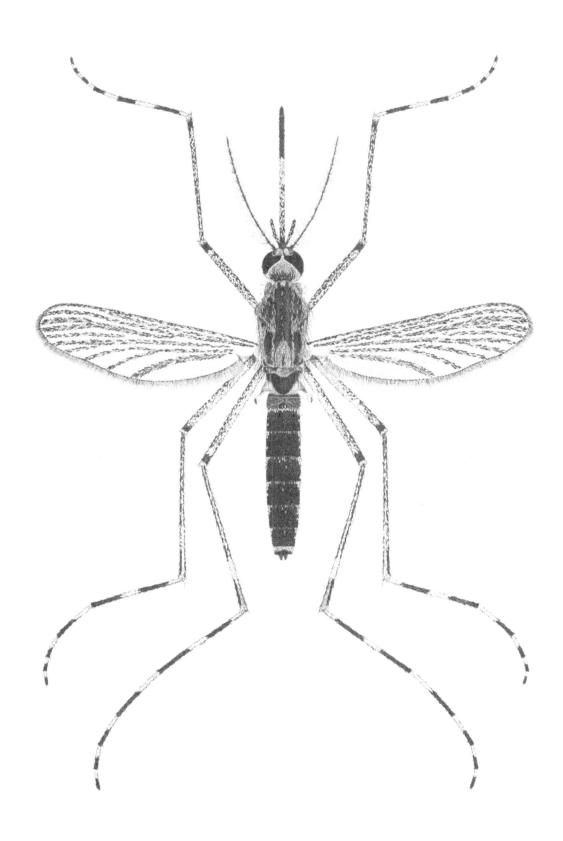

Plate 31. *Mansonia perturbans* (Walker), female.

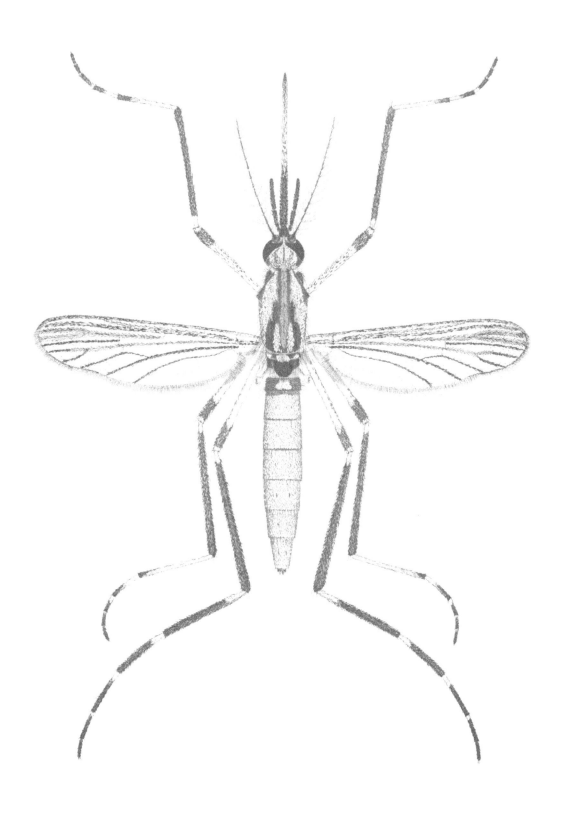

Plate 32. *Psorophora ciliata* (Fabricius), female.

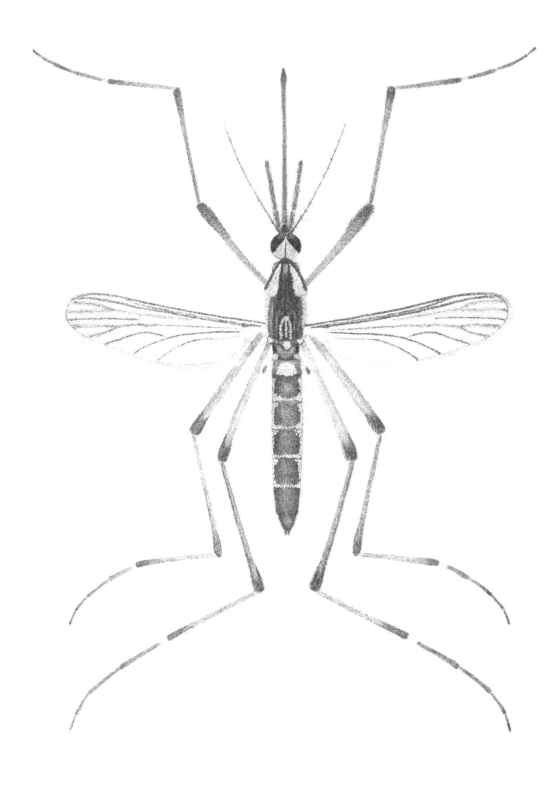

Plate 33. *Psorophora howardii* Coquillett, female.

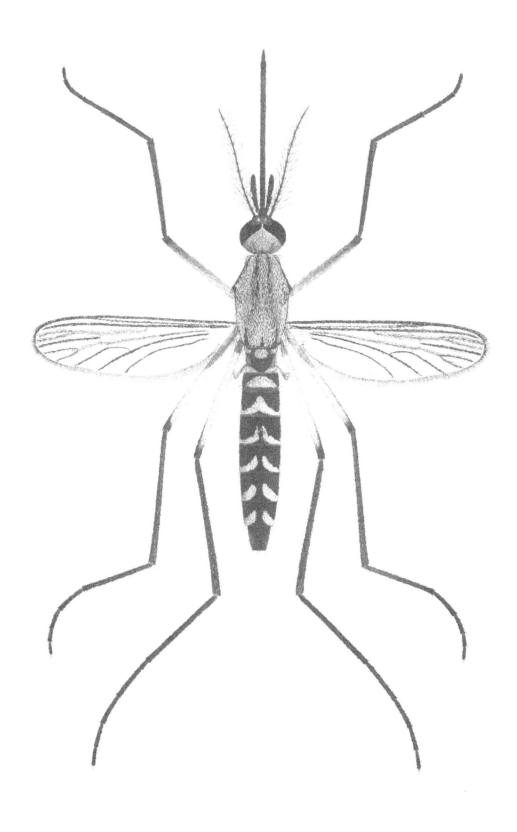

Plate 34. *Psorophora cyanescens* (Coquillett), female.

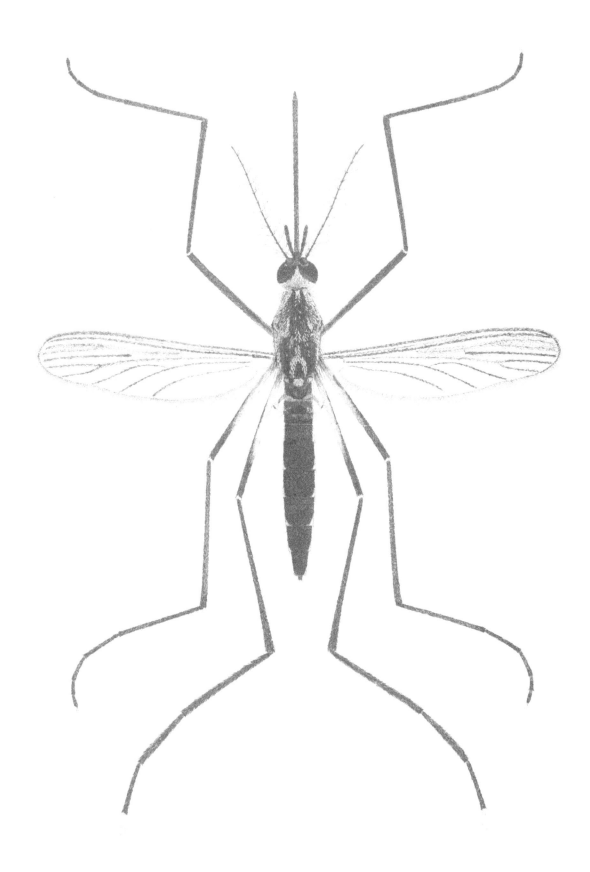

Plate 35. *Psorophora ferox* (Humboldt), female.

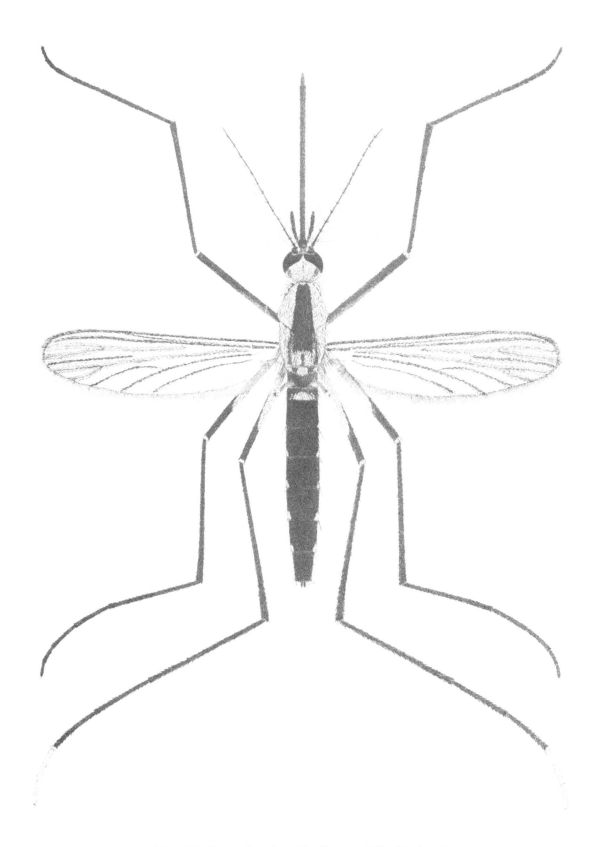

Plate 36. *Psorophora horrida* (Dyar and Knab), female.

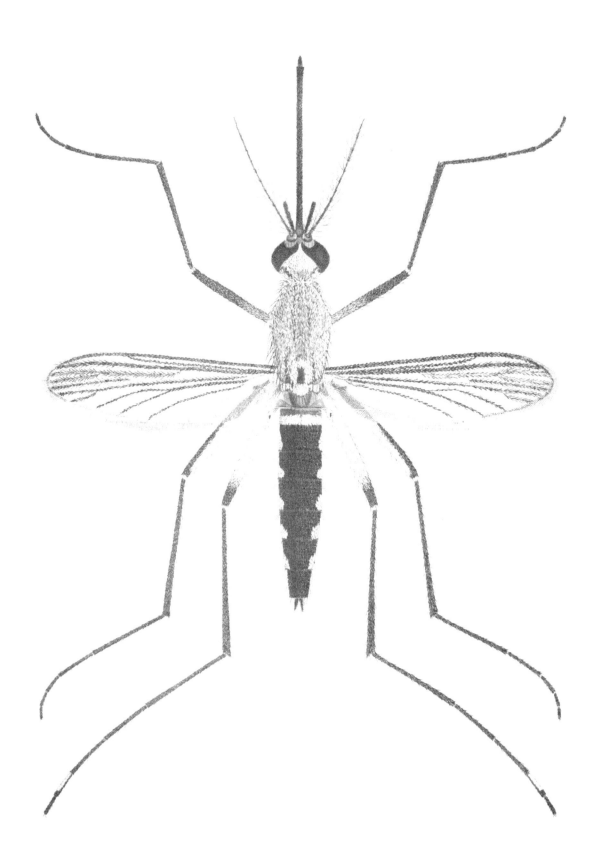

Plate 37. *Psorophora johnstonii* (Grabham), female.

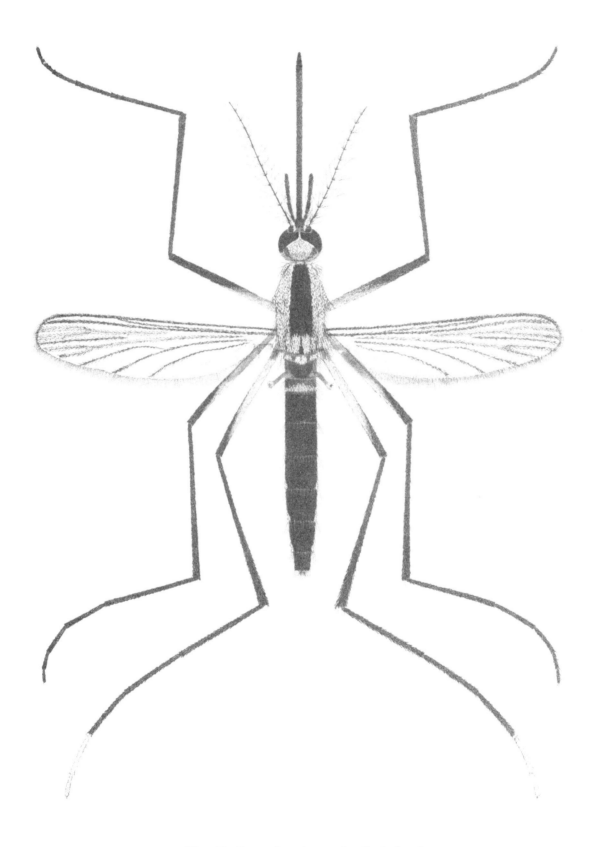

Plate 38. *Psorophora longipalpis* Roth, female.

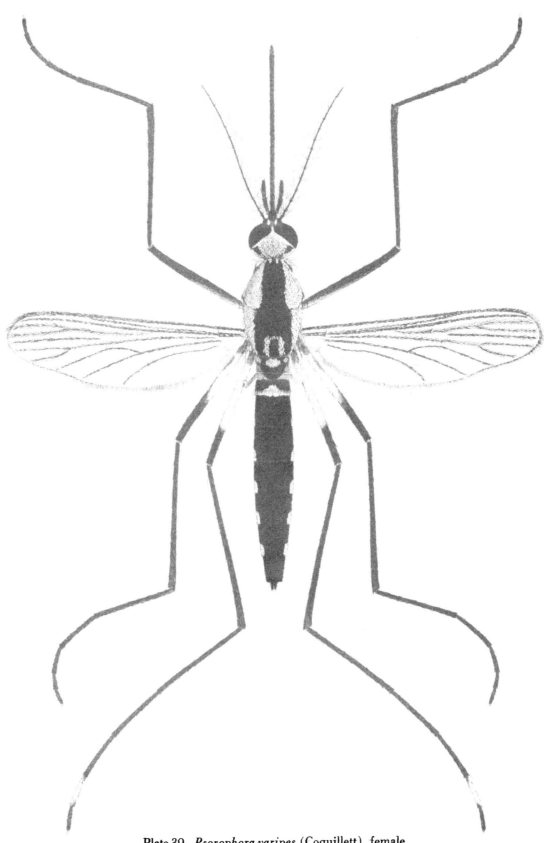

Plate 39. *Psorophora varipes* (Coquillett), female.

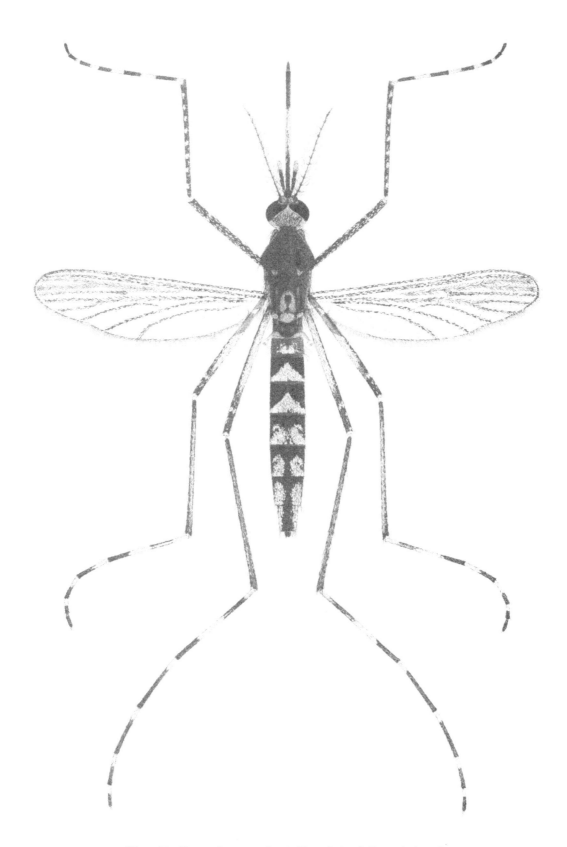

Plate 40. *Psorophora confinnis* (Lynch Arribálzaga), female.

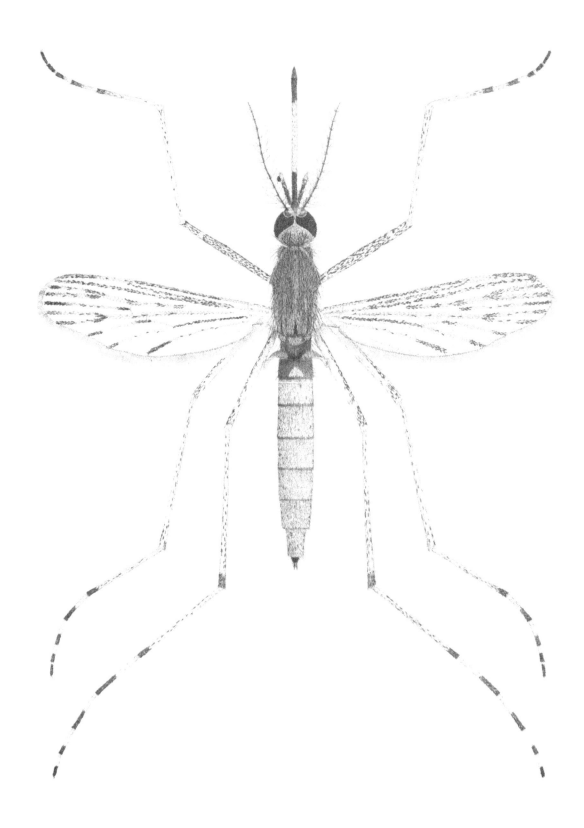

Plate 41. *Psorophora discolor* (Coquillett), female.

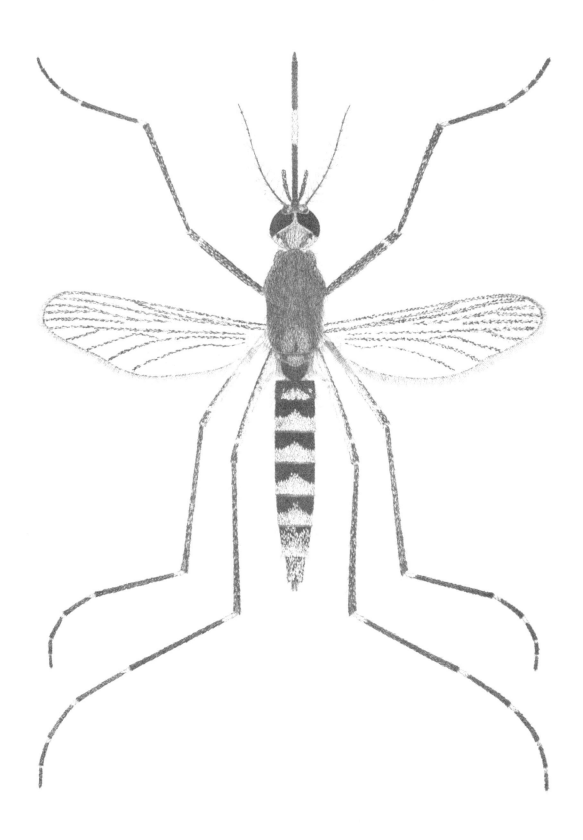

Plate 42. *Psorophora pygmaea* (Theobald), female.

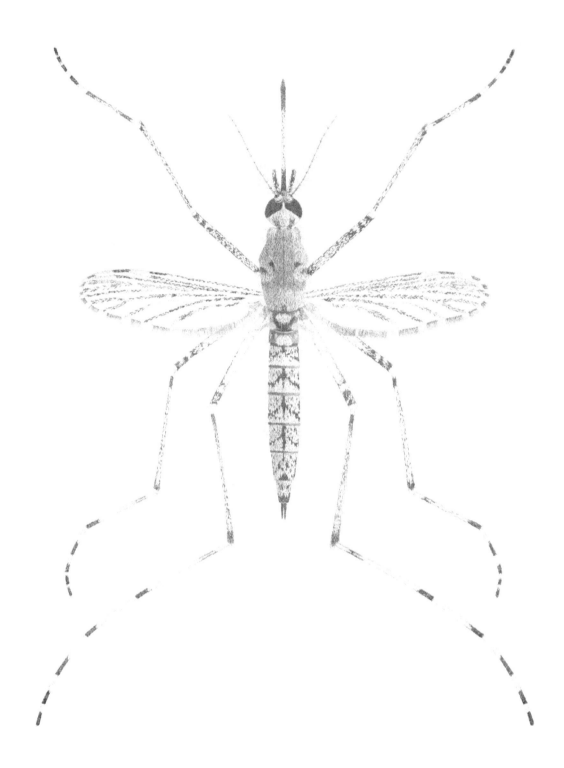

Plate 43. *Psorophora signipennis* (Coquillett), female.

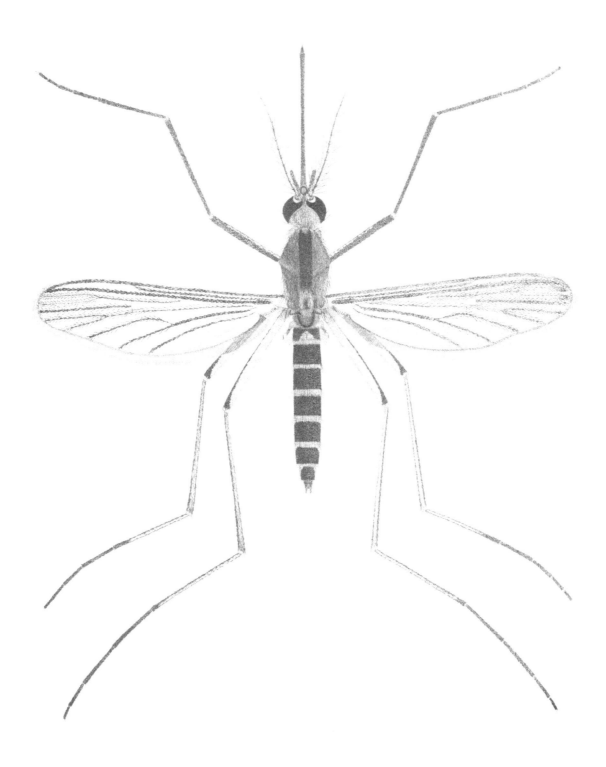

Plate 44. *Aedes aboriginis* Dyar, female.

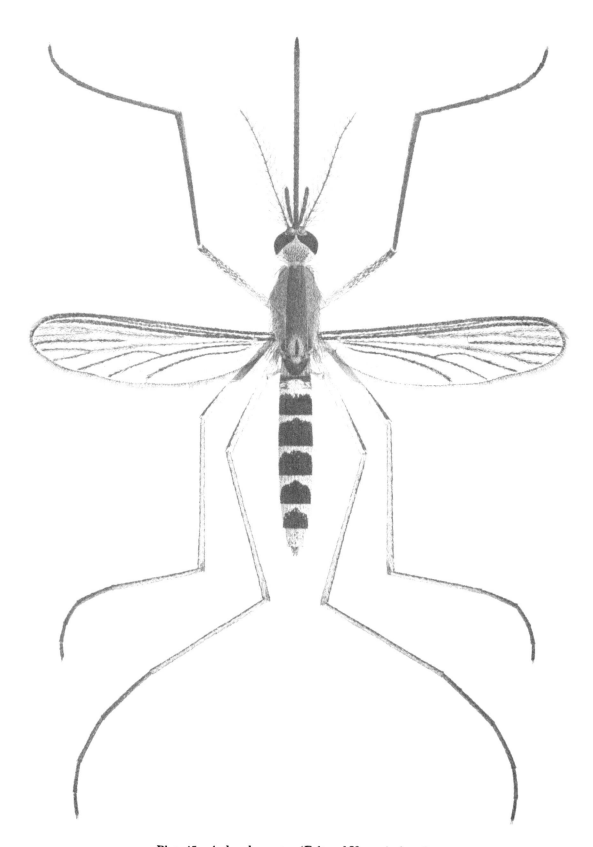

Plate 45. *Aedes abserratus* (Felt and Young), female.

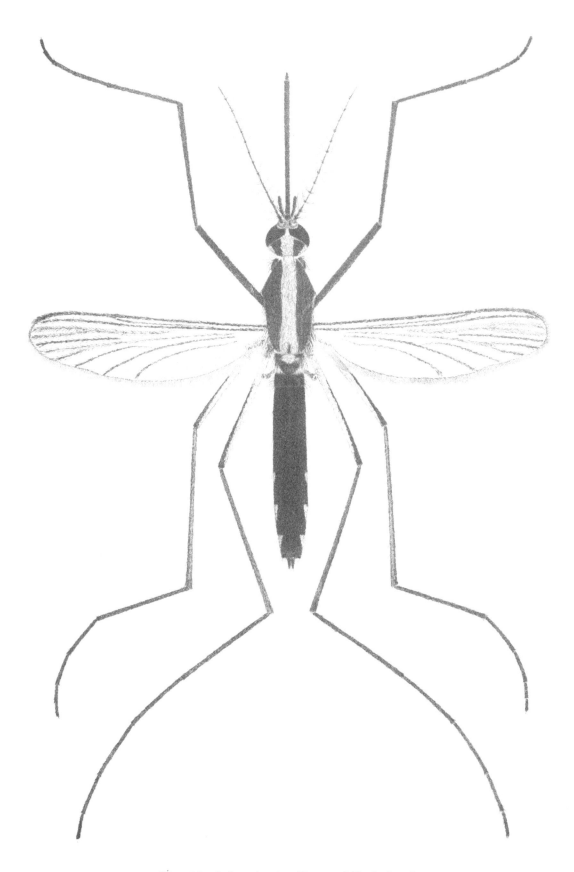

Plate 46. *Aedes atlanticus* Dyar and Knab, female.

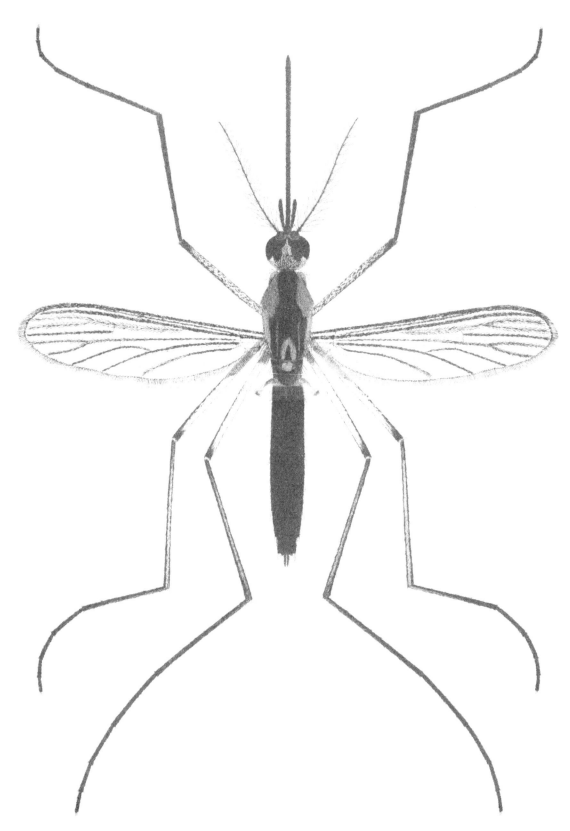

Plate 47. *Aedes aurifer* (Coquillett), female.

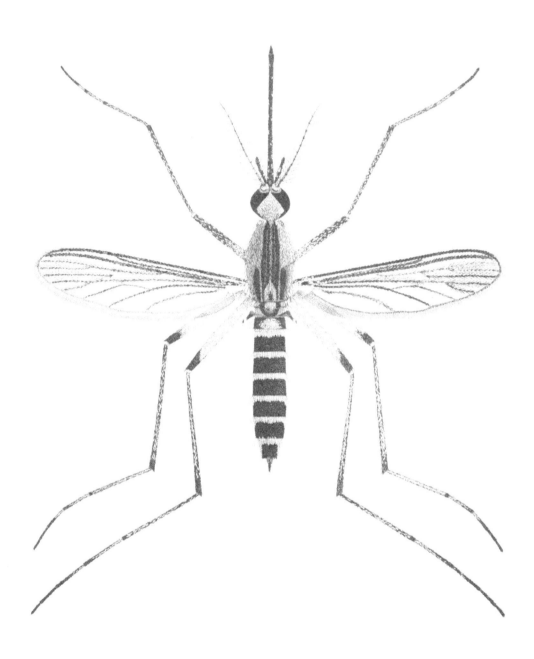

Plate 48. *Aedes bicristatus* Thurman and Winkler, female.

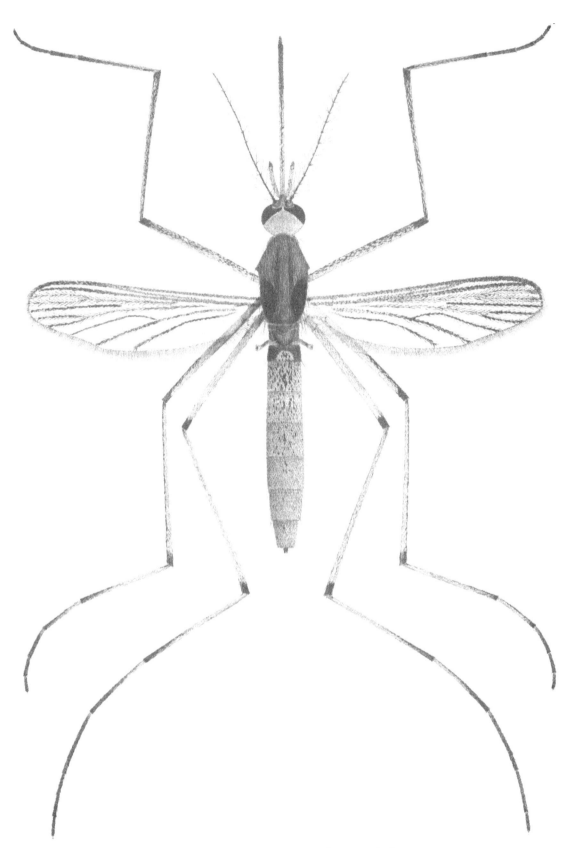

Plate 49. *Aedes bimaculatus* (Coquillett), female.

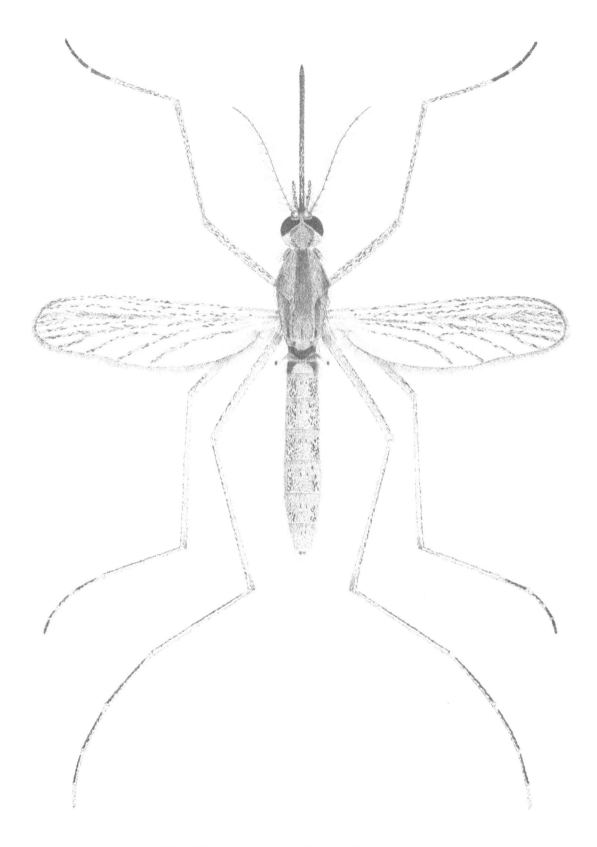

Plate 50. *Aedes campestris* Dyar and Knab, female.

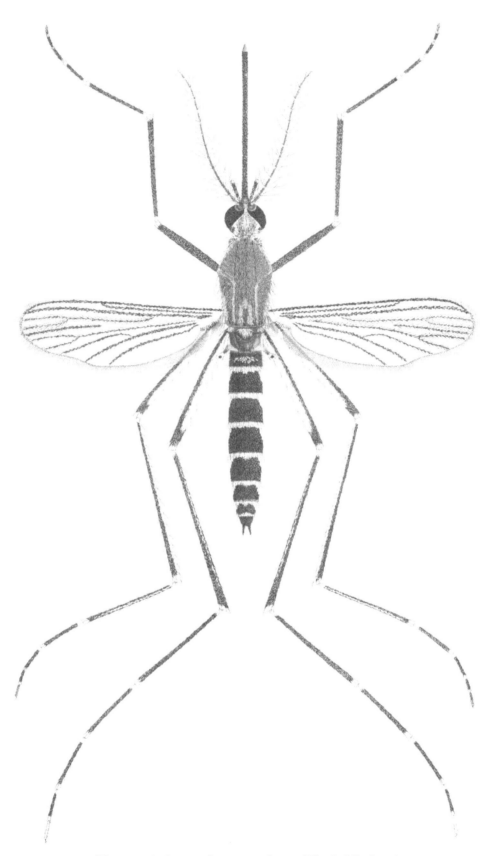

Plate 51. *Aedes canadensis canadensis* (Theobald), female.

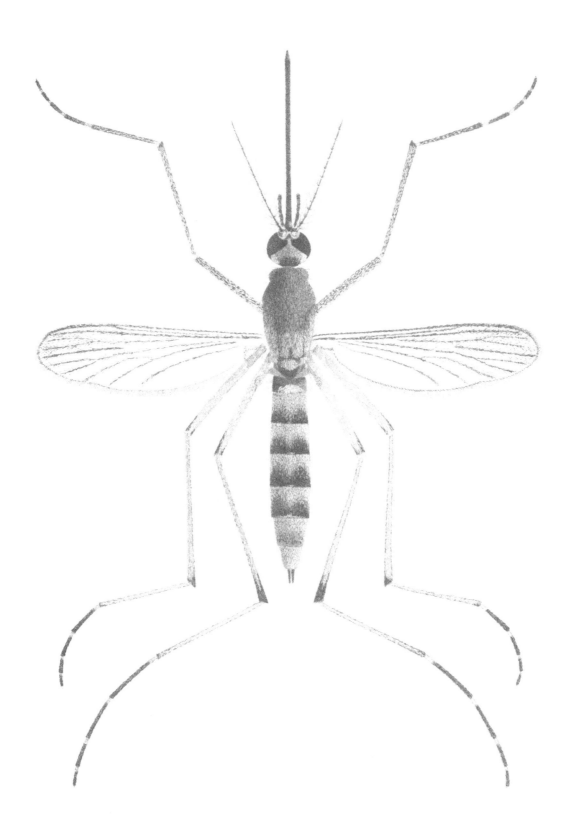

Plate 52. *Aedes cantator* (Coquillett), female.

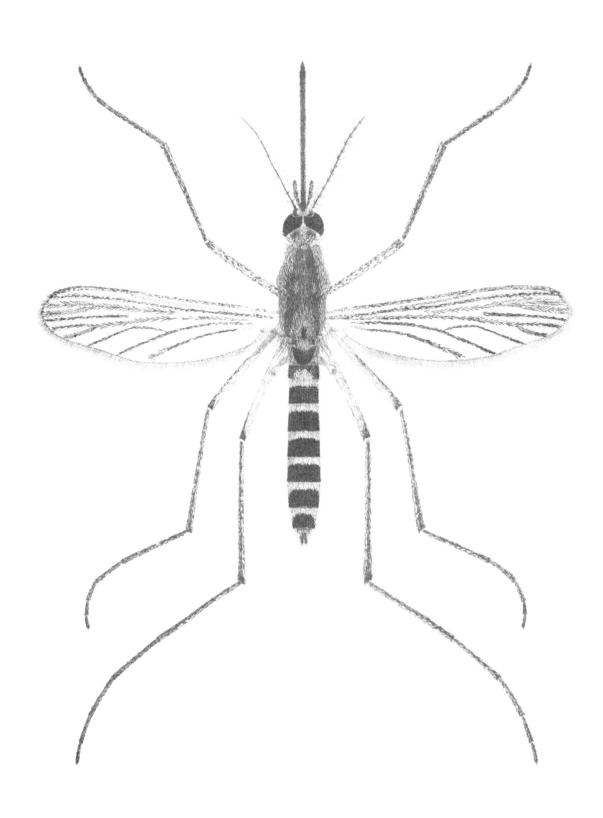

Plate 53. *Aedes cataphylla* Dyar, female.

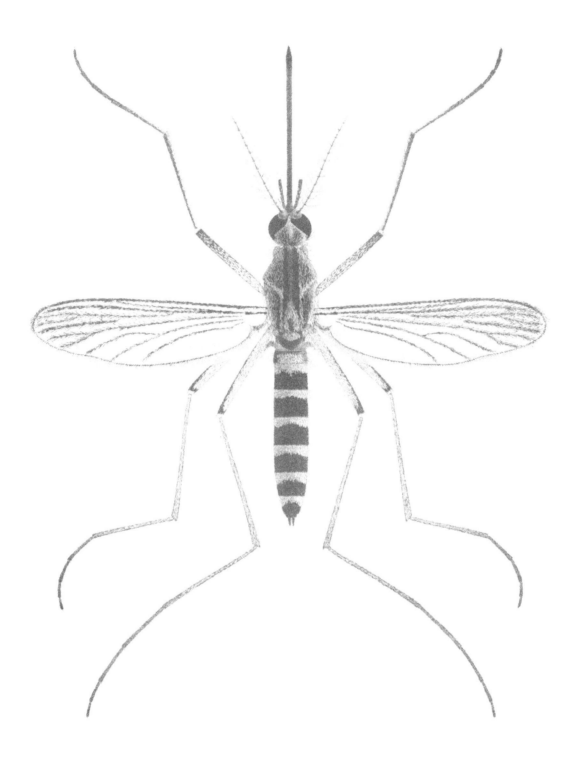

Plate 54. *Aedes communis* (De Geer), female.

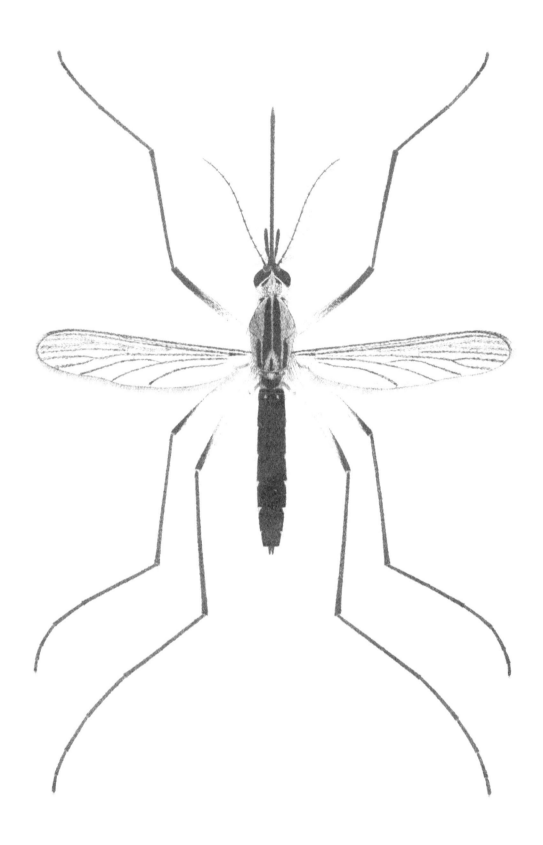

Plate 55. *Aedes decticus* Howard, Dyar, and Knab, female.

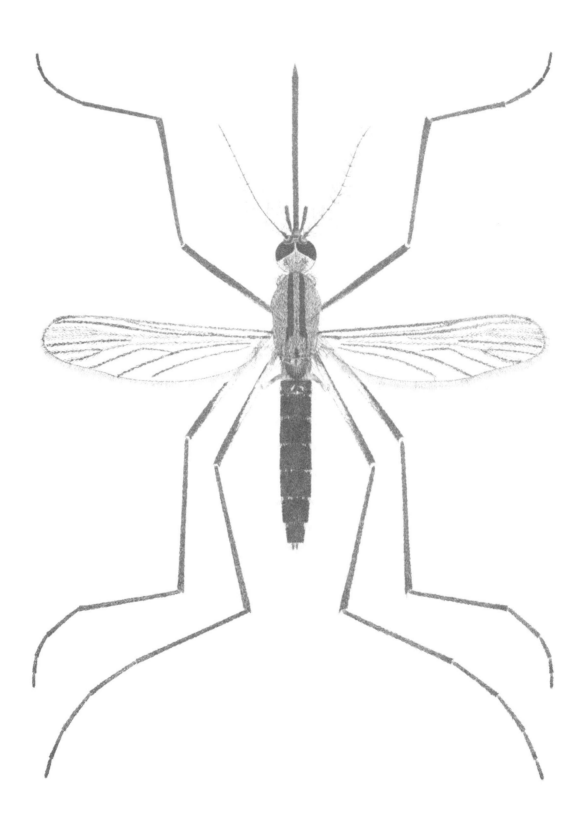

Plate 56. *Aedes diantaeus* Howard, Dyar, and Knab, female.

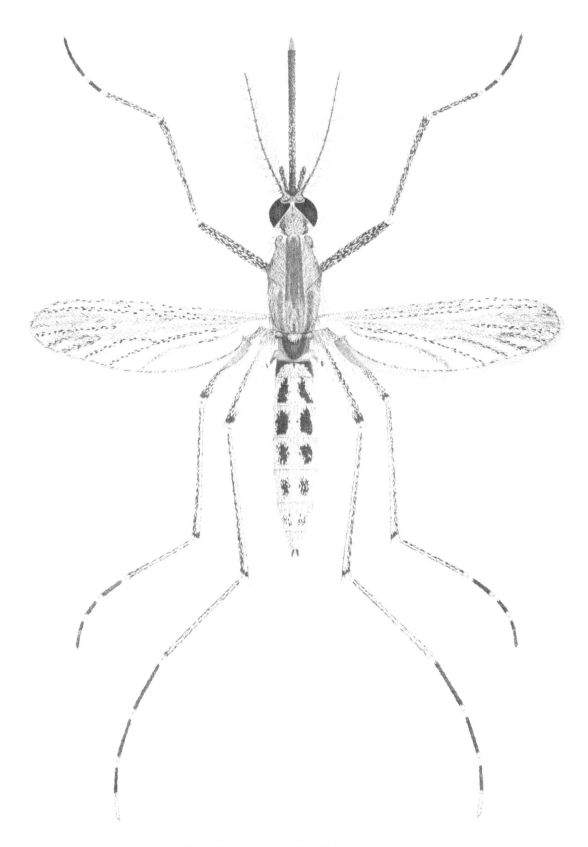

Plate 57. *Aedes dorsalis* (Meigen), female.

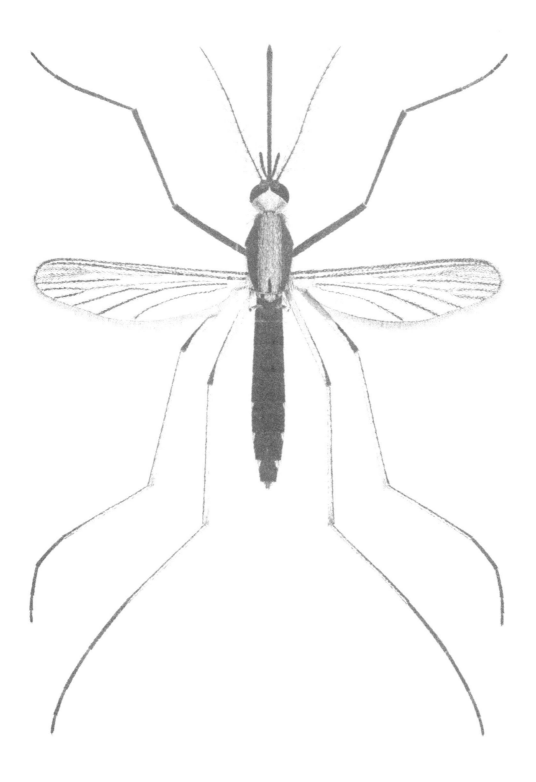

Plate 58. *Aedes dupreei* (Coquillett), female.

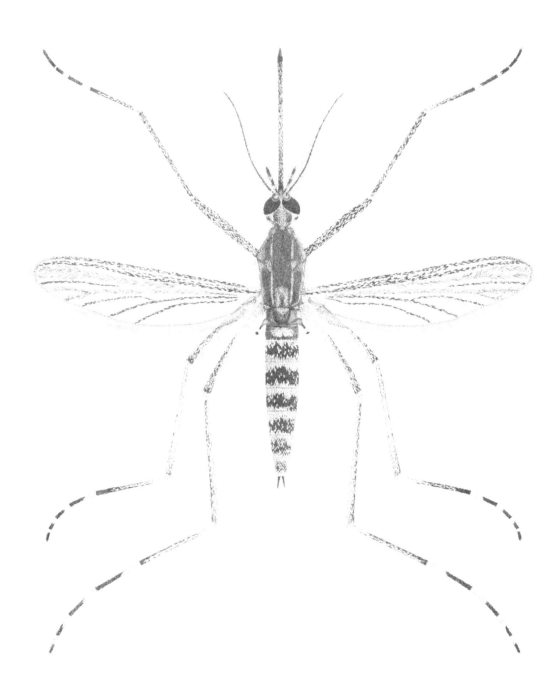

Plate 59. *Aedes excrucians* (Walker), female.

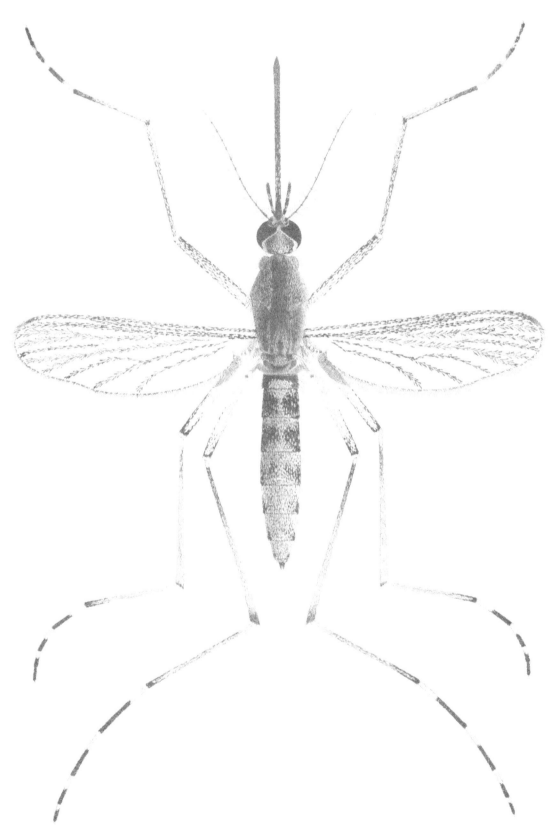

Plate 60. *Aedes fitchii* (Felt and Young), female.

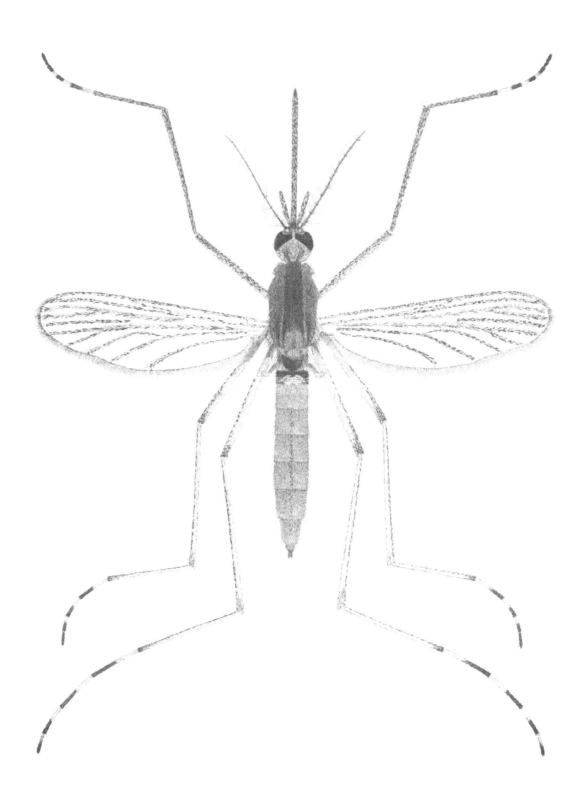

Plate 61. *Aedes flavescens* (Müller), female.

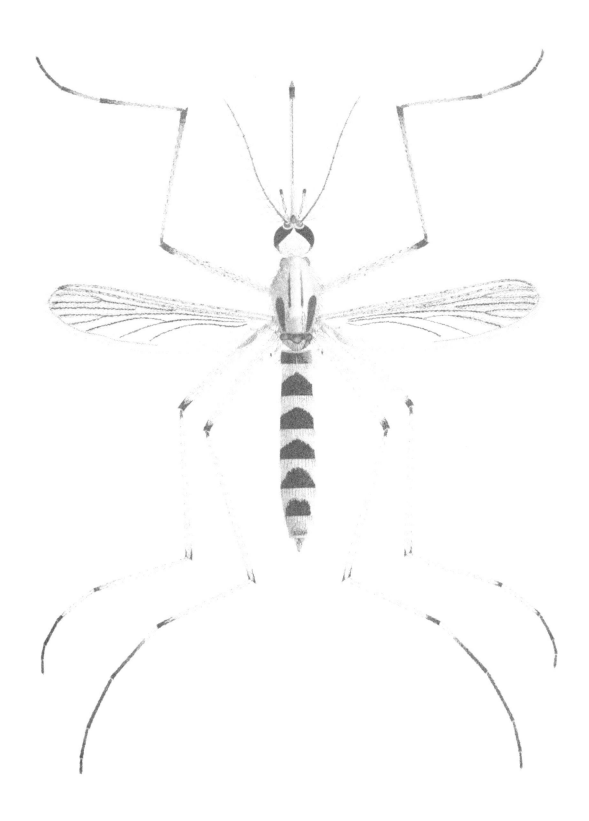

Plate 62. *Aedes fulvus pallens* Ross, female.

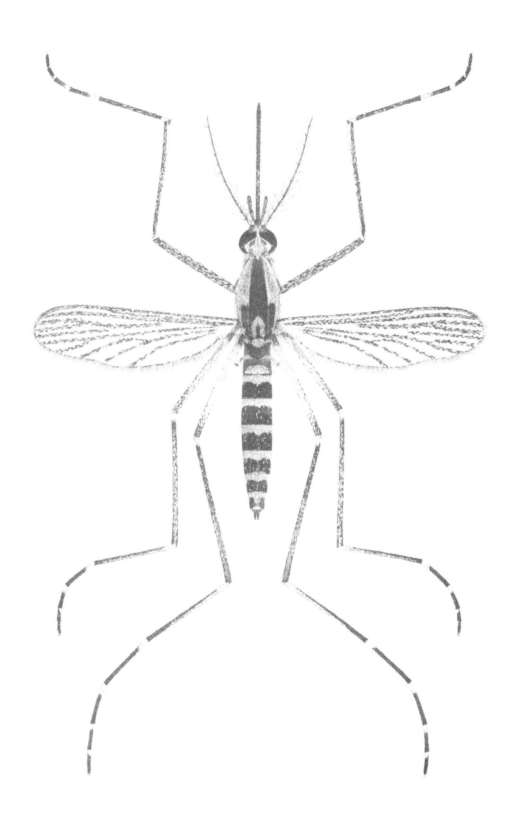

Plate 63. *Aedes grossbecki* Dyar and Knab, female.

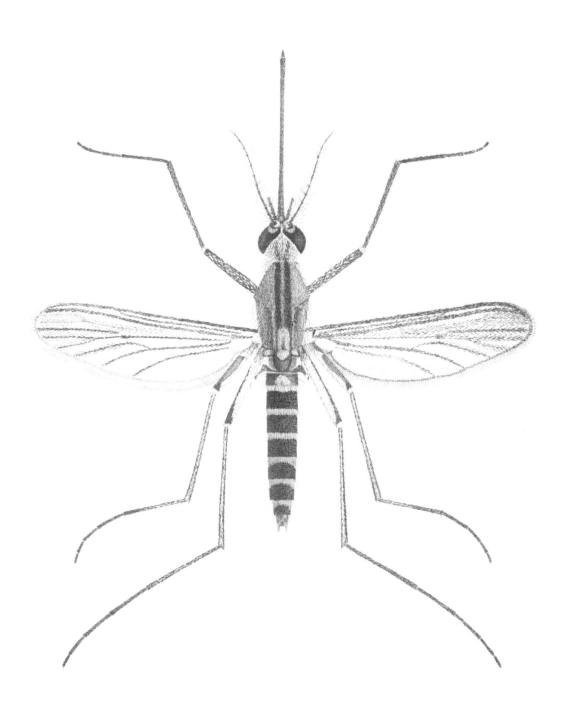

Plate 64. *Aedes hexodontus* Dyar, female.

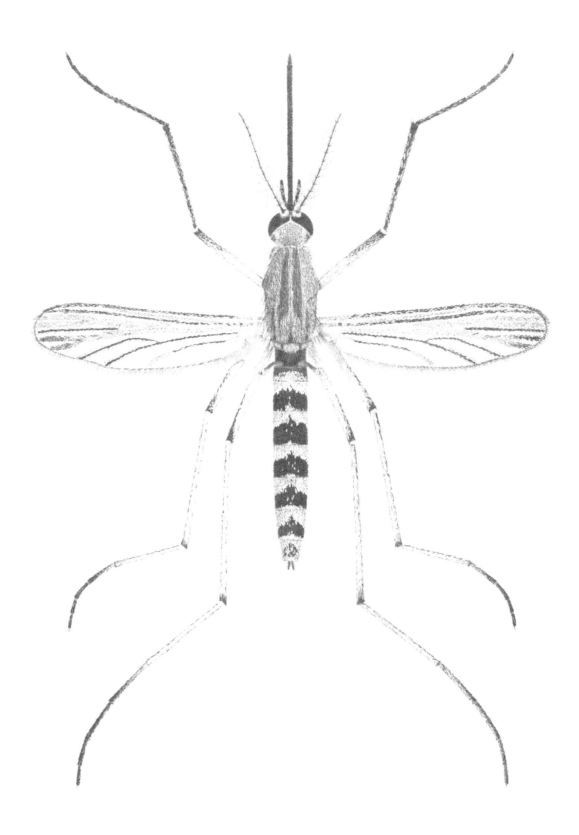

Plate 65. *Aedes idahoensis* (Theobald), female.

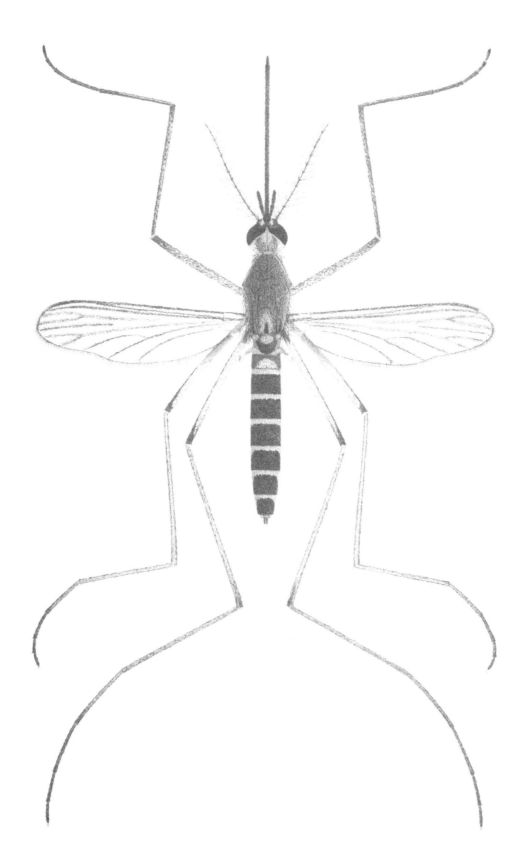

Plate 66. *Aedes impiger* (Walker), female.

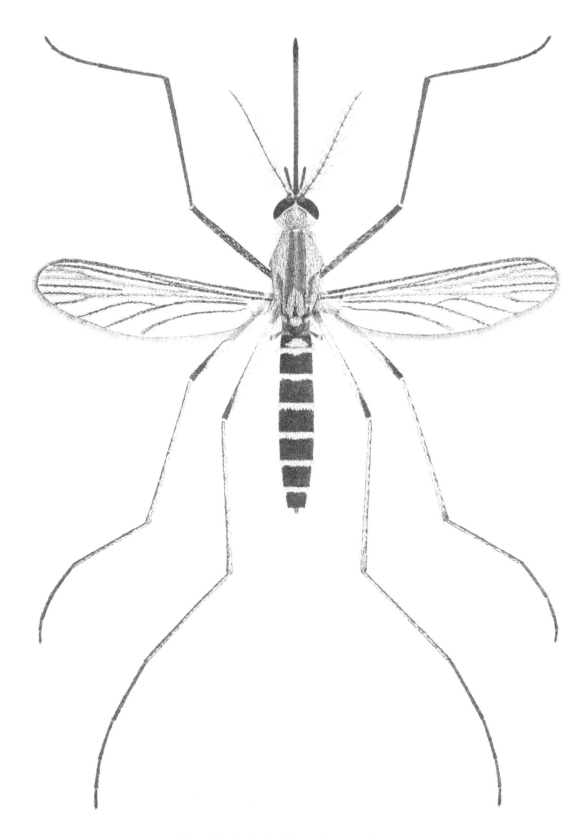

Plate 67. *Aedes implicatus* Vockeroth, female.

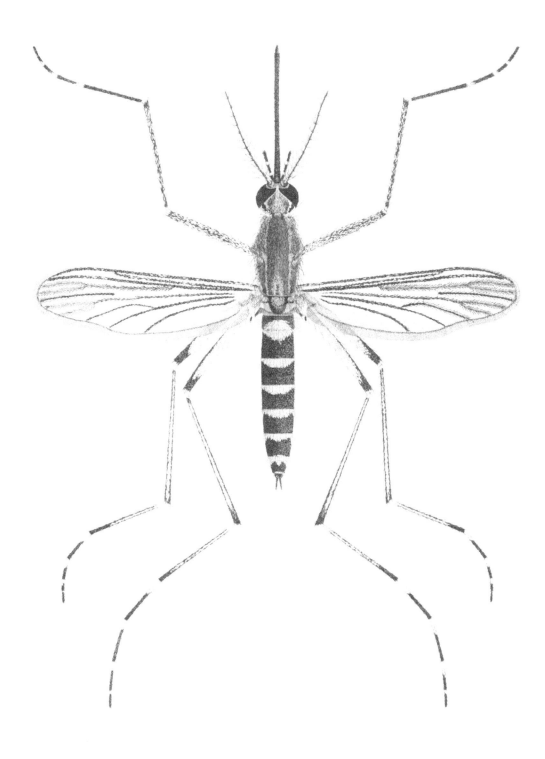

Plate 68. *Aedes increpitus* Dyar, female.

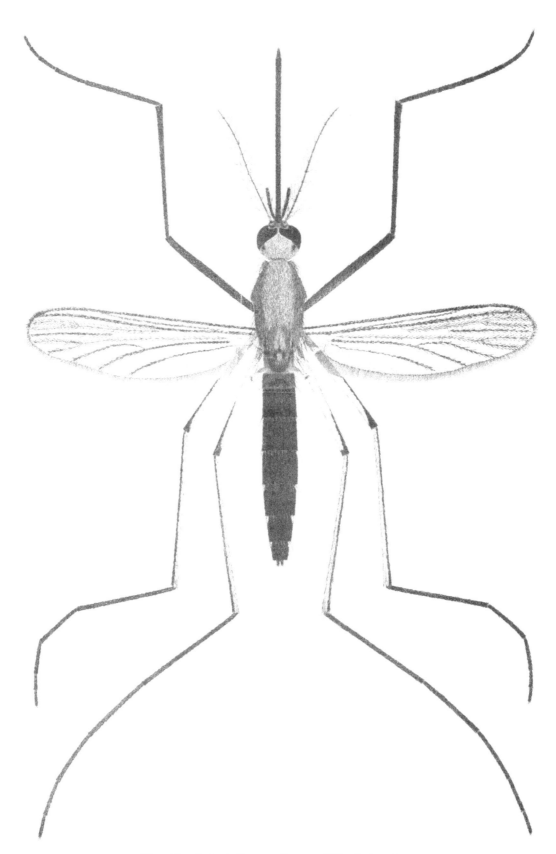

Plate 69. *Aedes infirmatus* Dyar and Knab, female.

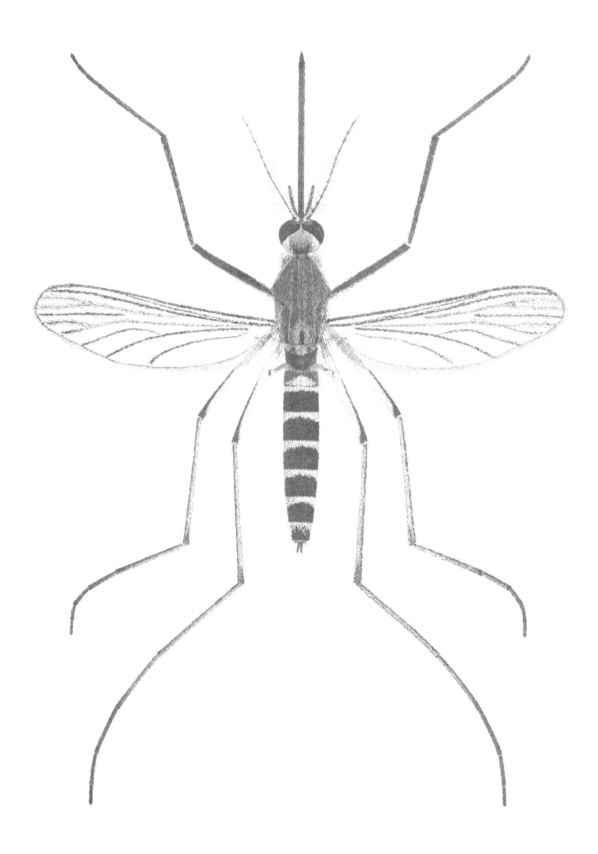

Plate 70. *Aedes intrudens* Dyar, female.

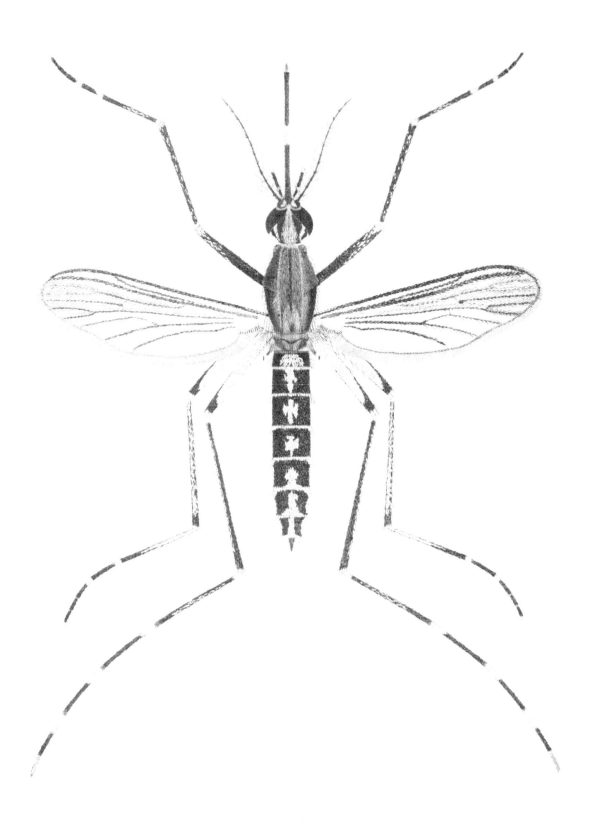

Plate 71. *Aedes mitchellae* (Dyar), female.

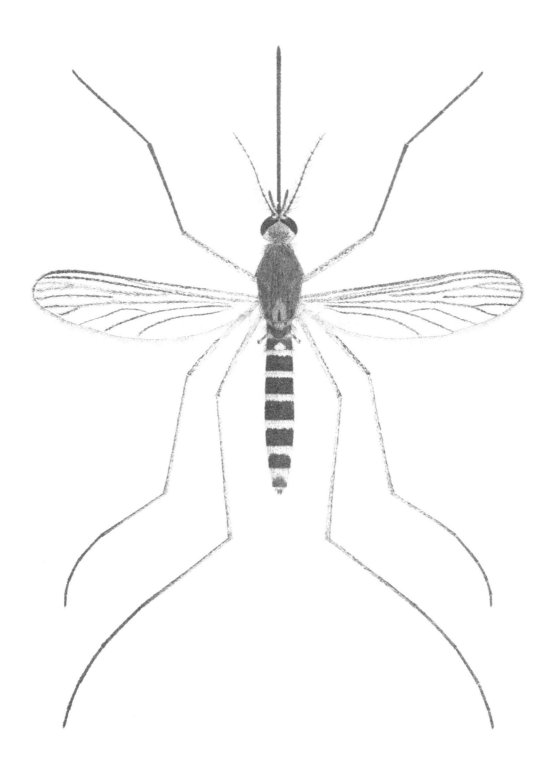

Plate 72. *Aedes nigripes* (Zetterstedt), female.

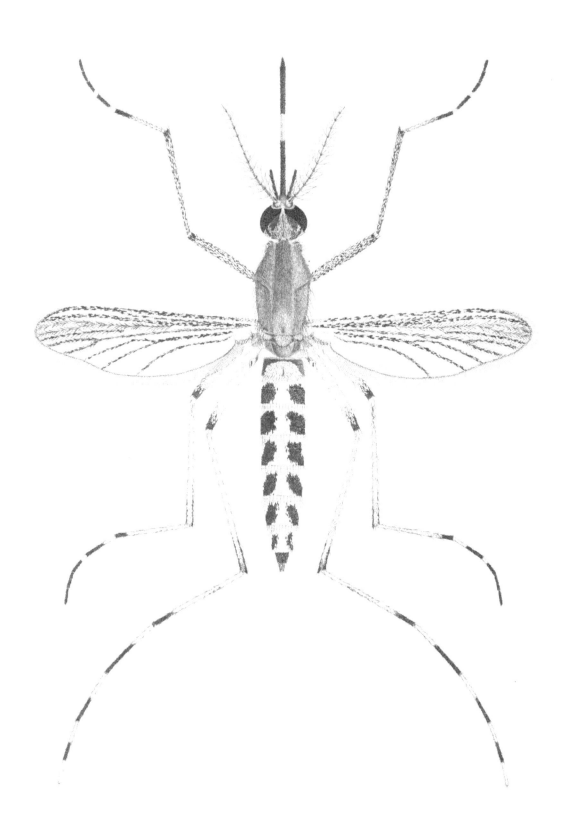

Plate 73. *Aedes nigromaculis* (Ludlow), female.

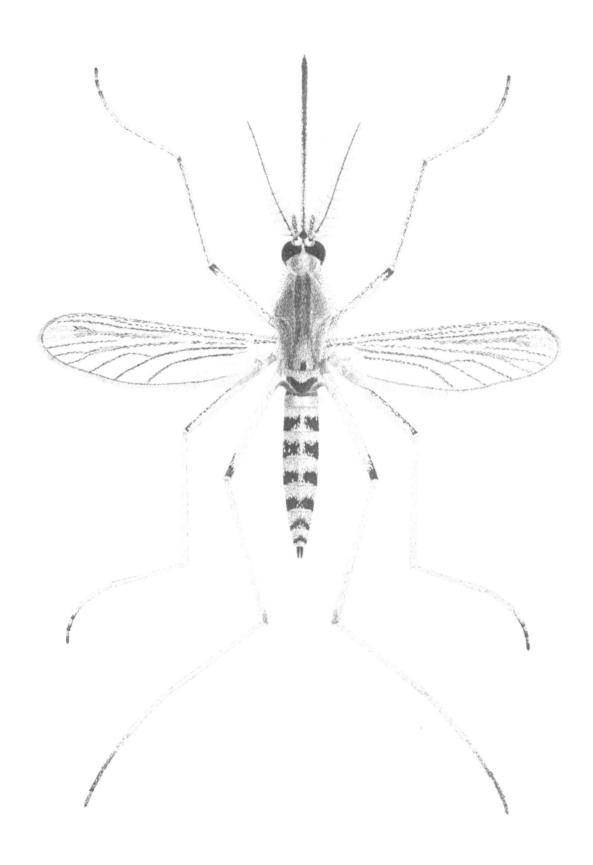

Plate 74. *Aedes niphadopsis* Dyar and Knab, female.

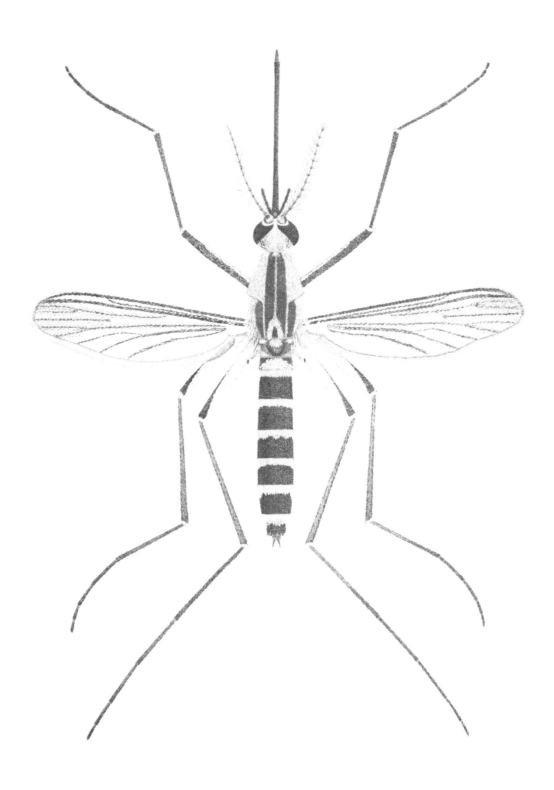

Plate 75. *Aedes pionips* Dyar, female.

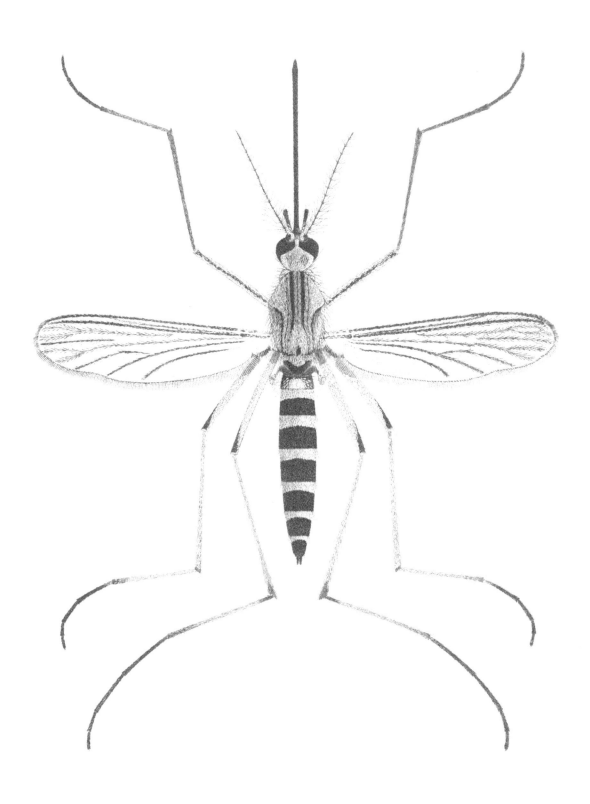

Plate 76. *Aedes pullatus* (Coquillett), female.

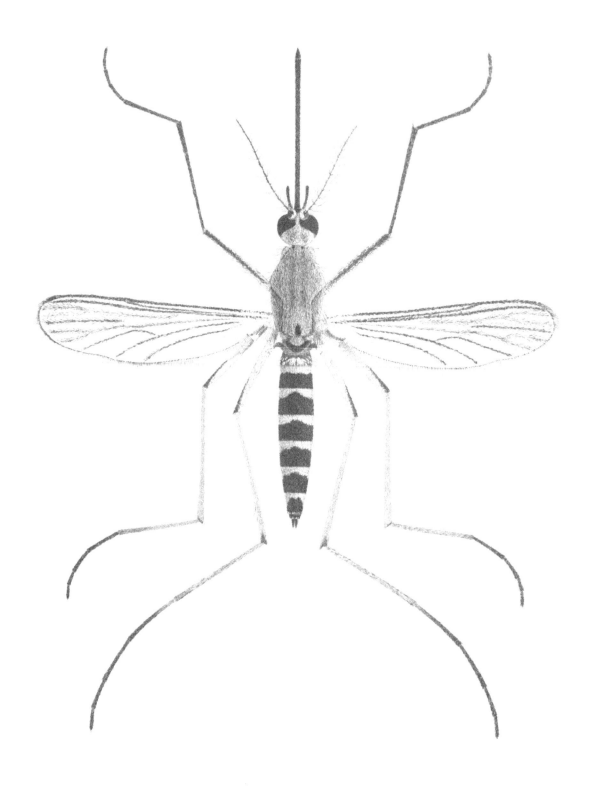

Plate 77. *Aedes punctor* (Kirby), female.

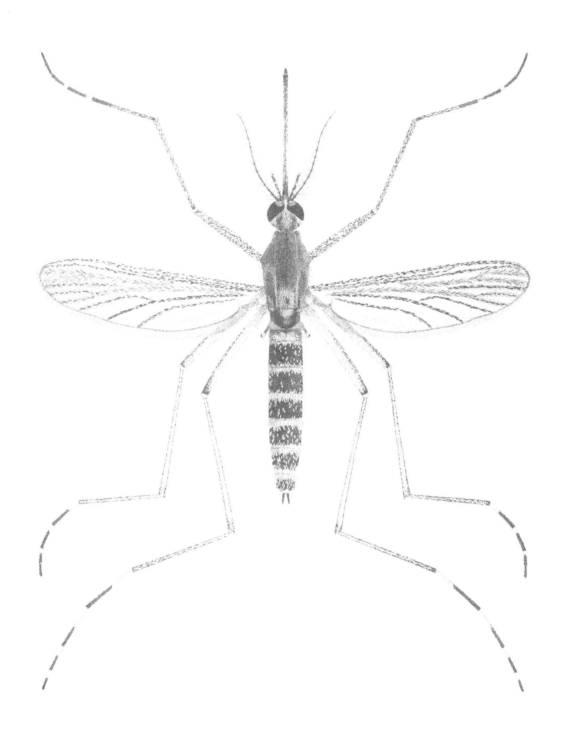

Plate 78. *Aedes riparius* Dyar and Knab, female.

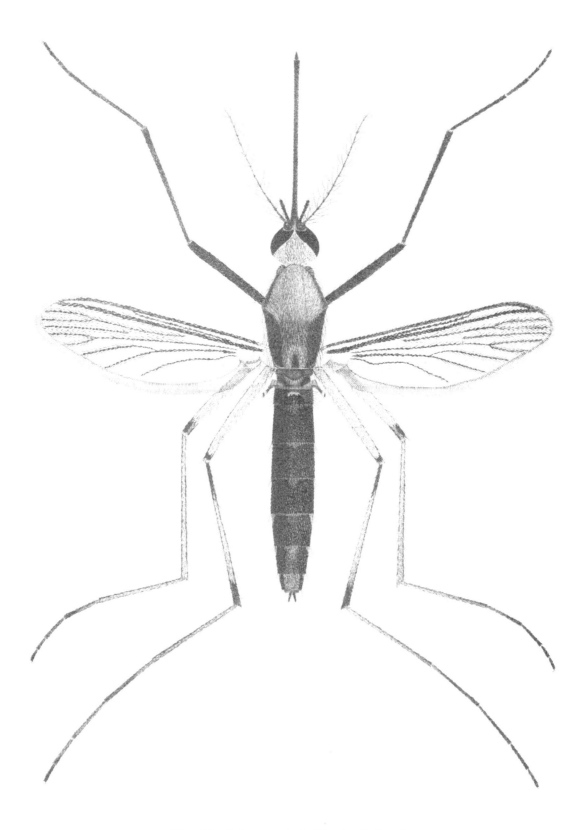

Plate 79. *Aedes scapularis* (Rondani), female.

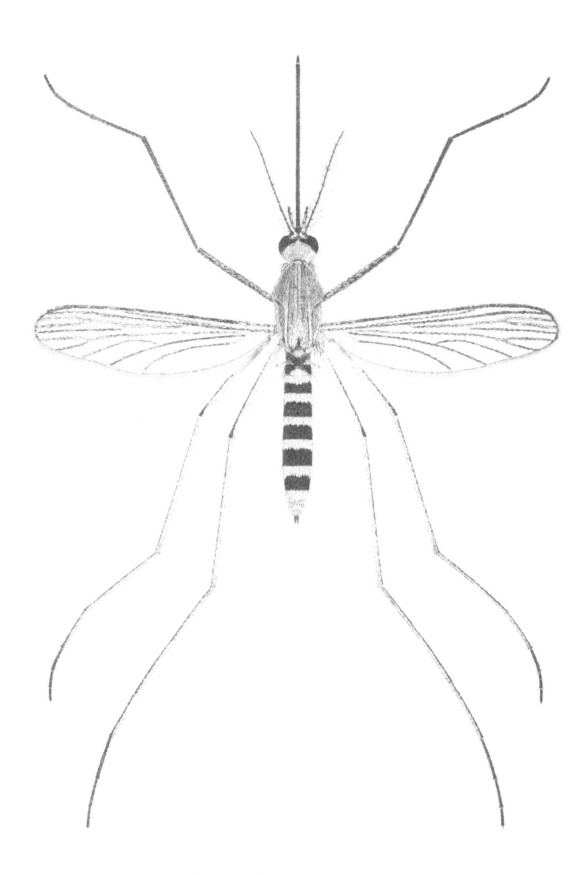

Plate 80. *Aedes schizopinax* Dyar, female.

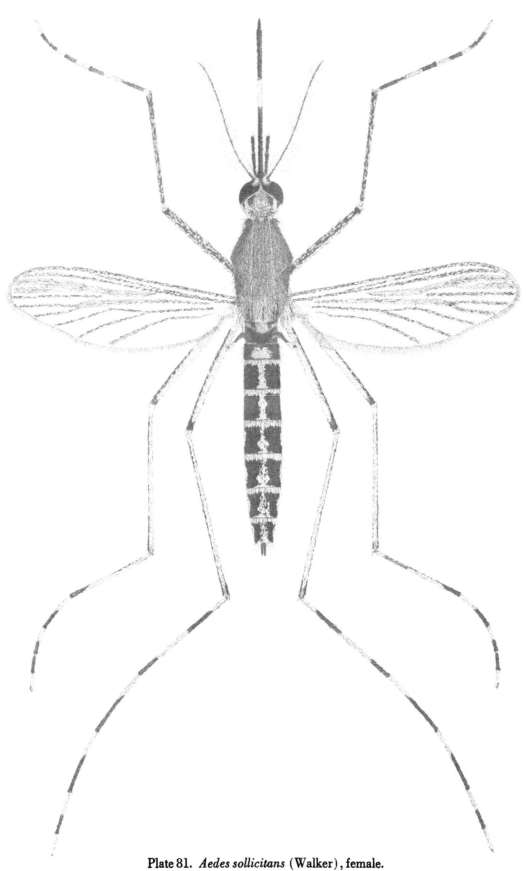

Plate 81. *Aedes sollicitans* (Walker), female.

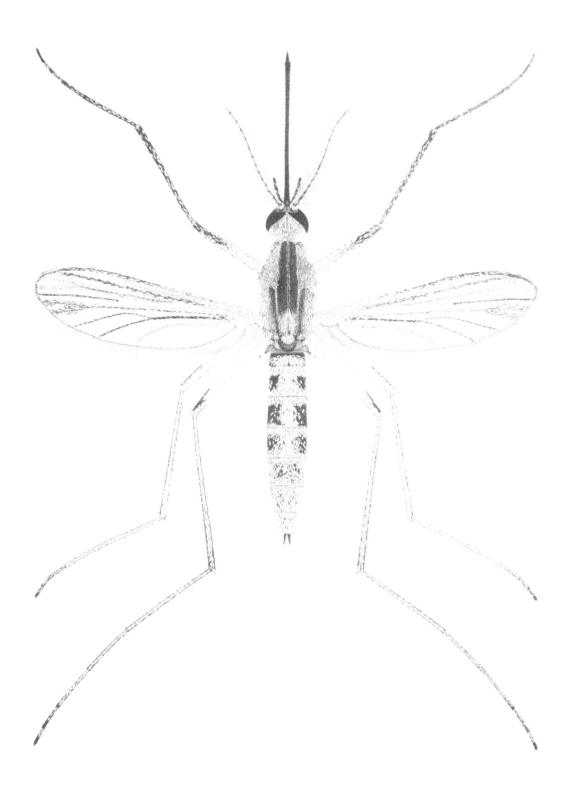

Plate 82. *Aedes spencerii* (Theobald), female.

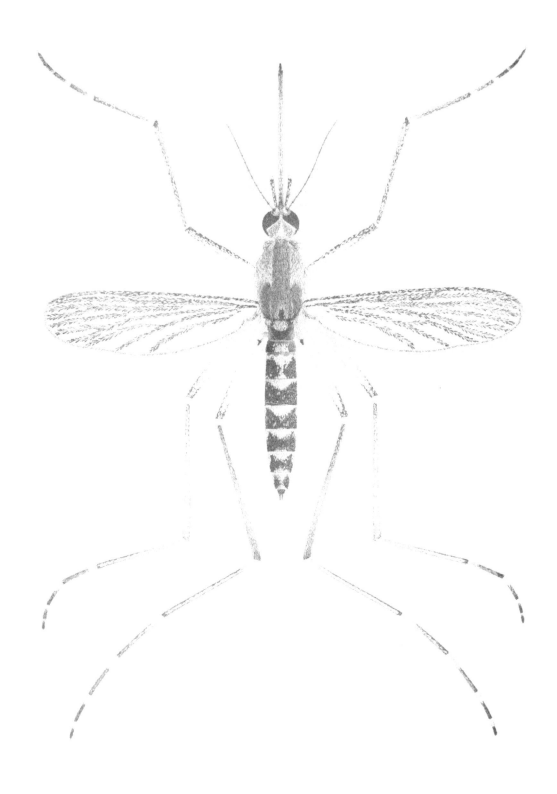

Plate 83. *Aedes squamiger* (Coquillett), female.

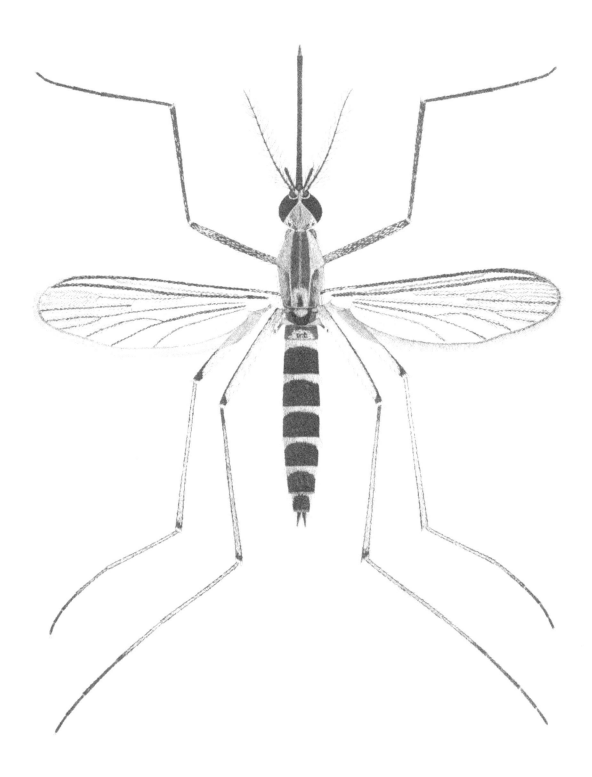

Plate 84. *Aedes sticticus* (Meigen), female.

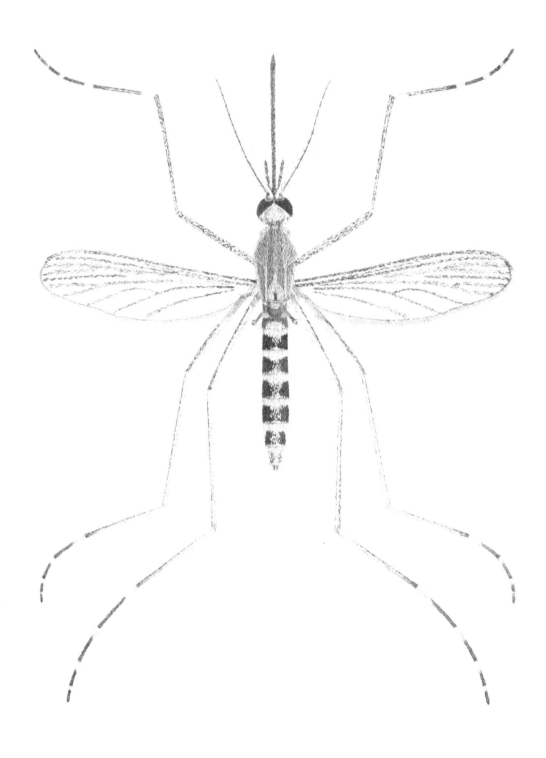

Plate 85. *Aedes stimulans* (Walker), female.

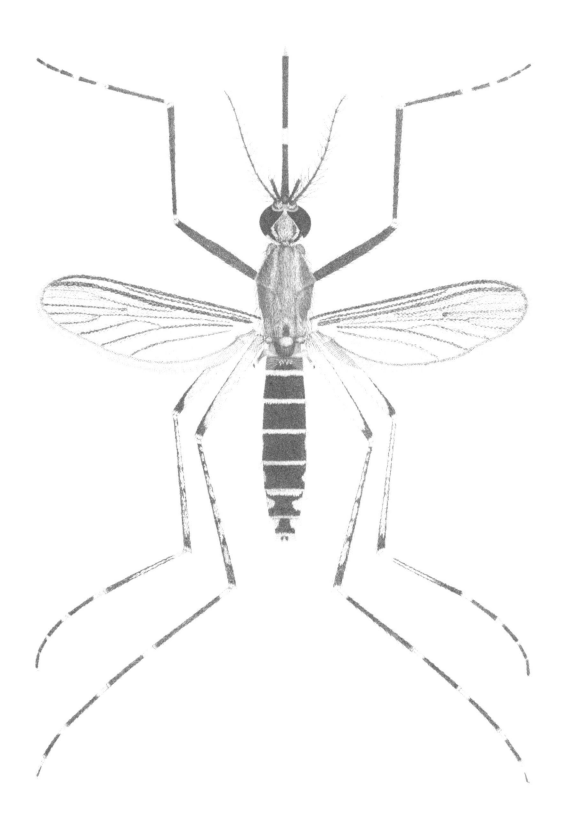

Plate 86. *Aedes taeniorhynchus* (Wiedemann), female.

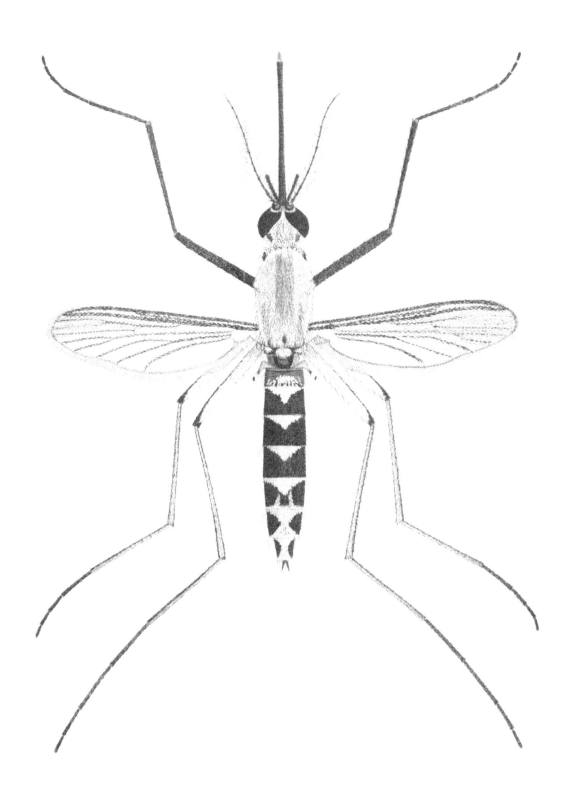

Plate 87. *Aedes thelcter* Dyar, female.

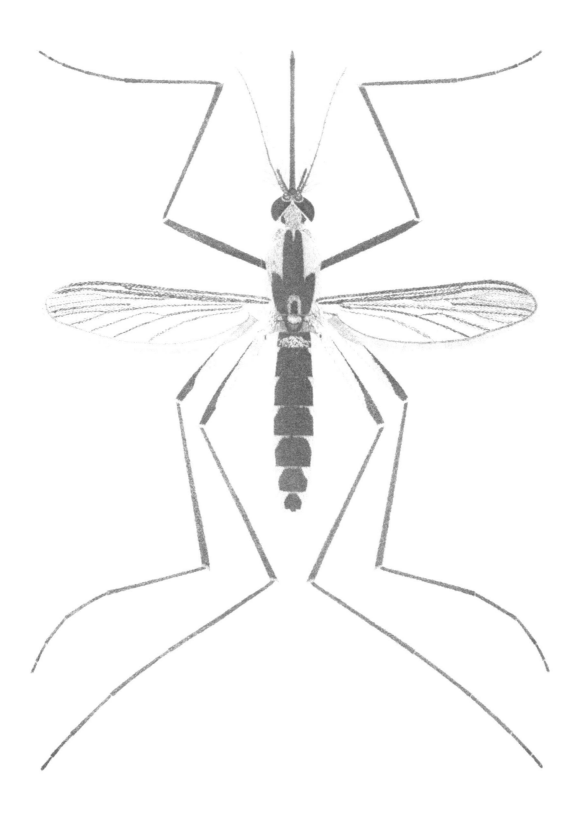

Plate 88. *Aedes thibaulti* Dyar and Knab, female.

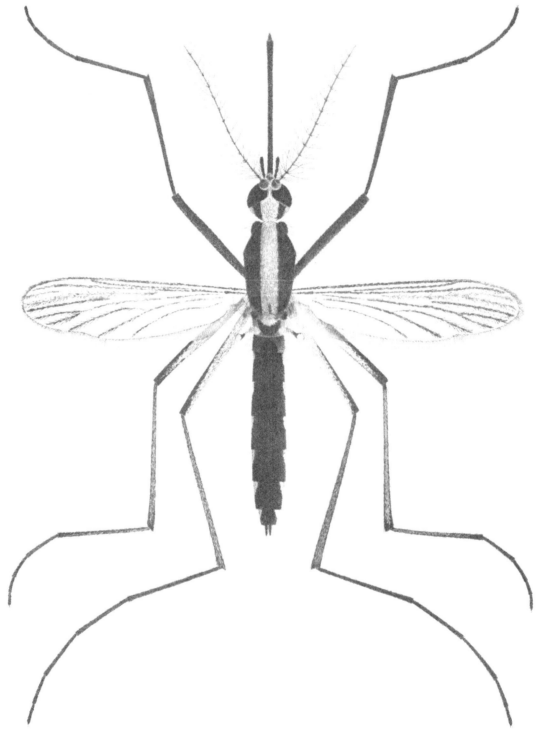

Plate 89. *Aedes tormentor* Dyar and Knab, female.

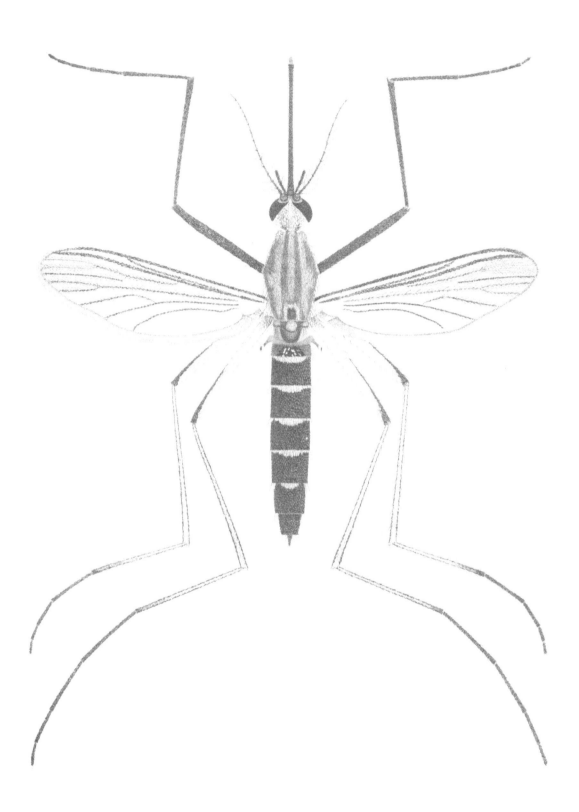

Plate 90. *Aedes tortilis* (Theobald), female.

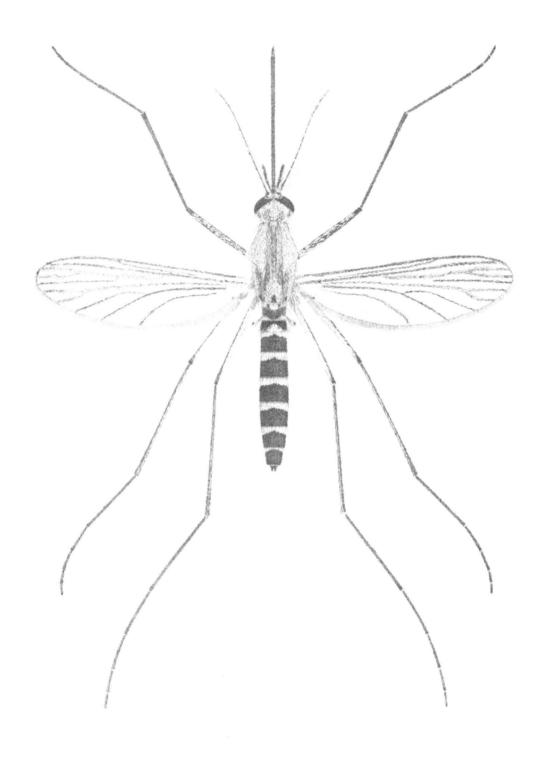

Plate 91. *Aedes trichurus* (Dyar), female.

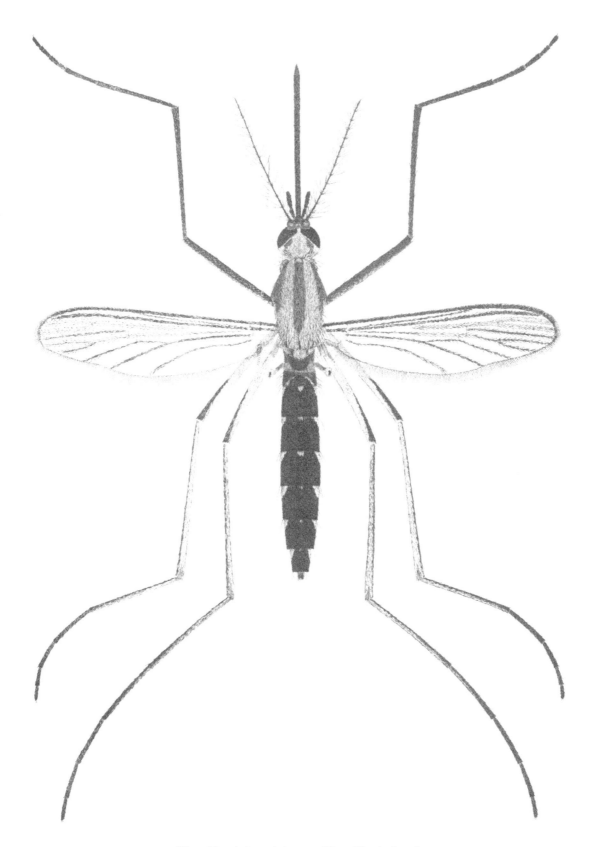

Plate 92. *Aedes trivittatus* (Coquillett), female.

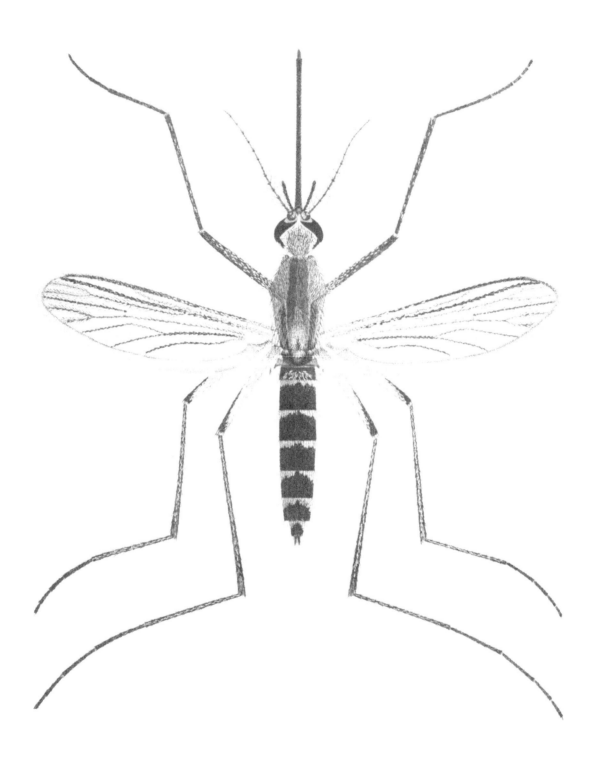

Plate 93. *Aedes ventrovittis* Dyar, female.

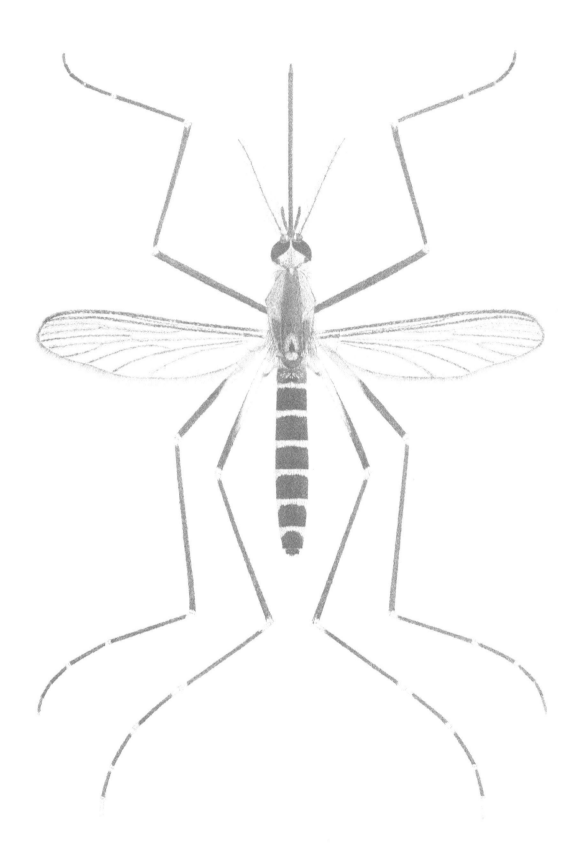

Plate 94. *Aedes atropalpus* (Coquillett), female.

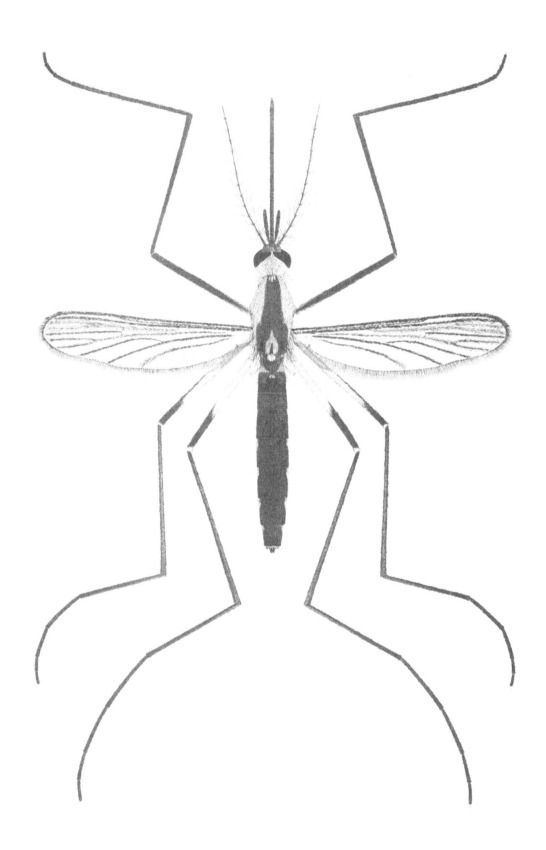

Plate 95. *Aedes triseriatus* (Say), female.

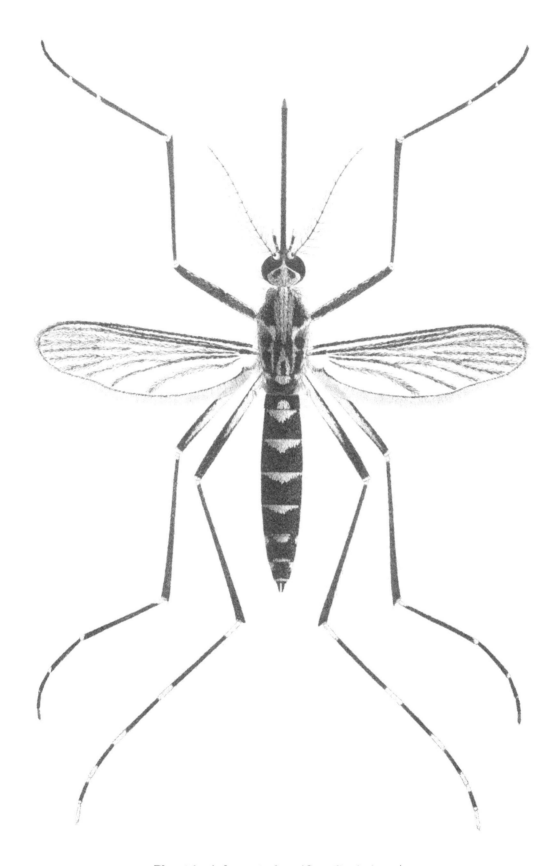

Plate 96. *Aedes varipalpus* (Coquillett), female.

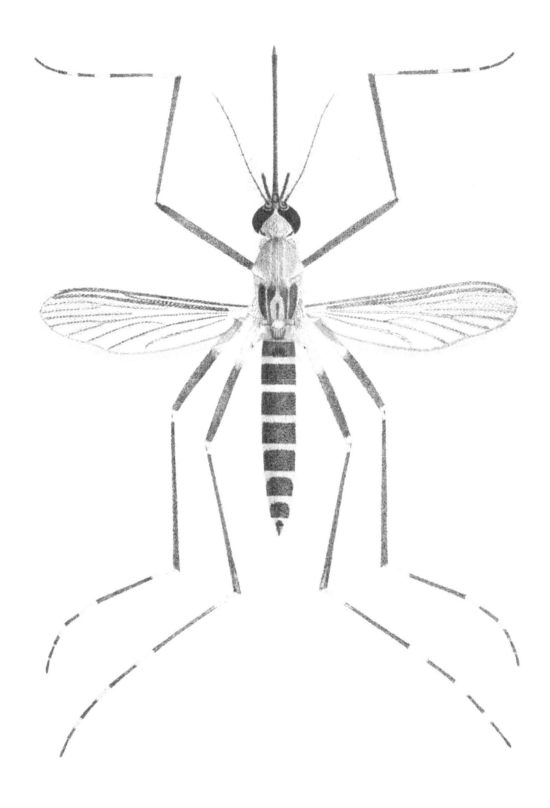

Plate 97. *Aedes zoosophus* Dyar and Knab, female.

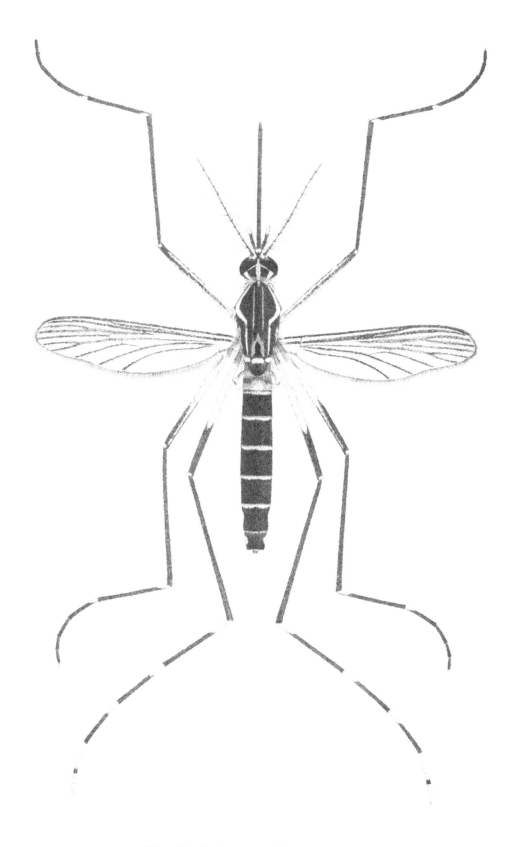

Plate 98. *Aedes aegypti* (Linnaeus), female.

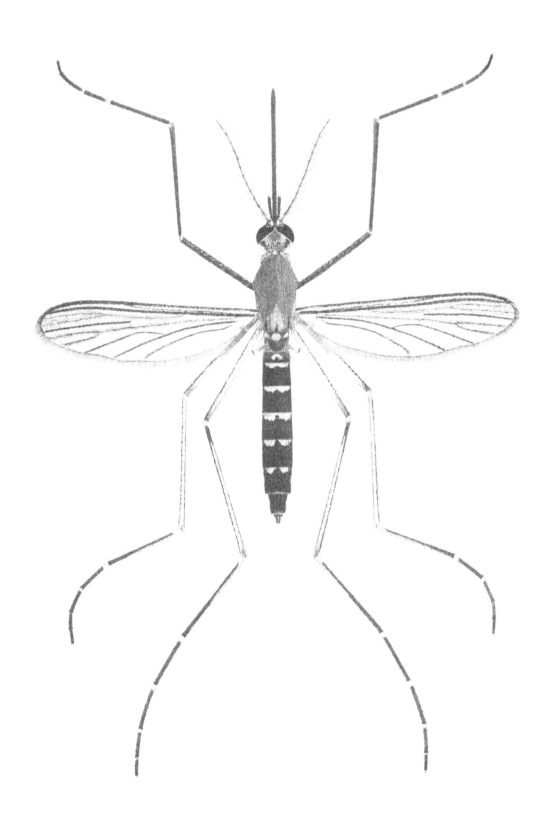

Plate 99. *Aedes vexans* (Meigen), female.

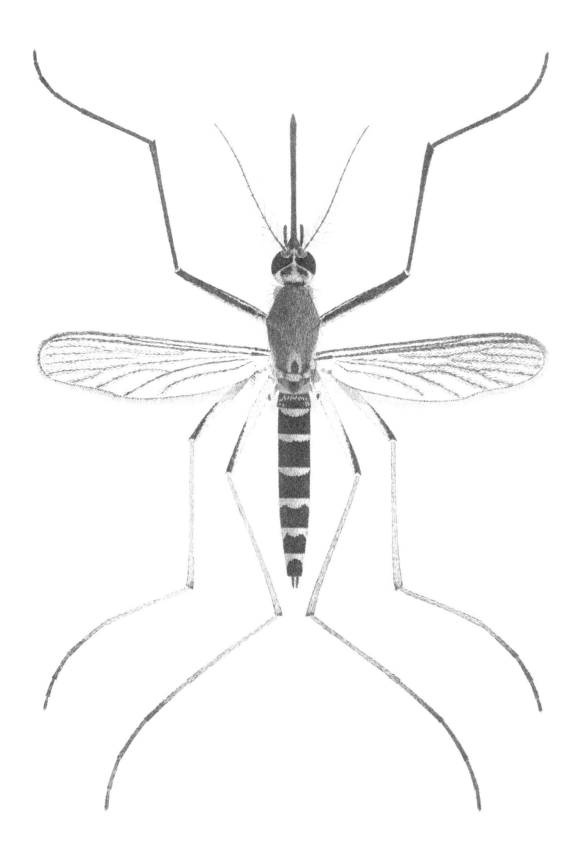

Plate 100. *Aedes cinereus* Meigen, female.

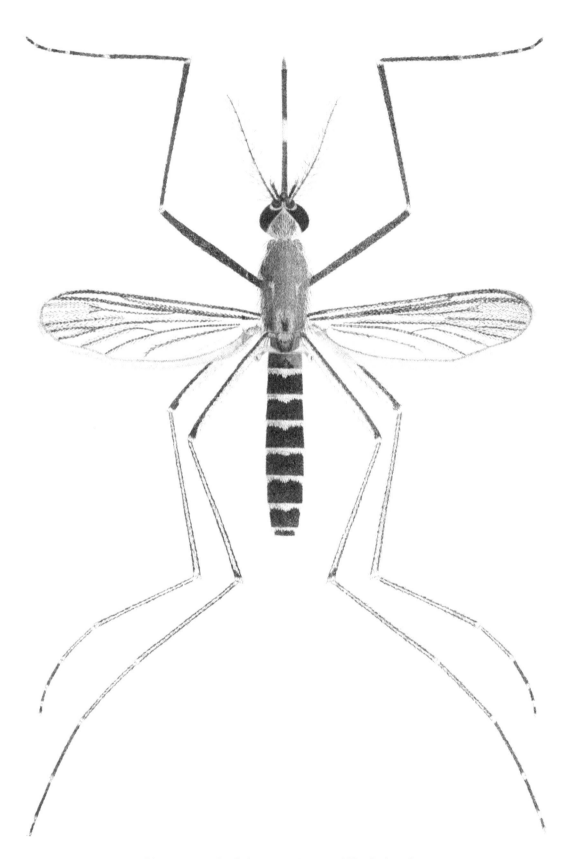

Plate 101. *Culex bahamensis* Dyar and Knab, female.

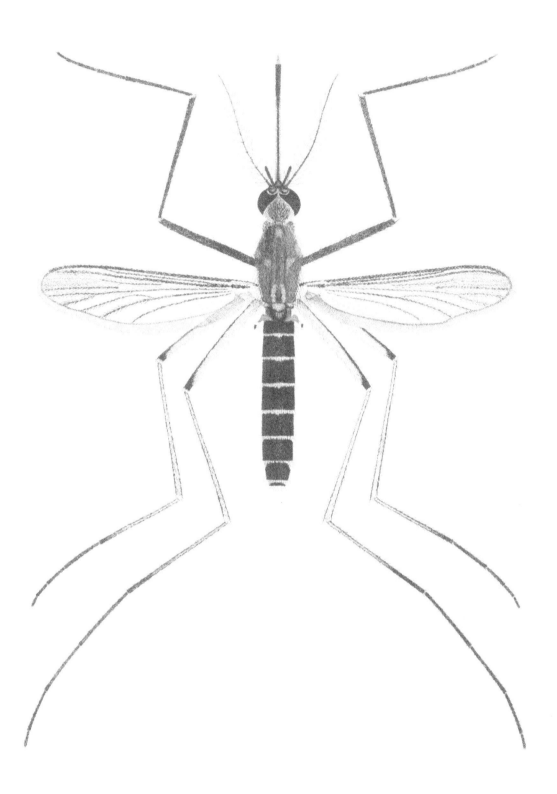

Plate 102. *Culex chidesteri* Dyar, female.

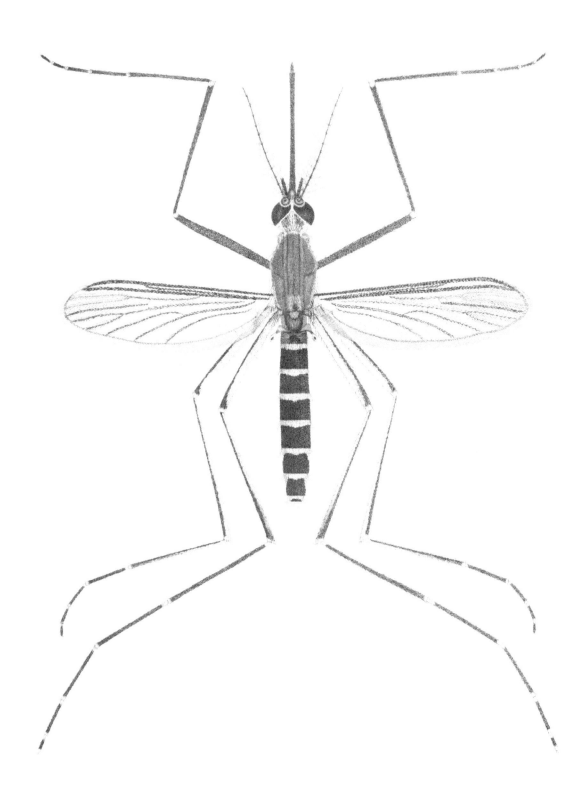

Plate 103. *Culex coronator* Dyar and Knab, female.

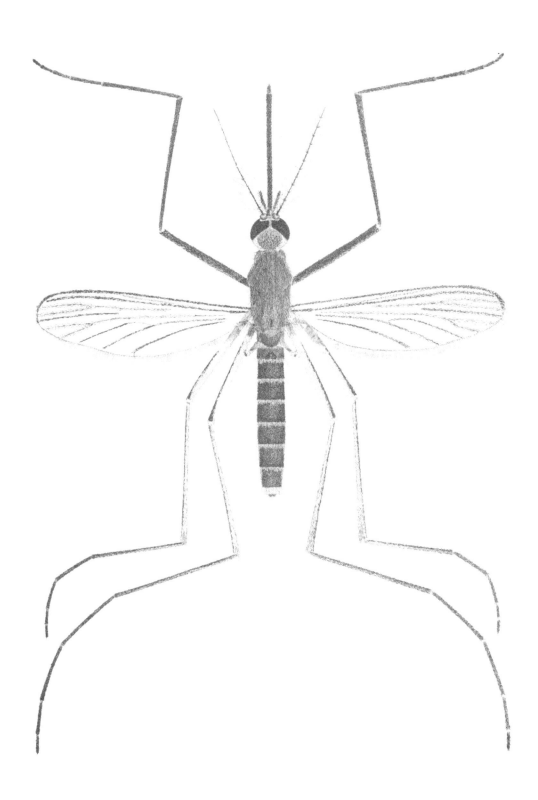

Plate 104. *Culex erythrothorax* Dyar, female.

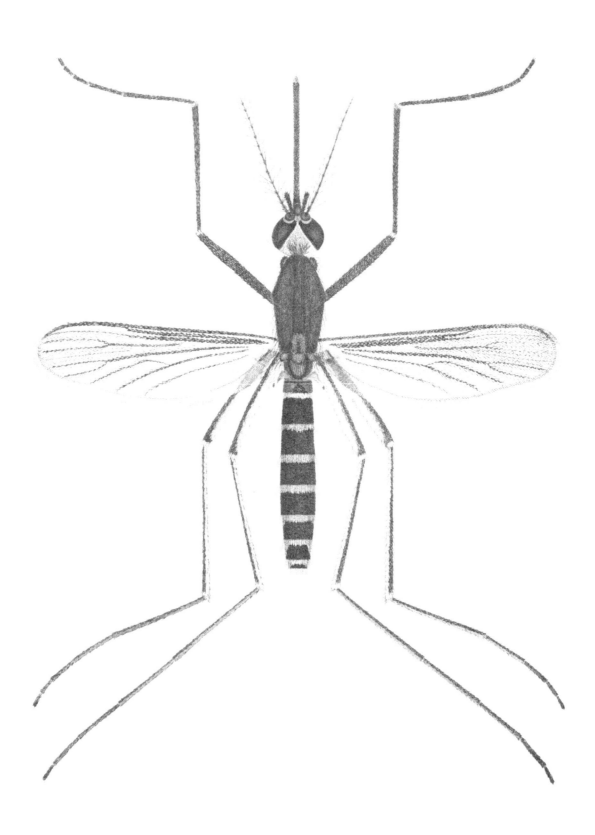

Plate 105. *Culex interrogator* Dyar and Knab, female.

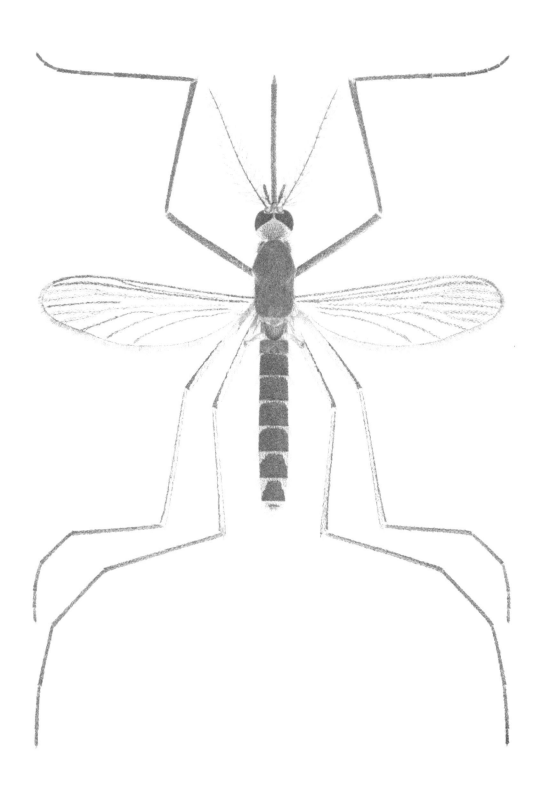

Plate 106. *Culex nigripalpus* Theobald, female.

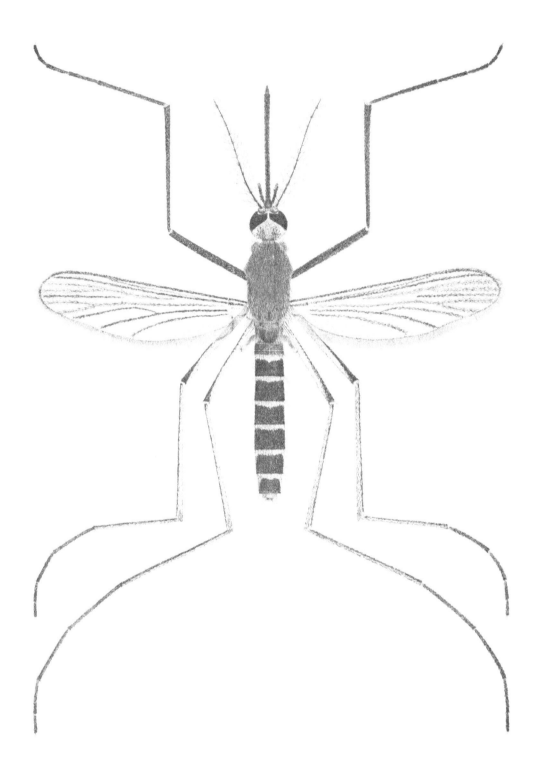

Plate 107. *Culex pipiens* Linnaeus, female.

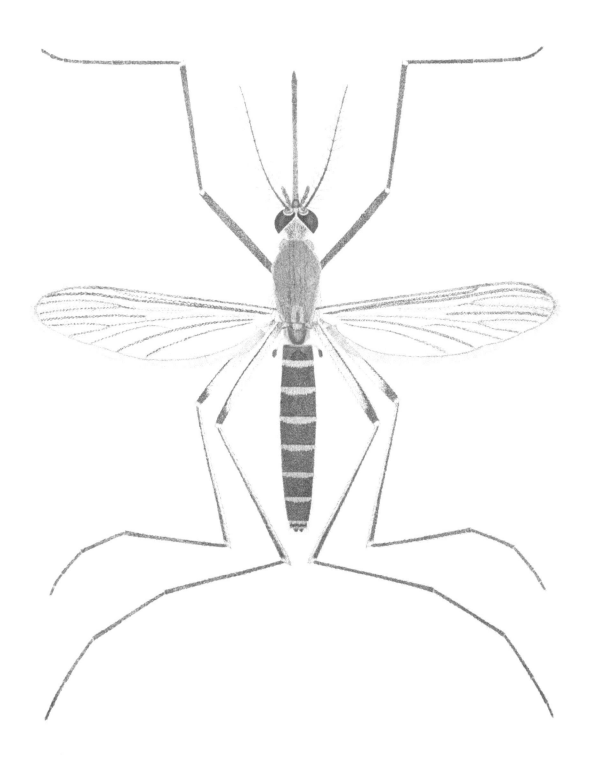

Plate 108. *Culex quinquefasciatus* Say, female.

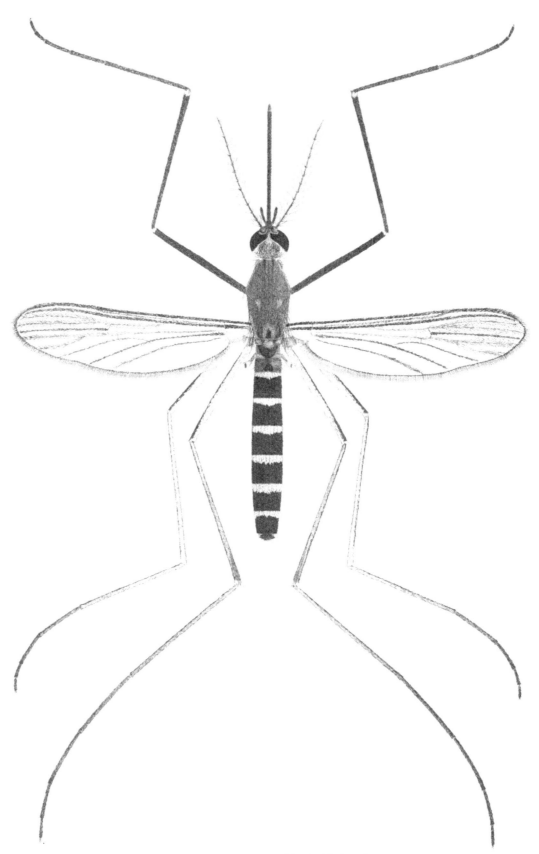

Plate 109. *Culex restuans* Theobald, female.

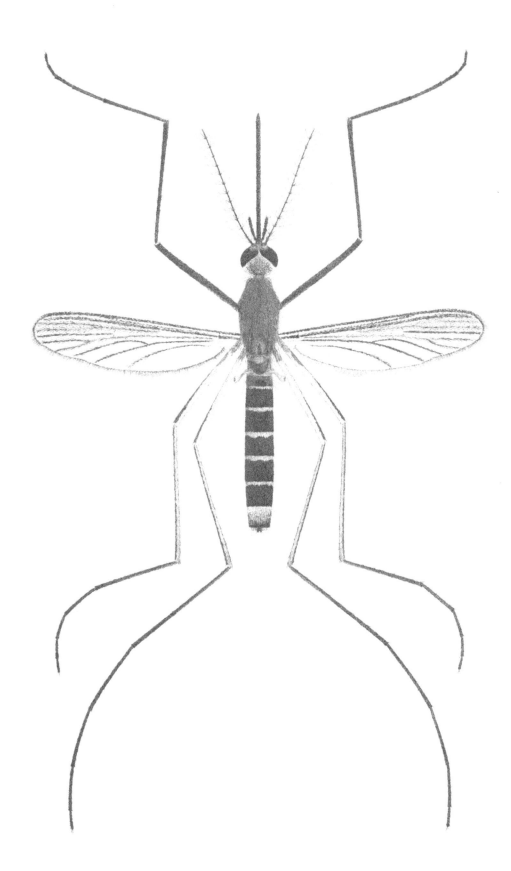

Plate 110. *Culex salinarius* Coquillett, female.

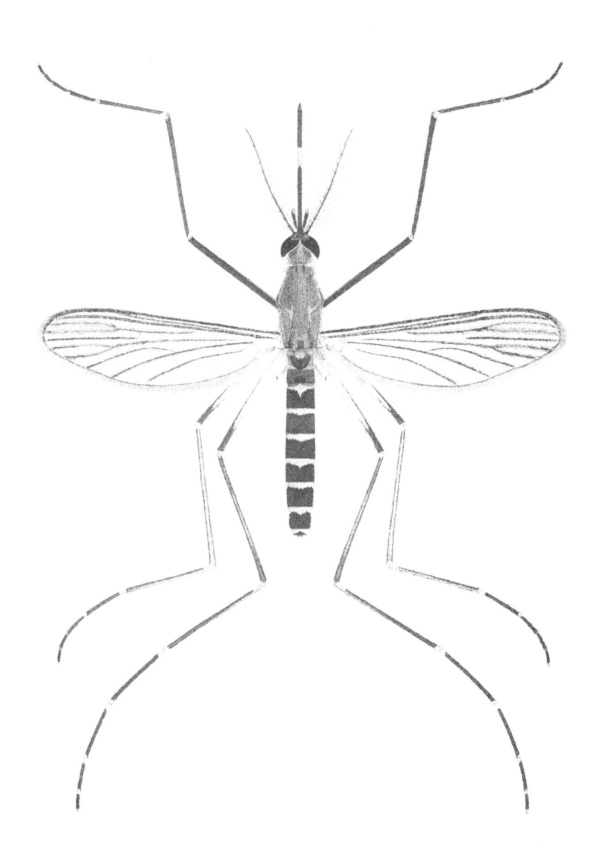

Plate 111. *Culex stigmatosoma* Dyar, female.

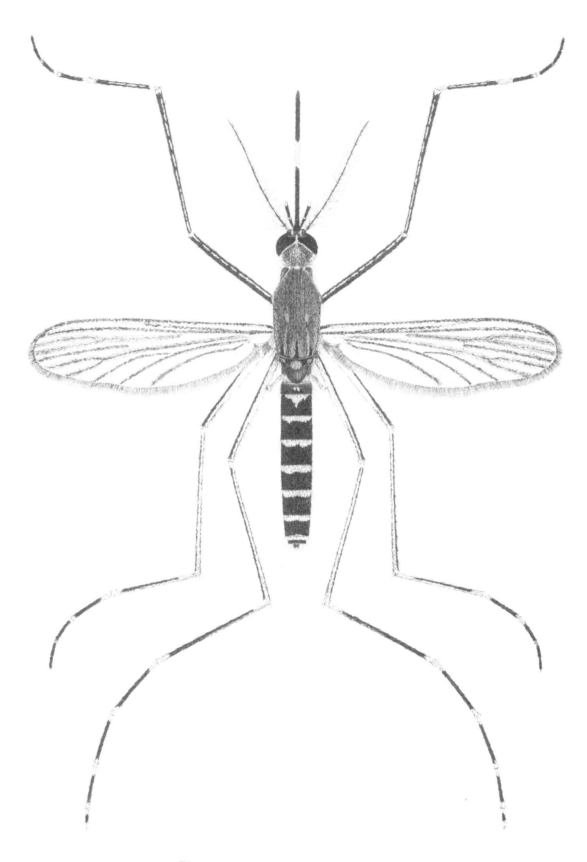

Plate 112. *Culex tarsalis* Coquillett, female.

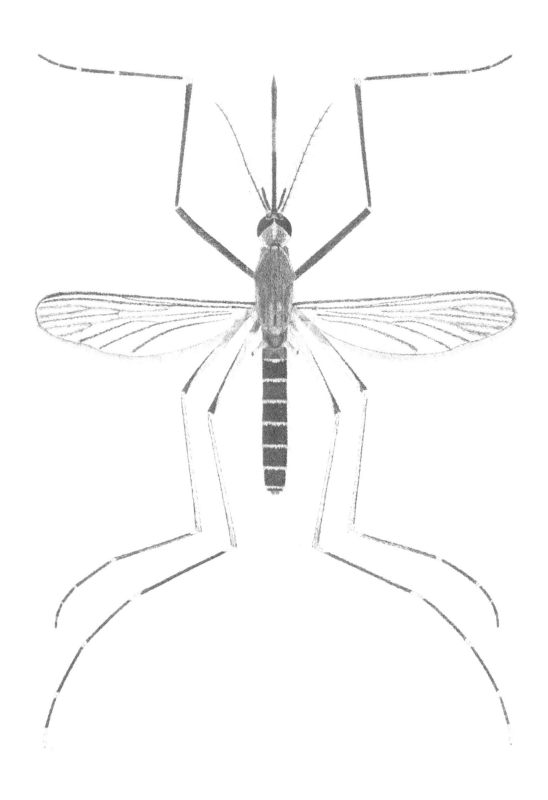

Plate 113. *Culex thriambus* Dyar, female.

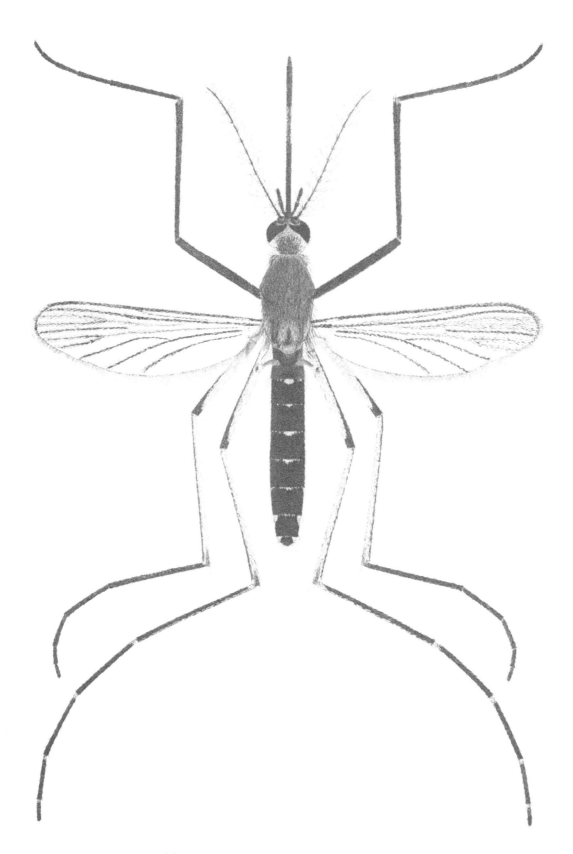

Plate 114. *Culex virgultus* Theobald, female.

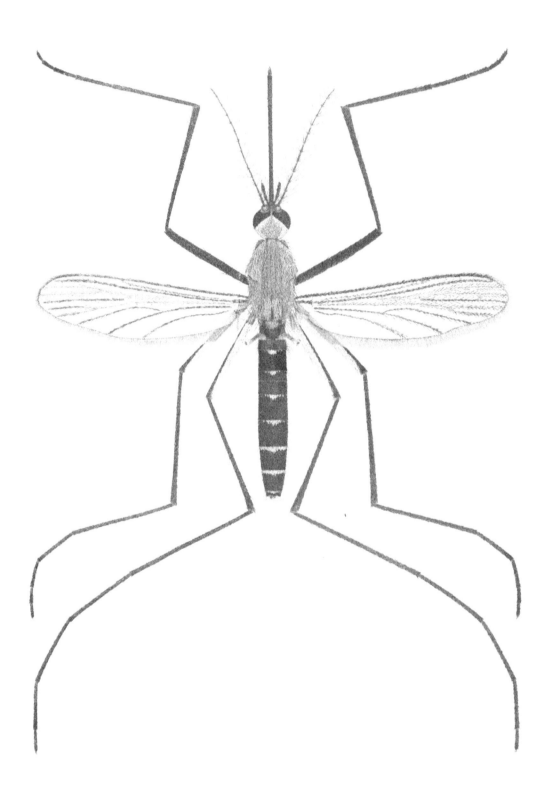

Plate 115. *Culex abominator* Dyar and Knab, female.

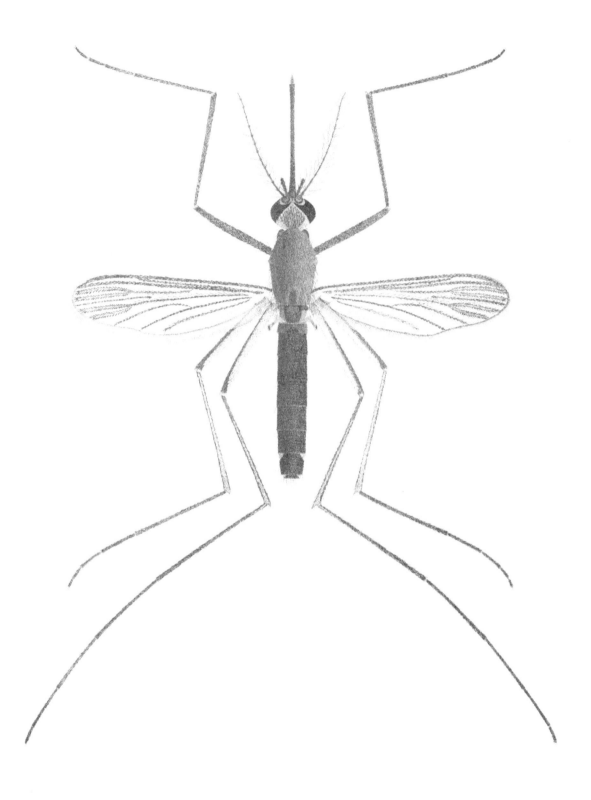

Plate 108. *Culex quinquefasciatus* Say, female.

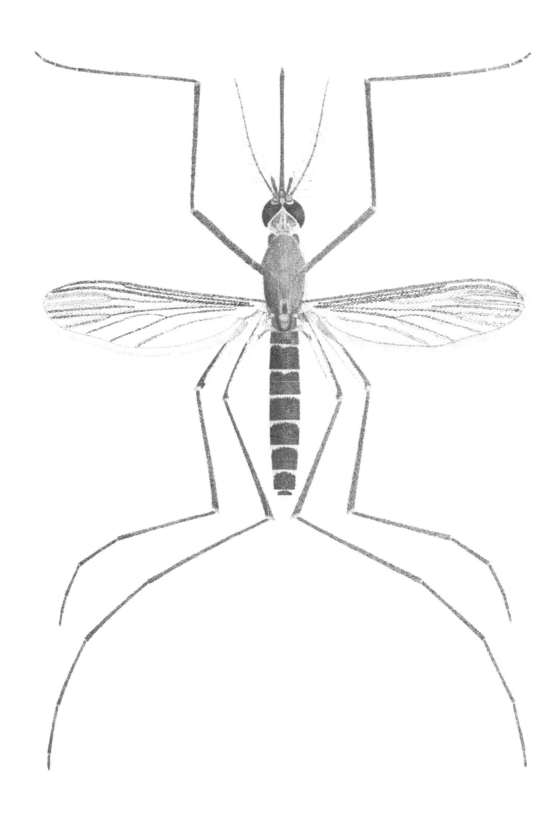

Plate 117. *Culex erraticus* (Dyar and Knab), female.

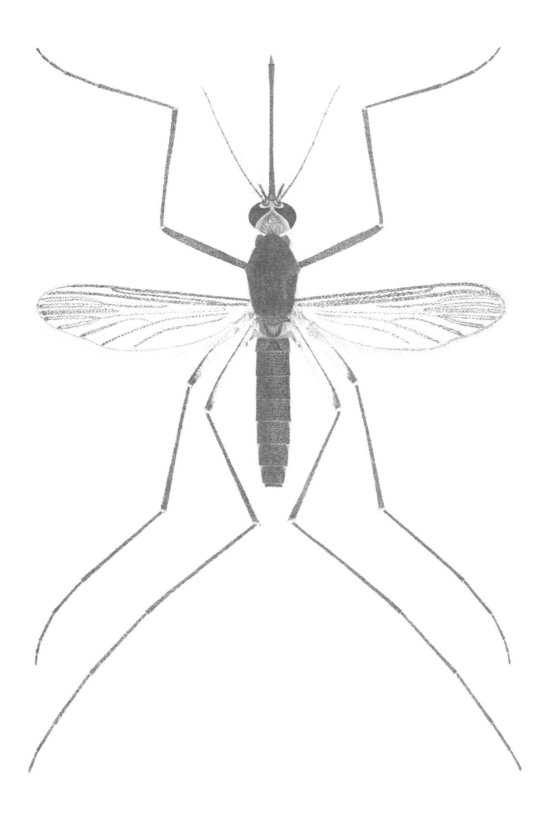

Plate 118. *Culex iolambdis* Dyar, female.

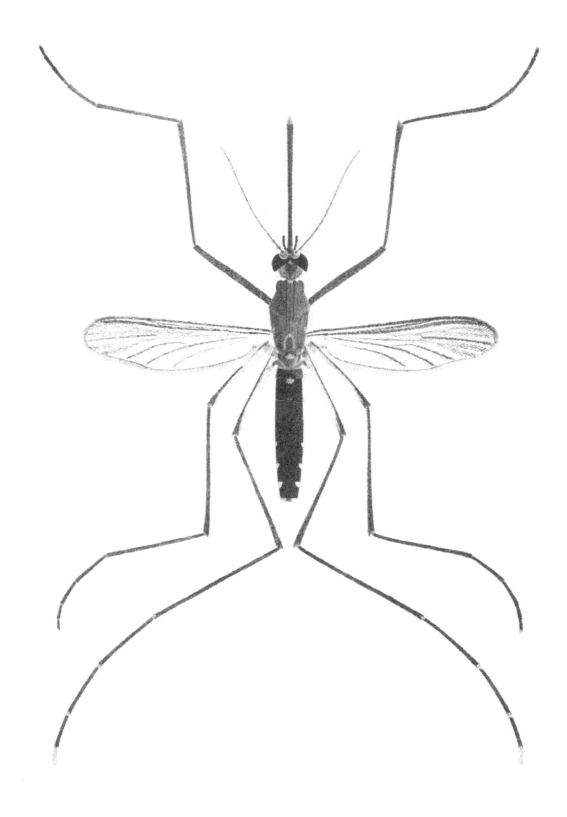

Plate 119. *Culex opisthopus* Komp, female.

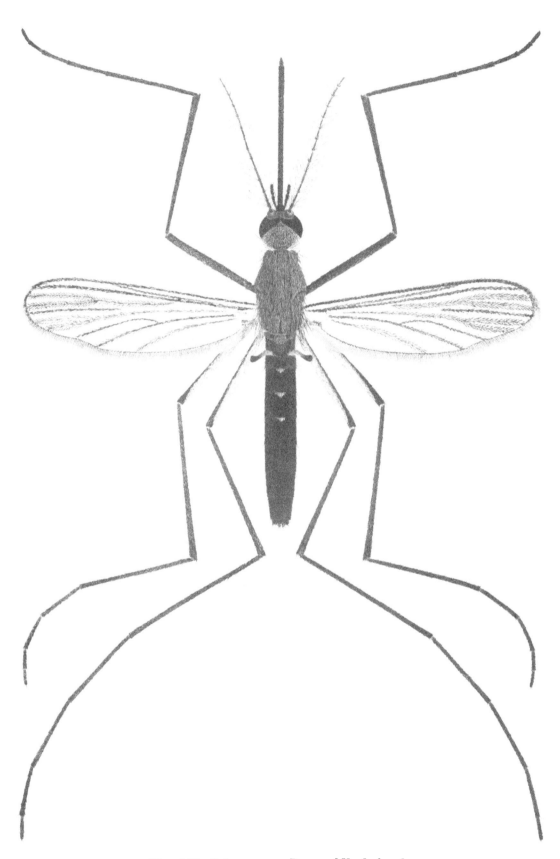

Plate 120. *Culex peccator* Dyar and Knab, female.

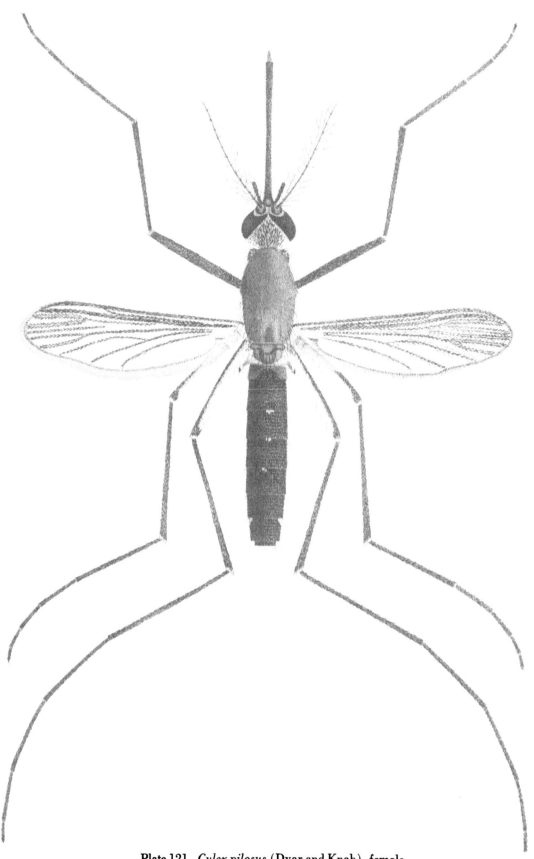

Plate 121. *Culex pilosus* (Dyar and Knab), female.

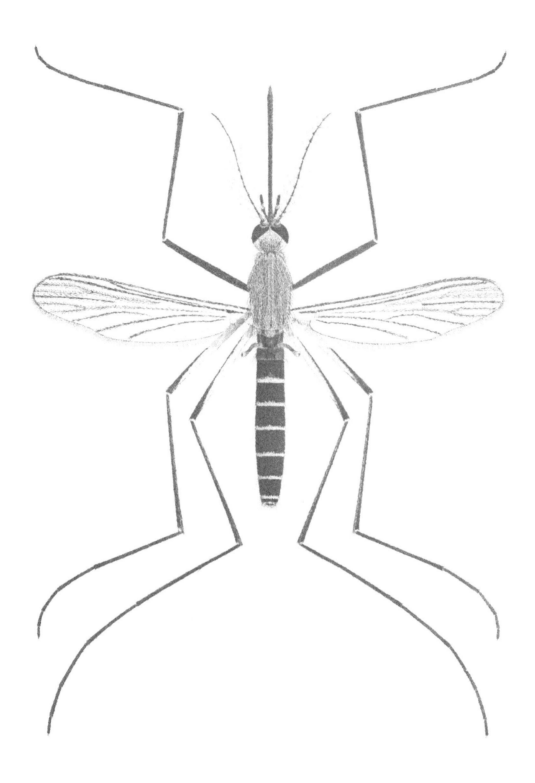

Plate 122. *Culex apicalis* Adams, female.

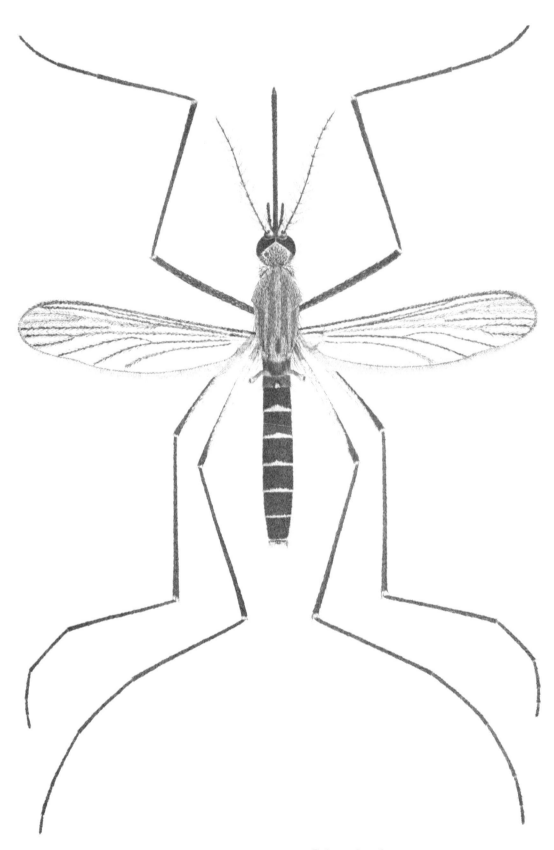

Plate 123. *Culex arizonensis* Bohart, female.

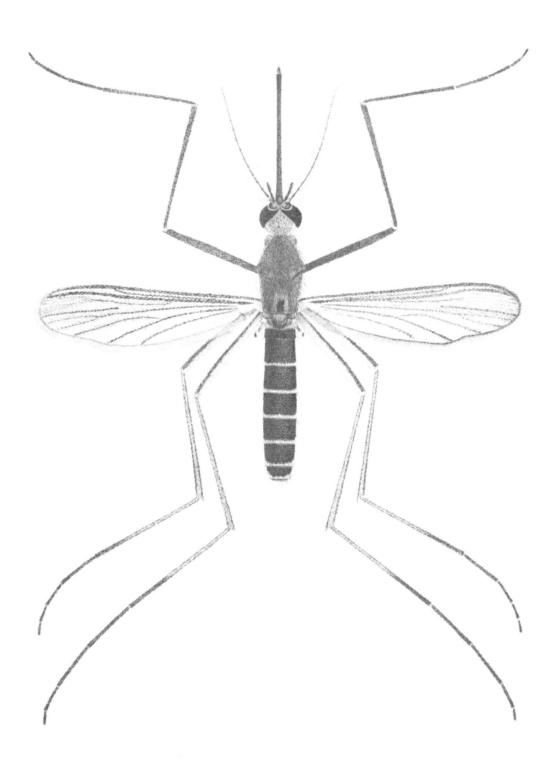

Plate 124. *Culex boharti* Brookman and Reeves, female.

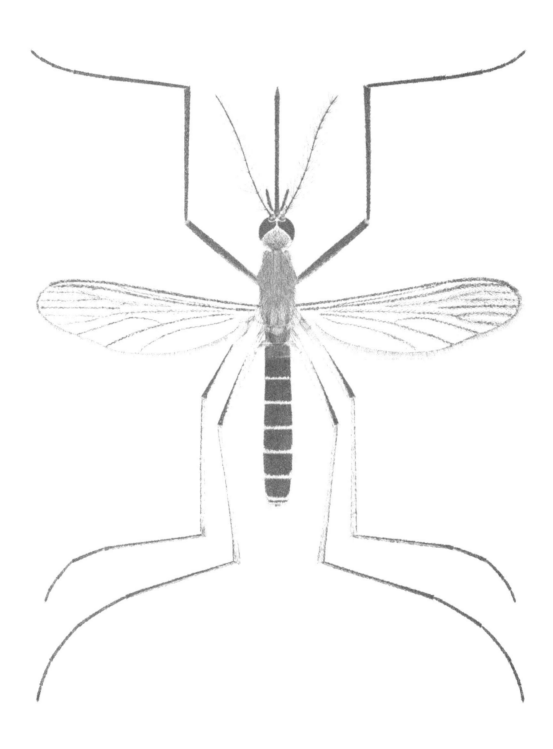

Plate 125. *Culex territans* Walker, female.

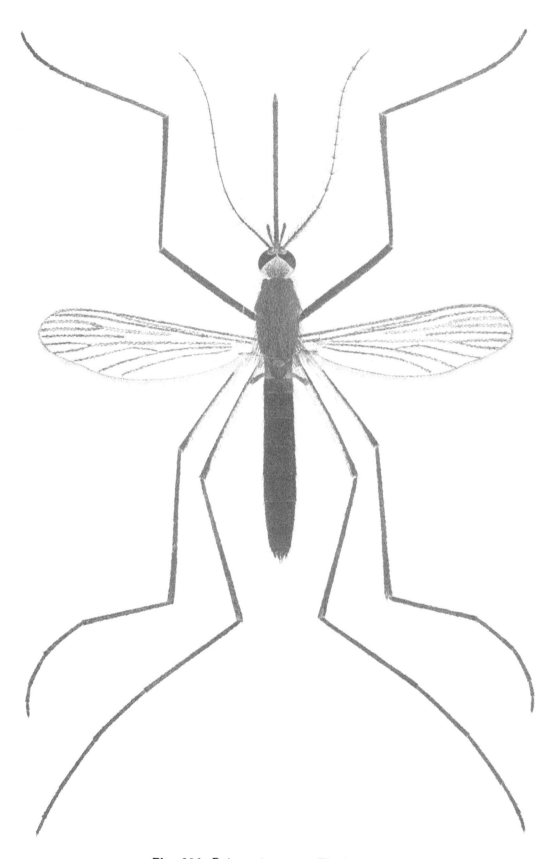

Plate 126. *Deinocerites cancer* Theobald, female.

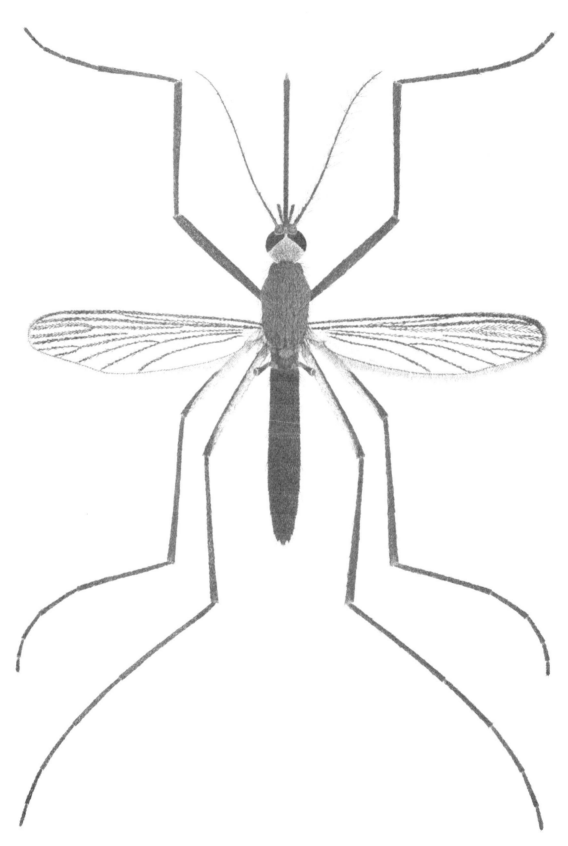

Plate 127. *Deinocerites spanius* (Dyar and Knab), female.

Milton Keynes UK
Ingram Content Group UK Ltd.
UKHW021910140724
445352UK00002B/38

9 780520 366619